ANIMAL COGNITION

Series Foreword

With this volume we initiate a series of books in comparative cognition and neuroscience. The presentations at the Harry Frank Guggenheim Conference, out of which the present volume grew, showed that this field of enquiry into cognitive functioning and its neural basis had reached maturity. That conference also crystallized the needs that the series is intended to meet.

The primary function of the series is to provide an outlet for further discussion and development of the field of comparative cognition and neuroscience. It is also a recognition of the need for and value of transdisciplinary dialogues about cognitive processes, neurophysiological mechanisms underlying those processes, and the evolutionary and ethological factors that shape them. The series' editorial board speaks to our commitment to that dialogue.

In view of the fact that the price of academic books has become a major hindrance to academic communication, the regular editions of all books published in this series will be accompanied by the publication of low-priced "student editions." Further, we are committed, whenever possible, to use computers to facilitate the publishing process. The price of the present volume is indicative of the initial effects of both of these two measures. We hope in these ways to return academic publishing to the service of academics.

We are proud to be able to include the present volume in the series even though it was substantially completed prior to the inception of the series. We are particularly grateful for the important contribution of Herbert Terrace in the organization of this conference and the production of this volume.

T. G. Bever
D. S. Olton
H. L. Roitblat

Editorial board:

Werner K. Honig   Klaus Immelman   Herbert M. Jenkins
Stephen Kosslyn   Larry R. Squire   Richard F. Thompson
Allan Wagner

ANIMAL COGNITION

**Proceedings of the
Harry Frank Guggenheim Conference
June 2-4, 1982**

Edited by

H. L. Roitblat
T. G. Bever
H. S. Terrace
Columbia University

LEA

LAWRENCE ERLBAUM ASSOCIATES, PUBLISHERS
1984   Hillsdale, New Jersey   London

Lawrence Erlbaum Associates, Inc., Publishers
365 Broadway
Hillsdale, New Jersey 07642

Library of Congress Cataloging in Publication Data

Harry Frank Guggenheim Conference (1982: Columbia
   University)
   Animal cognition.

   Bibliography: p.
   Includes indexes.
   1. Cognition in animals—Congresses.   I. Roitblat,
H. L.   II. Bever, Thomas G.   III. Terrace, Herbert S.,
1936-   .   IV. Title.   [DNLM: 1. Cognition—Congresses.
2. Psychology, Comparative—Congresses.   BF 311 H323a
1982]
QL785.H355   1982   591.51   83-20714
ISBN 0-89859-334-4
ISBN 0-89859-407-3 (pbk.)

Printed in the United States of America
10 9 8 7 6 5 4 3 2 1

CONTENTS

V. JUDGMENTS OF SIMILARITY AND DIFFERENCE

PREFACE

On June 2-4, 1982, the Harry Frank Guggenheim Conference on Animal Cognition was held at Columbia University. About 75 scholars discussed the new cognitive perspective on animal behavior and examined the changes that this perspective is producing.

An earlier conference, held in 1976, can be regarded as the modern birth of the cognitive approach to animal behavior. Many of the chapters in the book resulting from that meeting[1] begin with an attempt to justify the cognitive approach. Given the many years that operationalist behaviorism had dominated the study of learned behavior, some authors felt it necessary to *prove* that animals use internal cognitive processes to guide their behavior. Similar attempts at justification are notably absent from the present volume, suggesting that the approach has met with wide acceptance. If that conference marked the birth of a cognitive movement, we hope that this conference will mark its coming of age.

Several factors have influenced the development of animal cognition as a unique field. First, the widespread application of cognitive approaches to human behavior has shown how abstract models can account for behavioral data. Second, the discovery that there are species-specific constraints on learning called attention to the weaknesses of universal theories of learning such as behaviorism. Third, the simultaneous development of economic analyses of animal behavior in the fields of animal psychology and in behavioral ecology emphasized a view of animals as rational decision makers.

Animal cognition represents not only a distinct subject matter, cognitive processes in animals, but also a perspective from which to view organisms in relationship to their environment. No amount of data could ever prove this, or any other perspective, correct, but the chapters in the present volume document its usefulness as a framework for the study of animal behavior.

Animal cognition is concerned with explaining animal behavior on the basis of cognitive states and processes, as well as on the basis of observable variables such as stimuli and responses. For a time it appeared, at least to some, that discussion of cognitive states was not necessary, either because they were exhaustively determined by environmental events, or because they were epiphenomenal and without any causal force. In any case, it was assumed that a sufficiently detailed description of overt events would suffice for explanation. A great deal of the research into animal behavior has made it clear, however, that such cognitive states are real and necessary components of any adequate theory that seeks to

---

[1]Hulse, S. H., Fowler, H., & Honig., W. K. (Eds.) **Cognitive processes in animal behavior.** Hillsdale, N.J.: Erlbaum, 1978.

explain animal behavior.

An essential feature of the cognitive approach is an emphasis on the organism and its internal states and processes. Organisms are assumed to have internal cognitive structures which depend on their individual development as well as their evolution. External objects cannot enter directly into an organism's cognitive system, and so they must be internally encoded, that is, "represented." Accordingly, much cognitive research involves techniques for the study of the representations used by an organism, the processes that produce, maintain, and operate on the representations, and the environmental and situational factors that affect the representations.

The techniques used in the study of animal cognition are typically more complicated than are the methods of a few years ago. For example, the environments used to test animals include a relatively empty room with a sandy floor, well-controlled environmental chambers of various sorts, simulated natural habitats, and others. The stimuli utilized include colors, food presentations, letters of the alphabet, the sites of storage caches, photographic slides of moths, trees, fish, and other natural objects. This complexity is not the result of a belief that cognitive explanations are appropriate to complex behaviors but not to simpler behaviors. Rather, it is a reflection of the ideas that (a) the natural behaviors of a species determine many aspects of its cognitive capacities, (b) those capacities are best observed by utilizing tasks that have some ecological validity for the organism, and (c) even a complete understanding of simple phenomena may not be a sufficient basis for the understanding of complex phenomena.

Studies of animal cognition have been conducted with a large number of different species. Work described in the present volume, for example, includes pigeons, rats, bluejays, apes, Clark's nutcrackers, monkeys, early hominids, etc. Such diversity reflects the fact that animal cognition can be viewed as comparative in at least three senses. First, it inevitably concerns the evolution of cognition: Species may differ in their cognitive capacities as a result both of the phylogenetic histories and as a result of the particular ecological niche they exploit. Second, the field is comparative in the sense of comparing human and animal cognition in order to gain insight into both through an analysis of their similarities and differences. For example, in discussing the scientific analysis of human language, Miller[2] suggests:

> One way is to look at human language as a particular form of communication, one among many alternative varieties. If we can see it side by side with other forms of communication, we will have taken a first step toward putting it in scientific perspective (p. 11).

We think that the same principle holds for the study of all cognitive capacities of organisms.

Finally, animal cognition is comparative in the service of understanding the particular physiological mechanisms that underlie all cognitive processes and structures. Experimental investigations of the neurology underlying

---

[2]Miller, G. A. **Language and speech.** New York: Freeman, 1981.

behavior and thought are limited to the study of brain mechanisms in animals. The importance of such research for understanding human neurology will depend on sufficient understanding of the cognitive mechanisms instantiated in animals by those physiological mechanisms.

The chapters of this volume are indicative of the broad range of techniques being applied to the study of animal cognition. We hope that they stimulate further thought and research on the wide variety of topics that were discussed at the conference on which this book is based.

In addition to the financial support of the Harry Frank Guggenheim foundation, the Cognitive Neurosciences Institute and Columbia University also provided funds which we gratefully acknowledge. Many of our students helped in the conduct of the conference and in the preparation of this volume, Caroline Carrithers, Stephen Dopkins, Steven Jandreau, Nicholas Manna, Christine Moon, Scot Nourok, Kevin O'Connor, Bennett Pologe, and Robert Scopatz. We also wish to thank Dr. Howard Eskin, Dr. Bruce Gilchrist, and Mark Lerner of the Columbia University Center for Computing Activities for their support in preparation of this volume. Particular thanks are also due to Scott Strohm and Richmond Sweet who worked hard to format the book and set it in type using the Columbia University computer. Their efforts enabled the publisher to make the book available at a relatively inexpensive price.

The Editors
New York, August 27, 1983

I. COGNITION IN ANIMALS AND HUMANS

1   ANIMAL COGNITION

H. S. Terrace
Columbia University

I.   INTRODUCTION

After a number of false starts, animal cognition has taken root as a viable area of inquiry. Instead of sterile debates about the validity of behavioral, as opposed to cognitive, accounts of animal learning we have recently been witness to an impressive array of empirical and theoretical analyses of memory, expectancy and other cognitive processes in animals (see Hulse, Fowler, & Honig, 1978 and Griffin, 1982 for recent summaries).

As significant as such developments may be, it is important to understand how the current Zeitgeist, the widespread study of human cognition in particular, has contributed to the development of animal cognition. The purpose of this essay is· to call attention both to some features of recent research on animal cognition which distinguish it from earlier, less successful, efforts and to some basic differences between the assumptions underlying the study of animal and human cognition.

II.   SOME DISTINGUISHING FEATURES OF ANIMAL COGNITION

A. ANIMAL COGNITION IS NOT THE STUDY OF ANIMAL CONSCIOUSNESS

Many of the basic concepts of animal cognition were borrowed from human cognitive psychology. It is, therefore, hardly surprising to discover that a number of controversial issues that have been raised about the nature of human cognition have appeared in discussions of animal cognition. Two such issues which often obfuscate the nature of animal cognition and the rationale for its study are the purported introspective basis of human cognitive psychology and the validity of studying processes which cannot be observed directly.

The first of these issues, which is mainly of historical interest, is mentioned only to emphasize the unconscious nature of cognitive processes attributed to animals and to make clear that objections to animal cognition that are based upon its "mentalistic" nature are no longer relevant. It is, of course, true that early students of animal behavior, or "animal intelligence" as it was then called, speculated freely about the consciousness, feelings, ideas, images, etc. of animals [e.g., Romanes (1882), Morgan (1894, p. 77), Thorndike (1911, p. 16; also see Chapter 3, this volume, by Wasserman)]. Such speculations provoked two kinds of reactions from psychologists who argued against reference to cognitive entities in the analysis of animal behavior.

Watson (1914) dismissed them on the grounds that they could not be observed objectively. Accordingly, "psychology must discard all references to consciousness." Skinner (1938, 1950) simply questioned their explanatory value. Saying that an animal did X because it "believed" that X would lead to a reward was, for Skinner, a vacuous exercise in directing attention away from observable independent variables to an inner belief whose features coincided exactly with those needed to account for X.[1]

While questions about the nature of consciousness in animals still arise (e.g., Griffin, 1976, 1978), it has not been a major issue in the recent revival of interest in animal cognition. Just as the modern rationale for using human cognitive terms is not based upon arguments that appeal to consciousness or to introspective reports, the rationale for the study of cognitive processes in animals requires no reference to animal consciousness. Both in human and animal cognition it is assumed that the normal state of affairs is unconscious activity and thought. Indeed, additional processes must be postulated in order to account for one's consciousness of activities, perceptions, thoughts, and so on (cf. Pylyshyn, 1973, 1981; Skinner, 1969, Chapter 6). In this respect it is interesting to note a rare point of agreement between such dissimilar psychologists as Freud, Skinner, and modern investigators of human cognition.

Exorcising the ghosts of consciousness and introspection has proved to be a much easier task than demonstrating and defining the unconscious processes that constitute the subject matter of animal cognition. It is not too great an oversimplification to observe that many of the major recent advances in animal cognition are best characterized as convincing demonstrations of the inadequacies of so-called S-R models of animal behavior, -models which rely exclusively on observable (or potentially observable) stimuli and responses. On the other hand, less progress has been made in characterizing the nature of the cognitive processes whose existence has been demonstrated clearly by a wide variety of recent studies (see below).

B. A POSSIBLE DISCONTINUITY BETWEEN ANIMAL AND HUMAN COGNITION

One obvious obstacle to progress in defining the nature of animal cognition is the difficulty of inferring the critical features of unobservable processes, -a problem shared by the study of human cognition. Even though that difficulty has been reduced by focusing on unconscious processes and by the use of objective criteria for postulating their existence, our understanding of animal cognition has been impeded by an

---

[1]Unlike Watson, Skinner accepted introspective reports of consciousness (from humans) as a valid subject of study (Skinner, 1945, 1953, Chapter 17, 1969, Chapter 6). Indeed, he proposed a seldom recognized theory concerning the origins of so-called "private events." The basic features of Skinner's theory of private events have been adopted in toto by proponents of attribution theories of human beliefs and attitudes (e.g., Bem, 1967)).

uncritical application of concepts borrowed from the study of human cognition. This practice reveals clearly the influence of the shift in the study of human memory from S-R models of association to models of information processing. It fails, however, to take into account a fundamental difference between human and animal cognition.

Most models of human cognition rely on linguistically mediated processes. This is true even in the case of non-verbal stimuli, e.g., photographs and geometric shapes. The unavailability of language as a medium of cognition for animals dictates that models of animal cognition will differ in many important respects from their human counterparts.[2] When studying animal cognition, it seems prudent to keep in mind Piaget's many demonstrations of the differences between a child's and an adult's mode of thought. It would be surprising if the differences between animal and human cognitive processes prove to be less profound (Terrace, 1982).

III. THE ORIGINS OF CONTEMPORARY STUDIES OF ANIMAL COGNITION

In order to appreciate the kinds of phenomena that models of animal cognition will have to explain, it would be helpful to ask why recent demonstrations of cognitive processes in animals are more compelling than those advanced more than 50 years ago by Tolman and by other like-minded psychologists. Even though one can point to certain superficial similarities between early and contemporary studies of cognitive processes in animals, a failure to recognize their differences can only obscure the reasons for the success of recent investigations of animal cognition.

A. "S-R" MODELS OF BEHAVIOR

An S-R model of behavior seeks to explain any instance of learned or unlearned behavior by reference to associations between particular stimuli and particular responses. Such models come in many varieties. Unlearned behavior has been characterized as reflexes (e.g., Sherrington, 1906; Pavlov, 1927), tropisms (e.g., Loeb, 1900), orienting responses (e.g., Fraenkel and Gunn, 1961) and fixed-action-patterns (e.g., Tinbergen, 1951; Thorpe, 1963; Lorenz, 1950). Learned behavior has been characterized either as a new reflex that is established by pairing a conditioned stimulus with an unconditioned stimulus (e.g., Pavlov, 1927; Watson, 1916, 1919; Hull, 1943; Spence, 1950), or as a modification of existing behavior that results from positive or negative consequences of that behavior (e.g.,

---

[2]The ability of apes to learn signs of American Sign Language (Gardner & Gardner, 1971, 1978; Fouts, 1972; Patterson, 1978, 1979; Terrace, 1979a; Terrace, Petitto, Sanders, & Bever, 1979) or of artificial languages (Premack, 1970; Premack, 1976; Rumbaugh, 1977) may provide a basis for studying, in a non-human subject, the relationship between symbols, memory, and thought. With but one 1interesting exception (Rumbaugh, Rumbaugh, Smith, & Lawson, 1980), this topic does not appear to have been dealt with by projects that have studied the linguistic capacity of apes.

Thorndike, 1898; Skinner, 1938).

As disparate as these and related models may be, they share two features which have made them highly resistant to cognitive interpretations of behavior. Every S-R model claims to be more parsimonious than its cognitive counterpart and each claims to rely exclusively on observable (or potentially observable) stimuli and responses.

B. EXPERIMENTA CRUCIS AS "DISPROOFS" OF S-R THEORY

Cognitive analyses of animal behavior, which began to appear with increasing frequency in the thirties, sought to demonstrate the inability of S-R theory to account for various complex examples of learned behavior. Transposition, for example, was cited as a demonstration that an animal could learn the relative value of two discriminative stimuli, -as opposed to the absolute values stipulated by S-R theory (e.g., Kohler, 1955, Bingham, 1922). Studies of non-continuous learning claimed that an animal must form hypotheses about the relevant dimension(s) of a discriminative stimulus before it could actually learn the required discrimination (e.g., Krechevsky, 1938). Studies of latent learning questioned whether reinforcement was needed for learning (e.g., Tolman and Honzik, 1930).

Place learning was presented as evidence that animals learned spatial relationships (i.e., S-S associations) rather than specific responses (e.g., Tolman, Ritchie and Kalish, 1946a, b). In general, cognitive psychologists answered the question, "what is learned?" by reference to expectancies rather than to S-R associations.

S-R psychologists responded to experiments purporting to show the weakness of their theories by elaborating certain basic concepts (cf. Goldstein, Krantz, & Rains, 1965). Transposition was derived from interacting gradients of excitation and inhibition (Spence, 1937).

Demonstrations of non-continuous learning were attributed to poor stimulus control (e.g., Ehrenfreund, 1948; see Terrace, 1966, for a review). Latent learning was explained by appeal to reinforcement that derived from escaping from or exploring a maze (e.g., Thistlethwaite, 1951). Restle (1957) showed how place learning could be explained by models which predicted why, in certain types of mazes, rats came under the control of kinesthetic cues (response learning) and, in others, under the control of visual cues (place learning). Evidence purporting to reveal an animal's expectancy of a particular stimulus was explained by reference to fractional anticipatory goal responses (e.g., Spence, 1956; Amsel, 1958; Wagner, 1966).

The resiliency of S-R theory in answering the many purported *experimenta crucis* generated by cognitive psychologists enabled S-R theory to continue as the dominant theory of animal behavior through the early sixties. While occasional pockets of resistance appeared (e.g., Lawrence, 1955; Sutherland & Mackintosh 1971; Zeaman & House, 1963), there seemed little reason to argue for any major changes in the assumptions of S-R theory. Even when it appeared as if S-R theory might be forced to compromise some of its basic assumptions, it persisted, if for no other reason than that the cognitive point of view amounted to little more than a collection of provocative demonstrations. This state of affairs is a good illustration of Conant's incisive observation that discrepant facts don't suffice to defeat a theory. What is needed is a new theory.

The tug of war between S-R and cognitive theory was resolved by external developments rather than by a decisive breakthrough in either camp. One line of research questioned the universality of S-R laws of behavior, both within and between species. Another sparked a shift away from S-R theories of *human* learning toward cognitive models. It was this Zeitgeist that gave rise to modern concepts of animal cognition.

C. BIOLOGICAL CONSTRAINTS ON LEARNING

Recent advances in animal cognition notwithstanding, it would be erroneous to identify those advances as the sole reason for the loss of faith in S-R theory that we have witnessed during the last two decades. Extensive damage was also done by studies of so-called "biological constraints" of learning which have questioned a basic assumption of S-R theory: the arbitrary nature of the stimuli, responses, and reinforcers (e.g., Seligman & Haber, 1961; Garcia & Koelling, 1966; Brown & Jenkins , 1968). More so than any purported demonstration of an animal's cognitive ability, experiments which showed the effectiveness of particular combinations of stimuli, responses and reinforcers have undermined the generality of S-R laws of behavior, –laws which make no provision for such specificity.

While the study of biological constraints is of fundamental importance in understanding the nature of animal learning, it provides little insight into an animal's cognitive abilities. An animal's preparedness to make a particular response or to form associations between particular stimuli provides no impetus to turn to cognitive explanations of conditioned behavior. From the point of view of animal cognition, the main consequence of the discovery of biological constraints was to question the universality of S-R theory as an explanation of learned behavior.

D. HUMAN VERBAL LEARNING

The emergence of a new approach to the study of human verbal learning and memory was of much greater significance than the discovery of biological constraints in influencing the development of animal cognition. Ever since Ebbinghaus' classic studies (Ebbinghaus, 1885), association theory has dominated research on this problem. In paired-associate learning, for example, it is assumed that subjects learn to make particular responses to particular stimuli. In serial learning it is assumed that one response serves as a cue for the next. Transfer effects are explained by one or another version of interference theory, –a theory which makes ample use of such S-R concepts as excitation, inhibition, stimulus generalization and response generalization.

Just as in the case of animal learning, cognitively oriented psychologists who studied verbal learning tried to discredit S-R theory by various types of *experimenta crucis* (e.g., Rock, 1957). They achieved little success until a major weakness of S-R theory was revealed by experiments which demonstrated the existence of associations between non-adjacent items [e.g., experiments on free recall (Bousfield, 1953; Bousfield & Bousfield, 1966; Tulving, 1968)]. While various networks of associations might, in principle, explain the clustering of items on free recall tests, the nature of such networks is, of necessity, *ad hoc* and unparsimonious. Much greater success in explaining the results of such experiments was achieved by models of memory based upon the hierarchical organization of the items recalled (Bower, 1972; Mandler, 1967; Tulving, 1962).

Essential to the modern study of human memory are a variety of concepts which derive from computer programming: information processing, branching, executive control, and so on.   Others, such as representation, short- and long-term memory, rehearsal, retrieval and control processes were suggested by various empirical discoveries about human memory.   In a fairly short time, these and other concepts were integrated into an information processing model of human memory which has been widely accepted as a more powerful alternative to S-R models (cf. Anderson & Bower, 1973).

IV.   REPRESENTATION IN HUMANS AND ANIMALS

Representation is a central concept in both human and animal cognition.   In each instance an important basis for appealing to representations are demonstrations in which the subject makes a response without guidance from any physically present stimulus as, for example, when drawing something from memory, or when generating "missing" elements of a proximal stimulus in order to make an impoverished stimulus meaningful (cf. Gregory, 1975).

Almost 70 years ago, Hunter (1913), an early behaviorist, prescribed a similar logic for postulating representations in animals.

> . . .If comparative psychology is to postulate a representative
> fact,. . .it is necessary that the stimulus represented be absent
> at the moment of the response.   If it is not absent, the reaction
> may be stated in sensory-motor terms (Hunter, 1913, p. 21).

By stipulating that ". . .the stimulus represented be absent at the moment of the response," Hunter argued that the organism generates some representation of the absent stimulus.   That representation functions just as an exteroceptive stimulus might in evoking appropriate behavior.[3]

In order to grasp fully Hunter's view of an animal's ability to use a self-generated representation of some feature of its environment, it is important to ask why he did not see any need to argue for representation

---

[3]In the various types of delayed response apparatus that Hunter devised to study representation in animals, he took pains to rule out particular postures or orientations of the organism as mediators between the stimulus and the response recorded by the experimenter.   Hunter correctly rejected, as representations, kinesthetic feedback from such mediators because they could be construed as members of a covert chain of stimuli and responses.   Overall, Hunter found very little evidence that animals could represent features of their environment in ways that could not be explained in S-R terms.

in those instances in which a stimulus reliably precedes a response. Why not, for example, appeal to representations of a CS or of a US in a typical conditioning experiment?[4] Hunter's answer was the logic of parsimony. Like Skinner and other behaviorists, Hunter noted that our ability to predict or to explain behavior is not enhanced by appealing to a representation of a stimulus if that stimulus is available when the organism responds.

This is not to say that, somehow, an animal doesn't store memories of its experience. Quite obviously it must if it is able to react to a stimulus at $time_2$ in a manner similar to its reaction to that stimulus at $time_1$. It is, however, necessary to distinguish between an organism's ability to generate, or to at least maintain, a representation of some previously experienced stimulus that is present when the response in question occurs and its ability to respond when that stimulus is absent.

A. REPRESENTATIONS OF STIMULI THAT ARE ABSENT DURING THE EXECUTION OF A CONDITIONED RESPONSE

More so than any other type of study, recent experiments on animal memory provide compelling evidence that animals form representations. In each case the subject is required to recall certain features of one or more events and, in the absence of those events, to use that information as a basis for performing some response.

1. A pigeon's memory of which of two locations it pecked. In one study, pigeons were trained to report the spatial location of one of a series of three responses they made to either of two response keys (Shimp, 1976). During the first part of each trial, a white "X" appeared three times, in an irregular sequence, on one or the other key. Pecking the "X" caused it to disappear momentarily, after which it reappeared either in the same or in the other position. After the third "X" was pecked, the pigeon was queried as to which of the two keys it pecked at different stages of the three-peck sequence.

The retention test used three response keys. First, the center key was illuminated by either a red, blue or a white light. Pecking the center key turned it off and illuminated the side keys with the same color. Red side keys posed the question, "which key was pecked during the first presentation of the 'X'?" If, for example, the first "X" appeared on the right side key, a peck to the right side key resulted in reinforcement. Blue center and side keys posed the same question about the location of the second "X" while white center and side keys required the pigeon to report the location of the third "X." Shimp showed that pigeons could report, at greater than chance levels of accuracy, the location of each "X." As one might expect, however, accuracy was greatest in the case of reporting the location of the most recently presented "X." Overall, accuracy decreased as the time between the appearance of a particular "X" and the retention

---

[4]It is immaterial, for this argument, whether the representation is of the CS, the US, the UR, or of some S–S or S–R connection. The form of the representation is less critical than the fact that some type of representation exists.

test increased.

2. *A rat's memory of places in which it obtained food.* In Shimp's experiment, the sequence of "X'"s that the pigeon pecked and was required to remember was defined by the experimenter. A series of studies by Olton and his colleagues showed that rats can remember what they did in circumstances in which their behavior was less constrained by the experimental paradigm (see Chapter 10, this volume). Using a radial maze, in which 8 runways radiate from a common start point, Olton and Samuelson (1976) and Olton (1978, 1979) demonstrated that rats can remember which of the runways they entered while searching for food. In the basic paradigm, each runway was identical and was baited with equal amounts of food that could not be seen from the entrance to the runway. The variable of interest was how many runways were entered before all of the food was consumed. On the average, Olton's rats reentered less than one alley per trial.

Of greater interest was evidence that Olton's subjects did not show any consistent pattern in going from one runway to another and that they could remember which arms of the maze they had entered without the benefit of specific olfactory (or of any other exteroceptive) cues. In one instance a highly odorous substance was used to mask presumed olfactory cues. In another the olfactory nerve was cut. Neither procedure impaired the rat's highly efficient performance. In yet another experiment the position of the runways was rotated after 4 entries. This provided the rats with a choice between previously visited alleys (now situated in unvisited spatial locations) and unvisited alleys (now situated in previously visited spatial locations). The results demonstrated clearly that the rat responded on the basis of its memory of what *spatial* locations it had visited previously.

3. *Serial learning in the pigeon.* A third study which demonstrates an animal's ability to represent recent events used a variation of the conventional chaining paradigm. Conventional chaining paradigms do *not* require the organism to memorize the sequence that defines a particular chain. For example, in learning to run through a typical maze, a rat has only to learn what to do at various choice points. Since each choice point is discriminable, the rat's task can be characterized as learning to solve a set of discrimination problems in which the discriminative stimuli are encountered successively.

In a procedure they referred to as a "simultaneous chain" (in order to distinguish it from the conventional "successive chain"), Terrace and his collaborators showed that pigeons could learn a sequence of four colors (A->B->C->D) in which all of the colors (i.e., all of the choice points) are presented simultaneously (Terrace, Straub, Bever, & Seidenberg, 1977; Straub, Seidenberg, Bever, & Terrace, 1979; Straub & Terrace, 1981). The position of the four colors, each presented on a different response key, varied from trial to trial. The task was to peck the colors in the sequence A->B->C->D, irrespective of their positions. No differential feedback was provided after each response.

Pigeons learned to perform the A->B->C->D sequence on the arrays on which they were trained at levels of accuracy that exceeded 70%. No decrement in performance was observed even when the colors A, B, C and D were configured in novel arrays. This showed that the pigeon had not simply mastered a set of rotely learned response sequences to the arrays used during training. Thus, the only basis the pigeon had for choosing a particular color is its representation of what color it just pecked and what color should next be pecked.

A pigeon's ability to form a representation of the sequence was also demonstrated by its accurate performance on "sub-sets" of the original sequence (e.g., B and D, A and D, C and D, and so on). On arrays presenting B and D (in which the required sequence was B->D), the pigeon had neither the advantage of the normal starting color nor, having pecked B, the advantage of an adjacent element. Just the same, accuracy of performance on arrays requiring the sequence B->D was as great as it was on arrays requiring the sequence A->B.

Another study showed that the ordinal position of the middle element *per se* controlled performance on three-element sequences of colored elements (Terrace, 1983). Following acquisition of the sequence A->B->C, training commenced on one of three new "lists" of elements. All of the lists consisted of one old element, B, the middle color of the original list, and two new elements, X and Y, -a white vertical line and a white diamond (on black backgrounds), respectively. For one group, B retained its original position in the new sequence: X->B->Y. For both of the other groups, the position of B was shifted (with respect to A->B->C training). One group was trained on the sequence X->Y->B; the other on the sequence B->X->Y.

The subjects trained on the sequence in which B retained its original position (X->Y->B) learned that sequence more rapidly than the subjects trained on the sequences in which the position of B was changed (X->Y->B and B->X->Y). Further the subjects trained on the X->B->Y sequence mastered that sequence more rapidly than they did the original A->B->C sequence. Conversely, it took longer to acquire either the X->Y->B and the B->X->Y sequences than it did to acquire the original A->B->C sequence. These results show that in addition to mastering a three-element sequence of colors, pigeons also acquired knowledge about the ordinal position of the second element. That knowledge adds to the complexity of the representation that would be needed to account for performance on simultaneous chaining.

Yet another consequence of learning to produce a sequence is positive transfer to sequence recognition. In a study recently completed in my laboratory, sequence recognition was trained by a "yes-no" paradigm. On half of the trials the sequence A->B->C was presented on a single key. On the remaining trials A->B->C sequences of the elements A, B, and C were presented in other orders. Following each type of sequence, "test" stimuli were presented on keys to the left and the right of the key on which the sequence was shown. Food reward was provided for pecks to the left hand key following A->B->C sequences and to the right-hand key following non A->B->C sequences (see Dopkins, Scopatz, Roitblat, & Bever, 1983; Chapter 4, this volume, by Bever; and Weisman, Wasserman, Dodd, & Larew, 1980, for similar procedures for training sequence discriminations).

The acquisition of sequence discrimination was studied under three conditions. One group of pigeons first learned to produce an A->B->C sequence under the simultaneous chaining procedure described earlier. A second group was given the same amount of training on a *successive* training procedure in which the elements A, B, and C were presented one at a time. (A peck to A produced B and turned off A, a peck to B produced C and turned off B, and a peck to C turned off C and produced food reward.) The third group had no experience with the elements A, B, and C prior to sequence discrimination training. The first group learned to discriminate the A->B->C from the non A->B->C sequences much more rapidly than the subjects of the other two groups. Thus, the representation that a pigeon acquires when learning to produce a sequence facilitates learning to discriminate that sequence from other sequences containing the same elements.

B. REPRESENTATIONS OF STIMULI AVAILABLE TO THE ORGANISM DURING THE EXECUTION OF A CONDITIONED RESPONSE

Given the evidence that Hunter had available (indeed, given the evidence that was available during the next half century), it seemed reasonable to restrict reference to representation in animals to situations in which the occurrence of a learned response could not be predicted from any constellation of exteroceptive or proprioceptive stimuli that were available while the organism executed a conditioned response. In both the case of human and animal behavior, however, it has often been shown that one cannot predict the subject's response without knowledge of the *coded* form of the relevant stimulus. A clear example is Conrad's study of memory in humans (Conrad, 1964). Subjects who were asked to report visually presented letters, made errors which could only be predicted from the sound of the letters (e.g., following an *E*, an errant subject is more likely to say *B* or *D* rather than *F*).

A variety of recent studies of animal behavior, employing both classical and operant conditioning paradigms, provide evidence that an animal encodes stimuli that are available at the time it is performing a conditioned response. While it is too early to generalize about the nature of the circumstances which induce an organism to form representations of currently available stimuli, the variety of those circumstances suggests that representations may play a significant role in most conditioning experiments.

1. The role of the contingency between the conditioned and the unconditioned stimulus. In view of the central importance of classical conditioning to behavioral theories of animal behavior, one would expect that this paradigm would be the least likely one for uncovering evidence of an animal's ability to encode a conditioned stimulus. That ability, however, was revealed clearly in Rescorla's seminal paper on the distinction between the contiguity and contingency in classical conditioning (Rescorla, 1967; Rescorla, 1972). By showing that classical conditioning takes place only if $p(US|CS) > p(US|\overline{CS})$, Rescorla, and others, have demonstrated that an animal somehow notices whether the CS is predictive of the US. Any model which seeks to predict this outcome must postulate some mechanism for comparing a representation of the US, given a prior CS, with a representation of the US when it is not preceded by the CS. It is in this sense that Rescorla has argued that classical conditioning results in an "expectancy" of the US following presentations of the CS (also see Gibbon, Berryman, & Thompson, 1974).

It is important to note the difference between Tolman's and Rescorla's use of "expectancy." As Tolman himself acknowledged, his reference to expectancy was based on an anthropomorphic analysis of how the rat perceived its environment, specifically, how the rat perceived, "what leads to what." The weakness of this approach is that descriptions of the rat's expectancies varied with different experimenter's views of the rat's perceptions. Reference to "expectancy" in analyses of contingencies between the CS and the US are more objective in that they are based upon models of the predictive power of the CS.

2. Blocking. The phenomenon of blocking provides another familiar example of selective stimulus control in which mere contiguity between the CS and the US is not sufficient to produce conditioning (cf. Kamin, 1967).

In the basic blocking paradigm a compound conditioned stimulus (consisting of the elements A and B) regularly precedes the US. Prior to compound training, only A signals the occurrence of the US. After compound training, A and B are presented by themselves. Under these circumstances, it has been widely shown that A is more likely to evoke a conditioned response than B.

Both the role of contingencies and blocking demonstrate that the mere presence of a stimulus prior to the presentation of a US is not sufficient to establish a conditioned response. Stated more positively, these phenomena show that an organism somehow processes stimuli from its environment in a way that maximizes information (cf. Kamin, 1969). This view of contingencies and blocking has been formalized in the well known Rescorla-Wagner model of associative learning (Rescorla & Wagner, 1972; Wagner & Rescorla, 1972) a model which assumes the animal's ability to process information but which makes no attempt to explicate the nature of that process.

That step was taken by Wagner (1978, 1981) who argued that the strength of the association between a CS and a US depended upon the degree to which the CS and the US was "rehearsed" following a conditioning trial. The more strongly a stimulus is represented in short-term memory, the less likely it will be rehearsed. In the case of blocking, the subject would only rehearse B (the CS element added to A) when B was followed by a change in the value of the US (e.g., an increase in intensity or its elimination). This is not the case for control group subjects whose conditioning history began with the compound AB. Such subjects would have no reason to rehearse A more than B, or vice versa.

In the case of the contingency between the CS and the US, Wagner argues that the subject would be less likely to rehearse the US if it occurred both after the CS and during the inter-trial interval than if it occurred only after the CS. This follows from the fact that, when the US occurs between pairings of the CS and the US, the US was present in short-term memory at the start of a trial.

3. Does a pigeon encode the sample when performing the matching-to-sample task? Studies of contingencies and blocking provide evidence that animals process stimuli in the sense that they define circumstances under which a supra–threshold, but uninformative, stimulus fails to gain control over a response.   More positive evidence that animals encode stimuli has been obtained in studies which have used variations of the discriminative operant paradigm.   One study (Roitblat, 1980), focused on the kinds of errors pigeons made in mastering a delayed symbolic matching–to–sample task (see also Chapter 5, this volume, by Roitblat).   In this modification of the standard matching–to–sample task, the sample and choice stimuli were selected from different continua.   Unlike identity matching, in which the sample and the correct choice are identical, a symbolic matching task requires a subject to match a sample to an arbitrarily dissimilar choice (cf. Cumming & Berryman, 1965).   Most of the subjects in Roitblat's study were shown colored samples (blue, orange, and red), and were required to match a particular color to an achromatic line of a particular orientation (0°, 12.5°, and 90°).   Another subject, who was shown achromatic line samples of the three different line orientations, was required to match a particular line orientation to one of the three colors. Under both conditions, the selection of the sample and choice stimuli insured that similar comparison stimuli corresponded to dissimilar sample stimuli and that dissimilar comparison stimuli corresponded to similar sample stimuli.   This design allows one to compare two independent contributions to accuracy of choice: discriminability of the sample and discriminability of the choices.

Under each condition, confusions between similar *test* stimuli resulted in a greater decrement in performance than confusions between similar sample stimuli.   If, during the delay between the sample and the choice, the pigeon represents the sample in its original form, errors would be more likely to occur to the dissimilar choice stimuli.   This follows from the fact that similar choice stimuli correspond to dissimilar test stimuli.   Since Roitblat obtained the opposite pattern of confusions with respect to the test stimuli, he concluded that his subjects encoded the sample stimulus with respect to a particular test stimulus.

4. *Concept formation.*   The pigeons of Roitblat's study appeared to use a simple encoding process.   Each sample stimulus was encoded with respect to the test stimulus with which it was matched.   A much more complex type of stimulus encoding is evident in the case of pigeons that Herrnstein, and others, trained to learn "natural" concepts (Cerella, 1977, 1979; Herrnstein, 1979; Herrnstein, Loveland, & Cable, 1976; also see Chapter 14, this volume, by Herrnstein).

Concept formation was trained by a standard multiple schedule of reinforcement in which the discriminative stimuli were photographs.   On positive trials, transparencies showing exemplars of some concept were projected above the response key (trees, water, people, a particular person, an oak leaf, fish, to name a few).   On negative trials, transparencies were displayed which did not contain the exemplar in question.   The reliable outcome was a substantially higher rate of responding to positive than to negative exemplars.   Of greater significance is the fact that this difference was also observed on the first occasion on which novel transparencies were presented.

An important property of a natural concept, such as trees or a particular person, is that it cannot be characterized by any formal set of rules.   Unlike stimuli from such physical continua as wavelength and intensity, there is no known dimension, or set of dimensions, along which a natural concept can be defined.   As a consequence, pigeons can recognize trees or a particular person more reliably than any known artificial intelligence device.

In accounting for the pigeon's ability to form natural concepts one must postulate some encoding process that enables the pigeon to react in the same manner to transparencies whose only common property is the presence or absence of an exemplar of a particular concept. It has yet to be shown whether the pigeon abstracts a single feature, e.g., a "prototype," as has been hypothesized in the case of humans (Rosch, 1978), or multiple features, as has been suggested by Herrnstein and de Villiers (1980) and by Morgan *et al.* (1976) (also see Chapter 15, this volume, by Lea). What is clear is that the most detailed physical specification of a set of positive and negative exemplars cannot provide an adequate basis for predicting how the pigeon will generalize to novel exemplars.

V. THE ROLE OF REPRESENTATION IN THE ANALYSIS OF BEHAVIOR

The handful of studies of animal cognition we have considered hardly does justice to the variety of research that has been pursued during recent years.   Of necessity, some topics will have to be mentioned by name only: perception of sequences (Weisman & DiFranco, 1981; Weisman, Wasserman, Dodd, & Larew, 1980; also see Chapter 12, this volume, by Weisman and von Konigslow), serial pattern learning (Hulse, 1978; also see Chapter 11, this volume, by Hulse, Cynx, and Humpal) memory of recently presented photographs (Wright, Santiago, Urcuioli, Sands, 1981; also see Chapters 20 and 21, this volume, by Wright, Santiago, Sands, & Urcuioli and Sands, Urcuioli, Wright, & Santiago, respectively), directed forgetting (Rilling, Kendrick, & Stonebraker, 1984), causal inference (Premack & Woodruff, 1978), and analogical reasoning (Gillan, Premack, & Woodruff, 1981; Gillan, 1981).   These and other studies provide impressive evidence that an animal can form representations of its environment and that such representations often play an essential role in the explanation of animal behavior (cf. Roitblat, 1980).

Less clear is the nature of animal representations.   In studying this problem, however, it is helpful to observe that the concept of representation should not raise the spectre of dualism that provoked Watson and other behaviorists to banish it and other cognitive concepts from psychology.   As studies of human cognitive psychology have made clear, it is meaningful and essential to investigate representations as *psychological* phenomena in their own right.

A. REPRESENTATIONS AS A PROBLEM OF STIMULUS CONTROL

The fact that representations are generated within the organism may suggest that they should be regarded as mere physiological phenomena. Like all psychological phenomena, representations obviously have a physiological basis.   That truism should not, however, detract from the importance of specifying representations at a psychological level of analysis.   Indeed, without psychological specifications of representations, it is not clear how one can hope to discover anything about their physiological nature.

The examples we have considered suggest that the task of specifying the features of a representation overlaps, in many respects, with the task of specifying the stimulus control exerted by an exteroceptive stimulus.   In this sense a representation may be construed as an adjunct, albeit a fundamental adjunct, of an exteroceptive stimulus that is generated by the organism (cf.   Leahey, 1981; also see Chapter 16, this volume, by Blough).

When studying the stimulus control of an exteroceptive stimulus, it is usually assumed that variations in the value of the stimulus are sufficient to account for concomitant variations in the strength of some learned response (e.g., Terrace, 1966)   A common thread of the various studies reviewed earlier suggest that it is important to supplement one's specification of relevant exteroceptive stimuli with a description of the organism's representation of those stimuli.

Studies of contingencies and blocking indicate that an organism may selectively ignore particular suprathreshold stimuli that add no new information to the organism's store of knowledge.   Once selected, a stimulus can undergo a variety of transformations.   Radial maze studies show that the rat can organize its external environment into sites in which it has recently fed and sites in which it has yet to be fed.   Studies of symbolic matching—to—sample show that a sample stimulus is not retained as such but, instead, is encoded in a form that is determined by the test stimulus with which it is matched.   Studies of concept formation suggest that generalization of an exteroceptive stimulus cannot be predicted from any physical feature of the stimulus.   Studies of simultaneous chaining show that an organism can represent an arbitrary sequence of arbitrary elements, that it can do so even when normally adjacent elements are omitted and that the ordinal position of an element controls sequential performance.

Extending the study of stimulus control to stimuli that the organism generates poses a variety of technical problems.   However, from a conceptual point of view, there is no basis at present to distinguish qualitatively between environmental and self—generated stimuli.

B. "BEHAVIORAL" ALTERNATIVES TO THE CONCEPT OF REPRESENTATION

It has been observed that the processes whereby an organism evaluates or encodes a stimulus are simply instances of behavior that are reinforced by their outcome (cf.   Catania, 1979).   This argument is but one of many manifestations of resistance to the concept of a representation in the analysis of conditioned behavior.   Another is the view that a stimulus can act at a considerable temporal distance from the response it controls (Marr, 1983).   These and related "behavioral" formulations of representation beg the question of the nature of the processes that are suggested as alternatives.   They also imply, incorrectly, that an organism's behavior can be predicted from a sufficiently rich description of the organism's environment. In addition to such a description, it is necessary to determine the nature of the stimulus as processed by the organism.

Skinner's ambivalent discussions of "private events" provides a good illustration of the kind of difficulty one encounters by avoiding reference to an organism's representations of its external environment. Consider, for example, Skinner's analysis of thinking which makes reference to processes such as "conditioned seeing." Conditioned seeing gives rise to "images." Likewise, "conditioned hearing" gives rise to "internal voices" (e.g., Skinner, 1953, Chapter 16). When we can't remember a name, we presumably engage in behavior that is similar to a search for a lost object in our external environment. When looking for an object in our "mind's eye," we are engaged in "looking" behavior which will eventually result in the "appearance" of the sought after object.

So long as these process are characterized by non-observable "behavioral" responses such as "looking" or "imagining," and by their equally non-observable consequences, Skinner is content to include them in his conceptual system. It is not clear, however, how such processes can be distinguished from so-called cognitive processes. Just the same, when cognitive processes are couched in non-behavioral terms, objections are raised on the grounds that they refer to meaningless homunculi, quasi-neurological constructs, or subjective entities (Skinner, 1950). While once cogent, the relevance of such objections to the modern study of cognitive processes has yet to be demonstrated.

The recent variety of research on animal cognition questions the validity of another of Skinner's objections to cognitive psychology. Skinner observed that cognitive terms draw attention away from the important variables which control behavior: contingencies of reinforcement and exteroceptive discriminative stimuli. However important such variables may be, they provide no basis for suggesting the broad range of studies of animal memory and representation that have preoccupied students of animal learning for the past decade. If anything, a strong case can be made that exclusive concern with contingencies of reinforcement and discriminative operants has detracted from the study of cognitive processes in animals (see Chapter 3, this volume, by Wasserman; Wasserman, 1983).

C. DIFFERENCES BETWEEN HUMAN AND ANIMAL REPRESENTATIONS

Though our knowledge of animal representations is embarrassingly meager, we can be fairly confident that animal representations differ from those generated by humans in two important respects. Most studies of human memory use verbal stimuli. Even when non-verbal stimuli are used, memory may be facilitated by verbal mnemonics and control processes. In the absence of such mnemonics and control processes, it seems foolhardy to assume that animals rehearse stimuli the way humans do or that there is much overlap between animal and human encoding processes. It also seems clear that biological constraints may limit the generality of cognitive processes in animals more so than they do in humans.

In the case of the radial maze, for example, an important basis of the rat's ability to avoid previously visited alleys is an unlearned "win–shift" strategy that it follows when searching for food.   While a win–shift strategy is not sufficient to explain the highly efficient performance of Olton's rats, it appears to be a necessary condition.   This becomes evident when comparing the radial–maze performance of pigeons, who are "win-stay" organisms, with that of rats.   Pigeons appear to be considerably less efficient than rats in avoiding previously visited alleys (Bond, Cook, & Lamb, 1981).   Given the pigeon's ability to home, it seems more plausible to attribute their poor performance in the radial maze to its win–stay tendency than to a poorer ability to represent spatial locations.

Putting aside the contribution of a win–shift strategy, it is unclear that the rat's ability to perform efficiently in a radial maze has very much in common with such superficially similar human abilities as remembering elements of arbitrary lists, e.g., which of a group of people have yet to be called, which errands have yet to be performed, and so on.   At present, we have no basis for assuming that a rat's ability to keep track of alleys that it has visited could generalize to tasks which require other responses (e.g., bar–pressing) or other reinforcers (e.g., water) or to the many kinds of arbitrary non–spatial tasks that language makes possible in the case of humans.

Virtually all of the examples of representation in animals described earlier warrant similar caution when it comes to extrapolating to human cognitive processes.   Both Herrnstein and Lea have noted that the processes used by pigeons to form concepts may differ considerably from those used by humans (Herrnstein & DeVilliers, 1980; Lea & Harrison, 1978; Morgan, Fitch, Holman, & Lea, 1976).   In the case of a pigeon's ability to represent a group of elements in performing a serial learning task, it is unlikely that their representations of these elements has much in common with human representations of serially ordered elements.   Both involve representation and both involve sequences.   There is, however, good reason to assume that, unlike the pigeon, humans encode each element of the sequence verbally.

These, and other problems suggested by recent demonstrations of animal cognition, leaves us with a baffling but fundamental question.   Now that there are strong grounds to question Descartes' contention that animals lack the ability to think, it is appropriate to ask, how *does* an animal think?   In particular, how does an animal think without language?   Learning the answer to that question will provide an important biological benchmark against which to assess the evolution of human thought.

REFERENCES

Amsel, A. The role of frustrative nonreward in noncontinuous reward situations. *Psychological Bulletin*, 1958, *55*, 102–119.

Anderson, J. R., & Bower, G. H. *Human Associative Memory*. Washington, D. C.: V. H. Winston & Sons, 1973.

Bem, D. J. Self–perception: An alternative interpretation of cognitive dissonance phenomena. *Psychological Review*, 1967, *74*, 183–200.

Bingham, H. C. Visual perception in the chick. *Behavior Monograph*, 1922, *4(20)*, 1–104.

Bond, A. B., Cook, R. G., & Lamb, M. R. Spatial memory and the performance of rats and pigeons in the radial-arm maze. *Animal Learning and Behavior*, 1981, *9*, 575–580.

Bousfield, W. A. The occurrence of clustering in the recall of randomly arranged associates. *Journal of General Psychology*, 1953, *49*, 229–240.

Bousfield, A. K., & Bousfield, W. A. Measurement of clustering and of sequential constancies in repeated free recall. *Psychological Reports*, 1966, *19*, 935–942.

Bower, G. H. A selective review of organizational factors in memory. In E. Tulving & W. Donaldson (Eds.), *Organization of memory*. New York: Academic Press, 1972.

Brown, P. L., & Jenkins, H. M. Auto-shaping of the pigeon's keypeck. *Journal of the Experimental Analysis of Behavior*, 1968, *11*, 1–8.

Catania, A. C. *Learning*. Englewood Cliffs, N.J.: Prentice-Hall, 1979.

Cerella, J. Absence of perspective processing in the pigeon. *Pattern Recognition*, 1977, *9*, 65–68.

Cerella, J. Visual classes and natural categories in the pigeon. *Journal of Experimental Psychology: Human Perception and Performance*, 1979, *5*, 68–77.

Conrad, R. Acoustic confusions in immediate memory. *British Journal of Psychology*, 1964, *55*, 78–84.

Cumming, W. W., & Berryman, R. The complex discriminated operant: Studies of matching-to-sample and related problems. In D. I. Mostofsky (Ed.), *Stimulus generalization*. Stanford: Stanford University Press, 1965.

Dopkins, S. C., Scopatz, R. A., Roitblat, H. L., & Bever, T. G. *Encoding and decision processes in the discrimination of 3-item sequences*. Proceedings of the 54th annual meeting of the Eastern Psychological Association, Philadelphia, April, 1983.

Ebbinghaus, H. *Memory*. New York: Teachers College, 1885. Translated by H. A. Ruger & C. E. Bussenius, 1913.

Ehrenfreund, D. An experimental test of the continuity theory of discrimination learning with pattern vision. *Journal of Comparative and Physiological Psychology*, 1948, *41*, 408–422.

Fouts, R. S. The use of guidance in teaching sign language to a chimpanzee. *Journal of Comparative and Physiological Psychology*, 1972, *80*, 515–522.

Fraenkel, G. S., & Gunn, D. L. *The orientation of animals: Kineses, taxes and compass reactions*. New York: Dover Publications, 1961.

Garcia, J., & Koelling, R. A. Relation of cue to consequence in avoidance learning. *Psychonomic Science*, 1966, *4*, 123–124.

Gardner, B. T., & Gardner, R. A. Two-way communication with an infant chimpanzee. In A. M. Schrier & F. Stollnitz (Eds.), *Behavior of non-human primates*. (Vol. 4.) New York: Academic Press, 1971.

Gardner, R. A., & Gardner, B. T. Comparative psychology and language acquisition. In K. Salzinger & F. Denmark (Eds.), *Psychology: The state of the art*. New York: New York Academy of Sciences, 1978.

Gibbon, J., Berryman, R., & Thompson, R. L. Contingency spaces and measures in classical conditioning. *Journal of the Experimental Analysis of Behavior*, 1974, *21*, 585–605.

Gillan, D. J. Reasoning in the chimpanzee: II. Transitive inference. *Journal of Experimental Psychology: Animal Behavior Processes*, 1981, *7*, 150–164.

Gillan, D. J., Premack, D., & Woodruff, G. Reasoning in the chimpanzee: I. Analogical reasoning. *Journal of Experimental Psychology: Animal Behavior Processes*, 1981, *7*, 1–17.

Goldstein, H., Krantz, D. L., & Rains, J. D. *Century Psychology*. Volume 1: *Controversial issues in learning*. New York: Meredith, 1965.

Gregory, R. L. Do we need cognitive concepts? In M. S. Gazzaniga & C. Blakemore (Eds.), *Handbook of psychobiology*. New York: Academic Press, 1975.

Griffin, D. R. *The question of animal awareness: Evolutionary continuity of mental experience*. New York: Rockefeller University Press, 1976.

Griffin, D. R. Prospects for a cognitive ethology. *The Behavioral and Brain Sciences*, 1978, *1*, 527–538.

Griffin, D. R. (Ed.) *Life Sciences Research Report*. Volume 21: *Animal mind – Human mind*. Berlin: Springer–Verlag, 1982.

Herrnstein, R. J. Acquisition, generalization, and discrimination reversal of a natural concept. *Journal of Experimental Psychology: Animal Behavior Processes*, 1979, *5*, 116–129.

Herrnstein, R. J., & DeVilliers, P. A. Fish as a natural category for people and pigeons. In G. H. Bower (Ed.), *The psychology of learning and motivation*. (Vol. 14.) New York: Academic Press, 1980.

Herrnstein, R. J., Loveland, D. H., & Cable, C. Natural concepts in pigeons. *Journal of Experimental Psychology: Animal Behavior Processes*, 1976, *2*, 285–302.

Hull, C. L. *Principles of behavior*. New York: Appleton–Century–Crofts, 1943.

Hulse, S. H. Cognitive structure and serial pattern learning by animals. In S. H. Hulse, H. Fowler, & W. K. Honig (Eds.), *Cognitive processes in animal behavior*. Hillsdale, N.J.: Erlbaum, 1978.

Hulse, S. H., Fowler, H., & Honig, W. K. (Eds.) *Cognitive processes in animal behavior*. Hillsdale, N.J.: Erlbaum, 1978.

Hunter, W. S. The delayed reaction in animals. *Behavior Monograph*, 1913, *2*, 6.

Kamin, L. J. Selective association and conditioning. In N. J. Mackintosh & W. K. Honig (Eds.), *Fundamental issues in associative learning*. Halifax: Dalhousie University Press, 1969.

Kohler, W. Simple structural functions in the chimpanzee and in the chicken. In W. D. Ellis (Ed.), *A source book of Gestalt psychology*. New York: Humanities Press, 1955.

Krechevsky, I. A study of the continuity of the problem solving process. *Psychological Review*, 1938, *45*, 107–133.

Lawrence, D. H. The nature of a stimulus: Some relationships between learning and perception. In S. Koch (Ed.), *Psychology: A study of a science*. (Vol. 5.) New York: McGraw–Hill, 1963.

Lea, S. E. G., & Harrison, S. N. Discrimination of polymorphous stimulus sets by pigeons. *Quarterly Journal of Experimental Psychology*, 1978, *30*, 521-537.

Leahey, T. H. *The revolution never happened: Information processing is behaviorism*. Paper presented at the 52nd Annual meeting of the Eastern Psychological Association, April, 1981.

Loeb, J. *Comparative physiology of the brain and comparative psychology*. New York: Putnam, 1900.

Lorenz, K. Z. The comparative method in studying innate behavior patterns. *Symposia of the Society for Experimental Biology*, 1950, *4*, 221-268.

Mandler, G. Organization and memory. In K. W. Spence & J. T. Spence (Eds.), *The psychology of learning and motivation*. (Vol. 1.) New York: Academic Press, 1967.

Marr, J. Memory: Models and metaphors. *The Psychological Record*, 1983, *33*, 12-19.

Morgan, C. L. *An introduction to comparative psychology*. London: Walter Scott, 1894.

Morgan, M. J., Fitch, M. D., Holman, J. G., & Lea, S. E. G. Pigeons learn the concept of an "A". *Perception*, 1976, *5*, 57-66.

Olton, D. S. Characteristics of spatial memory. In S. H. Hulse, H. Fowler, & W. K. Honig (Eds.), *Cognitive processes in animal behavior*. Hillsdale, N.J.: Erlbaum, 1978.

Olton, D. S. Mazes, maps, and memory. *American Psychologist*, 1979, *34*, 583-596.

Olton, D., & Samuelson, R. J. Remembrance of places passed: Spatial memory in rats. *Journal of Experimental Psychology: Animal Behavior Processes*, 1976, *2*, 97-116.

Patterson, F. The gestures of a gorilla: Language acquisition in another pongid. *Brain and Language*, 1978, *5*, 72-97.

Patterson, F. G. *Linguistic capabilities of a lowland gorilla*. PhD Thesis, Stanford University, 1979.

Pavlov, I. P. *Conditioned reflexes*. London: Oxford University Press, 1927.

Premack, D. A functional analysis of language. *Journal of the Experimental Analysis of Behavior*, 1970, *4*, 107-125.

Premack, D. *Intelligence in ape and man*. Hillsdale, N.J.: Erlbaum, 1976.

Premack, D., & Woodruff, G. Does the chimpanzee have a theory of mind? *The Behavioral and Brain Sciences*, 1978, *1*, 515-526.

Pylyshyn, Z. W. What the mind's eye tells the mind's brain: A critique of mental imagery. *Psychological Bulletin*, 1973, *80*, 1-24.

Pylyshyn, Z. The imagery debate: Analogue versus tacit knowledge. *Psychological Review*, 1981, *88*, 16-45.

Rescorla, R. A. Pavlovian conditioning and its proper control procedures. *Psychological Review*, 1967, *74*, 71-80.

Rescorla, R. A. Informational variables in Pavlovian conditioning. In G. H. Bower (Ed.), *The psychology of learning and motivation*. (Vol. 6.) New York: Academic Press, 1972.

Rescorla, R. A., & Wagner, A. R. A theory of Pavlovian conditioning: Variations in the effectiveness of reinforcement and nonreinforcement. In A. H. Black & W. F. Prokasy (Eds.), *Classical conditioning II: Current research and theory*. New York: Appleton–Century–Crofts, 1972.

Restle, F. Discrimination of cues in mazes: A resolution of the "Place-vs.–Response" question. *Psychological Review*, 1957, *64*, 217–228.

Rilling, M., Kendrick, D. F., & Stonebraker, T. B. Stimulus control of forgetting: A behavioral analysis. In M. L. Commons, R. J. Herrnstein, & A. R. Wagner (Eds.), *Quantitative analyses of behavior: Vol. III, Acquisition*. Ballinger, 1984. In press.

Rock, I. The role of repetition in associative learning. *American Journal of Psychology*, 1957, *70*, 186–193.

Roitblat, H. L. Codes and coding processes in pigeon short-term memory. *Animal Learning and Behavior*, 1980, *8*, 341–351.

Roitblat, H. L. The meaning of representation in animal memory. *The Behavioral and Brain Sciences*, 1982, *5*, 353–406.

Romanes, G. J. *Animal intelligence*. London: Kegan Paul, 1882.

Rosch, E. Principles of categorization. In E. Rosch & B. B. Lloyd (Eds.), *Cognition and categorization*. Hillsdale, N.J.: Erlbaum, 1978.

Rumbaugh, D. M. (Ed.) *Language learning by a chimpanzee: The Lana project*. New York: Academic Press, 1977.

Rumbaugh, E. S., Rumbaugh, D. M, Smith, S. T., & Lawson, J. Reference: The linguistic essential. *Science*, 1980, *210*, 922–924.

Seligman, M. E. P., & Hager, J. L. (Eds.) *Biological boundaries of learning*. New York: Appleton–Century–Crofts, 1972.

Sherrington, C. S. *Integrative action of the nervous system*. Cambridge, England: Cambridge University Press, 1906.

Shimp, C. P. Short-term memory in the pigeon: Relative recency. *Journal of the Experimental Analysis of Behavior*, 1976, *25*, 55–61.

Skinner, B. F. *The behavior of organisms*. New York: Appleton–Century–Crofts, 1938.

Skinner, B. F. The operational analysis of psychological terms. *Psychological Review*, 1945, *52*, 270–277.

Skinner, B. F. Are theories of learning necessary? *Psychological Review*, 1950, *57*, 193–216.

Skinner, B. F. *Science and human behavior*. New York: Macmillan, 1953.

Skinner, B. F. *Contingencies of reinforcement: A theoretical analysis*. New York: Appleton–Century–Crofts, 1969.

Spence, K. W. The differential response in animals to stimuli varying within a single dimension. *Psychological Review*, 1937, *44*, 430–444.

Spence, K. W. Cognitive vs. Stimulus–Response theories of learning. *Psychological Review*, 1950, *57*, 159–172.

Spence, K. W. *Behavior theory and conditioning*. New Haven: Yale University Press, 1956.

Straub, R. O., & Terrace, H. S. Generalization of serial learning in the pigeon. *Animal Learning and Behavior*, 1981, *9*, 454–468.

Straub, R. O., Seidenberg, M. S., Bever, T. G., & Terrace, H. S. Serial learning in the pigeon. *Journal of the Experimental Analysis of Behavior*, 1979, *32*, 137–148.

Sutherland, N. S., & Mackintosh, N. J. *Mechanisms of animal discrimination learning*. New York: Academic Press, 1971.

Terrace, H. S. Stimulus control. In W. K. Honig (Ed.), *Operant behavior areas of research and application*. Englewood Cliffs, N.J.: Prentice Hall, 1966.

Terrace, H. S. *Nim*. New York: Knopf, 1979.

Terrace, H. S. Animal vs. human minds: Commentary on Roitblat's "The meaning of representation in animal memory". *The Behavioral and Brain Sciences*, 1982, *22*, 391-2.

Terrace, H. S. Simultaneous chaining: The problem it poses for traditional chaining theory. In M. L. Commons, R. J. Herrnstein, & A. R. Wagner (Eds.), *Quantitative analyses of behavior: Vol. III, Acquisition*. Cambridge, Mass.: Ballinger, 1983. In press.

Terrace, H. S., Petitto, L. A., Sanders, R. J., & Bever, T. G. Can an ape create a sentence? *Science*, 1979, *206*, 891-900.

Terrace, H. S., Straub, R. O., Bever, T. G., & Seidenberg, M. S. Representation of a sequence by a pigeon. *Bulletin of the Psychonomic Society*, 1977, *10*, 269.

Thistlethwaite, D. A critical review of latent learning and related experiments. *Psychological Bulletin*, 1951, *48*, 97-129.

Thorndike, E. L. Animal intelligence: An experimental study of the associative processes in animals. *Psychological Review, Monograph Supplement*, 1898, *2*, 8.

Thorndike, E. L. *Animal intelligence*. New York: Macmillan, 1911.

Thorpe, W. H. *Learning and instinct in animals*. London: Methuen, 1963.

Tinbergen, N. *The study of instinct*. London: Oxford University Press, 1951.

Tolman, E. C., & Honzik, C. H. Introduction and removal of reward, and maze performance in rats. *University of California Pub.*, 1930, *4*, 257-275.

Tolman, E. C., Ritchie, B. F., & Kalish, D. Studies in spatial learning. I. Orientation and the short-cut. *Journal of Experimental Psychology*, 1946, *36*, 13-24. (a)

Tolman, E. C., Ritchie, B. F., & Kalish, D. Studies in spatial learning. II. Place learning versus response learning. *Journal of Experimental Psychology*, 1946, *36*, 221-229. (b)

Tulving, E. Subjective organization in free recall of "unrelated" words. *Psychological Review*, 1962, *69*, 344-354.

Tulving, E. Theoretical issues in free recall. In T. R. Dixon & D. L. Horton (Eds.), *Verbal behavior and general behavior theory*. Englewood Cliffs, N.J.: Prentice-Hall, 1968.

Wagner, A. R. Frustration and punishment. In R. N. Haber (Ed.), *Current research in motivation*. New York: Holt, Rinehart and Wilson, 1966.

Wagner, A. R. Expectancies and the priming of STM. In S. H. Hulse, H. Fowler, & W. K. Honig (Eds.), *Cognitive processes in animal behavior*. Hillsdale, N.J.: Erlbaum, 1978.

Wagner, A. R. SOP: A model of automatic memory processing in animal behavior. In N. E. Spear & R. R. Miller (Eds.), *Information processing in animals: Memory mechanisms*. Hillsdale, N.J.: Erlbaum, 1981.

Wagner, A.R., & Rescorla, R.A. Inhibition in Pavlovian Conditioning: Application of a Theory. In R.A. Boakes & M.S. Holliday (Eds.), *Inhibition and learning*. New York: Academic Press, 1972.

Wasserman, E. A. Is cognitive psychology behavioral? *The Psychological Record*, 1983, *33*, 6–11.

Watson, J. B. *Behavior: An introduction to comparative psychology*. New York: Henry Holt, 1914.

Watson, J. B. The place of the conditioned reflex in psychology. *Psychological Review*, 1916, *23*, 89–116.

Watson, J. B. *Psychology from the standpoint of a behaviorist*. Philadelphia: Lippincott, 1919.

Weisman, R. G., & DiFranco, M. P. Testing models of delayed sequence discrimination in pigeons: Delay intervals and stimulus durations. *Journal of Experimental Psychology*, 1981, *7*, 413–424.

Weisman, R. G., Wasserman, E. A., Dodd, P. W. D., & Larew, M. B. Representation and retention of two-event sequences in pigeons. *Journal of Experimental Psychology: Animal Behavior Processes*, 1980, *6*, 312–325.

Wright, A. A., Santiago, H. C., Urcuioli, P. J., & Sands, S. F. Monkey and pigeon acquisition of same/different concept using pictorial stimuli. In M. L. Commons, R. J. Herrnstein, & A. R. Wagner (Eds.), *Quantitative analyses of behavior: Vol. III, Acquisition*. Cambridge, Mass.: Ballinger, 1984. In press.

Zeaman, D., & House, B. J. The role of attention in retardate discrimination learning. In N. R. Ellis (Ed.), *Handbook of mental deficiency*. New York: McGraw-Hill, 1963.

2 CONTRIBUTIONS OF ANIMAL MEMORY TO THE INTERPRETATION OF ANIMAL LEARNING

Werner K. Honig
Dalhousie University

I. INTRODUCTION

The processes of learning and memory in animals would, on the face of it, appear to be closely related. No subject will show the benefits of learning unless it can remember between trials what it has learned, and the memory of what has happened within a trial will presumably determine the course of learning itself. In fact, one could define learning as the encoding of information into memory. Yet, as Bolles (1976) has pointed out, students of learning have "gained very little from the work of their colleagues studying memory (p. 24)." He suggests that "The investigation of memory has produced many conceptual tools that could be of great value in working out a better understanding of learning in particular and behavior in general (p. 25)." In this essay I will try to show that Bolles' remarks were appropriate and prophetic.

To begin, it will be useful to distinguish between three conceptual categories for the study of animal memory (cf. Honig, 1978). Reference memory is the "long-term store" of environmental contingencies -- or of the appropriate responses which they engender -- acquired in the course of learning. Without reference memory, the subject would forget between trials what it has learned. Obviously, reference memory is involved in learning. Furthermore, some processes which are revealed by the study of reference memory may be important for acquisition of behavior as well. I will argue that retrieval plays such a role.

Working memory (which is often called "short-term" memory) is the memory for an event in a trial which governs behavior after a delay or memory interval later in the same trial. For example, in delayed matching-to-sample there is a delay between the sample, or initial stimulus, and the choice at the time of the test stimulus. Since the content of the memory changes from trial to trial, the subject may learn to "reset" working memory between trials. I have proposed for some time that working memory involves "prospective" as well as "retrospective" memory processes (Honig, 1978, 1981; Honig & Thompson, 1982).

Finally, there is associative memory, which has received rather little attention. It also involves a delay between important events in each trial, but this delay is in effect from the beginning of training. This delay or memory interval can be placed between an initial stimulus or cue, and the correct response. This is the old delayed response paradigm. The cue can also be the outcome of the prior trial, or the response on that trial, in which case the inter-trial interval serves as the memory interval. This often takes the form of a delayed alternation. Associative memory is also required when there is a delay of reinforcement or other trial outcome. This is often called "long-delay learning" (Lett, 1979). It is observed in classical conditioning, as in flavor-aversion learning, or in instrumental discriminations with delayed reinforcement.

Associative memory paradigms are, in principle, the same as other "simple" learning procedures in which there is no explicit memory requirement. However, even in these, the successiveness of significant events -- the cue followed by the response, or the response followed by the reinforcer -- requires the subject to remember the first event at the time of the second. The introduction of a delay makes the memory requirement explicit. This method may then serve as a sort of temporal "macroscope" to examine memory processes that may be operating, but may not be recognized, in associative memory, and thus enter into the establishment of reference memories.

II.   REFERENCE MEMORY AND RETRIEVAL

A retrieval process is identified when a cue from the original learning situation facilitates the performance of previously learned behavior, in conjunction with the discriminative stimuli that direct or control such behavior. In many situations, the discriminative stimulus may suffice as the retrieval cue. However, such cues are most clearly identified if these functions are separated -- if it is shown that the cue is necessary to reactivate a reference memory, but not sufficient to control the appropriate behavior. The reactivation allows further strengthening or modification of the reference memory in the course of learning.

The "domain of memory retrieval," as Spear (1981) has called it, is extensive and well documented. I will cite only one example from current research to illustrate the potency of retrieval cues. Thomas and his associates (Thomas, 1981; Thomas, McKelvie, Ranney, & Moye, 1981) trained pigeons in a successive operant discrimination in which responses to a spectral light of 530 nm (S+) on the key were reinforced, and responses to a light of 576 nm (S-) were not. The discrimination was then reversed during a single, extended training session in the same context. A post-discrimination spectral generalization test showed that the birds were controlled by the second discrimination. Other groups of birds learned the two discriminations in different contexts; one was houselight plus tone, while the other was no houselight plus noise. When the birds were then tested in the context that had been present in the first discrimination, they were controlled by the cues appropriate to that discrimination. If several reversals were carried out successively with different contexts, performance became dominated by the context to such a degree that contrasting gradients could be obtained more or less at will.

This research with a discrimination paradigm demonstrates the power of retrieval cues in a rather extreme form. Much of the other research in this area has been done with simple acquisition. The presentation of some aspect of a prior learning trial improves performance, even though this procedure is not sufficient to evoke the trained response. Retrieval cues are generally presented near the beginning of a trial; presumably they activate a memory of events from prior trials. If so, then the memory of past events must take the form of an anticipation or "representation" (Roitblat, 1982) of events which will occur in the current trial. I want to suggest in this essay that a similar process of retrieval and representation enters into acquisition, or associative memory, as well, particularly when there are delays between the important events within trials.

III. ASSOCIATIVE MEMORY MECHANISMS

A. DELAYED ALTERNATION

In Capaldi's work on delayed alternation of reward, rats learned to use the outcome of one trial as the cue for the reinforcement condition in the next trial. Capaldi and Stanley (1963) showed that rats learned to run quickly after an unrewarded trial (N) and slowly after an unrewarded one (R) in the NRNRNR. . .sequence. Performance was rather insensitive to the ITI, and an ITI of 20 min was mastered with little difficulty. The paradigm is not restricted to the alternation of reward and non-reward. Pschirrer (1972) ran rats with a fixed repeating sequence of milk, pellets, and non-reward. They learned to runs slowly only on those trials which followed pellets as the prior reward. Thus they learned to use the type of reward from a prior trial as a cue for non-reward in the current trial following a 15-min ITI. Petrinovich and Bolles (1957) showed that rats could learn to alternate the arms of a T-maze on successive runs during a session. Some rats could perform well with ITI's of a few hours.

It is of course possible that memories for appropriate events remain active between trials in situations such as these. But a retrieval mechanism provides a more credible account. Presumably the memory of each trial became "inactive" during the long ITI, and cues from the apparatus retrieve it at the beginning of the next trial. The rats would need to learn to retrieve the memory only of the preceding trial; otherwise interference would result. An interpretation in terms of retrieval receives further support from work by Capaldi (1971) on a double-alternation problem (NNRRNNRR. . .). In this procedure, non-rewarded trials follow rewarded and non-rewarded trials equally often. Rats find it almost impossible to learn this problem in the standard way. Capaldi overcame the difficulty by alternating the color of the alley (black or white) so that the problem was reduced to two independent single-alternation problems differentiated by color: BN, WN, BR, WR, BN, WN, BR, WR. . . Alley color presumably provided a retrieval cue which enabled the rat to remember the outcome of the previous trial with the same color. When alley color was randomized with respect to the double alternation, the rats did not learn the problem.

B. DELAY OF OUTCOME

Other associative memory procedures incorporate a delay between the initial cue or the response and the outcome of the trial. A familiar example is flavor-aversion learning. There may be an interval of hours between the time that a rat is exposed to a flavor and the time that it is poisoned, yet it develops an association between these two events. Clearly, learning theory should provide a mechanism for this association.

Flavor-aversion learning with long delays is not an isolated adaptation to "natural" contingencies (the delay between eating and suffering the consequences); long-delay learning is also demonstrated with discrete-trial instrumental procedures. Lett has worked extensively with delay of reward in a T-maze. Rats learned to choose one of two highly discriminable arms even if the reward was delayed by a number of minutes or even an hour (see Lett, 1979, for a review). The animals normally spent the memory interval in their home cage, and were given their food reward in the start box of the maze at the beginning of the next trial.

D'Amato and his associates carried out similar research with monkeys and rats (D'Amato, Safarjan, & Salmon, 1981; see also D'Amato & Buckiewicz, 1980). These authors studied "spatial attractions" by placing animals on one or the other arm of a T-maze, removing them, feeding them after a 30-min delay in the start box of the maze. Even within a few trials, the subjects began to spend most of their time in the arm of the maze which preceded reward. In spite of occasional difficulties in obtaining positive results, research with these paradigms indicates that animals can associate an initial event with a trial outcome over delays of several minutes and in some cases even hours.

In this sort of "long-delay learning" situation the subject presumably needs to remember, at the time of reinforcement, what it did (or consumed) earlier in the trial, rather than the outcome of the previous trial. Lett (1979) suggested that the memory of the earlier event becomes "inactive" during the delay and is retrieved at the time of the trial outcome. The problem is to identify the retrieval cues at the end of the trial. In the normal instrumental learning situation there are presumably many apparatus cues in common to the performance of the response and the subsequent delivery of reinforcement. In fact, according to Lett, these cues provide ample opportunity for interference which will prevent associations over a delay. Stimuli from the environment will reactivate memories of the initial stimulus during the memory interval, and these can then be associated with an improper "trial outcome" -- the subject's current behavior, or other aspects of the situation. For this reason, experiments with associative delays are generally most successful if the memory interval is spent in the home cage.

Lett supported her theory with an experiment in which rats were kept in the correct end-box for a portion of the memory interval. This should reduce the length of the memory interval and facilitate learning. Instead, it interfered with long-delay learning; rats which spent half of the memory interval in the correct-end box did not learn the discrimination, in contrast to rats who spent the entire memory interval in the home cage (Lett, 1975). In other studies (Barnes, 1978; cited by Lett, 1979), rats were put in the start box during the memory interval, where they eventually received reward, or into a facsimile of the start box in a different location. Again, learning was poorer than for a control group of rats which were kept in their home cages during the entire memory interval and then fed in the start box. These findings indirectly support Lett's retrieval hypothesis. They show that if the start box is "isolated" in the trial as a retrieval cue learning is enhanced.

Lett also tested a more direct deduction from her hypothesis, namely that long-delay learning is facilitated by providing retrieval cues from the maze at the time of the reinforcement. Following an unpublished study by Denny, she either rewarded rats in the start box (which would provide such cues) or in their home cages. She found that, in fact, rats did learn the correct choice even if rewarded in the home cage. This was confirmed by D'Amato et al. (1981), who rewarded their rats and monkeys outside the training situation. Lett suggests that the experimenter may have provided retrieval cues because he or she both placed the animal into the apparatus at the beginning of each trial and then provided the reward. Unfortunately, no one has tested this notion by providing the reward (outside the start box) by remote control.

While the retrieval hypothesis may be defensible for long-delay learning with instrumental tasks, it is not clear how it would work for flavor-aversion learning. What cues are common to the events that comprise the association of interest? Perhaps it is handling by the experimenter. But this does not happen to the rat in "real life." The rat simply eats a substance, becomes ill, and later avoids the same substance. Perhaps illness is a mechanism for retrieving the memory of foods that have recently been eaten. But this is ad hoc, and it violates the basis of the retrieval paradigm, namely that an initial event (the flavor) and the outcome (illness) must have some element in common.

My general conclusion from this research on long-delay learning is that retrieval cues are effective at the beginning of a trial if the discriminative cue occurred on a prior trial, as in alternation studies. In my view, it has yet to be demonstrated that retrieval cues are effective within trials -- that the reinforcer serves as a cue to retrieve the memory of a prior response or of the CS.

IV. WORKING MEMORY AND PROSPECTIVE PROCESSING

In working memory paradigms, delays occur within trials, normally between an initial discriminative stimulus (or "sample"), and test stimulus (or "comparison"). the problem is usually learned without a delay, so that the association is readily established. The memory test is a test of discriminative control, not of the ability to learn the problem.

The most familiar working memory paradigm is delayed matching-to-sample, where the sample provides the cue for the appropriate choice between the comparison stimuli after the memory interval. The decline of performance with increases in the memory is usually quite rapid -- within a few seconds when the pigeons are used (Roberts & Grant, 1976), and within a minute or two when monkeys or dolphins serve as subjects (D'Amato, 1973; Herman & Thompson, 1982). It is likely that memory in some form remains active within trials. A retrieval process is unlikely, because pigeons, at least, do no better on "true" matching to sample than they do on delayed conditional matching, where neither of the test stimuli is actually identical to the sample, and the "matches" are assigned by an arbitrary rule (also called symbolic matching). The specific dimensions, particularly of the initial stimuli, are much more important for the pigeon. Colors are better remembered than line orientations, no matter what the test stimuli are.

Research with monkeys and dolphin supports this general impression gained from the pigeon. D'Amato and Worsham (1974) explicitly compared performance on delayed matching–to–sample and delayed conditional matching with visual stimuli. While their monkeys learned the latter problem more slowly, loss of information over memory intervals ranging up to 2 min was very similar for both problems. Herman and Thompson (1982) obtained similar results with their dolphin. (But see Zentall, Hogan, & Edwards, Chapter 22, this volume, regarding transfer of the "identity concept" across different problems.)

If a retrieval mechanism is not at work in working memory, what other processes are available? Two kinds of retrospective processes have been suggested. One is the trace of the initial stimulus, which decays during the memory interval, and which controls choice of the correct test stimulus if it is sufficiently strong at the end of the memory interval. This hypothesis was presented and defended at length by Roberts and Grant (1976), although it has recently been called into question by Grant (1981). Another suggested process is temporal discrimination of the most recent initial stimulus (D'Amato, 1973). If the animal can remember which was the most recent initial stimulus at the time of the test stimulus, then he can make the correct response. There is some evidence in favor of this notion; in particular, monkeys do better with a large sample set of six or seven initial stimuli than with a small sample set of only two. In the latter case, the relative recency of the sample controlling the incorrect choice will generally be higher than in the former. It is hard to distinguish experimentally between this hypothesis and the trace theory, since they make similar predictions (but see Chapter 5, this volume, by Roitblat). With a small sample set it is more likely that there will be a residual trace from a prior trial which is "matched" by the incorrect choice on the current trial. However, as we shall see in the following discussion, it is possible that pigeons use a prospective memory process even in delayed matching–to–sample and delayed conditional matching procedures.

A. PROSPECTIVE MEMORY PROCESSES

A good deal of recent research and writing supports the notion of "prospective" rather than retrospective processes in working memory (Grant, 1981; Honig & Thompson, 1982). The hypothesis is that subjects learn to anticipate the correct response, the trial outcome, or some other event within the trial, and can remember that anticipation throughout the memory interval. The most relevant experimental paradigms are those in which sufficient information regarding the trial outcome or the correct response is presented early in the trial.

I (Honig, 1978) devised a procedure in which the initial stimulus indicated whether the trial would end with a relatively extended exposure to a favorable schedule (VI 30 sec) or to a period of extinction. Following the offset of the initial stimulus there was a delay condition, in which reinforcement was available at a low rate (VI 120 sec). The first part of the delay was also the memory interval. When this ended, the pigeon could procure the favorable outcome or choose to remain in the delay condition. The subjects showed the appropriate discrimination with rather little loss, even over fairly extended delays ranging up to 30 sec. This procedure is described in detail elsewhere (Honig, 1978, 1981; Grant, 1981). However, the method did not provide a direct comparison with a method where the trial outcome could not be readily anticipated by the subject.

This comparison is facilitated with a design carried out by Honig and Wasserman (1981). Each trial began with a color as the initial stimulus on the pigeon's key, and ended with a horizontal or vertical line. In the delayed simple discrimination, all trials beginning with one color ended with reinforcement for pecking to either line, while trials beginning with the other color ended in a time-out. In this delayed simple discrimination, the subjects could anticipate the trial outcome from the time of the initial stimulus. In the delayed conditional discrimination, the trial outcome was determined conjointly by the initial stimulus and the test stimulus, so that only certain sequences preceded reinforcement, while others did not. In the delayed conditional discrimination, then, the subject could not anticipate the trial outcome until the time of TS. When memory intervals were introduced after the initial stimulus, performance on the delayed simple discrimination was clearly superior as indicated by differential rates of responding to the test stimulus on positive and negative trials.

Even when the subject has to remember the initial stimulus, as in the delayed conditional discrimination, that memory can be enhanced if information regarding the nature of the trial outcome is provided during the memory interval. This is shown by the so-called differential outcome studies. DeLong and Wasserman (1981) trained pigeons with a delayed conditional discrimination similar to the one just described. However, for a differential outcome group, the two initial stimuli were associated with different probabilities of reward on positive trials (e.g., 1.0 for trials beginning with red and 0.2 for trials beginning with green). For the non-differential outcome group, the probability of reward on all positive trials was 0.6. Pigeons in the differential condition acquired the discrimination more readily, and performed more accurately with memory intervals of 5 and 10 sec. This experiment is a systematic replication of the work of Peterson and his associates (Peterson, Wheeler, & Trapold, 1980; Peterson & Trapold, 1980) who used traditional delayed matching procedures involving a choice between two test stimuli and rather different outcome dimensions (see Peterson, Chapter 8, this volume).

It seems clear, then, that pigeons can anticipate a trial outcome, and use the anticipation to enhance working memory. The question remains whether they can also encode and remember the appropriate response when all trials can in principle end with the same reinforcement. Grant (1981) supported the notion of the anticipation of the correct choice in a delayed conditional matching procedure. He presented two or three initial stimuli at the start of some trials. He observed that choice responding was enhanced when these "congruent" stimuli cued the same correct choice, compared to the standard procedure, when only one initial stimulus was presented. The improvement was the same whether the same initial stimulus was repeated, or whether congruent initial stimuli from different dimensions were presented. The initial stimulus dimensions were key color, food presentation vs. no food, and number of pecks required to terminate a white initial stimulus. Thus the repetition of three green initial stimuli had the same effect (compared to a single green initial stimulus) as the successive presentation of green, food, and the requirement of a single response. Likewise, interference was produced by two incongruent initial stimuli to the same degree, whether they came from the same or different dimensions. Similar findings were reported by Maki, Moe, and Bierly (1977).

Other evidence also supports the notion of prospective processing, such as the work on directed forgetting (Grant, 1981; Maki, 1981; Rilling, Kendrick, & Stonebraker, 1984), and on the "representation" of test stimuli (Roitblat, 1980, 1982, and Chapter 5, this volume). This material cannot be fully reviewed here. The important point is that working memory can be prospective in nature, and that trial outcomes and specific discriminative responses can be encoded and anticipated. But this does not exclude the possibility of retrospective processing. I have already indicated that in the delayed conditional discrimination the pigeon probably remembers the initial stimulus until the test stimulus is presented. In certain other paradigms, "short-term memory" in the more traditional sense is clearly at work. A straightforward example is the probe recognition of stimuli in a list (Thompson & Herman, 1976; Sands & Wright, 1980; see also Chapter 21, by Sands, Urcuioli, Wright, & Santiago, and Chapter 20, by Wright, Santiago, Sands, & Urcuioli in this volume).

V.   INTEGRATION: A REINTERPRETATION OF LONG-DELAY LEARNING

The foregoing material has provided the background for this section, in which I want to apply concepts derived from animal memory to the interpretation of animal learning. The discussion is focused upon long-delay learning because that procedure incorporates a memory interval. However, I think that my conclusions also apply to "regular" training, when there is no explicit delay between a response and its consequences. I want to interpret long-delay learning by applying the notion of retrieval, derived from reference memory, and the notion of prospective processing, derived from working memory.

In the course of training, animals are exposed to repeated sequences of events called "trials." These sequences establish temporal representations of the events, which correspond roughly to the representations of geographical arrays which have come to be called "cognitive maps." I have therefore called these representations "temporal maps" (Honig, 1981). The events within a trial are temporally spaced, but they are clearly separated by much larger intervals from prior and subsequent trials. At the beginning of a long–delay learning trial, the initial cues activate the appropriate temporal map, and thereby retrieve the memory of events from previous trials. This memory now determines in anticipatory fashion the subsequent choices or other behaviors on that trial. Trial outcomes have the effect of confirming (or disconfirming) expectancies generated by activation of the temporal map.

This retrieval process differs from the role assigned to it by Lett in the following way: Lett suggests that in long–delay learning, retrieval occurs at the end of the trial, when reinforcement is presented. At that time the cues retrieve the memory of the initial stimulus or of prior behavior on that particular trial, and this memory is then associated with the trial outcome. This is clearly retrospective. I suggest the opposite process: At the start of a trial, the retrieval cue establishes an anticipation which is subject to further confirmation later in the trial, following a memory interval in long–delay learning procedures. This anticipation controls behavior in the presence of the initial stimulus.

This general paradigm can be applied to specific cases. I will try to show that it provides a more adequate understanding of several long–delay learning phenomena than does an interpretation based on retrospective processing or retrieval at the time of reinforcement.

A. FLAVOR-AVERSION LEARNING

It is not known how illness can provide a retrieval cue for the initial flavor, unless a highly specific adaptive mechanism is postulated for this process. Flavor and illness need not have distinctive stimuli in common, such as handling by the experimenter or a common location. Let us assume instead that the subject does not retrieve the flavor cue when he becomes ill. Instead, suppose that when he encounters the flavor on the next occasion after poisoning, this retrieves the sequence of events following the prior encounter with the flavor. The anticipation of illness suppresses drinking. If the animal drinks again and is not poisoned, the representation of subsequent events is changed, and extinction will occur. But if retrieval occurred only at the time of poisoning, it would be difficult to explain extinction, since there would be no event to retrieve memory of the flavor to modify the association.

B. T-MAZE DISCRIMINATIONS

When the rat enters the start box of the maze, events from prior trials are retrieved. One set of events (going to the correct end box) is prospectively associated with reward; the other, with extinction. Lett (1979) suggested that feeding the animal in the start box after the intra-trial delay is important because the start box retrieves the memory of the choice from the same trial and this has to be associated with food. I maintain that this is irrelevant; feeding the animal only confirms an anticipation of reinforcement based upon prior trials. In fact, as we have seen (D'Amato et al., 1981), cues from the start box or the maze are not essential at the time of the reinforcement.

Lawrence and Hommel (1961) replicated a study by Grice (1948) who had shown that rats can suffer only a short delay of reinforcement in a black–white discrimination procedure. The rats had to choose a white or a black tunnel between a start box and a gray delay chamber. After the delay they were admitted to a goal box. Lawrence and Hommel provided distinctive goal boxes following the choice of the black or white tunnel. This greatly increased accuracy of choice with a delay of reinforcement.

These findings cannot be readily explained with the help of retrieval cues. There is no reason to suppose that increasing the discriminability of the "outcome chambers" would enhance the retrieval of the memory of running through a black or a white alley on the way to the holding chamber. In fact, in one of their conditions, the rat was fed in a black end box if the white alley was correct, and vice versa. If the color of the end box acted as a retrieval cue, this should have led to interference. Instead, facilitation was similar to that observed when the goal boxes differed on other dimensions (size and floor texture).

The results of this study do make sense within the prospective interpretation of learning proposed here. The effect of making the "correct" and the "incorrect" goal boxes discriminable is parallel to the differential outcomes effect described earlier, except that the stimuli accompanying reward and non-reward, rather than the trial outcomes themselves, are rendered more discriminable.

The role of retrieval cues at the end of the trial is also put in question by a T-maze study carried out by Grant (1980). On each trial, rats were forced to run into one arm of the maze as the initial cue (the "forcing" run). After a memory interval of 0 or 20 sec, they were given a test run with both arms open. The baited arm was always opposite to the arm which had been open during the forcing run. This whole sequence was at times preceded by an "interfering" run, in which the turn was also forced, but in the direction opposite to the forcing run. Such interference reduced the accuracy on the test run. In prior unpublished work, Roberts had shown that contextual cues can modulate this interference. When the contexts of the interfering and the forcing trials differed, the interference was reduced. This can be explained on the assumption that retrieval cues occur during the free-choice trial. When the context of the interfering run differed from the context of the forcing run and of the test run, the memory of the former was presumably less strongly activated during the test trial.

Grant tested this notion by running three kinds of sequences of the interfering, forcing, and test trials. In one, the contextual cues (room lights on or off) were all the same (AAA). In another, the contextual cue of the interfering trial differed from the others (BAA). In the third, the cues accompanying the interfering and the forcing trial differed from each other (ABA). Grant observed reliably less interference with BAA and ABA than with AAA. However, the two sequences involving different contexts for the interfering and the forcing trials (BAA and ABA) produced the same amount of interference. If retrieval cues had been effective, then performance should have been poorer with ABA than with BAA, since the cues at the time of the test run should have retrieved the memory of the interfering run more strongly. This was not the case, and Grant attributes the effect of different contexts to the enhanced discriminability of the forcing and the interfering trials prior to the choice. If the rat generates a temporal representation of the sequence, and can identify the second run as the relevant cue, then it would enhance the discriminability of that cue to distinguish it from the first run by providing different contexts.

C. PATTERNING OF REINFORCEMENT

In a prior section, I described patterning of reinforcement in the runway. The retrieval hypothesis seemed to account for single and double delayed alternation. The retrieval would have to occur at the beginning of the trial, when the subject needs a discriminative stimulus in order to direct its speed of running. This in keeping with the general role that I have suggested for the retrieval mechanism. My critique of the use of this concept has been limited to "long–delay" learning when the delay occurs between the correct response and reinforcement.

Nevertheless, the notion of prospective processing may also be important for the understanding of patterned reinforcement, where successive runs are required within trials, and there is a discriminable ITI which exceeds the interval between runs. Capaldi and Verry (1981) trained rats with such a procedure. There were five runs on each trial. If the first run provided reinforcement of 20 pellets, subsequent runs were not reinforced (the negative, or 20–0–0–0–0 sequence). If the first run was not reinforced, the last run provided 20 pellets (a positive, or 0–0–0–0–20 sequence). The rats eventually differentiated their running speed on the last run in each trial, running slowly in the former, and quickly in the latter sequence.

There is no simple way to account for this performance if the rats retrieved only the memory of the outcome of the prior run, since the outcome of the second, third, and fourth runs was always the same. Of course, the rat might have retrieved the memory of the outcomes of all of the prior runs in the course of a trial; then he would have had to recollect the number of prior unrewarded runs in order to make the discrimination. It is not more complex to suggest that the outcome of the first run, which follows a discriminable ITI, provides a cue for anticipating the rest of the trial, and the rat adjusts its running speeds accordingly.

It is perhaps of some interest that on positive trials the rats increased their running speed successively on the second, third, and fourth runs, even when they were very well trained. On negative trials, they ran slowly on all but the first run by the end of training. This is what one would expect if the subject anticipates reinforcement at the end of the positive trial, but is uncertain regarding the time of its presentation. A similar acceleration was observed by Heise, Keller, Khavarhi, and Laughlin (1969) in a study of discrete–trial alternation of reinforcement of bar pressing in the rat. In the pattern where three negative trials followed each rewarded trial (+,–,–,–,+,–,–,–,+,. . .) response rate increased in the course of the successive negative trials.

D. EXTINCTION AND LATENT INHIBITION

There are two procedures in which a CS (conditioned stimulus) is presented without reinforcement, namely extinction and "latent inhibition." In the former case, an established response declines; in the latter, the acquisition of a conditioned response is retarded. It is not altogether easy to account for either process by a retrospective model of learning. If behavior is maintained because the subject scans its recent past at the time of reinforcement, then such scanning would not occur when reinforcement is omitted. Thus the process which underlies response reduction or inhibition is not entirely clear. On the other hand, a prospective model permits the anticipation of reinforcement, and this is evoked by the CS at the beginning of a trial. Clearly, this anticipation will change when reinforcement is omitted, and behavior will be modified accordingly.

In the "latent inhibition" procedure, the CS is simply presented by itself in advance of acquisition. Such presentations generally retard subsequent acquisition with the same CS. This treatment does not turn the CS into a genuine inhibitor in the sense in which the term is normally used (Rescorla, 1971). The retardation is due to a loss of "associative potential" on the part of the CS. Since the CS is not followed by any event of significance when it is first presented, it would not be retrieved by a relevant subsequent event. ' There would be no opportunity for it to become associated with anything on a retrospective basis. A prospective interpretation, on the other hand, can handle latent inhibition quite well. The subject establishes a temporal representation in which the CS is followed by no consequences. This expectation has to be overcome when the reinforcer is then added during training trials.

VI. CONCLUSION

In this essay I have tried to show that concepts from the area of animal memory can help us to understand the acquisition of behavior, and particularly of discriminative behavior. If these concepts are to add anything to the many theoretical accounts of learning, we should be able to identify what is "new and different" in the present account. The invocation of retrieval cues is hardly a novel idea. However, I suggest that they do more than to facilitate appropriate learned response patterns. They also activate prospective representations of other trial events, particularly outcomes, and these are then subject to modification on the basis of the responding which is governed by these cues. The notion of prospective processing is perhaps less common, but it has recently gained a lot of support from studies of working memory (Grant, 1981; Honig, 1981) where, by and large, the problem is acquired before memory intervals are introduced. In this chapter I suggest that even in "long-delay" learning, where memory within trials is required in acquisition, prospective processing is maintained throughout the memory interval.

I do not intend to imply that associations are established only prospectively.   There is good evidence for retrospective processing in working memory (Honig & Thompson, 1982).   Thus, if working memory processes enter into "long–delay" associations, there is no reason to exclude retrospective processes in favor of prospective ones.   As we have seen, retrospective interpretations of the delay of reinforcement have been common since the time of Grice (1948), and are contained in systematic reviews of Revusky (1971) and Lett (1979). However, the present account, among others, suggests that prospective processes are more "robust" (that is, they decay less rapidly) than retrospective ones. This conclusion emerges from comparisons of delayed simple and delayed conditional discriminations , and from the facilitation of memory through the anticipation of differential outcomes.   Following the present line of reasoning, I would conclude that retrospective associative memory processing is more likely to occur in learning situations with relatively short intervals between the cue and the response, or the response and the trial outcome.

It would of course be valuable to obtain more direct evidence in favor of prospective processing in the establishment of associations.   My argument has been based upon analogies with memory processes that have emerged from the study of established associations.   The literature is curiously silent on the temporal "direction" of associations when they are formed.   However, we may note that after Hogan and Zentall (1977) taught pigeons a symbolic matching problem, they failed to obtain any evidence for "backward" associations when they interchanged the sample and comparison stimuli.   A retrospective process of acquisition should strengthen such associations.   However, Hogan and Zentall used a relatively direct transfer test, and a more complex design which controls for stimulus generalization decrement and similar procedural problems may be required to discover retrospective associations.

Otherwise, I can support the "prospective" view only intuitively.   First, it is more parsimonious to assume a single associative mechanism, rather than two, one for short and one for long within–trial delays.   Second, if retrieval cues can direct learned behavior (as I have suggested for delayed alternation experiments), then it is reasonable to suppose that they will facilitate prospective processing of other events, such as the outcomes of such behavior.   Finally, I think that behavior may have evolved as a "commitment to the future," as Ilan Golani has put it (personal communication).   Animals may quite "naturally" look ahead to the consequences of behavior, rather than looking back upon behavior when such consequences occur.

In any case, whatever the ultimate conclusion may be on the mechanisms of association, I think that this chapter has shown that the "conceptual tools" generated by systematic research in the area of animal memory can help us to "work out a better understanding of learning," as Bolles suggested. The procedures of learning entail processes of memory, both within and between the experimental units which we call "trials." These processes, behaviorally identified, contribute to the acquisition, modification, and maintenance of behavior.   Theoretical schemes, such as temporal or spatial "maps" and "representation," will, if they are valid, provide a conceptual framework within which a meaningful distinction between learning and memory would be hard to maintain.

ACKNOWLEDGMENTS

The preparation of this chapter, and the author's research described in it, were supported by grant no. AO-102 from the Natural Sciences and Engineering Research Council of Canada.

REFERENCES

Barnes, G. Place of delay and interference in long-delay learning. Master's thesis, Memorial University of Newfoundland, 1978.

Bolles, R. C. Some relationships between learning and memory. In D. L. Medin, W. A. Roberts, & R. T. Davis (Eds.), *Processes of animal memory.* Hillsdale, N.J.: Erlbaum, 1976.

Capaldi, E. J. Memory and learning: A sequential viewpoint. In W. K. Honig & P. H. R. James (Eds.), *Animal memory.* New York: Academic Press, 1971.

Capaldi, E. J., & Stanley, L. R. Temporal properties of reinforcement aftereffects. *Journal of Experimental Psychology,* 1963, *65,* 169-175.

Capaldi, E. J., & Verry, D. R. Serial order anticipation learning in rats: Memory for multiple hedonic events and their order. *Animal Learning and Behavior,* 1981, *9,* 441-453.

D'Amato, M. R. Delayed matching and short-term memory in monkeys. In G. H. Bower (Ed.), *The psychology of learning and motivation.* New York: Academic Press, 1973.

D'Amato, M. R., & Buckiewicz, J. Long delay, one-trial conditioned preference and retention in monkeys *(Cebus apella). Animal Learning and Behavior,* 1980, *8,* 359-362.

D'Amato, M. R., & Worsham, R. W. Retrieval cues and short-term memory in capuchin monkeys. *Journal of Comparative and Physiological Psychology,* 1974, *86,* 274-282.

D'Amato, M. R., Safarjan, W. R., & Salmon, D. Long-delay conditioning and instrumental learning: Some new findings. In N. E. Spear & R. R. Miller (Eds.), *Information processing in animals: Memory mechanisms.* Hillsdale, N.J.: Erlbaum, 1981, 113-142.

DeLong, R. E., & Wasserman, E. A. Effects of differential reinforcement expectancies on successive matching-to-sample performance in pigeons. *Journal of Experimental Psychology: Animal Behavior Processes,* 1981, *7,* 394-412.

Grant, D. S. Delayed alternation in the rat: Effect of contextual stimuli on proactive interference. *Learning and Motivation,* 1980, *11,* 339-354.

Grant, D. S. Short-term memory in the pigeon. In N. E. Spear & R. R. Miller (Eds.), *Information processing in animals: Memory mechanisms.* Hillsdale, N.J.: Erlbaum, 1981.

Grice, G. R. The relation of secondary reinforcement to delayed reward in visual discrimination learning. *Journal of Experimental Psychology,* 1948, *38,* 1-18.

Heise, G. A., Keller, C., Khavarhi, K., & Laughlin, L. Discrete-trial alternation in the rat. *Journal of the Experimental Analysis of Behavior*, 1969, *12*, 609-622.

Herman, L. M., & Thompson, R. K. R. Symbolic, identity, and probe delayed matching of sounds in the bottlenosed dolphin. *Animal Learning and Behavior*, 1982, *10*, 22-34.

Hogan, D. E., & Zentall, T. R. Backward associations in the pigeon. *American Journal of Psychology*, 1977, *90*, 3-15.

Honig, W. K. Studies of working memory in the pigeon. In S. H. Hulse, H. Fowler, & W. K. Honig (Eds.), *Cognitive processes in animal behavior*. Hillsdale, N.J.: Erlbaum, 1978.

Honig, W. K. Working memory and the temporal map. In N. E. Spear & R. R. Miller (Eds.), *Information processing in animals: Memory mechanisms*. Hillsdale, N.J.: Erlbaum, 1981.

Honig, W. K., & Thompson, R. K. R. Retrospective and prospective processing in animal working memory. In G. H. Bower (Ed.), *The psychology of learning and motivation*. (Vol. 16.) Academic Press, 1982.

Honig, W. K., & Wasserman, E. A. Performance of pigeons on delayed simple and conditional discriminations under equivalent training procedures. *Learning and Motivation*, 1981, *12*, 149-170.

Lawrence, D. H., & Hommel, L. The influence of differential goal boxes on discrimination learning involving delay of reinforcement. *Journal of Comparative and Physiological Psychology*, 1961, *51*, 552-555.

Lett, B. T. Long-delay learning in the T-maze. *Learning and Motivation*, 1975, *6*, 80-90.

Lett, B. T. Long-delay learning: Implications for learning and memory theory. In N. S. Sutherland (Ed.), *Tutorial essays in psychology: A guide to recent advances*. (Vol. 2.) Hillsdale, N.J.: Erlbaum, 1979.

Maki, W. S. Directed forgetting in pigeons. In N. E. Spear & R. R. Miller (Eds.), *Information processing in animals: Memory mechanisms*. Hillsdale, N.J.: Erlbaum, 1981.

Maki, W. S., Moe, J. C., & Bierley, C. M. Short-term memory for stimuli, responses, and reinforcers. *Journal of Experimental Psychology: Animal Behavior Processes*, 1977, *3*, 156-177.

Peterson, G. B., & Trapold, M. A. Effects of altering outcome expectancies on pigeons' delayed conditional discrimination performance. *Learning and Motivation*, 1980, *11*, 267-288.

Peterson, G. B., Wheeler, R. L., & Trapold, M. A. Enhancement of pigeons' conditional discrimination performance by expectancies of reinforcement and nonreinforcement. *Animal Learning and Behavior*, 1980, *8*, 22-30.

Petrinovich, L., & Bolles, R. Delayed alternation: Evidence for symbolic processes in the rat. *Journal of Comparative and Physiological Psychology*, 1957, *50*, 363-365.

Pschirrer, M. E. Goal events as discriminative stimuli over extended intertrial intervals. *Journal of Experimental Psychology*, 1972, *96*, 425-432.

Rescorla, R. A. Summation and retardation tests of latent inhibition. *Journal of Comparative and Physiological Psychology*, 1971, *75*, 77-81.

Revusky, S. H. The role of interference in association over a delay. In W. K. Honig & P. H. R. James (Eds.), *Animal memory*. Academic Press, 1971.

Rilling, M., Kendrick, D. F., & Stonebraker, T. B. Stimulus control of forgetting: A behavioral analysis. In M. L. Commons, R. J. Herrnstein, & A. R. Wagner (Eds.), *Quantitative analyses of behavior: Vol. III, Acquisition*. Ballinger, 1984. In press.

Roberts, W. A., & Grant, D. S. Studies of short-term memory in the pigeon using the delayed matching-to-sample procedure. In D. L. Medin, W. A. Roberts, & R. T. Davis (Eds.), *Processes of animal memory*. Hillsdale, N.J.: Erlbaum, 1976.

Roitblat, H. L. Codes and coding processes in pigeon short-term memory. *Animal Learning and Behavior*, 1980, *8*, 341–351.

Roitblat, H. L. The meaning of representation in animal memory. *The Behavioral and Brain Sciences*, 1982, *5*, 353–406.

Sands, S. F., & Wright, A. A. Serial probe recognition performance by a rhesus monkey and a human with 10- and 20-item lists. *Journal of Experimental Psychology: Animal Behavior Processes*, 1980, *6*, 386–396.

Spear, N. E. Extending the domain of memory retrieval. In N. E. Spear & R. R. Miller (Eds.), *Information processing in animals: Memory mechanisms*. Hillsdale, N.J.: Erlbaum, 1981.

Thomas, D. R. Studies of long-term memory in the pigeon. In N. E. Spear & R. R. Miller (Eds.), *Information processing in animals: Memory mechanisms*. Hillsdale, N.J.: Erlbaum, 1981.

Thomas, D. R., McKelvie, A. R., Ranney, M., & Moye, T. B. Interference in pigeons' long-term memory viewed as a retrieval problem. *Animal Learning and Behavior*, 1981, *9*, 581–586.

Thompson, R. K. R., & Herman, L. M. Memory for lists of sounds by the bottle-nosed dolphin: Convergence of memory processes with humans? *Science*, 1977, *153*, 501–503.

3   ANIMAL INTELLIGENCE: UNDERSTANDING THE MINDS OF ANIMALS THROUGH THEIR BEHAVIORAL "AMBASSADORS"

Edward A. Wasserman
The University of Iowa

I. INTRODUCTION

A century ago, publication of George J. Romanes' *Animal Intelligence* (1882) marked the beginning of the field of comparative psychology, a field that sought to define the dimensions of animal intelligence and to compare the intelligence of humans and animals. (For more on the origins of comparative psychology, see Jaynes, 1969, and Warden, 1928.) Romanes had been a close acquaintance of Charles Darwin and later published some of Darwin's notes posthumously in a second book, *Mental Evolution in Animals* (1883). Thus, in approaching the many questions of animal and human intelligence, Romanes accepted Darwin's evolutionary thesis and adopted Darwin's inductive approach, amassing voluminous reports relating exceptional feats of animal intelligence.

In the ensuing years, the Darwinian theory of evolution has achieved scientific respectability; however, reliance on largely unverifiable claims of animal genius has given way to exact experimental investigation under highly controlled conditions. Unfortunately, for many decades, those experimental methods have been rather little used in disclosing the limits of animal intelligence or drawing meaningful parallels between human and animal cognition. Instead, these techniques have been widely used in giving a momentary empirical edge to warring theoretical camps (see Bower & Hilgard, 1981) in often arcane debates, many proving to be of little enduring significance (see Skinner, 1950).

Recent interest in two-way communication between humans and animals via sign languages has brought the attention of the behavioral science community back to the central concerns of comparative psychology. (See Wasserman, 1981, 1982, 1983 for further discussion and analysis of this trend.) Accordingly, it is appropriate to reexamine the basic ideas and tools of comparative psychology before we embark on our second century of inquiry into this fascinating domain. Here, a careful look at Romanes' *Animal Intelligence* provides a useful means of considering many of the enduring problems and procedures of comparative psychology; such an examination also permits us to appreciate better Romanes' place in this important field of scholarship.

II. BRIEF BIOGRAPHY

The scientific world of Victorian England was astir with the theory of organic evolution.   Into that revolutionary period stepped George J. Romanes (1848–1894), a man well-prepared to make a significant contribution to evolutionary science (E. Romanes, 1898).   Romanes entered Cambridge in 1867 and after an abortive attempt at studying for the clergy, he turned his attention to the natural sciences.   In 1870, he passed his tripos along with Darwin's son, Francis.   Subsequent to his university studies, Romanes conducted empirical research in a number of areas, but primarily in the physiology of marine invertebrates.

During his early scientific career, Romanes received strong encouragement from C. Darwin, and a very strong personal and professional bond grew between them.   It was Darwin who first praised Romanes' work and who, along with T. H. Huxley and J. D. Hooker, proposed Romanes for the Linnean Society; and it was Romanes who bestowed on Darwin his most sincere affection, devotion, and gratitude. Romanes and Darwin defined disciple and master.

Perhaps because of his early interest in theology and philosophy, Romanes began studying animal intelligence in 1877.   Exploration into this area had, of course, become particularly imperative, owing to the evolutionary speculations of Darwin and Herbert Spencer, several years earlier.   If the human mind had evolved from earlier animal forms, then what different kinds of intelligence are shown by today's animals?   And if contemporary animals vary in their ancestry to us, then how do their cognitive capabilities compare to our own?   Answering these two questions were the main items on Romanes' agenda for comparative psychology.   Presenting evidence on the nature of nonhuman cognition was the prime aim of *Animal Intelligence*, whereas systematizing these data into a comparative psychology of cognition was the central concern of *Mental Evolution in Animals*.

Most students of comparative psychology are familiar with the evidential material contained in *Animal Intelligence*: copious cases of intelligent action in all of the major groups of animals, from protozoans to primates.   Yet the Preface and the Introduction to this volume reveal Romanes to be a thoughtful analyst of the fundamental problems and procedures of comparative psychology.   It is to these that we now turn.

III. COMPARATIVE PSYCHOLOGY

Romanes felt that the scientific world of 1882 was ready for the new field of comparative psychology.   The evolutionary thesis of mental continuity throughout the animal kingdom had been proposed with considerable fanfare, but it was in great need of extensive empirical substantiation.   Unlike the field of comparative anatomy, comparative psychology then lacked a sound data base from which clear evolutionary trends could be discerned.   Instead, only scattered tales or anecdotes were available to the classifier of animal intelligence, and these were penned by popular writers, not only careful students of animal behavior. Romanes wrote: "the phenomena of mind in animals, having constituted so much and so long the theme of unscientific authors, are now considered well-nigh unworthy of serious treatment by scientific methods (1977, p. vi)." Scientific respectability thus awaited controlled and reproducible behavioral observations, which Romanes encouraged.

IV. THE ANECDOTE

Although dismayed by the dearth of scientific data, Romanes assembled the most compelling of the available facts into his volume, *Animal Intelligence*. Contrary to widespread opinion, Romanes was acutely aware of the problems in basing important evolutionary conclusions upon the anecdotal reports of untrained observers. He thus took special care to include in his book only those reports that: (a) could be attributed to a particular individual, (b) were unlikely to involve a serious observational error, and (c) were corroborated by a number of independent observers.

Beyond this process of selection, Romanes begged the reader's pardon for having to rely on anecdotal evidence, and provided the following defense:

> If the present work is read without reference to its ultimate object of supplying facts for the subsequent deduction of principles, it may well seem but a small improvement upon the works of the anecdote-mongers. But if it is remembered that my object in these pages is the mapping out of animal psychology for the purpose of a subsequent synthesis [*Mental Evolution in Animals*], I may fairly claim to receive credit for a sound scientific intention, even where the only methods at my disposal may incidentally seem to minister to a mere love anecdote (p. vii).

This defense notwithstanding, history has judged Romanes by his actions, not by his intentions.

V. MIND IN OURSELVES AND IN OTHERS

Prior to considering mental phenomena throughout the animal kingdom, Romanes distinguished two ways by which evidence of mind may be obtained: subjectively and objectively. Subjective evidence of mind comes from our own private experiences and reflections -- our consciousness. In this inherently subjective domain, "we are restricted to the limits of a single isolated mind which we call our own, and within the territory of which we have immediate cognizance of all of the processes that are going on, or at any rate of all of the processes that fall within the scope of our introspection (p. 1)."

We gain evidence of mind in others through a rather different means: namely, through observation of their behaviors. "In our objective analysis of other. . .minds we have no. . .immediate cognizance; all our knowledge of their operations is derived, as it were, through the medium of ambassadors —— these ambassadors being the activities of the organism (p. 1)." Mind in others must be inferred because it cannot be directly known.

While not original, Romanes' distinction between the subjective and the objective material of psychology is important for the development of comparative psychology. Spencer had originally tied the notion of mental evolution to consciousness and Darwin had done little to separate private from public manifestations of mind (Magoun, 1960). From this vantage point, then, if our prime concern is with mental phenomena as directly known, a comparative psychology of mind or consciousness in others is precluded. A scientific comparative psychology must then consider as its subject matter mental phenomena as reflected in overt behavior.

VI. METHODS OF INFERRING MIND IN OTHERS

Romanes used, but did not explicitly distinguish, two methods of inferring mind from another's actions. First, reasoning by analogy, investigators might project onto another those inner thoughts and feelings that they themselves consciously experience under similar circumstances (called "subjective inference" . Second, reasoning by function, they might assume the operation of some capacity or process (like learning) that permits the observed organism to adjust to the specific contingencies of its environment (called "objective inference" by Mackenzie, 1977).

Romanes' own descriptions of these inferential methods are instructive. On reasoning by analogy: "Starting from what I know of the operations of my own individual mind, and the activities which in my own organism they prompt, I proceed by analogy to infer from the observable activities of other organisms what are the mental operations that underlie them (pp. 1–2)." On reasoning by function: "All, in fact, that in an objective sense we. . .mean by a mental adjustment is an adjustment of a kind that has not been definitely fixed by heredity as the only adjustment possible in the given circumstances of stimulation (p. 4)."

VII. MIND AS INFERRED BY ANALOGY AND BY FUNCTION

Reasoning by analogy and by function have played important parts in the history of comparative psychology. With the advent of experimental methods for studying psychological processes around the turn of the century, investigators developed the alternative to the rather sterile tactic of inferring mind by analogy. The behavioral analyses of J. Loeb and H. S. Jennings, plus the learning techniques of I. P. Pavlov and E. L. Thorndike contributed to a highly rigorous and objective school of human and animal psychology —— behaviorism. J. B. Watson was, of course, the school's most forceful proponent.

Some insight into this new approach and its relationship to analogical and functional inference can be gained if we look at the writings of Jennings (1904, 1906). A student of protozoan behavior, Jennings nevertheless recognized complexity in the actions of his lowly subjects:

> The writer is convinced, after long study. . ., that if Amoeba were a large animal, so as to come within the everyday experience of human beings, its behavior would at once call forth the attribution to it of states of pleasure and pain, of hunger, desire, and the like, on precisely the same basis as we attribute these things to the dog (1976, p. 336).

Yet the very ease with which we project our own private thoughts and feelings onto others should be a cause for great caution. Do we really know whether such analogical reasoning is merited?

> It is clear that objective evidence cannot give a demonstration either of the existence or of the non-existence of consciousness, for consciousness is precisely that which cannot be perceived objectively. No statement concerning consciousness in animals is open to refutation by observation and experiment (pp. 335-336).

Because the concept of consciousness had no scientific utility, Jennings rejected it and all other subjective accompaniments of action -- in animals and humans. "There are no processes in the behavior of organisms that are not as readily conceivable without supposing them to be accompanied by consciousness as with it (p. 336)."

Despite Jennings' clear logic, throughout the formative years of behaviorism, efforts nonetheless continued to interpret the behavior of animals in light of human thoughts and emotions. Examples of such efforts come from some expected and some surprising sources. For instance, C. L. Morgan -- often acclaimed as the man who delivered comparative psychology from the darkness of subjectivity fostered by Romanes -- advanced the following in his 1894 text, An Introduction to Comparative Psychology:

> In the systematic training of the comparative psychologist, the subjective aspect is not less important than the objective aspect (1896, p. 50).
>
> The wise and cautious student never forgets that the interpretation of the facts in psychical terms is based upon the inductions he has reached through introspection. The facts are objective phenomena; the interpretation is in terms of subjective experience (1896, p. 47).

Consider also the case of Thorndike. A reading of his attack on the anecdote (1911, pp. 22-26) soon discloses that he disdained at least some aspects of Romanes' comparative psychology. Yet the man who pioneered the laboratory study of instrumental conditioning in America did not disavow the importance of subjective phenomena in animal behavior:

> Behavior includes consciousness and action, states of mind

and their connections (p. 15).

The student of behavior, by. . .studying the animal in action as well as in thought, is surer of getting an adequate knowledge of whatever features of an animal's life may finally be awarded the title of mind (p. 16).

Indeed, in explaining his preference for experimental over anecdotal evidence, Thorndike offered that his new method for studying the intelligence of animals, "will not only tell more accurately *what they do*, and give the much-needed information *how they do it*, but also inform us *what they feel* while they act (p. 26)."

And take M. F. Washburn.   A student of the introspectionist E. B. Titchener, Washburn (1926) expectedly emphasized the necessity of considering the subjective experience of animals:

. . .we   are   obliged   to   acknowledge   that   *all   psychic interpretations of animal behavior must be on the analogy of human experience*. . .   Whether we will or no, we must be anthropomorphic in the notions we form of what takes place in the mind of an animal (p. 12).

It is particularly noteworthy that she was professedly, "interested in what animals do largely as it throws light upon what they feel (p. 22)."

Thus, many who came after Romanes failed to see the error of their way and continued to infer mental activities in animals by the method of analogy, even those who disagreed with him on different issues.   Although additional factors may have contributed to continuation of this practice (see Mackenzie, 1977), Romanes' own justification of analogical inference could have provided workers with a cogent defense against those who balked at attributing human thoughts and feelings to animals:

. . .skepticism of this kind is logically bound to deny evidence of mind, not only in the case of the lower animals, but also in that of the higher, and even in that of men other than the skeptic himself.   For all objections which could apply to the use of this criterion of mind in the animal kingdom would apply with equal force to the evidence of any mind other than that of the individual objector.   This   is   obvious,   because. . .the   only evidence we have of objective mind is that which is furnished by objective activities; and as the subjective mind can never become assimilated with the objective so as to learn by direct feeling   the   mental   processes   which   there   accompany   the objective activities, it is clearly impossible to satisfy any one who may choose to doubt the validity of inference, that in any case other than his own mental processes ever do accompany objective   activities. . .   Common   sense,   however,   universally feels that analogy here is a safer guide to truth than the skeptical demand for impossible evidence. . . (pp. 5-6).

Beyond this defense, Romanes knew that some analysts selectively applied the analogy to animals similar to human beings; as the organism's physical resemblance to humans decreased, so did application of the analogy (also see Yerkes, 1905).   Romanes could not abide this inferential selectivity, for the process of reasoning by analogy was logically the same whether one was considering the behavior of the next door neighbors or that of their child's pet turtle:

Therefore, having full regard to the progressive weakening of the analogy from human to brute psychology as we recede through the animal kingdom downwards from man, still, as it is the only analogy available, I shall follow it throughout the animal series (p. 9).

It was, of course, possible that trying to comprehend the minds of animals with our own mind would seriously distort the character of their mental life. But what else could be done, other than to abandon the endeavor altogether?

The mental states of an insect may be widely different from those of a man, and yet most probably the nearest conception that we can form of their true nature is that which we form by assimilating them to the pattern of the only mental states with which we are actually acquainted (p. 10).

The rise of behaviorism did, ultimately, occasion a decline in the use of analogical inference. By 1928, C. J. Warden was able to write that, "success of the objective movement . . . meant overthrow of the principle of anthropomorphic analogy (p. 513)." Coupled with the descent of analogical inference was growth in the use of functional inference. Perceptual, associative, and motivational processes obtained operational definitions and were intensively and experimentally examined in the laboratories of C. L. Hull, K. W. Spence, E. C. Tolman and many others.

Tolman (1936), in particular, contributed to the analysis of mind through functional inference with his concept of the intervening variable: a theoretical mediating process that connected the initiating causes of behavior, on the one hand, with the final resulting behavior, on the other. The intervening variable was then a purely hypothetical contrivance of the theorist, and had no surplus meaning or independent status outside of the environment–behavior relationships that it comprised. Psychological concepts that were intervening variables were thus inferred from behavior in a very different way than were those inferred by analogy, although the same term –– for example, "expectancy" –– might be used in both cases.

The movement away from subjectivity continued with B. F. Skinner. Whereas Tolman considered intervening cognitive and motivational variables important elements in his behavior theory (see MacCorquodale & Meehl, 1954), Skinner tried to construct a behavior system that eschewed mediating mental or physiological processes (see Verplanck, 1954). Analogical inferences of mind were thus disparaged because they were based upon unverifiable reports of self–observation and because they perpetuated various pre–scientific, animistic accounts of the causation of behavior. Functional inferences of mind, too, were discouraged; even though cognitive terms could be made objective by careful operational definition, their subjective connotations persisted nonetheless. Further, Skinner argued, the invention and theoretical manipulation of intervening variables tended to divert the energies of experimental psychologists away from what should have been their main aims: namely, the identification of environmental variables of which behavior is a function, and the induction of a limited number of descriptive laws which adequately comprise those environment-behavior interrelations.

Skinner's "radical" behaviorism has of course joined Tolman's "operational" behaviorism as a major school of modern psychology (for more on these approaches, see Wasserman, 1981, 1982, 1983). Much of Skinner's appeal has come because his approach has led to important practical advances in psychopathology, psychopharmacology, and education. Skinner has also proven to be an effective advocate for his ideas -- both practical and theoretical.

Recently, however, Skinner (1977) has tended to equate advocacy of radical behaviorism with opposition to operational behaviorism.   This equation is unfortunate for at least two reasons.   First, Skinner's main initial dissatisfaction with intervening variables was not that they were inherently subjective , but that, no matter how careful the theorist was, it was impossible to stop others from imbuing terms like perception, memory, and expectancy with subjective referents.   Second, in his criticism of cognitive intervening variables, Skinner seems to have confused the methods of analogical and functional inference.   Thus, Skinner provided the following characterization of cognitive animal psychology:

> It would be as extensive as the science of behavior because there would presumably be a feeling for each act. But feelings are harder to identify and describe than the behavior attributed to them, and we should have abandoned an objective matter in favor of one of dubious status, accessible only through necessarily defective channels of introspection (p. 3).

Were he criticizing psychologists such as Romanes, Morgan, Thorndike, and Washburn for their interest in studying the subjective or conscious experiences of animals, Skinner's remarks would have some force. However, as applied to the more modern researchers of Tolman, Hull and Spence and their use of functionally inferred mental processes, Skinner's complaints miss the mark.   Even more puzzling is Skinner's chastising others for their concern with subjective experience.   Although not widely appreciated, Skinner himself (1945, 1964) has for some time been interested in the analysis of private events (also see Terrace, Chapter 1, this volume).   Whether this analysis falls within the proper domain of radical behaviorism is not clear; nor is it obvious whether the analysis has any testable implications for comparative psychology (see Wasserman, 1982, 1983).

VIII.   FURTHER REFLECTIONS ON ANALOGICAL AND FUNCTIONAL INFERENCE

That psychological terms often have both objective and subjective meanings is a very bothersome point, one noted very early on by Morgan: "it is practically impossible to describe mental processes in their primitive unanalysed modes of occurrence without using phrases which are analytic in form (1896, p. 88)." How can we keep straight when use of a particular term, such as perception, is based on analogy with our own private experiences or based on function through the operation of some adaptive process or capability? One way is to continue to use the same term in both senses, but to clarify its precise meaning in a given situation with additional defining terminology. This tactic is very commonly employed. It has the advantage of providing a ready, common sense definition of the term, and it can make for rather easy communication among researchers. However, it has the serious disadvantage noted earlier by Skinner of never clearly separating the term's objective and subjective meanings. This semantic ambiguity can occasionally lead to argument and acrimony.

Another tactic is to associate the subjective and objective meanings with different terms, an approach associated with the 1899 paper by Beer, Bethe, and von Uexkull, "Proposals for an Objectifying Nomenclature in the Physiology of the Nervous System." Thus, as subjective terms, hearing, smell, and sight might continue to be used; but for objective purposes, the respective terms phono-reception, stibo-reception, and photo-reception should be substituted. The after effect of a stimulus in its subjective aspect might still be called memory; but in its objective aspect, it should be renamed resonance.

Despite the many merits of introducing new, neutral terms for psychological processes, the proposals of Beer, Bethe, and von Uexkull were not adopted in experimental psychology. Novel or unusual terms simply do not stand a good chance of being incorporated into an existing scientific lexicon. Even when new terminologies are adopted, another problem can arise. "There is the danger that the use of a terminology might put up the appearance of a theory when there is but an uncommon descriptive term inserted for a common term (Lewin, 1943, p. 289)."

Such may have been the case, for example, when Skinner (1938) substituted the terms respondent and operant behaviors for what had earlier been called reflexive and voluntary behaviors (Skinner, 1931).

We have yet to resolve the dual meanings of psychological terms. Inferences of mental processes by analogy and by function do seem to involve distinguishable methods. Yet the same behavioral "ambassador" may lead one to infer mind by different means, and the same term can then be applied to both inferences.

IX. THE RESURRECTION OF SUBJECTIVE INFERENCE

Despite Warden's earlier claim that behaviorism had overthrown the anthropomorphic analogy, the subjective inference of mind and consciousness has persisted from Romanes' time to today. We have already seen that Warden's contemporary, Washburn, considered comparative psychology properly to comprise studies of the animal mind, "the inner aspect of the behavior of animals (1926, pp. 21–22)." Later, Bierens de Haan (1946) continued in the same vein to use "sympathetic intuition" to interpret animal behavior. The primacy of subjective inference is obvious in the following quotations of the latter author:

*The object of animal psychology. . .*is not the animal soul but *the psychic phenomena in animals* (p. 11).

Only the man who tries to explain the actions of a fellow creature, be it a man or an animal, in terms of. . .subjective experience, may rightly call himself a psychologist (p. 10).

Both Washburn and Bierens de Haan felt that animal psychology could never prove to be as complete as human psychology because animals lacked the necessary communicative capabilities for fully relating to us the nature of their private experiences:

To this fundamental difficulty of the dissimilarity between animal minds and ours is added. . .the obstacle that animals have no language in which to describe their experience to us. . . the higher vertebrates could give us much insight into their minds if they could only speak (Washburn, 1926, p. 3).

Many people. . .doubt the possibility of a science of psychic phenomena in animals. Human psychology, they say is built up on two foundations: the introspection into one's own inner experiences and the communication in human language of that which another man experiences at a given moment or under given conditions. It will be clear that both methods fail us when we try to study the animal mind (Bierens de Haan, 1946, p. 11).

Given current advances in the study of animal communication, Griffin (1978) has recently proposed that communicative behavior may soon provide us with that long sought "window" through which to peer into the minds of animals:

to the extent that language is the best available window through which we learn about human thinking, why not open the window somewhat wider by recognizing that nonverbal communication can serve the same basic function as words and sentences, and that it could be used to gather information about the thoughts of other species, as well as those of our fellow men (p. 531)?

One can hardly question that studying the communicative behaviors of animals may greatly expand our understanding of their actions. However, one can question Griffin's rather blithe acceptance of communicative behavior as prima facie evidence of another's private experiences, as well as his conviction that the prospects of successfully understanding the subjective mental experiences of animals -- their sensations, feelings and intentions -- is any better now than in the days of Darwin and Romanes (1981, p. 27).

Quite apart from the criticisms that Skinner and others have made of the merits of a science of animal feelings, one can ask why Griffin, and his predecessors Washburn and Bierens de Haan put so much faith in communicative behaviors as "ambassadors" of mind. Surely other behaviors, even those commonly studied in learning laboratories, can be used to infer the presence of thoughts, memories, and emotions in animals. Consider these astute observations of E. G. Boring (1950) as they relate to analogical and functional inferences based upon, not obviously communicative, but conditioned responses:

> You can work out the laws of color vision easily with a human subject who is not color—blind, who uses your own language, who is intelligent and honest. He "describes his consciousness" to you, that is to say, he tells you what he sees. An animal can also tell you what he sees, but you have to build up with him a special language of conditioned responses before he can report to you (p. 621).
>
> When an animal learns to avoid. . .shock, he is using his new—found language of conditioned response —— not purposely, it is true, yet he is using it —— to tell you that he does not like the shock, that he knows it is coming, that he knows how to avoid it. His movements are a one—way language. His discriminations are his words (p. 621).
>
> Conditioning is an objective substitute for introspection, a form of language which enables an experimenter to know what discrimination an animal can make, what it does and does not perceive. Conditioning is, in fact, a kind of language, which the experimenter provides so as to enable an animal to communicate with him, but the phenomena of communication occur entirely on the objective level of stimuli, nerve—action and secretion, without any need for assuming consciousness as an entity (p. 637).

As comparative psychologists, we must be prepared to receive any and all of those behaviors that are the "ambassadors" of animals' minds. We must recognize that these behavioral "ambassadors" exist entirely in the world of objectivity; yet they may be interpreted in objective terms or in terms of our own subjective experiences. Finally, these "ambassadors" —— be they salivary secretions, button presses, waggle dances, or gestural signs —— are all exceedingly valuable in helping us to determine the nature of animal intelligence and to compare that alien intellect to our own (also see Mason, 1976; Ullman, 1978). I see no strong justification for considering some kinds of behavioral evidence to be special or superior to others (see Griffin, 1981, pp. 166–167 for the contrary opinion).

X. PHYSICAL VERSUS PSYCHICAL EVOLUTION

Romanes was adamant in his belief that physical and mental evolution had occurred and were still occurring. In *Animal Intelligence*, he stated that "there must be a psychological, no less than a physiological, continuity extending throughout the length and breadth of the animal kingdom (p. 10)."

Recognizing phylogenetic trends in existing animal species, however, raised a difficult problem in relating animals' psychological affinities to their structural similarities:

> If the animal kingdom were classified with reference to Psychology instead of with reference to Anatomy, we should have a very different kind of zoological tree from that which is

now given in our diagrams.   There is . . . a general and . . .
most important parallelism running through the whole animal
kingdom between structural affinity and mental development; but
this parallelism is extremely rough and to be traced only in
broad outlines. . . (p. ix).

There being no clear isomorphism between anatomical and
psychological evolution, Romanes judged that comparative psychology was
best advised to stick to its own subject matter without too great a
concern with the details of other disciplines:

although it is convenient for the purpose of definite
arrangement to take the animal kingdom in the order presented
by zoological classification, it would be absurd to restrict an
inquiry into Animal Psychology by any considerations of the
apparently disproportionate length and minute subdivision with
which it is necessary to treat some of the groups.
Anatomically, an ant or a bee does not require more
consideration than a beetle or a fly; but psychologically there is
a need for as great a difference of treatment as there is in the
not very dissimilar case of a monkey and a man (pp. ix–x).

Much of the recent upheaval (e.g., Hodos & Campbell, 1969; Lockard,
1971; Tobach, Adler, & Adler, 1973) concerning the proper foundations of
comparative psychology seems to stem from workers' misunderstanding
that our field has its own evidential base, and that it is not just a
subsidiary of other, more "scientific" disciplines.   The last word here goes
to King and Nichols (1960), who concurred with Romanes' concern over
the nonisomorphism between physical and psychical evolution: "It can be
said with some justification. . .that the zoological taxonomic system does
not furnish a sound basis for the fruitful organization of the data with
which comparative psychology must deal (p. 35)."

XI.   ROMANES' PLACE IN COMPARATIVE PSYCHOLOGY

His data were anecdotal, his interpretations were anthropomorphic and his
orientation was mentalistic.   For these reasons, Romanes has been judged
harshly by historians of comparative and biological psychology.   However,
such harshness seems not to provide an appropriate appreciation of
Romanes' scholarly work (also see Robinson, 1977). We can better evaluate
Romanes' place in comparative psychology by considering the following
contributions:

1. He helped to found the field of comparative psychology and
   set as its main goals the delimitation of animal intelligence and
   the comparison of intelligence in humans and animals.   These
   aims remain at the forefront of contemporary comparative
   psychology.

2. He clearly separated the subjective and objective data of

psychology, and put forth the two different ways that we use the behaviors of organisms as the "ambassadors" of their mental experiences and capabilities.   Despite a century of research and thinking, comparative psychologists are still debating the merits of analyzing the private experiences of others.

3. By placing the issues of psychology within the framework of organic evolution, he, along with Spencer and Darwin, helped to move the study of behavior from philosophy to biology, where it remains today.

4. By using anecdotal evidence to support claims of higher mental functions in nonhuman animals, he inspired the experimental study of animal intelligence under highly controlled conditions.   Experimentation rather than naturalistic observation continues to be the method of choice in the study of animal cognition.

Lest this list be construed as constituting uncritical acceptance of all his teachings, I note in conclusion that we can also learn a great deal from Romanes' errors.   First, we must be much more careful than Romanes in talking about mental phenomena.   To speak of behaviors as "ambassadors" of the mind strongly smacks of dualism -- however unintended. Contemporary workers in cognitive animal psychology do not generally believe that nonphysical forces impel and direct behavior (see Terrace, Chapter 1, this volume; Wasserman, 1983).   Yet the very language of cognitive psychologists often elicits strong aversive responses in others, (e.g., Skinner, 1977) and leads these critics to neglect the very important methods and data of cognitive psychology.   Perhaps as cognitive psychology continues to evolve in connection with mechanical computing devices, the language of cognition will move away from mentalistic terminology.

Second, unlike Romanes, we should not consider ourselves to be advocates for the intelligence of animals.   Indeed, the very reason that Morgan proffered his canon of parsimony was the predilection of Romanes and his followers to accept gossamer evidence as persuasive of the highest of intellectual achievements.   We can do no better than to embrace Morgan's canon while exploring the limits of animal intelligence, as Romanes urged.

And third, we must not confuse the descriptions and the interpretations of our research.   Romanes and his contemporaries were rightly criticized for not clearly distinguishing what they observed from what they inferred (again, see Thorndike, 1911).   In today's world of controversy over "symbolic communication" (Epstein, Lanza, & Skinner, 1980; Savage-Rumbaugh, Rumbaugh, & Boyson, 1978) and "self-awareness" (Epstein, Lanza, & Skinner, 1981; Gallup, 1970) this distinction is particularly important if we are to concentrate on substantive rather than semantic issues.

REFERENCES

Beer, T., Bethe, A., & Uexkull, J. V. Vorschlage zu einer objectivirenden Nomenclatur in der Physiologie des Nervensystems. *Biologisches Zentralblatt*, 1899, *19*, 517–521.

Bierens de Haan, J. A. *Animal psychology: Its nature and its problems.* London: Burrow's Press, 1946.

Boring, E. G. *A history of experimental psychology.* New York: Appleton–Century–Crofts, 1950.

Bower, G. H., & Hilgard, E. R. *Theories of learning.* Englewood Cliffs, N.J.: Prentice–Hall, 1981.

Epstein, R., Lanza, R. P., & Skinner, B. F. Symbolic communication between two pigeons *(Columba livia domestica). Science*, 1980, *207*, 543–545.

Epstein, R., Lanza, R. P., & Skinner, B. F. "Self–awareness" in the pigeon. *Science*, 1981, *212*, 695–696.

Gallup, G. G., Jr. Chimpanzees: Self–recognition. *Science*, 1970, *167*, 86–87.

Griffin, D. R. Prospects for a cognitive ethology. *The Behavioral and Brain Sciences*, 1978, *1*, 527–538.

Griffin, D. R. (Ed.) *The question of animal awareness.* New York: Rockefeller University Press, 1981.

Hodos, W., & Campbell, C. B. G. Scala naturae: Why there is no theory in comparative psychology. *Psychological Review*, 1969, *76*, 337–350.

Jaynes, J. The historical origins of "ethology" and "comparative psychology". *Animal Behavior*, 1969, *17*, 601–606.

Jennings, H. S. *Behavior of the lower organisms.* Bloomington, Indiana: Indiana University Press, 1976. Originally published in 1904, 1906.

King, J. H., & Nichols, J. W. Problems of classification. In R. H. Waters, D. A. Rethlingshafer, & W. E. Caldwell (Eds.), *Principles of comparative psychology.* New York: McGraw–Hill, 1960.

Lewin, K. Remarks to Mr. Hull's supplementary note. *Psychological Review*, 1943, *50*, 288–290.

Lockard, R. B. Reflections on the fall of comparative psychology: Is there a message for us all? *American Psychologist*, 1971, *26*, 168–179.

MacCorquodale, K., & Meehl, P. E. Edward C. Tolman. In W. K. Estes, S. Koch, K. MacCorquodale, P. E. Meehl, C. G. Mueller, W. N. Schoenfeld, & W. S. Verplanck (Eds.), *Modern learning theory.* New York: Appleton–Century–Crofts, 1954.

Mackenzie, B. D. *Behaviourism and the limits of scientific method.* Atlantic Highlands, N.J.: Humanities Press, 1977.

Magoun, H. W. Evolutionary concepts of brain function following Darwin and Spencer. In S. Tax (Ed.), *Evolution after Darwin: The evolution of man.* (Vol. II.) Chicago: University of Chicago Press, 1960.

Mason, W. A. Windows on other minds. *Science*, 1976, *194*, 930–931.

Morgan, C. L. *An introduction to comparative psychology.* London: Walter Scott, 1896. Originally published in 1894.

Robinson, D. N. Preface to Romanes' *Animal intelligence.* In D. N. Robinson (Ed.), *Significant contributions to the history of psychology 1750-1920. Series A: Orientation. Vol. VII: G. J. Romanes.* Washington, D.C.: University Publications of America, 1977.

Romanes, E. *The life and letters of George John Romanes.* London: Longmans, Green, 1898.

Romanes, G. J. *Animal intelligence.* New York: Appleton, 1883. (Reprinted in D. N. Robinson (Ed.), *Significant contributions to the history of psychology 1750-1920. Series A: Orientation. Volume VII: G. J. Romanes.* Washington, D.C.: University Publications of America, 1977) (Originally published in 1882.)

Romanes, G. J. *Mental evolution in animals.* New York: Appleton, 1898. Originally published in 1883.

Savage-Rumbaugh, E. S., Rumbaugh, D. M., & Boyson, S. Symbolic communication between two chimpanzees (*Pan troglodytes*). *Science,* 1978, *201,* 641-644.

Skinner, B. F. The concept of the reflex in the description of behavior. *Journal of General Psychology,* 1931, *5,* 427-458.

Skinner, B. F. *The behavior of organisms.* New York: Appleton-Century-Crofts, 1938.

Skinner, B. F. The operational analysis of psychological terms. *Psychological Review,* 1945, *52,* 270-277.

Skinner, B. F. Are theories of learning necessary? *Psychological Review,* 1950, *57,* 193-216.

Skinner, B. F. Behaviorism at fifty. In T. W. Wann (Ed.), *Behaviorism and phenomenology.* Chicago: University of Chicago Press, 1964.

Skinner, B. F. Why I am not a cognitive psychologist. *Behaviorism,* 1977, *5,* 1-10.

Thorndike, E. L. *Animal intelligence.* New York: Macmillan, 1911.

Tobach, E., Adler, H. E., & Adler, L. L. Comparative psychology at issue. *Annals of the New York Academy of Sciences,* 1973, *223.*

Tolman, E. C. Operational behaviorism and current trends in psychology. In E. C. Tolman (Ed.), *Behavior and psychological man: Essays in motivation and learning.* Berkeley: University of California Press, 1966. Paper originally presented at Proceedings of the Twenty-fifth Anniversary Celebration of the Inauguration of Graduate Studies at the University of Southern California, 1936.

Ullman, S. Mental representations and mental experiences. *The Behavioral and Brain Sciences,* 1978, *4,* 605-606.

Verplanck, W. S. Burrhus F. Skinner. In W. K. Estes, S. Koch, K. MacCorquodale, P. E. Meehl, C. G. Mueller, W. N. Schoenfeld, & W. S. Verplanck (Eds.), *Modern learning theory.* New York: Appleton-Century-Crofts, 1954.

Warden, C. J. The development of modern comparative psychology. *Quarterly Review of Biology,* 1928, *1,* 486-522.

Washburn, M. F. *The animal mind.* New York: Macmillan, 1926. Third edition.

Wasserman, E. A. Comparative psychology returns: A review of Hulse, Fowler, and Honig's "Cognitive processes in animal behavior". *Journal of the Experimental Analysis of Behavior,* 1981, *35,* 243-257.

Wasserman, E. A. Further remarks on the role of cognition in the comparative analysis of behavior. *Journal of the Experimental Analysis of Behavior*, 1982, *38*, 211–216.

Wasserman, E. A. Is cognitive psychology behavioral? *The Psychological Record*, 1983, *33*, 6–11.

Yerkes, R. M. Animal psychology and criteria of the psychic. *Journal of Philosophy, Psychology, and Scientific Methods*, 1905, *2*, 141–149.

4 THE ROAD FROM BEHAVIORISM TO RATIONALISM

Thomas G. Bever
Columbia University

I. ON THE ROAD

Most of the articles in this book reflect some dissatisfaction with the behaviorist approach to animal activity. During the last 70 years behaviorism has appeared in many guises – for the present discussion I will assume that a behaviorist framework follows two principles;

Physicalism:
> every term in a description must be based on a physically definable entity.

Associationism:
> distinct descriptive terms can be related to each other only by undifferentiated association.

Many scholars now doubt that certain aspects of animal behavior can be described by a model which strictly obeys such postulates. This doubt underlies what can be taken as the current "cognitive revolution" in the study of animal behavior. In this essay, I contrast two directions that this revolution can now take; one that continues the emphasis on learning particular behaviors for which the behaviorist framework still appears to be adequate, and a new direction towards rationalism – the unashamed description of animal minds.

The inadequacy of behaviorism as a complete account of behavior has in fact been known for a long time. The demonstration of perceptual gestalten simultaneously invalidated the physicalist and associationist principles, by proving the existence of internal plans of perception (Wertheimer, 1923); rapidly-executed and species-specific behaviors indicated the presence of internal plans of action (Tinbergen, 1951); maze learning in rats was shown to be like acquiring a "map" of the maze rather than a series of responses (Tolman, 1948); the conceptual unity of reasoning processes in higher primates inexplainable by any compilation of behaviorist principles (Kohler, 1925). Thus, a definitive crisis in animal psychology should have occurred many years ago. It is instructive to consider why it did not.

How should psychologists have responded to crucial demonstrations of behaviorism's inadequacies? There were two choices: one could reject behaviorism entirely, or one could retain it for those phenomena for which it was not yet invalidated.   The latter choice required two related assumptions, one about descriptive efficiency, the other about organismic efficiency.   I will call these assumptions Descriptive and Representational Reductionism.

Descriptive Reductionism:
> Science best proceeds by applying only the weakest descriptive device until forced to apply a stronger one.

Representational Reductionism:
> An animal organizes each newly acquired behavior with the most concrete mechanisms available.

The appeal of these assumptions explains why gestalt and ethological demonstrations of abstract perceptual and conceptual units had little impact on the behaviorist program.   The first assumption allowed behaviorism to continue as the theory of non-gestalt phenomena; the second assumption rationalized the application of behaviorist descriptions to behaviors of higher animals whenever possible.   This scientific posture was aided further by the fact that the simple gestalt and ethological phenomena were taken to be innate and not a problem for learning theory; the learned skills that were also complex were limited to a few higher organisms, such as apes and humans.

Descriptive reductionism is a doctrine from the philosophy of science, whose acceptance depends on its overall efficacy – yet, there are notorious counterexamples to the view that science best proceeds from the simple to the complex.   Often a crucial scientific revolution is characterized by the replacement of an unwieldy and complex model with a much simpler one that makes different assumptions.   I do not wish to belabor this point, since the general topic is more appropriate for a philosophical context.   The primary moral for our considerations here is that descriptive reductionism is not always the correct move: therefore, it was not necessary to maintain behaviorism as true of some phenomena, once it was proven to be inadequate for other phenomena.

Descriptive reductionism has another argument that can be made in its favor – if representational reductionism is true, then descriptive reductionism will be most likely to lead us to correct theories of how new skills are learned.   This raises a question that is potentially empirical – is representational reductionism true?

A. REPRESENTATIONAL REDUCTIONISM IS NOT TRUE OF AT LEAST ONE ANIMAL

To examine the validity of representational reductionism we must first define a concrete/abstract continuum for representations; such a continuum is required to give content to the claim that one kind of representation is more concrete than another.   The definition of such a continuum is surely a central theoretical problem for psychological science, one which I will not solve here. Rather, I will rest my argument on some intuitive notions of this continuum and attempt to use only clear relative cases along it, cases that will hold up even when a precise scientific specification of the continuum has been developed. The concrete/abstract dimension I have in mind may be interpreted as one of functional specificity – concrete representations are relatively close to specific sensori–motor phenomena, while abstract representations draw more on internal codes, and can be applied to a wider variety of sensori–motor phenomena (see also Gallistel, 1980).

How do humans first represent a new problem to themselves? In particular, do we follow the representational reductionist principle, always representing a new problem at the most concrete level, and resorting to a more abstract level only when we must? I will review three kinds of evidence that support a negative answer to this question – anecdotal, as in learning a new physical skill; observational, as in mastering a game; and experimental, as in retaining a list of stimuli. In each case, there is a contrast between the initial approach to the task and its ultimate organization. Characteristically, humans develop a concrete, automatic representation only after they have started out with a relatively abstract and conscious representation of a new problem.

1. Learning to ski. Skiing well is a physical skill that relies on developed "instincts," i.e., rapid movements that are appropriate to the slope's surface and the directional intent of the skier. The main problem is learning to turn (and thereby to stop) with a minimum of effort and a maximum of grace. Clearly the component acts of this skill, once developed, are largely unconscious – a good skier has internalized a set of general movements and slope–appropriate parameters that he or she can adjust without direct conscious thought about them. Once learned, it would indeed seem that whenever possible this behavior is organized at a concrete level of representation.

In this essay I am concerned with the problem of learning itself as well as the end product of learning, so the pertinent question is, how does the apprentice skier represent the problem during the early stages of acquisition? The anecdotal evidence is that all too often, novice skiers think explicitly about their individual body movements, and even the physical theory underlying how turns are made. The result of such conscious activity is usually snow–encrusted chaos. This fact exemplifies two points at once; first, we often do not start out conceptualizing a problem at its most concrete level; second, we initially depend on an abstract level of representation that may be dysfunctional.

2. Learning to play chess. The behavior of chess–players has received considerable attention within the contemporary framework for the study of cognition (De Groot, 1965). One reason is that there is wide variation in the skill level among players: this highlights a distinction between formal knowledge and perceptual heuristics, rather like the knowledge/behavior–program distinction in skiing. Here, too, we see a course of learning that moves from the abstract to the concrete – novice chess players concentrate on stringing legal moves together in order to increase the chance of checkmating an opponent's king. Skilled chess players have developed perceptual schemata that signal situation–

appropriate strategies.   The latter skill depends in part on acquiring set sequences of moves, and in part on nonliteral perceptual organizations of the situation.   These relatively concrete perceptual processes lead to improved performance not only in the game, but in such arbitrary tasks as remembering the positions of all the pieces at a particular point in a game. Conversely, skilled chess players recall the location of *fewer* pieces than unskilled players, when the pieces are randomly arranged on the board (Chase & Simon, 1972).   That is, skilled players have started from an abstract representation of the board and legal moves, and developed a set of perceptual schemata that categorize important aspects of the configuration.

3. Learning lists of things.   The study of cognitive psychology in humans has often been taken to be tantamount to the study of memory. Ebbinghaus (1885) was not only the first to develop the study of memory via the learning of arbitrary sequences; he noticed early in his research on his own memorization of lists of real words, that he could not avoid utilizing the meanings of the words as part of his memorization process. This interfered with his scientific goal, to study pure memory – specifically, how sets are memorized "by rote," without the aid of any internal relations among the components, and without any prior associations.   To this end, he invented the "nonsense syllable" as a meaningless stimulus that can be used to probe the learning and recall processes themselves.   He used this kind of stimulus to explore basic parameters of learning and forgetting, for example the effects of repeated presentations of different kinds.

The succeeding century of research has gradually instructed us that subjects generally attempt to avoid rote learning: they use abstract codes as the basis for their memorization of lists, even lists of nonsense syllables.   Despite the psychologist's attempt to force subjects to learn the stimuli as unorganized and association-free entities, subjects impose abstract analyses on the stimuli whenever possible (Anderson & Bower, 1973).   Such reanalyses make certain kinds of syllables selectively easier to recall than others.   For example, pronounceable syllables are more easily learned than unpronounceable ones; syllables that have a high "meaning" associate (either via synesthetic content or by being phonetically close to a real word) are easier to recall; syllables or words that are easily imagable are easier to recall; finally, lists of words that can be interrelated propositionally are easiest of all to recall (see Glass, Holyoak, & Santa, 1979, for a review).

Relative ease of recall has become a tool in cognitive psychology to demonstrate the existence of a particular kind of representational code.   In fact, much current research on memory is specifically addressed to the interactions among different kinds of representations and methods of presentation (see, e.g., Norman & Rumelhart, 1975).   In this way, cognitive science has made a virtue out of necessity.   The original goal was to study memory with association-free stimuli so that differential representational mechanisms would not obscure the properties of a "pure" memory process.   Human subjects resist that goal whenever they can.

B. REPRESENTATIONAL ABSTRACTION IN HUMANS - CONCLUSION

The above examples support several principles. First, humans do not always practice representational reductionism when presented with new tasks; rather, they would seem initially to represent a new task at the most abstract level possible. Second, fully mastered skills may depend on more concrete representations than they first receive; these representations themselves are relatively specialized, "automatic" and inflexible.

If humans are typical in this regard, then the principle that is true for the initial representation of new problems is:

Representational Abstraction:
    an animal represents a newly acquired behavior at the most abstract level of which it is capable.

A moment's thought shows that such a principle is neither gratuitous nor dysfunctional, despite its inconvenience for behaviorism. The first stage of solving a new task is to figure out what kind of task it is, or is best represented as. The more abstract the level of representation, the less prematurely committed the organism is to a particular mode of representation. Thus, even on evolutionary and functional grounds, the principle of representational abstraction makes good sense. This point is similar to that concerning the functional value of consciousness in animals (Griffin, 1978). A conscious representation of the world may be the very best way to negotiate through variations in it. In this sense, consciousness is the ultimate expression of the principle of representational abstraction. We know this principle to characterize at least part of human behavior; a conservative assumption is that other animals follow the principle as well.

II. SERIAL LEARNING IN PIGEONS

The mastery of a rapid sequence of behaviors precludes simple explanation in terms of isolated stimulus-response associations (Lashley, 1950).

Accordingly, we have studied the acquisition and maintenance of both production and discrimination of sequences. The fact that pigeons, for example, can learn to peck a particular sequence of colors and correctly discriminate a presented sequence would seem to force us to accept the view that the pigeons can "represent" a sequence — in that sense it would seem that we are forced to study avian "cognition." I shall show that the issue does not concern whether there is a representation, but what the nature of the representation is, and most important, how the representation is performed.

A. SEQUENCE PRODUCTION

Consider first the production of a sequence. In a typical paradigm, birds are required to peck four simultaneously–presented colored buttons in a particular order of colors (the colors are presented in randomly distinct physical locations) – the task is of interest particularily because correctly ordered pecks within the sequence are not individually reinforced once the skill is acquired. The animal is positively reinforced only after an entire correct order is produced. (In these experiments, however, negative reinforcement – in the form of trial termination – immediately follows an incorrectly ordered peck. (Terrace, Straub, Bever, & Seidenberg, 1977; Straub, Seidenberg, Bever, & Terrace, 1979; Straub, 1979; Straub & Terrace, 1981; Bever, Straub, Terrace, & Townsend, 1980). How do we assess the competence that an animal has when it has attained this skill? From the perspective of the most obvious behaviorist model, the question is whether the animal has learned an unordered group of adjacent pair-wise associates, which can be performed in only one order, or whether the animal has learned a particular ordinal position for each response. The latter representational form would appear to be difficult to represent within the usual behaviorist framework.

1. Generalization to subsequences. One way to determine this is to present deformed versions of the problem to. see if the animal is sensitive to relative order. Suppose buttons are arbitrarily labelled in the order, ABCD. To see if the animal learned the sequence with each item in a particular ordinal position, one could observe its behavior when only a subset of the 4 buttons is available to peck, e.g., AB, AC, BD, etc. My colleagues have shown that such transfer can occur from a full sequence to subsequences of it (Straub & Terrace, 1981; Terrace, 1983). This proves that the animal indeed mastered the sequence, but the mechanisms of its representation remain undetermined. If the animal pecks the correct relative order that the subsets have in the complete sequence, we know only one thing: the animal can apply its complete serial skill to deformations of the original serial problem. This could be because the original skill is itself composed out of a complex of both adjacent and remote ordering relations, or because the animal can transform the sequence it has learned into a new sequence; it could also be because the animal treats the generalization subset as a full stimulus in which some of the buttons are missing (e.g., it "air–pecks" the missing colors). One thing is certain from such results, however: the animals are capable of relatively abstract operations on the skill they have mastered, operations that are inconsistent with physicalism. Transforming the sequence into subsequences is surely such an abstraction, as would be "air–pecking" responses to physically absent colors.

The generalization technique exemplifies negative reasoning: a particular example of the behaviorist framework is invalidated by a particular experiment; other versions of behaviorist models will have to be systematically disqualified as they appear. To further narrow down alternative models and emphasize the claim that the animal must have a non–behaviorist "representation" of the learned sequence, Terrace (1983; Chapter 1, this volume) has reported several other studies in which learning to produce a sequence transfers to other sequences.   To show that pigeons learn the ordinal position of each item in a sequence, he trained subjects on ABC and tested their generalization to XBY (in which X and Y are new colors).   The subjects learned that new sequence more quickly than BXY or XYB, presumably because XBY maintains the ordinal position of B. Like the earlier generalization studies, there are many alternate interpretations of what this means about the original learning.   A similar point applies to Terrace's demonstration (see Chapter 1, this volume) that learning to discriminate a sequence is facilitated by having already learned to produce the same sequence.   None of the generalization studies rules out the possibility that the subjects learn and represent the sequence according to adjacent pair-wise associations, and map what they learn onto new situations by some other, more abstract, processes.   Indeed, following the principle of scientific reductionism, this is the required interpretation.

   This emphasizes the futility of trying to disqualify behaviorism by proving that an animal has learned a non-behaviorist "representation" of a skill.   Every internalized skill presumes a representation of some kind.   The problem is that we must also have a theory of how the representation is *performed* before we can test its validity. (See Roitblat, 1982, and Anderson, 1978.)

B. MODELS OF THE ANIMAL'S MIND

A positive technique to explore the nature of the mastery of a sequence is to construct a model that describes as many observed parameters as possible. One can then take the features of the model to be hypotheses about the animal's processes – the model is of the behaving *animal*, not the behavior.   Such practice in the study of human cognition has flourished under the rubric of "information-processing" automata (Newell & Simon (1972), recently revised into a system of "productions": conditional statements on internal operations).   I constructed such a model for the 4-color sequence production experiment by Straub (1979), which is shown in Figure 4.1.   This model has two temporal parameters, the time to move along the sequence for each new peck position and the time to rehearse each item in the sequence up to the new peck position.   It is striking that the two parameters predict more than 99% of the (mean) variance observed in the latencies to the 16 independent kinds of responses.   The model also allows for the description of the two main types of out-of-order pecks that occur; repeat pecks to a just-pecked key occur when a false negative occurs in response to condition 1 ("are there positions to the right"); keys are skipped when the same conditional is answered with a false positive.   (There are 4 correct responses, A, B, C, D; 3 "repeat pecks" on keys A, B, or C; 6 forward errors that skip over a key; and 3 backwards responses, e.g., pecking A, B, C, A; see Bever *et al.*, 1980, for a more complete description of how the model operates.)

   The model in Figure 4.1 describes the results for a study by Straub in which pigeons were pre-trained on the full 4-color sequence only in a left-right, "forwards" manner.   That is, after pre-training on individual colors presented alone, they were presented with overlapping subsequences building from left to right; first, AB, then ABC and finally

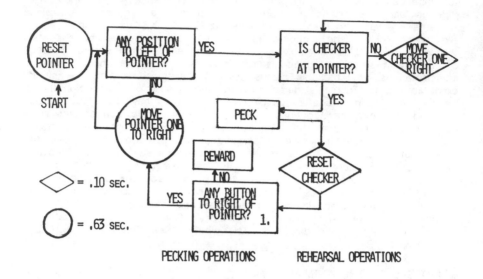

FIG. 4.1. Information-processing model of serial pecking in pigeons with initial forward training. (Adapted from Bever et al., 1980.)

ABCD. In the model this property corresponds to the fact that after each independent peck, the checking operation resets to the beginning of the sequence, and the sequence is "rehearsed" from the beginning until the new peck position.

In another experiment (Straub *et al.*, 1979), the subjects were initially focused on the final color of the sequence before forwards training was initiated (so-called "backwards" training). The animals were exposed repeatedly to the last color alone, D, then the last two, CD, and so on. (This pretraining was ultimately halted in favor of forwards training, because the subjects almost always ignored the early colors and pecked the last one.) The best-fit model of the type in Figure 4.1 for this experiment has *exactly* the same temporal parameters as the forwards model, and one change in the model's operation – the checking operation resets to the *end* of the sequence after each peck; the sequence is, in effect, rehearsed from the end to the beginning.

Each model accounts for more than 99% of the variance in the observed latencies in the experiment it describes. (One model's predictions account for only 58% of the variance in the other model's predictions, demonstrating that the models are distinct from each other.) Thus, the two experiments confirm both the types of operations of the models and their temporal parameters. Such models are also consistent with successful transfer to sequence subsets, at least because of the indeterminant meaning of such transfer that I noted above. The models isolate two principles governing the serial skills of our subjects.

1. Attach a peck response to the end of what has just been rehearsed.

2. After each peck, go to a constant reference point and rehearse the sequence up to the next color where a peck can be attached.

There are many ways to instantiate each of those principles. (1) tells us that our subjects pecked the colors in order by running through the sequence before each peck. (2) reflects the fact that in order to produce ordered behavior, the animal must have some way of running through it to keep track of where it is in the sequence. The variation in the reference-point according to the early pre-training demonstrates the way in which critical early exposure to a problem can set a parameter that dominates its subsequent organization.

At first blush, the empirical success of these models is impressive. But the models have several limitations as well. First, they are models of the acquired skill, far transcending any direct theory of how they are acquired. Second, no matter how successful, they must be taken as initial metaphors in relation to actual physiological processes. The main potential virtue of behaviorist descriptions is that they are simultaneously models of learning and directly interpretable as hypotheses about physiological mechanisms. An information processing model of a skill does not dictate how it is learned nor how it is physically instantiated. It is a model of what is learned and instantiated. Finally, the information-processing systems that they exemplify are arbitrarily powerful, setting no constraints on possible conditions and operations. Before we can argue that they have explanatory power, we must show that they deploy formally constrained mechanisms. We remain afflicted by a basic scientific dilemma. We are between the Scylla of behaviorist descriptive limits, and the Charybdis of information processing models' unconstrained descriptive power.

C. SEQUENCE DISCRIMINATION

When theory is inconclusive or too rich, one instinctively tries to enrich the data and restrict the theory. To achieve this, we have explored the mastery of the perceptual discrimination of sequences. The sequence discrimination paradigm has its own properties and exigencies that have lead us in some new directions as well as supplementing what we have learned from the study of sequence production (Roitblat, Dopkins, Scopatz, & Bever, 1983; Dopkins, Scopatz, Roitblat, & Bever, 1983; Bever, Scopatz, Dopkins, & Roitblat, 1983). I will argue that the best way to understand the data is to assume that the pigeons divide the problem of sequence discrimination into terms of the same two mechanisms isolated in the study of sequence production, rehearsal and pecking.

The first step is to show that the birds can learn a sequential discrimination at all. We extended a paradigm of Weisman and his colleagues (Weisman, Wasserman, Dodd, & Larew, 1980) to the discrimination of a sequence of three colors, (A, B, C) presented on the same key (see Roitblat *et al.*, 1983, for a more complete description; also see Chapter 12, this volume, by Weisman & von Konigslow). Birds are reinforced contingent on at least one peck following presentation of a correct color sequence ("ABC"); there are no consequences of a peck following a negative sequence; the measure of discrimination is the ratio of response rate following a positive sequence to the rate following both kinds of sequences; most sequences were presented at about 1 stimulus/sec, which corresponds to the rate of correctly ordered pecks in the production experiments.   In the first study, the negative sequences were all possible orders of the same three colors (except one order, CCC, which was used as a (successful) post-training generalization test).   After initial pretraining to establish an orienting and responding sequence of behaviors to a white key, sequences of 3 stimuli were presented without any pretraining on shorter sequences.   With about 60 sessions, the birds mastered the discrimination to a reasonable criterion level.

1. The uniqueness of the sequence-final color.   Our procedure naturally prompted us to consider the stages of acquisition imposed by the subjects during their acquisition of the skill.   One question is, what does the animal learn first? – to which kinds of negative sequences does the subject first stop responding?   Our initial analysis suggests that responses to incorrect sequences ending in something other than the last correct color, C, drop out first.   That is, the subjects appear spontaneously to learn the perceptual task by first discriminating the final position.   The next-acquired position tends to be the initial one, A, rather than the medial one, B.

We have examined the special importance of the final color in two other ways, first by selectively withholding certain negative sequences.   In particular, one group of subjects received no trials with a negative sequence that ended in C, while another group received no negative sequence beginning in A. Informationally, the two paradigms are exactly equivalent.   Yet, the group with no negative sequences ending in C discriminated the positive sequence much more quickly.   Apparently, an unambiguously correct C in the final position is much more conducive to learning to discriminate the correct sequence than is an unambiguously correct A in initial position.

In a third set of paradigms, the subjects were presented within the same session with correct sequences of length 3, 2, and 1 color.   The sequences of length 2 and 1 were either successive overlapping subsets from the beginning or end of the complete 3-color sequence.   (The forwards overlapping set was ABC, AB, A; the backwards overlapping set was ABC, BC, C). In each case, the corresponding negatives of each length were also presented.   The results are dramatic: subjects exposed to the backwards overlapping set learned the 3-color discrimination sequence quickly, while subjects exposed to the forwards overlapping set barely learned the 3-color sequence at all.   In brief, the experimental set that provided multiple instances of C as the final color was much easier to learn than the set that provided multiple instances of A as the initial color. This too, highlights the importance of isolating the final color.

The emphasis on discrimination of the final color might seem consistent with the principle of "backward spread of reinforcement."   On this view, later items are discriminated first by virtue of their relative proximity to reinforcement. This view is inconsistent with our findings in a number of ways. The second color to be discriminated was not clearly the next-to-last (B) but tended to be the first (A). The gradient hypothesis does not explain why the sequence set with missing negative initial A sequences is so difficult to learn: it would be expected on a reinforcement theory of any kind that systematic reduction of the alternative negatives should facilitate learning to some extent. Finally, the backwards set with overlapping positive sequences might be easier to learn, if distance from the reinforcement inhibits discrimination — the birds were presented with pure examples of the most distant color from reinforcement, A, which might be expected to facilitate its isolation and discrimination.

D. A MODEL OF SEQUENCE DISCRIMINATION

There appears to be something categorically unique about discriminating the last color in a sequence. We can explain this by constructing a model of the discrimination process for the normal conditions, with all negative sequences included. The subjects do learn to discriminate the sequence under such conditions. This means that they must have some mechanism for tracking the correct sequence; they must also have the means to attach the contingently necessary response to the end of a correct sequence. In brief, a consistent model of the serial discrimination task must include two processes, rehearsing the sequence internally to an external match, and discriminative pecking at the end of the sequence. (We have not yet constructed a detailed model like that for sequence production because we are still analyzing the data.) It is natural (though not formally necessary) to think of these two sub-mechanisms as the same as in the production model. (The sequence discrimination model may also require a color-matching procedure — unnecessary in the case of sequence production, since all colors were simultaneously present.)

According to the model, the animal divides the discrimination problem into two sub-problems, as in our model for sequence production — rehearsing the sequence internally (and matching each stage in the sequence to the input) and attaching a peck to the end of what is rehearsed. The key to the special status of the final color in sequence discrimination may be that the contingently important peck is always after the 3-color sequence is complete. Accordingly, the animal is not guided by intermediate training (as it is in the production studies) to attach pecks to the intermediate points in the 3-color sequence. We postulate that C is uniquely linked to a response because of its final position as the animal "rehearses" the 3-color sequence as part of tracking the input color sequence.

This interpretation of the special status of the final color in the original experiment with most negative sequences present as due to to dividing the task into internal tracking and pecking, can explain the results from the other studies. In the missing negatives set with initial A correct, attention may be called to the initial A, but the animal cannot link a discriminative peck response because (on the model) it has not yet internally rehearsed the sequence when it has perceived A. The same model can explain the difficulty of learning the forwards overlapping–positives set: A and B are sometimes the final color requiring attachment of the single contingent response to three different colors; in the alternative backwards set, C is always the final color available to attach to a peck.

III. BEHAVIORISM AND RATIONALISM - BACK AGAIN

What can we conclude about the pigeon's mind from these sequence production and discrimination studies? The production and perception experiments support the same conclusion via quite distinct theoretical and empirical routes. The fact that the sequence tasks are successfully learned demonstrates that the animal has a mechanism that keeps track of the ongoing sequence. The animal also has a distinct mechanism that links the final element in a just–rehearsed sequence to a peck. That is, a parsimonious explanation of the animal's behavior via information processing models differentiates two processes – internal rehearsal while tracking the sequence and pecking when the sequence is successfully rehearsed internally. The proposal that the animal keeps track of the sequence with one mechanism and pecks with another is not astounding. But it is important in light of the incompatibility of our models with several of the behaviorist principles outlined above.

Suppose the models are correct. First, they falsify the principle of representational reductionism. There is no a priori basis on which the sequence production problem is most simply represented in terms of rehearsal and pecking at the end of each sub–rehearsal; indeed, given the possibility of adjacent pair–wise associates (since the mean response rates are so slow) the animal should use them if representational reductionism were true. Similarily, the discrimination problems would be most simply represented as isolated positive and negative sequences. In fact, the comprehensive bipartite analysis that the animals apply (according to the model) is positively dysfunctional in certain cases. (e.g., the forward overlapping positive set).

Second, the internal workings of the models are not associationistic. No doubt, there are information processing models which can be intentionally constrained to have only associative relations among their components: but there hardly seems any reason to exert this constraint without the physicalist constraint as well – it is the latter that makes important the proposal that there is only one kind of mentally internal relation – association.

This leads to a further point – any controversy between physicalist and cognitive processes turns out to be totally irrelevant to our descriptions and discoveries. The argument here does not turn on the existence of a so–called "cognitive representation": the representation itself of the model could in fact be motoric, sensory, or an abstract structure (see also Roitblat, 1982, for a full discussion of this issue). Rather, the argument for internal, non–associationistic and non–reductionistic structure depends on the internal properties of the models that describe how the behaviors are organized and executed. The question of "cognition" in its everyday sense is irrelevant, since it does not involve what the animal may *know*. As in complex human behavior, the models describe distinct systems interrelated in aid of solving particular problems.

This is an important result of our rationalist attempt to describe the "whole bird" with one set of mechanisms, rather than constructing a different set for each behavior. In modelling the animal rather than the behavior alone, we may make discoveries about the internal mechanisms it has at its disposal. The different behaviors result from assembling the same mechanisms in different configurations (see Gallistel, 1980, for a general discussion of this approach).

The broadest question germane to this book is whether there is such a thing as animal cognition. The study of cognition in humans has involved a pun, which is tolerable when it is among us humans, but intolerable when applied to other species, who cannot talk back when a theory is obviously silly. There are two areas of study which currently go under the name "human cognition"; the nature of knowledge, and the mechanisms whereby we represent and act on it. In studying the latter in humans we often fudge the true nature of the former by common agreement that it is unnecessary initially to decide whether we "know" what we think we do, or only believe it. It is harder to justify this heuristic when dealing with organisms with whom we cannot communicate easily. It seems reasonable, however, to assume that whether animals know that they know what they know about such matters, that like humans, they have mechanisms for representation and behavior. (See also Chapter 13, this volume, by Shimp.)

There is every reason to believe further that the scientific study of these mechanisms in the human species is typical of what we will achieve in the study of other species.

ACKNOWLEDGMENTS

The main theme of this paper was stimulated by preparing the concluding presentation of the conference on which this book is based. I am grateful to my co-conferees for many discussions about these issues, both during the conference and afterwards. Preparation of this paper was in part supported by a grant from NIMH to Columbia University, "Acquisition and representation of stimulus sequences," Number RO1-MH-37070.

REFERENCES

Anderson, J. R. Arguments concerning representations for mental imagery. *Psychological Review*, 1978, *85*, 249–277.

Anderson, J. R., & Bower, G. H. *Human Associative Memory.* Washington, D. C.: V. H. Winston & Sons, 1973.

Bever, T. G., Scopatz, R. A., Dopkins, S. C., & Roitblat, H. L. The beginning and end of sequence discrimination. Paper presented at a meeting of the Psychonomic Society. November, 1983.

Bever, T. G., Straub, R. O., Terrace, H. S., & Townsend, D. J. Study of serial behavior in humans and non-humans. In P. Jucsyk & D. Klein (Eds.), *The nature of thought: Essays in honor of D. O. Hebb.* Hillsdale, N.J.: Erlbaum, 1980.

Chase, W. G. & Simon, H. A. The mind's eye in chess. In W. G. Chase (Ed.), *Visual information processing.* New York: Academic Press, 1972.

De Groot, A. D. *Thought and choice in chess.* The Hague: Mouton, 1965.

Dopkins, S. C., Scopatz, R. A., Roitblat, H. L., & Bever, T. G. *Encoding and decision processes in the discrimination of 3-item sequences.* Proceedings of the 54th annual meeting of the Eastern Psychological Association, Philadelphia, April, 1983.

Ebbinghaus, H. *Memory.* New York: Teachers College, 1885. Translated by H. A. Ruger & C. E. Bussenius, 1913.

Gallistel, C. R. *The organization of action: A new synthesis.* Hillsdale, N.J.: Erlbaum, 1980.

Glass, A. L., Holyoak, K. J., & Santa, J. L. *Cognition.* Reading, Mass.: Addison-Wesley, 1979.

Griffin, D. R. Prospects for a cognitive ethology. *The Behavioral and Brain Sciences,* 1978, *1,* 527-538.

Kohler, W. *The mentality of apes.* New York: Kegan Paul, Trench, Trubner, 1925.

Lashley, K. S. In search of the engram. *Symposia of the Society for Experimental Biology,* 1950, *4,* 454-482.

Newell, A. Physical symbol systems. In D. A. Norman (Ed.), *Perspectives on cognitive science.* Norwood, N.J.: Ablex, 1981.

Newell, A., & Simon, H. *Human problem solving.* Englewood Cliffs, N.J.: Prentice-Hall, 1972.

Norman, D. A., Rumelhart, D. E., & The LNR Research Group. *Explorations in cognition.* San Francisco: Freeman, 1975.

Roitblat, H. L. The meaning of representation in animal memory. *The Behavioral and Brain Sciences,* 1982, *5,* 353-406.

Roitblat, H. L., Dopkins, S. C., Scopatz, R. A., & Bever, T. G. Discrimination of 3-item sequences. Paper presented at a meeting of the Psychonomic Society. November, 1983.

Straub, R. O. *Serial learning and representation of a sequence in the pigeon.* PhD Thesis, Columbia University, 1979.

Straub, R. O., & Terrace, H. S. Generalization of serial learning in the pigeon. *Animal Learning and Behavior,* 1981, *9,* 454-468.

Straub, R. O., Seidenberg, M. S., Bever, T. G., & Terrace, H. S. Serial learning in the pigeon. *Journal of the Experimental Analysis of Behavior,* 1979, *32,* 137-148.

Terrace, H. S. Simultaneous chaining: The problem it poses for traditional chaining theory. In M. L. Commons, R. J. Herrnstein, & A. R. Wagner (Eds.), *Quantitative analyses of behavior: Vol. III, Acquisition.* Cambridge, Mass.: Ballinger, 1983. In press.

Terrace, H. S., Straub, R. O., Bever, T. G., & Seidenberg, M. S. Representation of a sequence by a pigeon. *Bulletin of the Psychonomic Society,* 1977, *10,* 269.

Tinbergen, N. *The study of instinct.* London: Oxford University Press, 1951.

Tolman, E. C. Cognitive maps in rats and men. *Psychological Review*, 1948, *55*, 189–208.

Weisman, R. G., Wasserman, E. A., Dodd, P. W. D., & Larew, M. B. Representation and retention of two–event sequences in pigeons. *Journal of Experimental Psychology: Animal Behavior Processes*, 1980, *6*, 312–325.

Wertheimer, M. Untersuchungen zur Lehre von der Gestalt: II. *Psychologische Forschung*, 1923, *4*, 301–350. Abridged translation by M. Wertheimer, 1958. Principles of perceptual organization. In D. C. Beardslee & M. Wertheimer (Eds.), *Readings in perception*. Princeton: Van Nostrand.

II.   WORKING MEMORY

5 REPRESENTATIONS IN PIGEON WORKING MEMORY

H. L. Roitblat
Columbia University

I. INTRODUCTION

In a quest for general and fundamental cognitive mechanisms the "minds " of humans, computers and animals have been explored. The problems faced by each of these approaches are similar, and revolve around the difficulty of inferring the workings of a system to which we have no direct access. One of the complicating factors in the study of human cognition is our intimate familiarity with at least one example of the system--ourselves. The lack of such intimacy in the case of computer or animal "minds" means that it is harder to conceal tacit assumptions and justifications derived from our own awareness of our own minds.

One result of this foreignness is that students of animal cognition have tended to be more skeptical in their demand for justification of cognitive conceptions, tending toward a mechanistic view. The behaviorist revolution took this tendency to an extreme by accepting a version of logical positivism as the ideal form of scientific justification. Although a strong positivist position has been largely rejected (e.g., Watkins, 1978) its influence is still apparent (also see Roitblat, 1982; Weimer, 1979). The result has been a cautious development of a comparative study of animal cognition.

One of the central concerns in any study of cognition is the means by which information is represented. If experience is to have an effect on future behavior then it is necessary that the experience produce some change that endures afterwards and thereby mediates its later effects. That change, in the context of its mediational role, is a representation (Roitblat, 1982).

The approaches to representation in animals within the behaviorist tradition have concentrated on the acquisition of habits and the like. This interest has its cognitive counterpart in the work of Wagner (1981) and others, for example, on cognitive interpretations of classical conditioning (see also Dickinson, 1980). Other investigations of representation in animals have concentrated on a group of tasks known collectively as delayed response tasks in which the subject is required to make a discriminative response on the basis of stimuli (discriminative cues) that are no longer present. The best known example of this kind of task is delayed matching-to-sample in which a sample stimulus is presented for a certain duration and followed by a retention interval and the simultaneous presentation of two or more comparison stimuli. A response to one of these comparison stimuli is rewarded, dependent on the identity of the sample stimulus which preceded it on that trial. The usual procedure is to test within-trial retention of the discriminative stimulus after acquisition is basically complete (although it is certainly possible to study acquisition).

To choose the correct comparison stimulus at the end of the retention interval the animal must remember information about the identity of the sample. This information is held in working memory (Honig, 1978; Chapter 2, by Honig, and Chapter 7, by Maki, this volume). In general working memory retains information about properties of the task that change from trial to trial (e.g., the sample in delayed matching-to-sample) and are important in determining the correct response for that particular trial (e.g., the choice of the correct comparison stimulus). Because different sample stimuli are presented, and different responses are correct on each trial, the information in working memory is typically of no use to and can interfere with later performance; hence the optimal strategy is to "flush," "reset," or forget the contents of working memory at the end of each trial.

In addition to the transient information held in working memory, certain transformational rules are needed to specify the relationships between samples and correct choices (e.g., choose the matching, rather than the different comparison stimulus; see Chapter 22, this volume, by Zentall, Edwards, & Hogan). These rules or mappings are constant from day to day and are held in reference memory (Honig, 1978). Reference memory contains the subject's knowledge base concerning the relationships among stimuli, the outcome of particular stimulus-behavior combinations, the rules of the task, etc. The subject must know both how to determine the correct choice, given the discriminative cue and the identity of the cue for adequate performance in delayed response tasks. Hence both working and reference memories are necessary.

The distinction between working and reference memory systems is functional. It may not be clear in all cases, but failure to consider it in others leads to obvious contradictions. For example, it is often found that pigeons perform delayed matching-to-sample more accurately when long as opposed to brief sample durations are used, even within the same session (Maki & Leith, 1973; Roberts, 1972; Roberts & Grant, 1974; Roitblat, 1980). Failure to distinguish between the two memory systems might force the conclusion that pigeons know how to perform the task with long sample durations but forget how to perform when short durations are used. Furthermore, although the functional distinction between working and reference memory does not, by itself, require that the two systems be anatomically distinct, a number of studies do indicate a dissociation between working and reference memory systems. For example, Olton and Papas (1979) found that sectioning the fimbria-fornix region of rats' brains temporarily interfered with both working and reference memory but reference memory recovered (cf. Nadel & MacDonald, 1980).

My research has focussed on the working memory system. In the rest of this paper I describe some attempts to understand the representations used by pigeons in their performance of delayed matching-to-sample.

There are a number of models consistent with the general finding that choice accuracy improves with increases in sample duration (Maki & Leith,

1973; Roberts, 1972; Roitblat, 1980). These models can be divided into two general classes: those based on the assumption that encoding is an all-or-none process and those based on the assumption that encoding is a gradual process. The available evidence indicates that during delayed matching-to-sample performance, the pigeon's working memory can best be conceptualized as a gradually changing memory process (cf. Grant, 1981).

In the usual two-alternative delayed matching-to-sample task there is no opportunity to distinguish between the various coding models because there is only one informative choice available on any trial. All models predict that the probability of an error on this choice should decline exponentially with increasing sample durations. The gradual models assert that the decline is a result of growing strength of the representation, the all-or-none models assert that this change is due to an increasing probability that the information has been encoded.

In order to explore differences among these models, I modified the delayed matching-to-sample task so that three comparison stimuli are available on every trial (instead of the usual two), each corresponding to one of three possible sample stimuli. The presence of a third alternative means that on those trials on which the first choice is an error, a second choice could be made between the remaining two stimuli. Second choice (as well as first choice) accuracy was found to improve in these circumstances with increasing sample durations (Roitblat, 1980) indicating that the birds have access to some information about the sample when making their second choices. These data thereby rule out a simple all-or-none model in which the information about the sample is either present, and a correct choice results, or absent, in which case the bird guesses.

A more complicated all-or-none model is based on the assumption that choice accuracy is controlled by two all-or-none processes. Encoding occurs in an all-or-none fashion, but on some trials, even when encoding has occurred, the bird does not base its choice on the information it possesses, perhaps because of elicitation processes resulting from the pairing of the comparison stimuli with reinforcement. In order to test this model the pattern of first-choice errors was examined (Roitblat, 1980). Contrary to the prediction of this model, the distribution of first-choice errors was found to depend on the sample; first choice errors thus contain information about the sample and are not produced solely by sample independent processes.

II. THE N-STATE MODEL

The presence of sample information in first-choice-errors is entirely consistent with a gradual encoding process but it raises the possibility that the original finding--above chance second-choice accuracy--may have been the result of choices made in response to first-choice-errors. For example, following a first-choice-error to one (e.g., the blue) comparison stimulus, the subject could still make a second-choice with above chance accuracy based on no other information about the sample except that errors to blue are more likely when green is the correct choice than when red is the correct choice (e.g., that responses to green following first-choice-errors to blue are more likely to be reinforced than are responses to red). This hypothesis corresponds to a discrete state model (called the N-state model) which contains one state for each of N sample (or comparison) stimuli.

The state of the system, at any point in time, depends on the identity of the sample if it is present and on the similarity among the comparison stimuli. While the sample is present the system is most likely to occupy the state corresponding to the correct choice. With some probability, however, it might change to the state corresponding to another stimulus. The probability of such a change depends on the similarity between the stimuli. The more similar the stimuli, the higher the likelihood of a state change between them. These state changes continue without further input from the sample during the retention interval, but, at any point in time, the system is in only one state and is unaffected by any previous state it might have occupied. According to this model, second-choices depend solely on the first-choice-error (more specifically on the state corresponding to the first-choice) and therefore contain no additional information about the sample than did the first-choice (Clarke, 1964). The pattern of second-choices made following each sample should, according to the N-state model, be fully described by the sequence of transitions from samples to first-choice-errors, and from first-choice-errors to second-choices. Four birds were tested (Roitblat & Scopatz, 1983) with two sample durations and a number of retention interval durations. Each combination of sample and retention interval duration produced three 3 x 3 matrices, an example of which is presented in Table 5.1. The sample/first-choice matrix enumerates the frequency with which each choice was made following each sample. The major diagonal of this matrix represents correct first-choices so those cells are set by definition to 0 in the sample/first-choice-error matrix. The sample/second-choice matrix enumerates the frequency with which each second-choice occurred following each sample. The first-choice/second-choice matrix enumerates the number of times each second-choice followed each first-choice. Because a first-choice to a given stimulus caused it to be darkened and therefore, no longer available, the cells along the major diagonal in this matrix are set to 0 by design.

The expected first-choice-error distributions were computed on the assumption that first-choice errors are distributed independently of the sample (Roitblat, 1980). To generate the expected sample/second-choice distribution, the counts in the sample/first-choice-error and in the first-choice-error/second-choice matrices were each converted to row proportions and the two matrices were multiplied. The resulting matrix, converted back to counts, was then compared with the obtained sample/second-choice distribution. A significant difference between the two indicates that second-choices contained information about the sample that was not contained in the first-choice-errors.

Figure 5.1 displays the results of the comparisons between observed and expected distributions in the form of Phi-prime scores (square root of chi-squared). This figure shows that sample information is present in first-choice-errors, but that amount of information is not sufficient to account for the distribution of second-choices. These data are

TABLE 5.1
Example matrices used to compute
expected distributions for the N-state model

Sample or choice	Obtained						Expected		
Sample	First-choice			First-choice error			First-choice error		
	R	G	B	R	G	B	R	G	B
R	45	14	1	0	14	1	.00	7.23	7.27
G	16	25	19	16	0	19	21.98	.00	13.02
B	20	12	28	20	12	0	19.64	12.36	.00

First choice	Second-choice		
	R	G	B
R	0	27	9
G	17	0	9
B	9	11	0

Sample	Second-choice			Second-choice		
	R	G	B	R	G	B
R	13	0	2	9.60	.55	4.85
G	8	26	1	8.65	22.45	4.00
B	5	12	15	7.85	15.00	9.15

Note-- Obtained data are from bird 453 tested with a 5000-msec sample and a 2800-msec retention interval. χ^2 (first choice) = 14.888, (second choice) = 11.635, both significant at .05. R = red, G = green, and B = blue.

incompatible with the N-state model or any other model which asserts that sample-independent error processes apply on some trials while "true" memory processes control responding on the rest. Instead, any discrete state model proposed in the future must include some mechanism for generating sample-dependent error patterns while simultaneously allowing for more information to be present in second-choices than would be transmitted from first-choice-errors. The N-state model included the former property in the state transition mechanism based on interstimulus similarity, but it could not account for the latter. Accurate delayed matching-to-sample performance appears to depend on a non-instantaneous coding process.

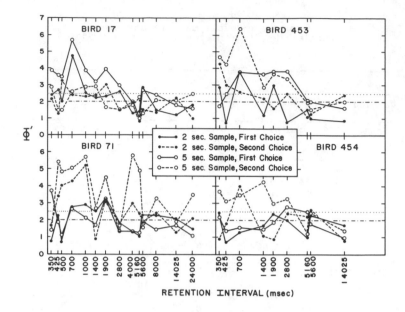

FIG. 5.1. The difference between observed and expected first-choice errors (solid lines) and second-choices distributions (dashed lines) as a function of sample duration and delay. Phi-prime is computed as the square root of chi-squared; critical values for significance (.05) are represented as the dash/dot line (first-choice-error) and the dotted line (second-choices).

III. PROSPECTIVE CODES

Theoretical accounts of delayed matching-to-sample have tended to emphasize the process suggested by its name as the basis of accurate performance namely that choices are determined by matching the observed comparison stimuli to a trace of the sample currently in memory (Roberts & Grant, 1976). Two assumptions are central to this hypothesis. The first is that subjects base their decisions on physical similarity between the remembered sample and the comparison stimuli. Reference memory, then, might contain a general rule of the form "choose the comparison stimulus that most closely resembles the remembered sample stimulus." The second assumption is that working memory contains a "copy" of the physical features of the sample. The gradual process apparently underlying pigeon delayed matching-to-sample performance on this view involves the slow development of this copy.

The first assumption receives some support from investigations of "concept formation" in pigeons. The evidence indicates that pigeons are capable of using a general rule concept (Herrnstein, Loveland, & Cable, 1976; Wright, Santiago, Urcuioli, & Sands, 1981; Zentall & Hogan, 1978,

Zentall, Edwards, Moore, & Hogan, 1981; in this volume: Chapter 14 by Herrnstein, Chapter 20 by Wright, Santiago, Sands, & Urcuioli, and Chapter 22 by Zentall, Edwards, & Hogan), but that it is not sufficient to account for all pigeon delayed matching–to–sample performance. If performance is based on a general rule, there should be positive transfer to new sets of stimuli as long as the task requirements are consistent. Zentall and his associates have found such transfer when familiarity with the stimuli was controlled, but the magnitude of this transfer was typically relatively small. In a number of experiments using matching–to– or oddity–from–sample tasks (e.g., Zentall & Hogan, 1974, 1978), subjects were trained on either oddity or matching problems and then tested with either the same or the complementary task. Choice accuracy was found typically to be about 88% at the end of training on either task. Following transfer, choice accuracy dropped to 60% correct if tested with the same task (e.g., transferred from matching to matching or vice versa), or to 40% when tested with the complementary task (e.g., when transferred from oddity to matching or vice versa). The difference between positive and negative transfer (i.e., to the same or complementary task) indicates that a general rule did play a role in controlling performance, but the difference between choice accuracy before and after transfer indicates that something more was involved. Another possible limitation of this evidence is that it is all obtained with simultaneous matching or oddity in which the sample and comparison stimuli are all present at the same time and there are few, if any, demands on working memory. The processes controlling delayed matching–to–sample may be very different from those controlling simultaneous matching or oddity.

Indeed, similarity between sample and correct comparison stimuli is not necessary to delayed matching–to–sample performance; a high level of accuracy is still obtained using symbolic delayed matching–to–sample in which the relation between the sample and the correct comparison stimulus is arbitrary. For example, two line orientation comparison stimuli might be presented after a red sample with a choice of "vertical" designated correct (Carter & Eckerman, 1975; Cohen, Looney, Brady, & Aucella, 1976; Maki, Moe, & Bierley, 1977; Roitblat, 1980). For pigeons, there appears to be little difference between symbolic and identity delayed matching–to–sample in terms of speed of acquisition or asymptotic level of performance (Carter & Eckerman, 1975; Cohen, Looney, Brady, & Aucella, 1976).

The alternative to the exclusive use of a general matching concept is a set of sample specific rules of the form "choose comparison stimulus 'X' after sample stimulus 'A'" (Carter, 1971; Carter & Eckerman, 1975; Carter & Werner, 1978; Cumming & Berryman, 1965). Use of these sample specific rules implies that the subject identifies the sample and uses that information as input to a transformation resulting in the correct choice when the alternative becomes available. This transformation could occur at any time between the onset of the sample and the choice. The available evidence suggests that it occurs relatively early and that the information about the sample is maintained during the retention interval in a form similar to the correct comparison stimulus. In a variation of the delayed matching–to–sample task (Roitblat, 1980), color sample stimuli were followed by line orientation comparison stimuli or, for a different bird, line orientation samples were followed by color tests. Two of the colors were very similar (red and orange) relative to the third color (blue). Three line orientation stimuli were also used: vertical (0o), a slant (12.5o), and horizontal (90o). For the birds trained with color samples the vertical stimulus was designated as the correct choice following blue samples, the slant was correct following orange, and the horizontal orientation was correct following red samples. One sample and all three comparison stimuli occurred on every trial. Note that two similar sample stimuli (red and orange) correspond to a dissimilar pair of comparison stimuli (horizontal and 12.5o slant, respectively), and that two similar comparison stimuli (vertical and 12.5o slant) correspond to two dissimilar sample stimuli (blue and orange). The same interstimulus relationships held for the bird trained with line orientation samples and color comparison stimuli.

The confusability of the similar comparison stimuli increased faster than did the confusability of the less similar comparison stimuli (see Figure 3 in Roitblat, 1982; also see Gaffan, 1977, for a similar experiment). On the assumption that the greater the similarity between memory items the greater the likelihood that they would be confused (Beals, Krantz, & Tversky, 1968), these results can be interpreted to mean that the birds used a prospective memory code (Honig & Wasserman, 1981), maintaining the information about the identity of the sample in a form resembling the comparison stimuli. I should emphasize that the change in confusability, not the absolute level of confusions is important in attributing performance to a prospective code. A higher rate of confusion between similar than between dissimilar comparison stimuli, alone, might indicate nothing more than the difference in discriminability. The change in confusability with increasing retention interval, however, cannot be similarly attributed since the stimuli themselves did not change during the delay; rather something in the bird's memory did. As the memory was maintained longer, more opportunities for confusing the sample's representation with its alternatives were available (Roitblat, 1982). Prospective memory codes receive further support from a variety of other experiments (e.g., see Chapter 8, this volume, by Peterson).

Honig and Wasserman (1981) found that delayed conditional discriminations were made as accurately as delayed simple discriminations when 0 sec intervened between sample and test, but that accuracy fell more rapidly with increasing delay for the delayed conditional than for the delayed simple discrimination. Since in both tasks a single sample is presented on each trial, Honig and Wasserman reasoned that the two tasks differed mainly in terms of their prospective memory requirements. The delayed simple discrimination task required only that the bird remember a decision (whether to respond to the test stimulus), while the delayed conditional discrimination task required a decision as well as memory for the identity of the appropriate stimulus (e.g., peck if red, not if blue).

Peterson and Trapold (1980, see also Chapter 8, this volume, by Peterson) found that pigeons utilized the outcome of particular delayed matching-to-sample trial sequences as part of the representation. Correct choices on one sample-comparison pair (in a two-alternative task) were reinforced with access to food, the other with presentation of a tone. Following a double reversal, the same sample-comparison pair was still correct, but followed by a disparate outcome. Choice accuracy following this reversal was initially good, but quickly declined, and eventually recovered, indicating that the outcome of a trial had a greater effect than simply rewarding correct choices. Trial outcome appears to be an intimate component of the working memory representation used by pigeons. Evidence for prospective memory processes has also been found in experiments on delayed sequence discrimination (Wasserman, Nelson, & Larew, 1980; Weisman & Dodd, 1979; Weisman, Wasserman, Dodd, & Larew, 1980) as well as in other experiments on delayed matching-to-sample (Grant, 1981; Roberts, 1980; Roitblat & Scopatz, 1983).

IV. THE DRIFT MODEL

One way of conceptualizing a gradual prospective memory process is by utilizing a spatial metaphor. The drift model is a conceptualization of this type which I have simulated on a computer. In this model, each of the comparison stimuli is represented as a point in "memory space". The distances between points represent the dissimilarities between the corresponding stimuli--more similar stimuli are located nearer each other than more dissimilar stimuli. To model the three-alternative delayed matching-to-sample tasks described above (Roitblat, 1980; Roitblat & Scopatz, 1983), I have chosen a two dimensional space, though other, more complex spaces may be required for other, more complex sets of stimuli. Each trial begins with a "pointer" located somewhere in the memory space. During the sample presentation the bird employs an analysis process to identify the sample stimulus through usual signal detection processes.

The output of the identification process drives the pointer on a random walk through memory space more or less in the direction of the point corresponding to the correct comparison stimulus for the identified sample. Following the termination of the sample presentation the pointer begins to drift. When the actual comparison stimuli are presented, the bird chooses the alternative corresponding to the point closest to the current position of the pointer. If that choice is incorrect, the comparison stimulus corresponding to the point next closest to the pointer is chosen.

The model, as it is presently described, can account quite well for the choices made on individual delayed matching-to-sample trials. The distance of the pointer from a central origin within the memory space corresponds to what might be called trace strength, which tends to increase during sample presentations. The model also makes clear, however, that the accuracy of performance does not depend solely on a change in this unidimensional variable. The use of three alternatives requires at least two dimensions to properly characterize the dimensional structure of the memory system as manifested in the patterns of choices made. The model also suggests that forgetting during the retention interval is not simply an undoing of the encoding, though it too is a gradual process. Forgetting in this model is simply a matter of gradually substituting one alternative for another with or without a modification of trace strength (i.e., the distance from a neutral origin).

In addition to the analysis of delayed matching-to-sample performance on isolated trials, trace strength theory has also been applied to performance obtained on sequences of trials (Grant, 1975; Roberts & Grant, 1974; Roberts & Grant, 1976; Grant & Roberts, 1973).

A number of studies have found general proactive interference (PI) effects in the form of less accurate performance at shorter than at longer intertrial intervals (Grant, 1975; Herman, 1975; Jarrard & Moise, 1971; Maki, Moe, & Bierley, 1977; Roberts, 1980; Roitblat & Scopatz, 1983). In addition, more specific proactive interference effects have been observed when the stimulus corresponding to the incorrect choice in two-alternative delayed matching-to-sample was the correct choice on the previous trial relative to cases in which that stimulus did not appear (Grant, 1975; Moise, 1976; Worsham, 1975). Trace strength theory interprets these results as the product of competition between the trace of the sample from the previous trial and the trace from the present trial, both of which are decaying with the passage of time. Roberts and Grant have developed trace theory into a competition/decay model in which the traces from each trial decay independently over time and compete for control of responding. These notions are paralleled in the drift model by an additional assumption that the pointer does not reset (i.e., does not return to the neutral origin) between trials but tends to remain in the general direction of the choice made.

The competition decay model attributes errors to two possible sources—decay of the trace from the present trial and intrusion or competition from the trace remaining from the previous trial (Roberts & Grant, 1976). The drift model combines these two processes into a single one mediated by the location of the pointer. Alternatively, there may be separate factors mediating the two kinds of proactive interference observed. The general effects of intertrial interval duration may be due to fatigue, temporary satiation, etc., independent of the specific effects of competition and intrusions of choices from the previous trial. Because only two alternatives are typically used in delayed matching-to-sample, it is difficult to discriminate between these alternatives but some information is available suggesting that even in this task the two effects may be separately mediated.

If relatively strong mismatching memories from two successive trials can combine to produce interference, then matching memories should combine to produce higher matching accuracy. This follows from either of two possible combination/competition rules. According to one rule, the stronger the memory, the higher the likelihood that it and not its competitor will control the choice made. According to the other rule, the two memories algebraically (or vectorially) combine their strength to produce the choice made. Choice accuracy, therefore, is predicted by the competition/decay model to be highest when short intertrial intervals separate trials involving matching memories. To the best of my knowledge, this finding has never been obtained. For example, Roberts (1980) tested pigeons at 1 and 20 sec intertrial interval (ITI) durations. He classified each trial according to the sample presented on the previous trial (the presample) and whether or not the correct choice had been made. His results are made more clear (because of a peculiar interaction he found and for theoretical reasons to be made clear shortly) by collapsing his classification according to the agreement between the choice made on the previous trial (prechoice) and the sample appearing on the later trial. Agreement between the prechoice and sample resulted in 73% correct choices at the 1 sec intertrial interval and about 80% correct choices following a 20 sec intertrial interval. Disagreement between prechoice and sample resulted in about 67% correct choices at the shorter intertrial interval and about 77% correct choices at the longer. While these results are consistent with the previously reported findings of higher accuracy at longer intertrial intervals and (after averaging together the relative frequencies of correct and incorrect prechoices) higher accuracy following intertrial agreement between samples, they are incompatible with the competition/decay model because performance was uniformly better at long than at short intertrial interval durations independent of the samples or choices made.

We (Roitblat & Scopatz, 1983) recently completed a series of experiments in which we explored this problem in greater depth using three alternative delayed matching-to-sample. Colors served as both samples and comparison stimuli. Pigeons were tested at three intertrial interval durations: 1, 5, or 30 sec, with 2 sec sample and 2 sec retention interval durations throughout. One intertrial interval duration was used between all trials on a particular day. Only one bird showed an overall effect on choice accuracy of intertrial interval duration, performing with lower accuracy at the shorter duration (cf. Riley & Roitblat, 1978). No effects of intertrial interval were found for any bird on second-choice accuracy.

Trials were examined in running dyads consisting of trials 1 & 2, 2 & 3, etc. The first trial in each pair is called the previous trial, the second is called the current trial. Each response was classified according to the sample presented on the previous trial (*presample* symbolized by P), previous choice (*prechoice* symbolized by R), current sample (symbolized by S) and current choice (symbolized by C) and The occurrence of each combination of these variables was counted (e.g., the number of times the red choice was made following a blue sample, green prechoice, and green presample) and the resulting 3 x 3 x 3 x 3 contingency tables were subjected to a log-linear analysis (Bishop, Fienberg, & Holland, 1975; Colgan & Smith, 1978).

Log-linear analyses have a function similar to that of analysis of variance. The latter technique partitions the variability of a set of observations (on a continuous scale) according to a model consisting of a set of additive components attributable to systematic effects of independent variables and chance factors. An F-test is used to determine whether the structure derived from the model is significantly different from the "structure" expected wholly on the basis of chance.

A log-linear analysis examines categorical variables cast into a multidimensional (multiway) table with each variable (e.g., sample, prechoice, etc.) corresponding to one dimension of the table (e.g., rows, columns, layers, etc.). It partitions the variability among the counts in the cells of the table according to a model whose components are linearly additive in the logarithmic scale (hence the name log-linear model) and are attributable to systematic effects of variables and chance factors. A chi-squared test is used to compare the observed distribution of counts (cell frequencies; the number of times an observation was made with that particular combination of categorical variables) with the distribution expected on the basis of the model (the goodness of fit), therefore, the logic of the test statistic is somewhat complementary to that used in analysis of variance. Each model tested represents a hypothesis about the structure of the data, which is then tested against other models and the data. The usual procedure is to try various models in order to find the one model that gives the best account of the structure of the data. The preferred model is usually the one that gives the best fit with the fewest number of free parameters. Interesting models contain fewer free parameters than the number of cells in the table. The parameters are estimated in these circumstances from the marginal tables which are subtables formed by collapsing over one or more variables. For example, in a two-way table, the column sums are one marginal table and the row sums are another. A given model is specified when its relevant marginal tables are listed in any order enclosed (in the present paper) by braces. Separate marginal tables (more properly the parameters specified by those marginal tables) exert independent influence. So, for example, a model specifying that choices (C) were made independent of the sample (S) presented would be symbolized {S,C} or {C,S}. On the other hand, the {CS} model specifies an interaction between C and S (i.e., that the proportion of counts at each level of S differed at the different levels of C). That is, that these two variables are statistically associated. The model therefore derives its parameters from the CS marginal table (the frequency with which each choice followed each sample), summed over the remaining variables—P and R—to determine the expected distribution. Only hierarchical models are tested, meaning that inclusion of an interaction term in a model implies that the lower level terms are also included in the model. Specification of the {CS} model, for example, implies that the C marginal table (the distribution of choices) and the S marginal table (the frequency with which each sample occurred) are both included. Any variables not explicitly or implicitly specified are assumed to have equal counts in each category and to have parameter values of 0 (Bishop, Fienberg, & Holland, 1975).

A summary of the best fitting models from this experiment is presented in Table 5.2. The test statistic is the likelihood ratio chi-squared test (LRX2), the degrees of freedom are those remaining after estimating all the used parameters, and the probability is the likelihood that the observed and expected distributions were drawn from the same population. A lower LRX2 value and a higher probability indicate, all other things being equal, a better fitting model and a better account of the data.

If there were no proactive interference, then each contingency table would be best fit by either the {CS,P} or the {CS,R} model in the presample and the prechoice tables respectively. According to these models, the distribution of responses in the three-way table depends on the joint distribution of choices and samples (CS) and on the frequency with which each presample or prechoice occurred. Presamples were each presented with equal frequency, however, so equivalent models involving prechoices would default to {CS}. In contrast, significant proactive interference is indicated by the presence of either (CP) or of (CR) in the best fitting models. Examination of table 5.2 shows the presence of proactive interference in every case, but two, however, this proactive interference took the form of a (statistical) association between the

TABLE 5.2
Best Fitting Models

Bird	ITI	Model	df	LRX2	Probability
Previous sample (P) × Sample (S) × Choice (C)					
17	1	CS	18	23.35	.18
	5	CS	18	14.13	.72
	30	CSP	0	0.00	1.00
90	1	CS,CP	12	10.65	.66
	5	CS	18	19.56	.36
	30	CS	18	13.04	.79
453	1	CS	18	26.32	.09
	5	CS	18	24.25	.15
	30	CS,CP	12	7.65	.81

Bird	ITI	Model	df	LRX2	Probability
Previous Choice (P) × Sample (S) × Choice (C)					
17	1	CS,CR,SR	8	7.66	.47
	5	CS,CR	12	14.58	.27
	30	CS,CR,SR	8	4.29	.83
90	1	CS,CR	12	12.94	.37
	5	CS,CR	12	8.13	.77
	30	CS,CR	12	10.85	.54
453	1	CS,CR,SR	8	6.14	.63
	5	CS,CR	12	18.25	.11
	30	CS,CR,SR	8	10.21	.25

Note-- ITI = intertrial interval; LRX2 = likelihood ratio chi-squared.

prechoice and the choice and not an association between the presample and the choice. This relationship is illustrated in Figure 5.2 in the form of a path diagram. It shows that the choice made on a given trial was due to two independent causal influences: That due to the sample and that due to the choice made on the previous trial. In other words, proactive interference consisted explicitly of intrusions of the choice made on the previous trial, not of the sample presented on that trial. This is the same effect reported by Roberts (1980) which, I argued, accounts for the interfering effect of intertrial sample disagreement as due to the nonrandom association between previous samples and previous choices not to a memory for the sample per se. This finding is entirely consistent with those obtained from the symbolic delayed matching–to–sample confusion experiment (Roitblat, 1980, Experiment 3) indicating prospective memory codes. Furthermore, with the exception of bird 17 at the longest intertrial interval, the proactive interference took the form of an association between previous choices and present choices and an association between present samples and present choices, not higher order combinations of the three. Hence, though both samples and prechoices affect current choices, they do so independently. On this basis it is possible to conclude that choices are remembered from one trial to the next and combine with the independent influence of the sample to determine the choice made.

PROACTIVE INTERFERENCE

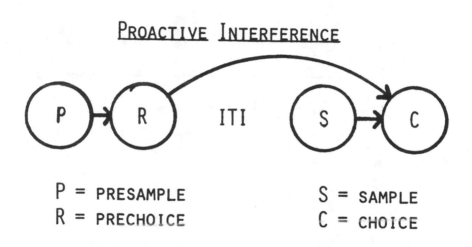

P = PRESAMPLE
R = PRECHOICE

S = SAMPLE
C = CHOICE

FIG. 5.2. Causal relationships obtained from log–linear analysis.

These relations were further explored in an experiment with the same three birds (Roitblat & Scopatz, 1983) that factorially combined two sample durations (0.5 or 2 sec), two retention interval durations (0.5 or 2 sec) and two intertrial interval durations (1 or 30 sec). All variables were manipulated between sessions and were constant within a session. The percentage correct responses of this experiment were analyzed using an analysis of variance which revealed that first– and second–choice accuracy improved for all birds with longer sample durations. One bird showed a decline in matching accuracy with longer retention interval durations.
As in the previous experiment, the response frequencies were cast

into separate three way prechoice (RSC) and presample (PSC) tables and again analyzed using log-linear analyses. With one exception, all tables broken down by presample were fit by the {CS} or simpler models. In contrast, most of the tables broken down by prechoice did show evidence of proactive interference by the prechoice. In this way this experiment replicated the findings of the previous one (see Roitblat & Scopatz, 1983).

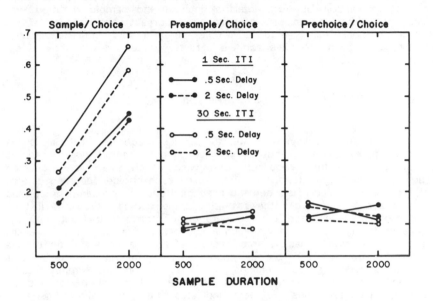

FIG. 5.3. Measures of Association as a function of intertrial interval, sample duration and retention interval duration. Reprinted from Roitblat & Scopatz, 1983. Copyright 1983 by the American Psychological Association.

The tables were further analyzed for the degree of association obtained between (a) presamples and choices, (b) prechoices and choices, and (c) between samples and choices. Associations (indexed by PHI, though other indices show the same pattern) between samples and choices varied as a function of sample duration and intertrial interval duration but no changes were detected as a function of delay nor were any interactions found (see Figure 5.3). Similarly, no changes in the degree of association between presamples and choices or between prechoices and choices were found as a function of any variable. This was a very surprising finding from the standpoint of the competition/decay model because it indicates that intertrial interval duration does indeed affect choice accuracy, but not by manipulating the magnitude of the competion, intrusion, or association from the previous trial. Longer intertrial interval durations result in stronger associations between the choice and the sample occurring within the same trial. The magnitude of the association between events from the previous trial and choice remained the same at all combinations of intertrial interval, sample duration and retention interval duration.

One possible way in which intertrial interval duration could affect the association between sample and choice might be to increase the efficiency (the rate of change in the probability of a correct choice per unit sample duration) of encoding mechanism (e.g., the rate at which the pointer moves in memory space). The data do not support this hypothesis, however, since no interaction between intertrial interval and sample duration was found using either the analysis of percentage correct responses or of the association measures. Instead, sample duration and intertrial interval duration were found to have additive effects, suggesting that they affect separate mechanisms.

V.   RETURN OF THE DRIFT MODEL

To summarize the findings to date, the choice made on any trial is the result of (a) the overall frequency with which each sample was presented (sample bias), (b) the frequency with which each prechoice occurred (prechoice bias), (c) the frequency with which each choice occurred (choice bias), (d) the association between the prechoice and the present choice, and (e) the association between the sample and the choice. Only the last of these factors is affected by the duration of the intertrial interval, the sample or the retention interval, each acting independently. Thus there are really two separate proactive interference effects, a general interference effect that depends on the duration of the intertrial interval and a prechoice-specific effect that is, at least within the limits tested, independent of the intertrial interval duration. The failure to find an interaction between these effects indicates that they affect separate mechanisms. This finding is inconsistent with the trace theory interpretation of delayed matching-to-sample, and requires some modification of the drift model.

Recall that the drift model involves two separate processes for encoding sample information. The analysis process determines the identity of the sample appearing on a trial. Its output then drives the random walk memory process. The general interference effect may be due to fatigue of dimension specific analyzers (e.g., Sutherland & Mackintosh, 1971). Utilization of an analyzer on one trial causes it to become refractory for some amount of time, dependent on the duration of the intertrial interval, or slows its rate of operation thereby delaying an identification. For example the proportion of time (e.g., of a session) during which an analyzer is in use may control the duration of this refractory period. Once this analyzer identifies the sample, however, it continues to drive the memory pointer at a fixed rate unaffected by the intertrial interval duration. Earlier identification of the sample results in more steps being taken, but does not affect their size. The pointer drifts during the retention interval, again at a rate independent of any of the variables investigated. Finally, during the intertrial interval, the pointer resets, not to the center of the memory space but to some point a fixed distance from the center toward the choice actually made.

ACKNOWLEDGMENTS

This research was supported by Grant BNS 79-14212 and Grant BNS 82-03017 from the U.S. National Science Foundation. I thank Robert Scopatz for his continuing contribution to the ideas contained here.

REFERENCES

Beals, R., Krantz, D. H., & Tversky, A. Foundations of multidimensional scaling. *Psychological Review*, 1968, *75*, 127-142.

Bishop, Y. M. M., Fienberg, S. E., & Holland, P. W. *Discrete multivariate analysis: Theory and practice*. Cambridge, Mass.: MIT Press, 1975.

Carter, D. E. *Acquisition of a conditional discrimination: A comparison of matching-to-sample and symbolic matching*. PhD Thesis, Columbia University, 1971.

Carter, D. E. & Eckerman, D. A. Symbolic matching by pigeons: Rate of learning complex discriminations predicted from simple discriminations. *Science*, 1975, *187*, 662-664.

Carter, D. E., & Werner, T. J. Complex learning and information processing by pigeons: A critical analysis. *Journal of the Experimental Analysis of Behavior*, 1978, *29*, 565-601.

Clarke, F. R. Confidence ratings, second-choice responses, and confusion matrices in intelligibility tests. In J. A. Swets (Ed.), *Signal detection and recognition by human observers*. New York: Wiley, 1964.

Cohen, L. R., Looney, T. A., Brady, J. H., & Aucella, A. F. Differential sample response schedules in the acquisition of conditional discriminations by pigeons. *Journal of the Experimental Analysis of Behavior*, 1976, *26*, 301-314.

Colgan, P. W., & Smith, J. T. Multidimensional contingency table analysis. In P. W. Colgan (Ed.), *Quantitative ethology*. New York: Wiley, 1978.

Cumming, W. W., & Berryman, R. The complex discriminated operant: Studies of matching-to-sample and related problems. In D. I. Mostofsky (Ed.), *Stimulus generalization*. Stanford: Stanford University Press, 1965.

Dickinson, A. *Contemporary animal learning theory*. Cambridge, England: Cambridge University Press, 1980.

Gaffan, D. Response coding in recall of colours by monkeys. *Quarterly Journal of Experimental Psychology*, 1977, *29*, 597-605.

Grant, D. S. Proactive interference in pigeon short-term memory. *Journal of Experimental Psychology: Animal Behavior Processes*, 1975, *1*, 207-220.

Grant, D. S. Short-term memory in the pigeon. In N. E. Spear & R. R. Miller (Eds.), *Information processing in animals: Memory mechanisms*. Hillsdale, N.J.: Erlbaum, 1981.

Grant, D. S., & Roberts, W. A. Trace interaction in pigeon short-term memory. *Journal of Experimental Psychology*, 1973, *101*, 21-29.

Herman, L. M. Interference and auditory short-term memory in the bottle nosed dolphin. *Animal Learning and Behavior*, 1975, *3*, 43–48.

Herrnstein, R. J., Loveland, D. H., & Cable, C. Natural concepts in pigeons. *Journal of Experimental Psychology: Animal Behavior Processes*, 1976, *2*, 285–302.

Honig, W. K. Studies of working memory in the pigeon. In S. H. Hulse, H. Fowler, & W. K. Honig (Eds.), *Cognitive processes in animal behavior*. Hillsdale, N.J.: Erlbaum, 1978.

Honig, W. K., & Wasserman, E. A. Performance of pigeons on delayed simple and conditional discriminations under equivalent training procedures. *Learning and Motivation*, 1981, *12*, 149–170.

Jarrard, L. E., & Moise, S. L. Short-term memory in the monkey. In L. E. Jarrard (Ed.), *Cognitive processes of nonhuman primates*. New York: Academic Press, 1971.

Maki, W. S., & Leith, C. R. Shared attention in pigeons. *Journal of the Experimental Analysis of Behavior*, 1973, *19*, 345–349.

Maki, W. S., Moe, J. C., & Bierley, C. M. Short-term memory for stimuli, responses, and reinforcers. *Journal of Experimental Psychology: Animal Behavior Processes*, 1977, *3*, 156–177.

Moise, S. L. Proactive effects of stimuli, delays, and response position during delayed matching from sample. *Animal Learning and Behavior*, 1976, *4*, 37–40.

Nadel, L., & MacDonald, L. Hippocampus: Cognitive map or working memory? *Behavioral and Neural Biology*, 1980, *29*, 405–409.

Olton, D. S., & Papas, B. C. Spatial memory and hippocampal function. *Neuropsychologia*, 1979, *17*, 669–682.

Riley, D. A., & Roitblat, H. L. Selective attention and related cognitive processes in pigeons. In S. H. Hulse, H. Fowler, & W. K. Honig (Eds.), *Cognitive processes in animal behavior*. Hillsdale, N.J.: Erlbaum, 1978.

Roberts, W. A. Short term memory in the pigeon: Effects of repetition and spacing. *Journal of Experimental Psychology*, 1972, *94*, 74–83.

Roberts, W.A. Distribution of trials and intertrial repetition in delayed matching to sample with pigeons. *Journal of Experimental Psychology: Animal Behavior Processes*, 1980, *6*, 217–237.

Roberts, W. A., & Grant, D. S. Short term memory in the pigeon with presentation time precisely controlled. *Learning and Motivation*, 1974, *5*, 393–408.

Roberts, W. A., & Grant, D. S. Studies of short-term memory in the pigeon using the delayed matching-to-sample procedure. In D. L. Medin, W. A. Roberts, & R. T. Davis (Eds.), *Processes of animal memory*. Hillsdale, N.J.: Erlbaum, 1976.

Roitblat, H. L. Codes and coding processes in pigeon short-term memory. *Animal Learning and Behavior*, 1980, *8*, 341–351.

Roitblat, H. L. The meaning of representation in animal memory. *The Behavioral and Brain Sciences*, 1982, *5*, 353–406.

Roitblat, H. L., & Scopatz, R. A. Sequential effects in pigeon delayed matching-to-sample performance. *Journal of Experimental Psychology: Animal Behavior Processes*, 1983, *9*, 202–221.

Sutherland, N. S., & Mackintosh, N. J. *Mechanisms of animal discrimination learning*. New York: Academic Press, 1971.

Wagner, A. R. SOP: A model of automatic memory processing in animal behavior. In N. E. Spear & R. R. Miller (Eds.), *Information processing in animals: Memory mechanisms*. Hillsdale, N.J.: Erlbaum, 1981.

Wasserman, E. A., Nelson, K. R., & Larew, M. B. Memory for sequences of stimuli and responses. *Journal of the Experimental Analysis of Behavior*, 1980, *34*, 49–59.

Watkins, J. The Popperian approach to scientific knowledge. In G. Radnitzky & G. Andersson (Eds.), *Progress and Rationality in Science*. (Vol. 58.) Dordrecht, Holland: D. Reidel, 1978.

Weimer, W. B. *Notes on the methodology of scientific research*. Hillsdale, N.J.: Erlbaum, 1979.

Weisman, R. G., & Dodd, P. W. D. The study of association: Methodology and basic phenomena. In A. Dickinson & R. A. Boakes (Eds.), *Mechanisms of learning and motivation: A memorial volume to Jerzy Konorski*. Hillsdale, N.J.: Erlbaum, 1979.

Weisman, R. G., Wasserman, E. A., Dodd, P. W. D., & Larew, M. B. Representation and retention of two–event sequences in pigeons. *Journal of Experimental Psychology: Animal Behavior Processes*, 1980, *6*, 312–325.

Worsham, R. W. Temporal discrimination factors in the delayed matching–to–sample task in monkeys. *Animal Learning and Behavior*, 1975, *3*, 93–97.

Wright, A. A., Santiago, H. C., Urcuioli, P. J., & Sands, S. F. Monkey and pigeon acquisition of same/different concept using pictorial stimuli. In M. L. Commons, R. J. Herrnstein, & A. R. Wagner (Eds.), *Quantitative analyses of behavior: Vol. III, Acquisition*. Cambridge, Mass.: Ballinger, 1984. In press.

Zentall, T. R., & Hogan, D. E. Memory in the pigeon: Proactive inhibition in a delayed matching task. *Bulletin of the Psychonomic Society*, 1974, *4*, 109–112.

Zentall, T. R., & Hogan, D. E. Same/different concept learning in the pigeon: The effect of negative instances and prior adaptation to the transfer stimuli. *Journal of the Experimental Analysis of Behavior*, 1978, *30*, 177–186.

Zentall, T. R., Edwards, C. A., Moore, B. S., & Hogan, D. E. Identity: The basis for both matching and oddity learning in pigeons. *Journal of Experimental Psychology: Animal Behavior Processes*, 1981, *7*, 70–86.

6   REHEARSAL IN PIGEON SHORT-TERM MEMORY

Douglas S. Grant
University of Alberta

I. INTRODUCTION

The notion that animals postperceptually process, or rehearse, information in short-term memory has been prominent in recent theoretical treatments of both associative learning (Wagner, 1976, 1978, 1981) and working memory (Grant, 1981a; Maki, 1981). The present chapter addresses the implications of the notion that animals rehearse for research and theory in the area of pigeon short-term memory. Considered initially is whether rehearsal has the same referent in the two contexts. To anticipate the subsequent discussion, it is argued that the characteristics attributed to rehearsal by Wagner differ markedly from those attributed to rehearsal by Grant and by Maki. It is suggested that this state of affairs arose because rehearsal in animals is not a unitary process. It is argued that it is appropriate to distinguish conceptually between two qualitatively different types or forms of rehearsal, types referred to as associative rehearsal and maintenance rehearsal.

With a distinction between associative rehearsal and maintenance rehearsal in hand, evidence consistent with the operation of each type of rehearsal in pigeon short-term memory is reviewed. Considered finally are some directions for further research suggested by the associative-maintenance rehearsal distinction. It is suggested that the conduct of such research may facilitate the development of more powerful theoretical models in which the role of rehearsal in associative learning and working memory is more adequately specified.

II. TYPES OF REHEARSAL

In this section, the conception of rehearsal offered by Wagner (1981) is contrasted with that offered by Grant (1981a) and by Maki (1981). It is suggested that the contrasting properties attributed to rehearsal by these authors stems from the different empirical domains addressed by their theories. It is argued that a distinction between two different forms of rehearsal, associative and maintenance, is intuitively plausible. Considered in a later section is whether one can go beyond intuition in supporting the validity of such a distinction.

A. ASSOCIATIVE REHEARSAL

Wagner (1981) has offered a formal model of Pavlovian conditioning in which the notion of rehearsal assumes a central role.   According to Wagner, a representation may be said to be in rehearsal in short–term memory when that representation is highly active.   It would thus appear that for Wagner rehearsal is more appropriately characterized as a state of activation than as a process responsible for, or necessary to, a state of high activation.   For present purposes, however, whether rehearsal is conceptualized better as a state or as a process is not a central concern. Rather, what is important in the present context is a delineation of the properties or characteristics ascribed to rehearsal by Wagner.

Perhaps the most fundamental property of rehearsal involves an influence on the formation of permanent memories.   Specifically, Wagner suggests that post–trial rehearsal is necessary to associative learning. Given the typical Pavlovian conditioning preparation in which a conditioned stimulus (CS) is followed shortly by an unconditioned stimulus (US), the amount learned about the CS–US relation on any particular trial is directly proportional to the length of time during which the representations of the CS and US are conjointly in rehearsal in short–term memory.   Because of the dependence of associative learning on this type of rehearsal, it is appropriate to refer to rehearsal in Wagner's sense as "associative rehearsal."

Wagner identifies two further properties of associative rehearsal important to the present treatment.   First, he refers to rehearsal as a "standard operating procedure," by which he emphasizes the notion that rehearsal is an inherent property of the information processing system. That is, the provocation of rehearsal is held to be determined solely by invariant operating characteristics of the information processing system. The second property, which may be viewed as a corollary of the first, is that once initiated, associative rehearsal proceeds mechanically, and thus independently of voluntary control.

The sense of these latter two properties may be captured by referring to associative rehearsal as an automatic, rather than controlled, process. Caution must be exercised, however, in that associative rehearsal in Wagner's scheme does not share all of the characteristics identified as those of automatic processes by Shiffrin and Schneider (1977).   One of the more obvious discrepancies concerns the relationship between automatic information processing and attentional demands.   Shiffrin and Schneider suggest that automatic processing proceeds without attention and thereby does not tax the limited processing capacity of short–term memory.   In contrast, Wagner holds that associative rehearsal does engage or occupy processing capacity (in fact, this proposition formed the basis for one of the earlier studies designed to implicate the operation of rehearsal in associative learning, see Wagner, Rudy, & Whitlow, 1973). However, if the caveat is kept in mind, it may be useful to refer to associative rehearsal as an automatic process in that such reference emphasizes the mechanical, invariant features of this type of rehearsal.

B. MAINTENANCE REHEARSAL

Grant (1981a) and Maki (1981) have offered models of working memory in which the notion of rehearsal has also figured prominently. As discussed below, the conception of rehearsal endorsed by these authors differs fundamentally from Wagner's view of associative rehearsal. Before considering these points of contrast, it is necessary to consider the general nature of the preparations used to obtain the data about which Grant and Maki theorized.

Most studies of short-term memory in animals have employed a task which may be characterized as a working memory preparation. According to Honig (1978), working memory preparations are those in which "different stimuli govern the criterion response on different trials, so that the cue that the animal must remember varies from trial to trial. Therefore, working memory situations often involve conditional discriminations (p. 213)." The delayed matching to sample task conforms to Honig's definition of a working memory preparation and has been the assessment procedure of choice in the experimental analysis of pigeon short-term memory. In delayed matching, a sample stimulus (usually a colored field or a line orientation) is presented and then withdrawn prior to illumination of two comparison stimuli. A response to the comparison matching the sample is reinforced, and a response to the alternate (nonmatching) comparison is nonreinforced. Memory for the sample is indexed by the accuracy of matching at various retention intervals (intervals between sample offset and comparison onset).

Grant (1981a) and Maki (1981) have each suggested that pigeons rehearse the sample memory during the retention interval. In contrast to rehearsal referred to here as associative, rehearsal in their schemes is viewed as promoting temporary maintenance of information only. That is, while rehearsal enhances short-term memory for the sample, it has no effect on the formation of permanent memories. Thus, as it appeared appropriate to refer to rehearsal in the sense intended by Wagner (1981) as associative rehearsal, it appears appropriate to refer to rehearsal in the sense intended by Grant and by Maki as maintenance rehearsal.

Not only do associative and maintenance rehearsal differ in terms of putative function, but in two other ways as well. First, recall that associative rehearsal was viewed as a standard operating procedure in the information processing system. In contrast, maintenance rehearsal is conceived by Grant and Maki as a process which develops with training. That is, pigeons learn the conditions under which rehearsal is appropriate, and may even learn to rehearse with increased efficiency as training proceeds. Second, whereas Wagner refers to rehearsal as an automatic process, Grant and Maki refer to maintenance rehearsal as a controlled process. Thus, associative rehearsal may be viewed as proceeding mechanically and invariably once initiated, whereas maintenance rehearsal may be viewed as a flexible process which remains under the organism's control.

C. CONCLUSION

The properties of rehearsal suggested by Wagner (1981) differ fundamentally and importantly from those suggested by Grant (1981a) and by Maki (1981). The present suggestion that the former and latter authors were considering two qualitatively different types of rehearsal, associative and maintenance, has some intuitive appeal. In the Pavlovian preparation, the organism is exposed to a relatively stable relationship between environmental events. On the other hand, the reinforcement contingencies in a working memory preparation such as delayed matching are dynamic. Viewed in this way, it is conceivable that the Pavlovian preparation might invoke modes of information processing fundamentally different from those invoked in a working memory preparation such as delayed matching. Considered in the next section is whether empirical evidence from studies of pigeon short-term memory may be viewed as consistent with the postulation of two different forms of rehearsal.

III.   EVIDENCE OF REHEARSAL IN PIGEON SHORT-TERM MEMORY

It was suggested above that associative rehearsal is invoked in associative learning tasks whereas maintenance rehearsal is invoked in working memory tasks. If so, it is understandable that different properties have been attributed to rehearsal within models of Pavlovian conditioning and delayed matching. Although the present distinction between two qualitatively different types of rehearsal is thus plausible, it is somewhat disquieting to base the distinction solely upon theoretical notions postulated to account for such markedly different phenomena as Pavlovian conditioning in rabbits and delayed matching in pigeons. The distinction would have greater force if it could be shown that the two types of rehearsal are operative in a single species tested under comparable conditions and across the same retention intervals.

It is the thesis of the present section that such evidence is already in hand from studies of pigeon short-term memory. It is of course possible to assess short-term memory using preparations other than delayed matching. The defining characteristic of a short-term memory preparation is the use of retention intervals of seconds or minutes, and not the use of a working memory task. Although a working memory preparation such as delayed matching is both a convenient and profitable tool for the investigation of short-term memory, its use is not obligatory. Of particular relevance to the present discussion is the fact that short-term memory can be investigated using associative memory tasks. For example, one could present a CS followed by a US and subsequently probe for memory of the CS–US episode at a variety of intervals, ranging from say 0 to 20 sec, after trial offset. The preparation would thus be procedurally identical to typical multi-trial Pavlovian conditioning except that the intertrial interval would be manipulated systematically to permit a short-term memory function to be plotted. Unfortunately, studies illuminating the processes of short-term memory in such preparations have not been performed with pigeons.

However, pigeon short-term memory has been investigated in preparations combining aspects of associative learning and working memory tasks. Studies of this general form provide evidence, although admittedly tentative, that a type of rehearsal having the characteristics of associative rehearsal operates in such a preparation. Prior to considering that evidence, data from studies of pigeon short-term memory using the delayed matching task are shown to support the proposition that maintenance rehearsal operates in a typical working memory preparation.

A. EVIDENCE FOR MAINTENANCE REHEARSAL

In this section, data consistent with the proposition that rehearsal in delayed matching is appropriately characterized as having the properties attributed earlier to maintenance rehearsal is reviewed. Specifically, considered initially is the present suggestion that sample rehearsal should not result in the establishment of a permanent memory. The discussion then turns to a consideration of the putative flexibility of rehearsal in delayed matching. Data consistent with this view suggest that pigeons learn to engage in rehearsal and that with continued training learn to do so with greater efficiency and/or probability. Also consistent with the flexibility property are findings suggesting that pigeons highly practiced in rehearsal can learn to terminate that process.

1. Rehearsal and long-term memory. The present view is that once the contingencies of delayed matching have been represented permanently in long-term memory (a process presumably requiring associative rehearsal), rehearsal of the sample memory will then be that characterized as maintenance rehearsal. The influence of the stimulus presented as the sample on trial n upon responding on trial n+1 should therefore be minimal when intertrial intervals are sufficiently long to permit the trial n sample memory to be lost from short-term memory. In reviewing the evidence on this issue, Roberts and Grant (1976) concluded that "the analysis of performance on delayed matching-to-sample trials preceded by other trials designed to provide conflicting information gives little evidence of long-term interference effects and suggests that long-term memories of events occurring on delayed matching-to-sample trials are not established in the pigeon (pp. 89-90)." Although recent studies of delayed matching in pigeons reveal a rather more complex view of sequential effects than that held in 1976 (Roberts, 1980; Roberts & Kraemer, 1982; Roitblat & Scopatz, 1983), the author is unaware of any findings which invalidate the view that presentation of a sample stimulus fails to provoke processes necessary for that stimulus to be represented permanently in long-term memory.

2. Learning to rehearse. If, as suggested earlier, presentation of a sample does not provoke associative rehearsal after the delayed matching task has been learned, then the question arises as to whether the sample is rehearsed in any manner. Two sets of findings, one rather casual and the other more formal, support the proposition that pigeons learn to rehearse the sample memory.

First, it is a well known part of the laboratory lore surrounding delayed matching that early delayed testing should employ relatively short retention intervals (say up to 4 or 5 sec). Once performance stabilizes, intervals of slightly longer duration may be successfully introduced, and so on. The reason for gradually increasing the length of retention intervals is that pigeons better tolerate (i.e., match more accurately at) longer intervals if such testing is preceded by a period of practice at shorter intervals. Perhaps the most dramatic instance of this improvement comes from a study by Grant (1976) in which highly practiced pigeons matched at above chance levels at a 60-sec retention interval. Most noteworthy from the present perspective is the finding that such progressive improvement in accuracy is confined primarily to performance at longer retention intervals (compare the control curve in Figure 6.1 from Grant, 1975 with the 4-sec presentation time curve in Figure 6.1 from Grant, 1976). In other words, the primary effect of continued training appears to be a reduced rate of forgetting and not a general enhancement of matching accuracy. Although this effect may be accounted for in a number of ways, the interpretation favored here is that pigeons learn to engage in maintenance rehearsal with increased efficiency as delayed matching testing progresses.

More definitive evidence that pigeons learn to rehearse is provided by a recent study by Maki (1979a). Pigeons were trained initially on a symbolic matching task in which a sample of food or no food was followed by one red and one green comparison stimulus. Responding to red was reinforced on food-sample trials, and responding to green was reinforced on no-food-sample trials. The delays were adjusted for each bird to maintain matching accuracy at about 88% correct. At this point, training began on a simple successive discrimination between a vertical line (S+) which always terminated in food and a horizontal line (S-) which always terminated in no-food. Following acquisition of the simple discrimination, the birds were infrequently tested on complex trials in which a simple discrimination trial preceded a matching trial in such a way that the outcome event of the simple trial also served as the sample for the matching test. Such probe trials were of two types; congruent trials in which S+ preceded a food sample and S- preceded a no food sample, and incongruent trials in which S- preceded a food sample and S+ preceded a no food sample.

The finding of greatest relevance to the present discussion is that accuracy of matching was higher on incongruent probes than on congruent probes, but only during the later sessions of testing. Maki argued that retention was equivalent on the two types of probes early in testing because the birds did not expect a matching test on trials beginning with either S+ or S-, and hence did not rehearse the outcome/sample. Maki suggested, however, that with extended training the birds learned that the configurations S- followed by food and S+ followed by no food signalled a forthcoming retention test, and hence rehearsed the outcome/sample on incongruent probes. On the other hand, the configurations S+ followed by food and S- followed by no food were eight times more likely to terminate in an intertrial interval than in a matching test. Thus, the birds may have failed to learn to rehearse the outcome/sample on congruent probe trials.

3. Learning not to rehearse.   The notion that pigeons maintain memory for the sample through a flexible, controlled process of maintenance rehearsal suggests that (a) pigeons should learn to rehearse under appropriate training conditions and (b) once trained to rehearse, it should be possible for pigeons to learn not to rehearse.   Evidence consistent with proposition "a" was considered in the previous section, and evidence consistent with proposition "b" is considered in the present section.

Several studies of directed forgetting have shown that instructional stimuli presented after sample offset can influence matching accuracy (Grant, 1981b; Kendrick, Rilling, & Stonebraker, 1981; Maki & Hegvik, 1980; Maki, Olson, & Rego, 1981; Stonebraker & Rilling, 1981; Stonebraker, Rilling, & Kendrick, 1981).   The basic training procedure involves the designation of forget and remember cues from a dimension orthogonal to that of the sample and comparison stimuli.   On remember—cued trials, the remember cue is presented during the retention interval and the trial terminates in the standard test of memory for the sample.   On forget—cued trials, the forget cue is presented during the retention interval and the trial terminates without a memory test.   After extensive exposure to these conditions of training, memory for the sample on forget—cued trials is tested on occasional probe trials.

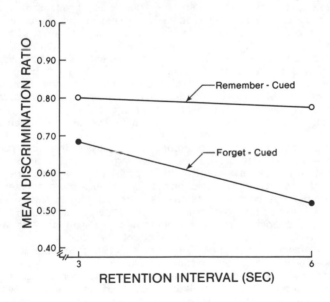

FIG. 6.1. Proportion of total responding directed toward matching comparison stimuli on forget—cued and remember—cued trials as a function of retention interval. Data are from Grant (1981b).

The typical outcome of such testing is shown in Figure 6.1 (Grant,

1981b). A successive matching task was employed in which only a single comparison stimulus is presented on each trial. If the comparison matches the sample, responding is reinforced on a fixed interval 5 sec schedule, and if the comparison does not match the sample, the comparison terminates in nonreinforcement after 5 sec. The index of matching accuracy employed in the successive procedure is the discrimination ratio which is the proportion of responses which are emitted to test stimuli matching the sample. The higher the discrimination ratio the more accurate is the performance. Examination of the figure reveals that accuracy was lower on forget–cued probe trials than on remember–cued trials. More importantly, the difference in accuracy on the two types of trials increased as retention interval was lengthened from 3 to 6 sec. Stated another way, rate of forgetting was faster on forget–cued probe trials than on remember–cued trials. The latter finding, which has also been obtained by Maki and his associates (Maki & Hegvik, 1980; Maki, Olson, & Rego, 1981), is exactly in accord with the notion that presentation of a forget cue reduces, or perhaps terminates, rehearsal of the sample memory.

Other findings from studies employing this preparation have provided additional support for the view that presentation of a forget cue reduces maintenance rehearsal. For example, Grant (1981b) presented a 1–sec forget cue during either the initial, middle, or final one–third of a 3–sec retention interval. Matching accuracy was lowest (and hence forget cue effectiveness was maximal) when the cue was presented immediately following sample offset, and accuracy increased progressively as the sample–cue interval increased. In a conceptually similar study, Stonebraker, Rilling, and Kendrick (1981) manipulated the inter–cue interval in a double cuing preparation. In the condition of interest, an initial forget cue presentation was followed after intervals of variable length by presentation of a remember cue. They found that presentation of the remember cue produced the greatest restoration of accuracy when the remember cue was presented immediately following forget cue offset.

In the studies cited above, forget cue effectiveness has been assessed by occasionally presenting a matching test on probe trials. As argued above, lower accuracy on forget–cued probe trials than on remember–cued trials may be interpreted as reflecting reduced memory for the sample on the former trial type. It is possible, however, that reduced accuracy might result because the animal is not expecting a retention test. In other words, the bird may remember the sample equally well on the two types of trials but perform less accurately on forget–cued trials because it is unprepared for a matching test. Rilling and his associates have suggested recently one such account of the directed forgetting phenomenon (Kendrick, Rilling, & Stonebraker, 1981; Stonebraker & Rilling, 1981; Stonebraker, Rilling, & Kendrick, 1981).

A necessary prerequisite for viewing directed forgetting as support for the postulation of a controlled process of maintenance rehearsal is convincing evidence that the forget cue reduces sample memory. Thus required is a preparation in which forget cue effectiveness could not result from the animal being unprepared for a retention test. Such a preparation has been developed recently in my laboratory. An intratrial interference preparation was employed in which two samples are presented successively prior to the choice test on interference trials. The first or interfering sample corresponds to the incorrect comparison and the second or target sample corresponds to the correct comparison. Several studies have demonstrated reduced matching accuracy on interference, trials relative to that on control trials in which the interfering sample is not presented (Grant, in press; Grant & Roberts, 1973; Zentall & Hogan, 1974, 1977).

Consider now a situation in which either the interfering or the target sample is followed by a forget cue (the alternate sample being followed by a remember cue). If presentation of a forget cue reduces sample rehearsal, then it should be possible to decrease interference by presenting a forget cue following the interfering sample, and to increase interference by presenting a forget cue following the target sample. On the other hand, if presentation of the forget cue reduces preparatory behaviors necessary to accurate test responding, presentation of a forget cue could not act to selectively reduce memory for one of two samples presented on a trial.

These predictions were evaluated recently (Grant, in press) using the same birds as had served in an earlier study of directed forgetting (Grant, 1981b). Control trials involved presentation of the target sample only, which was followed by a remember cue, and interference trials involved presentation of an interfering sample and a target sample. In the first experiment, the target sample was followed by a cue to remember on all interference trials, and the interfering sample was followed by a remember cue or by a forget cue equally often. The results of the experiment were unambiguous; matching accuracy was only 73.1% on interference trials in which the interfering sample was followed by a remember cue, increased to 81.7% when the interfering sample was followed by a forget cue, and was still higher at 92.6% when the interfering sample was not presented (control trials). A final experiment in the series demonstrated the symmetry of the selective action of a forget cue by reversing the cuing conditions. That is, the interfering sample was always followed by a cue to remember, and the target sample was followed by a cue to remember or by a cue to forget equally often. As anticipated by the rehearsal account, accuracy was lower when the target sample was followed by a forget cue (60.8%) than when it was followed by a remember cue (70.2%). These results suggest strongly that memory for a sample followed by a forget cue is reduced relative to memory for a sample followed by a remember cue, an effect viewed here as being mediated by the influence of such cues on processes of maintenance rehearsal.

B. EVIDENCE FOR ASSOCIATIVE REHEARSAL

It has been argued here that a controlled process of maintenance rehearsal operates in pigeon short—term memory when a working memory preparation is the assessment procedure. The argument was supported by considering data suggesting that rehearsal of the sample (a) does not promote the formation of a permanent representation, (b) develops during the course of training, and (c) remains under the organism's control after initiation. Considered in the present section is evidence consistent with the remaining half of the argument ——that an automatic process of associative rehearsal operates in pigeon short–term memory when the assessment technique involves an associative learning preparation.

The evidence considered below suggests that rehearsal of associative information has properties opposite those identified in points "b" and "c" above. More specifically, the data suggest that rehearsal of associative information is invoked automatically (does not develop with training) and, once initiated, proceeds automatically according to invariant properties of the system (is not under the organism's control). The present position cannot, of course, take particular comfort in the finding that permanent memories are established in associative learning tasks; only the converse, that maintenance rehearsal does not establish permanent memories, has meaning in the present context.

Studies implicating associative rehearsal in pigeon short–term memory have involved probing for memory of the outcome event of an associative episode presented seconds earlier. In order to conduct such memory tests, pigeons have been trained initially in delayed matching using samples identical to the outcome events of the associative episodes (food and no food). In this way, short–term memory for the outcome event of an associative episode may be assessed easily by presenting a matching test following the associative episode. Notice, however, that such a preparation is likely to invoke maintenance rehearsal in addition to associative rehearsal. That is, if an animal has been trained to remember the occurrence of a food sample, the occurrence of that event should be followed by maintenance rehearsal. However, if that event is also the outcome event of an associative episode, the present position anticipates that its occurrence would also provoke associative rehearsal. Interpretation of performance in this preparation is likely to be made more difficult by the operation of each type of rehearsal, possibly operating on independent representations. Further interpretive difficulties arise because we do not yet know how memory of an outcome event is translated into performance on a matching test. For example, might an animal remember that, say, a red light was followed by food, but demonstrate only marginal matching accuracy because the representation of "food" as an outcome in an associative episode differs from the representation of "food" as a sample stimulus?

Although a detailed discussion of these complications is beyond the scope of the present treatment, raising these issues may suggest one reason why the evidence for associative rehearsal in pigeons is less compelling than that for maintenance rehearsal. The primary concern is, however, whether such studies provide evidence for associative rehearsal. The present position is that they do in fact provide such evidence.

1. *Automatic triggering of rehearsal.*   Central to Wagner's (1981) model of Pavlovian conditioning is the notion that presentation of a surprising US provokes rehearsal of its representation, whereas presentation of an expected US does not provoke rehearsal. This view anticipates that a pigeon's short–term memory for the occurrence (or nonoccurrence) of a US should depend upon the surprisingness of that US occurrence (or nonoccurrence). Specifically, an event should be forgotten more slowly if it is surprising than if it is expected.

A recent series of experiments conducted in my laboratory (Grant,

Brewster, & Stierhoff, 1983) was designed to determine whether surprising events are more memorable than expected events (for an earlier, conceptually similar study, see Terry & Wagner, 1975). In the most definitive experiment of the series, pigeons were trained initially in a symbolic matching task employing samples of food and no food. Each sample presentation was preceded by a 10-sec "x" which terminated equally often in a 2-sec access to grain (food sample) and a 2-sec period of darkness (no food sample). A choice test between red and green comparison stimuli was presented after the retention interval. Choice of red was reinforced on food sample trials and choice of green was reinforced on no food sample trials.

Following acquisition of matching, the birds were trained on a simple Pavlovian successive discrimination between a large (CS+) and a small (CS-) triangle. Each 10-sec presentation of CS+ terminated in reinforcement (2-sec access to grain) and each 10-sec presentation of CS- terminated in nonreinforcement (2-sec of darkness). Following acquisition of the Pavlovian discrimination, the procedure was modified to permit an assessment of the effect of sample surprisingness on short-term memory. This was accomplished by chaining the two types of trials on probe trials. Probes were identical procedurally to the matching trials of training with the exception that a 10-sec presentation of CS+ or CS- replaced the standard 10-sec presentation of the "x." Thus, a sample of food was expected if preceded by CS+ and a sample of no food was expected if preceded by CS-. Similarly, preceding food by CS- and no food by CS+ rendered the sample surprising.

Performance on expected and surprising probe trials is shown in Figure 6.2. Although substantial forgetting is evident across the 10-sec retention interval when the sample was expected, there is no evidence of forgetting when the sample was surprising. Importantly, this finding was obtained from the outset of testing on probe trials. These data are thus consistent with the notion that evocation of enhanced rehearsal is an inherent property of surprising events. If so, rehearsal on probe trials does not have the properties identified in the working memory preparation as those of controlled, maintenance rehearsal. On the other hand, both the nature of the preparation (retention of an outcome) and the data obtained (surprising events evoking enhanced processing from the outset of testing) suggest that rehearsal in this preparation is that characterized as automatic, associative rehearsal.

Although there are many ways in which one might attempt to assess further the notion that surprising samples provoke associative rehearsal, perhaps the most intriguing involves the use of remember and forget cues. Suppose that effective remember and forget cues are established in a matching task employing samples of food and no food. Suppose we now assess the effectiveness of those cues on trials in which the sample of food or no food is made surprising. Because associative rehearsal is viewed as proceeding mechanically and automatically, accuracy on remember-cued and forget-cued trials should be approximately equivalent when the sample is surprising.

2. *Mechanical continuation of rehearsal.* In contrast to maintenance rehearsal, associative rehearsal is viewed as an automatic process which

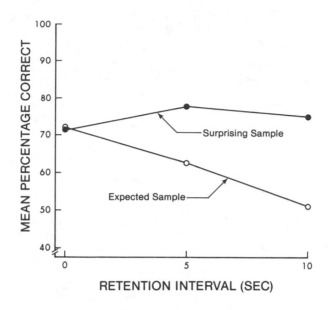

FIG. 6.2. Percentage of correct responses on surprising and expected sample trials as a function of retention interval. Data are from Grant, Brewster, & Stierhoff (1983).

proceeds mechanically once initiated.   Therefore, although associative rehearsal can be disrupted following initiation (Wagner, Rudy, & Whitlow, 1973), the present view suggests that maintenance rehearsal should be more easily disrupted.   At the very least, then, we might anticipate that certain treatments which disrupt maintenance rehearsal should have little, if any, effect upon ongoing associative rehearsal.

Maki (1979b) has obtained evidence consistent with this prediction. Pigeons were trained initially to "match" using samples of food (red correct) and no food (green correct).   In the test phase, food samples were preceded equally often by red and vertical (S+'s), and no food samples were preceded equally often by green and horizontal (S–'s). The first peck after 10-sec in the presence of an S+ produced a food sample, and a no food sample was presented after 10-sec in the presence of an S–. Notice that this preparation differs from the "surprisingness" preparation in two ways.   First, the sample was never surprising because it was never signalled inappropriately.   Second, the discriminative stimuli were not pretrained and instead acquired predictive significance on matching trials.   Thus, Maki's procedure permits the concurrent measurement of the development of associative learning (differential responding to the discriminative stimuli) and short-term memory for the outcome event (matching accuracy).

The manipulation of interest was the presence or absence of houselight during the retention interval (interval between termination of the outcome event and presentation of the matching test). For half of the birds, houselight was present during the retention interval of trials beginning with colored discriminative stimuli (red/food and green/no food). For the remaining half of the birds, houselight was present on trials beginning with line orientation discriminative stimuli (vertical/food and horizontal/no food). Consistent with prior findings (e.g., Grant & Roberts, 1976; Maki, Moe, & Bierley, 1977; Roberts & Grant, 1978), matching accuracy was reduced on trials in which the houselight was illuminated during the retention interval. However, there was no evidence that associative learning proceeded more slowly for the stimulus–outcome pair which was followed by houselight illumination.

Although Maki's findings are open to several interpretations, his data are consistent with the view that maintenance rehearsal is more easily disrupted than is associative rehearsal. The interpretation favored here is that houselight disrupted maintenance rehearsal (and hence reduced matching accuracy), but did not disrupt associative rehearsal (and hence did not retard associative learning). If this interpretation is to be viable, it must be suggested further that two independent representations of the outcome event were established on each trial; one involved in the associative representation (and which provoked associative rehearsal) and one which controlled performance on the matching test (and invoked maintenance rehearsal). If so, the associative representation of food apparently did not control matching performance because accuracy was at a chance level (50%) on houselight–present trials in spite of the fact that processing of the associative representation was presumably unaltered by that treatment.

An interesting paradox arises when Maki's results are compared to those of Grant et al., (1983). It was suggested that in Maki's study the associative representation of food did not control matching performance. In the Grant et al. study, however, it apparently did control performance. That is, if surprising samples provoke enhanced associative rehearsal, and if matching accuracy is higher on such trials, then the associative representation must have controlled performance on the matching test. The apparent paradox may be resolved by considering the conditions of prior training in the two studies. Briefly, it may be that training on matching trials employing the "x" as the sample signal mimicked sufficiently closely the conditions of probe trial testing that the birds effectively translated memory for the outcome event into accurate test responding. On the other hand, Maki's birds had had no such training, and may therefore have not used the associative representation of the outcome event to control matching performance.

IV. CONCLUSION

In considering directions for further research on directed forgetting in pigeons, Maki (1981) suggested that we can proceed to ask "whether the post–sample processes we study in tasks like delayed matching to sample and call rehearsal are homologous to those posttrial processes studied in conditioning preparations (p. 222)." The primary theme of the present chapter is that this question can be profitably addressed, at least in a preliminary way, through a consideration of available data and theory, and that such a consideration leads to a negative answer.   Specifically, the postsample processes operative in delayed matching are held to be those of a flexible, controlled process of maintenance rehearsal.   In associative learning tasks, the posttrial processes are held to be those of a mechanical, automatic process of associative rehearsal.   In preparations which combine associative learning and working memory tasks (such as those employed by Grant et al., 1983, and Maki, 1979a), both types of rehearsal are likely operative.

In offering a distinction between maintenance and associative rehearsal, the present treatment may stimulate research designed to further our understanding of the properties of each type of rehearsal, the conditions under which each is operative, and the ways in which the two processes interact.   Beyond this general prescription for further research, two related, and more specific, recommendations are offered.   First, given the paucity of available data, research concerned with the characteristics of associative rehearsal in pigeons is particularly warranted.   Such research need not, of course, involve some combination of associative learning and matching tasks.   In fact, research on short-term retention of incompletely learned associative tasks may prove more definitive because the operation of processes of maintenance rehearsal is effectively precluded.

A second issue for further research is to illuminate the ways in which associative and maintenance rehearsal interact in short-term memory.   This issue, which can be addressed only in preparations tapping processes of both associative learning and working memory, may not be resolved easily. The complexity of this issue is suggested by an apparent paradox arising from a comparison of the findings of Maki (1979b) and Grant et al. (1983).   In the former study, associative rehearsal of the outcome event (sample) was apparently not sufficient to support an above chance (50%) level of matching accuracy on houselight trials.   However, the latter study demonstrated that surprising samples are more memorable than expected samples, presumably because the former provoke enhanced associative rehearsal.   If so, them we are faced by the paradox that in one case an event receiving associative rehearsal supports matching, and in another case it does not.   It is suggested above that the difference may be attributed to conditions of prior training, but this assertion is post hoc and requires empirical assessment.

According to the present position, the ability to retain information for brief periods of time is intimately involved in both associative learning and working memory preparations.   This is not to say, however, that the processes whereby information is so retained are identical in the two cases; in fact, the fundamental proposition of the present viewpoint is that qualitatively different modes of information processing (referred to here as associative and maintenance rehearsal) are involved in the two preparations. Perhaps in tacit recognition of this point, theoretical treatments of processes of short-term memory have, to date, been confined to these processes as they operate either in associative learning (e.g., Pearce & Hall, 1980; Wagner, 1981) or in working memory (e.g., Grant, 1981a; Honig, 1978; Maki, 1981; Roberts & Grant, 1976).   It may now be appropriate, however, to begin the task of developing more powerful models of short-term memory, models capable of accommodating and integrating data from both associative learning and working memory tasks. A necessary prerequisite to such an effort may be explicit recognition of a distinction between associative rehearsal and maintenance rehearsal.

ACKNOWLEDGMENTS

Preparation of this chapter was supported by Grant A0443 from the Natural Sciences and Engineering Research Council of Canada.

REFERENCES

Grant, D. S. Proactive interference in pigeon short-term memory. *Journal of Experimental Psychology: Animal Behavior Processes*, 1975, *1*, 207-220.

Grant, D. S. Effect of sample presentation time on long-delay matching in the pigeon. *Learning and Motivation*, 1976, *7*, 580-590.

Grant, D. S. Short-term memory in the pigeon. In N. E. Spear & R. R. Miller (Eds.), *Information processing in animals: Memory mechanisms.* Hillsdale, N.J.: Erlbaum, 1981. (a)

Grant, D. S. Stimulus control of information processing in pigeon short-term memory. *Learning and Motivation*, 1981, *12*, 19-39. (b)

Grant, D. S. Intratrial proactive interference in pigeon short-term memory: Manipulation of stimulus dimension and dimensional similarity. *Learning and Motivation*, 1982, *13*, 417-433.

Grant, D. S. Directed forgetting and proactive interference in pigeon short-term memory. *Canadian Journal of Psychology*, in press.

Grant, D. S., & Roberts, W. A. Trace interaction in pigeon short-term memory. *Journal of Experimental Psychology*, 1973, *101*, 21-29.

Grant, D. S., & Roberts, W. A. Sources of retroactive inhibition in pigeon short-term memory. *Journal of Experimental Psychology: Animal Behavior Processes*, 1976, *2*, 1-16.

Grant, D. S., Brewster, R. G., & Stierhoff, K. A. "Surprisingness" and short-term retention in pigeons. *Journal of Experimental Psychology: Animal Behavior Processes*, 1983, *9*, 63-79.

Honig, W. K. Studies of working memory in the pigeon. In S. H. Hulse, H. Fowler, & W. K. Honig (Eds.), *Cognitive processes in animal behavior.* Hillsdale, N.J.: Erlbaum, 1978.

Kendrick, D. F., Rilling, M., & Stonebraker, T. B. Stimulus control of delayed matching in pigeons: Directed forgetting. *Journal of the Experimental Analysis of Behavior*, 1981, *36*, 241-251.

Maki, W. S. Discrimination learning without short-term memory: Dissociation of memory processes in pigeons. *Science*, 1979, *204*, 83-85. (a)

Maki, W. S. Pigeon's short-term memories for surprising vs. expected reinforcement and nonreinforcement. *Animal Learning and Behavior*, 1979, *7*, 31–37. (b)

Maki, W. S. Directed forgetting in pigeons. In N. E. Spear & R. R. Miller (Eds.), *Information processing in animals: Memory mechanisms*. Hillsdale, N.J.: Erlbaum, 1981.

Maki, W. S., & Hegvik, D. K. Directed forgetting in pigeons. *Animal Learning and Behavior*, 1980, *8*, 567–574.

Maki, W. S., Moe, J. C., & Bierley, C. M. Short-term memory for stimuli, responses, and reinforcers. *Journal of Experimental Psychology: Animal Behavior Processes*, 1977, *3*, 156–177.

Maki, W. S., Olson, D., & Rego, S. Directed forgetting in pigeons: Analysis of cue functions. *Animal Learning and Behavior*, 1981, *9*, 189–195.

Pearce, J. M., & Hall, G. A model for Pavlovian learning: Variations in the effectiveness of conditioned but not of unconditioned stimuli. *Psychological Review*, 1980, *87*, 532–552.

Roberts, W.A. Distribution of trials and intertrial repetition in delayed matching to sample with pigeons. *Journal of Experimental Psychology: Animal Behavior Processes*, 1980, *6*, 217–237.

Roberts, W. A., & Grant, D. S. Studies of short-term memory in the pigeon using the delayed matching-to-sample procedure. In D. L. Medin, W. A. Roberts, & R. T. Davis (Eds.), *Processes of animal memory*. Hillsdale, N.J.: Erlbaum, 1976.

Roberts, W. A., & Grant, D. S. An analysis of light-induced retroactive inhibition in pigeon short-term memory. *Journal of Experimental Psychology: Animal Behavior Processes*, 1978, *4*, 219–236.

Roberts, W. A., & Kraemer, P. J. Some observations of the effects of intertrial interval and delay on delayed matching to sample in pigeons. *Journal of Experimental Psychology: Animal Behavior Processes*, 1982, *8*, 342–353.

Roitblat, H. L., & Scopatz, R. A. Sequential effects in pigeon delayed matching-to-sample performance. *Journal of Experimental Psychology: Animal Behavior Processes*, 1983, *9*, 202–221.

Shiffrin, R. M., & Schneider, W. Controlled and automatic human information processing: II. Perceptual learning, automatic attending, and a general theory. *Psychological Review*, 1977, *84*, 129–190.

Stonebraker, T. B., & Rilling, M. Control of delayed matching-to-sample performance using directed forgetting techniques. *Animal Learning and Behavior*, 1981, *9*, 196–201.

Stonebraker, T. B., Rilling, M., & Kendrick, D. F. Time dependent effects of double cuing in directed forgetting. *Animal Learning and Behavior*, 1981, *9*, 385–394.

Terry, W. S., & Wagner, A. Short-term memory for "surprising" vs. "expected" unconditioned stimuli in Pavlovian conditioning. *Journal of Experimental Psychology: Animal Behavior Processes*, 1975, *104*, 122–133.

Wagner, A. R. Priming in STM: An information processing mechanism for self-generated or retrieval-generated depression in performance. In T. J. Tighe & R. N. Leaton (Eds.), *Habituation: Perspectives from child development, animal behavior, and neurophysiology.* Hillsdale, N.J.: Erlbaum, 1976.

Wagner, A. R. Expectancies and the priming of STM. In S. H. Hulse, H. Fowler, & W. K. Honig (Eds.), *Cognitive processes in animal behavior.* Hillsdale, N.J.: Erlbaum, 1978.

Wagner, A. R. SOP: A model of automatic memory processing in animal behavior. In N. E. Spear & R. R. Miller (Eds.), *Information processing in animals: Memory mechanisms.* Hillsdale, N.J.: Erlbaum, 1981.

Wagner, A. R., Rudy, J. W., & Whitlow, J. W. Rehearsal in animal conditioning. *Journal of Experimental Psychology,* 1973, *97,* 407–426.

Zentall, T. R., & Hogan, D. E. Memory in the pigeon: Proactive inhibition in a delayed matching task. *Bulletin of the Psychonomic Society,* 1974, *4,* 109–112.

Zentall, T. R., & Hogan, D. E. Short-term proactive inhibition in the pigeon. *Learning and Motivation,* 1977, *8,* 367–386.

7 SOME PROBLEMS FOR A THEORY OF WORKING MEMORY

William S. Maki
North Dakota State University

I. INTRODUCTION

For the last dozen years I (and my colleagues and students) have been investigating an assortment of problems in areas germane to the main topic of this book (cognitive processes and information processing in animals). Our initial efforts were shaped in part by two themes. First, we were impressed by how successfully the influential variable-reinforcement theory of Rescorla and Wagner (1972) coped with certain associative learning phenomena (e.g., blocking) that had been explained in more cognitive terms (Kamin, 1969). It appeared that traditional techniques used to study animal learning might leave us without decisive answers to questions about cognitive processes in animals (see Riley & Leith, 1976). Second, we were persuaded by Broadbent's (1961) argument that much could be learned about the nature of cognition and learning in animals by attending to the developing field of information processing in humans. The result was that we were led away from the tools of associative learning and instead adopted steady-state techniques that we saw as more compatible with method and theory favored by researchers of human information processing. We thus hoped to discover and understand the general attentional and memorial capacities of our animal subject in the context of the then-developing field of cognitive psychology.

Guided by these methodological and theoretical trends, my students and I have been investigating the nature of what's been called working memory in animals for the last few years. In this chapter I'll first describe current methods and theory connected with the study of working memory in animals and then present some new data I see as challenging our present notions of working memory and related processes in animals.

II. WORKING MEMORY IN ANIMALS

Miller, Galanter, and Pribram (1960) coined the term "working memory" to refer to a flexible, scratchpad-like memory that was conceived to be a temporary store in which "plans" could be maintained while being constructed and executed. Although we might use somewhat different terminology now, we are reminded of their conception by more recent discussions of both human and animal memory (e.g., Baddeley & Hitch,

117

1974; Honig, 1978; Olton, 1978).

A. WORKING MEMORY: METHODS

By far the most frequently used procedures in the investigation of working memory in animals are variations of the delayed matching-to-sample (DMTS) task. Usually a limited number of stimuli make up a set of "samples." On each trial, one sample is presented to the animal and later, after a delay (retention interval), that sample and another member of the set (the "comparison stimuli") are displayed. Choice of the sample is rewarded. Many (20 - 250) such trials occur within an experimental session separated by intertrial intervals (ITIs) normally longer than delays. To the extent that performance remains accurate in spite of a delay between the sample and comparison stimuli, one can conclude that the animal remembers the sample (or an associate of it; cf. Roitblat, 1980).

The delayed matching-to-sample task is a special case of the more general delayed conditional discrimination task. The delayed conditional discrimination and delayed matching-to-sample tasks are the same except for the "symbolic" nature of the conditional task, in which the sample and comparison sets are physically different but associatively related by the reinforcement contingencies chosen by the experimenter. For example, one might choose two forms (triangle and circle) as samples and two colors (red and green) as comparisons and define choice of red (green) as being correct after a sample of the triangle (circle).

Successive matching-to-sample is another task closely related to delayed matching-to-sample. In this task too, a trial is begun with a sample stimulus, but the trial ends with only one comparison -- either the sample or another member of the set. responses to the comparison that is identical to the sample are rewarded while responses to the nonmatching comparison are not. If the identity of the sample is remembered to the end of the delay then there should be a substantial difference between rates of responding to matching and nonmatching comparison stimuli.

There are probably numerous other kinds of tasks that could be mentioned as appropriate to the study of working memory in animals. Two will be of later concern in this chapter; both have been used to investigate spatial memory. The simpler of the two is the delayed alternation task, most often done with a T-maze. Both right and left alleys of the maze end in goal boxes, both of which are baited with rewards at the start of the trial. Trials then consist of two parts. The animal is first "forced" to choose either the right or left alley in which a reward is consumed. After a delay, the animal is returned to the maze and offered a "free" choice between alleys. Correct performance consists of choosing the alley opposite to the forced choice (in which an unconsumed reward remains). The other task makes use of the radial-arm maze (e.g., Olton & Samuelson, 1976). Here there are more (often eight) alternative alleys radiating from the central platform. Rewards are placed at the end of each alley and the animal continues to choose until all rewards are consumed. Repetitions of previously chosen alleys result in nonreward and are thus scored as errors. In both instances it is presumed that the animal must remember where it has recently been in order to avoid repeating choices.

It will be important later to note the many procedural differences between delayed matching–to–sample (and related tasks) and the two spatial tasks just described. Clearly the nature of the to–be–remembered events are different. So are the environmental conditions of testing; delayed matching–to–sample experiments are usually conducted within the confines of an operant conditioning chamber whereas the spatial memory experiments are conducted in larger experimental enclosures (whole rooms) in which the mazes are housed. And, whereas delayed matching–to–sample experiments employ many trials per session with rather short intertrial intervals, the spatial experiments employ few trials per session with longer intertrial intervals (sometimes only one trial per day).

B. WORKING MEMORY: A SAMPLE MODEL

An exemplary sequence of processes incorporating a working memory is listed in Table 7.1. The model is applied specifically to the two–choice delayed conditional discrimination problem (of which the well–known delayed matching–to–sample problem is an instance). At the onset of the trial, a to–be–remembered (TBR) event occurs and is soon represented (stored) in a working memory (WM). At the same time, or shortly afterward, the to–be–remembered event acts on some longer–term memory as a retrieval cue prompting activation of associates of that event which are in turn stored in working memory. After termination of the to–be–remembered stimulus, the contents of working memory are maintained during a retention interval (delay). After the delay, comparison stimuli are presented which support a decision process culminating in choice of one of those stimuli. Finally, maintenance of working memory ceases after the trial is over.

It will be recognized that the model is skeletal, lacking in details in most places and based on arbitrary theoretical choice in others. For example, "encoding" processes presumed to be involved in development of the to–be–remembered representation in working memory are ignored and one could certainly construct a variety of possible decision models (e.g., Wright & Sands, 1981). However, the model in Table 7.1 is not unrepresentative of the *kinds* of models that would be (and have been) generated by those of us who have investigated and speculated about the nature of working memory and related memory processes. Although this model was devised with pigeons, operant chambers, and delayed matching–to–sample all in mind, it is quite similar in major respects to Olton's (1978) flowchart of memory processes underlying a rat's performance of a spatial memory task in a radial–arm maze. Moreover, the tactical choices of what boxes to include, with what labels, and in what order were based on available data. First, there is ample evidence that pigeons and rats represent end–of–trial events at or shortly after the to–be–remembered stimulus; for example, pigeons have been reported to code to–be–remembered stimuli prospectively in terms of comparison stimuli (or responses to them; Honig, 1981; Roitblat, 1980) and also to code aindex<prospective memory codes> to–be–remembered events in terms of different end–of–trial reinforcers associated with them (see Peterson &

TABLE 7.1
A model of memory processes responsible for
performance of delayed conditional discriminations

Events in task	Memory and decision processes
TBR stimulus	* store TBR stimulus in WM * retrieve associates of TBR stimulus * store associates in WM
Delay-interval stimuli	* maintain contents of WM
Comparison stimuli	* inspect comparison #1 * respond to #1 if match to TBR stimulus * respond to #1 if match to associate * otherwise respond to comparison #2
End-of-trial events	* cease maintenance of WM

Note-- TBR = to be remembered, WM = working memory.

Trapold, 1980 and Chapter 8, by Peterson, this volume). Second, nearly flat retention gradients are sometimes observed, reflecting little loss of memory over the delay. These gradients can be taken as a sign of an active maintenance process (e.g., "rehearsal"); evidence adduced in support of an active maintenance process includes apparent attentional distraction by changes in delay-interval stimuli (e.g., Grant & Roberts, 1976; Maki, Moe, & Bierly, 1977) and our apparent ability to gain stimulus control over the hypothesized maintenance process by analogs of directed forgetting techniques used with humans (see Grant, 1981 and Chapter 6, by Grant, this volume; Maki, 1981). Third, there is the evidence already mentioned supporting the case for functional control of delayed matching performance by associates of nominal to-be-remembered events and also evidence in other experimental situations for control by a representation of a to-be-remembered event itself (Santi, Tombaugh, & Tombaugh, 1982).

Fourth, the cessation of maintenance processes after trial termination seems widely accepted (e.g., Honig, 1978; Maki, 1981; Olton, 1978) but there is little directly supportive evidence; however, we do have those data from directed forgetting analogs in which stimuli associated with premature ends of trials are reported to disrupt performance.

III. NEW STUDIES OF MAINTENANCE OF WORKING MEMORY

The main concern of this chapter will be with particular characteristics often attributed to working memory in animals (rather than with the hows and whys of encoding and decision processes implied in Table 7.1). In many respects, conceptions of working memory in animals have not been that far removed from characterizations of what was once called short-term memory in animals (e.g., D'Amato, 1973; Maki et al., 1977). First, the most common assumption about working memory is that it is used on demand as a holding place for information of temporary value to the organism (Miller et al., 1960) as in the conditional discrimination tasks described above (Honig, 1978). Second, working memory is often assumed to be of limited capacity, although the limits have yet to be properly defined (Olton, 1978; Roberts, 1979, and Chapter 24, by Roberts, this volume). Third, working memory, like short-term memory, has been acknowledged to last for relatively long periods of time, ranging from tens of seconds (Honig, 1978) to several minutes (D'Amato, 1973). Fourth, memories exhibited by animals in the tasks typically used have been viewed as based on "active" processes (Grant, 1981) as opposed to more traditional, passive accounts just incorporating simple notions of trace strength and decay (Roberts & Grant, 1976) and theories based on simpler conditioning preparations (Wagner, 1981).

We have been mostly interested in how animals hold temporarily useful information in working memory throughout a retention interval. Over the past few years an active-process, rehearsal model has been developed. The original model was quite simple. The rehearsal hypothesis advanced by Grant and Roberts (1976) and Maki et al. (1977) assumed a rehearsal-like process in animals that served to keep the memory for a to-be-remembered event (sample) active throughout a delay interval. Rehearsal was thus seen to protect the sample representation from the "trace decay" that would otherwise occur (cf. Roberts & Grant, 1976). The model can easily be quantified. Assume a sample memory to be established at trace strength S_0. When rehearsal occurs (with a probability of r), the trace is preserved intact. In the absence of rehearsal, the trace decays such that the fraction of trace strength remaining is $1-d$ momentarily and $(1-d)^T$ after T units of time. The remaining trace strength at time T is thus predicted to be

$$S_T = S_0[r + (1 - r)(1 - d)]^T, \qquad (1)$$

where both r and d range from 0 to 1. The most important aspect of the model is how trace decay is treated. Unlike in other "passive" accounts (e.g., Roberts & Grant, 1976; Wagner, 1981), a progressive diminution of trace strength over time is not inevitable in the rehearsal model. Instead, decay only occurs in the absence of active maintenance. Thus, as r varies from 0 to 1, a family of retention curves is predicted. At $r = 0$, S_T declines and approaches 0 at large values of T. At $r = 1$, S_T remains at S_0 and is independent of the passage of time as reflected by its constancy over variations in T.

The simple model is attractive for reasons of parsimony but leaves unanswered important questions about the nature of the rehearsal process. Consequently, our research became focused on revealing the characteristics of rehearsal. Rehearsal was seen to be a learned (Maki, 1979b) attentional process, influenced by the presence of distracting stimuli (Maki et al., 1977). It was thought to be analogous to human "maintenance rehearsal," having little if any influence on the state of associative information in long-term memory (Maki, 1979a). Our more recent empirical efforts have been devoted to a more detailed examination of effects of supposed attentional distractors and generalizability of this view of maintenance of working memory outside the narrow confines of the conditioning chamber.

A. RECOVERY FROM RETROACTIVE INTERFERENCE

One of the most pervasive findings in the literature on animal short-term memory is the fact that changes in incidental stimuli during a retention interval severely disrupt performance. The first reported effects were found with monkeys using delayed matching-to-sample; monkeys were reported to perform much better when the retention interval was spent in darkness than when a houselight was turned on in the chamber during the interval (D'Amato, 1973). Since then, the phenomenon has also been revealed in other species. Grant and Roberts (1976) reported that pigeons too performed delayed matching-to-sample worse as a consequence of delay-interval illumination, and Herman (1975) reported that dolphins performed an auditory short-term memory task worse when an irrelevant sound occurred during the delay than when the delay was spent in silence. In addition to its cross-species generality, this retroactive interference effect has considerable procedural generality. It has been demonstrated with different kinds of to-be-remembered events (visual stimuli, behaviors, and reinforcers; Maki et al., 1977) and with both delayed conditional discrimination and successive matching-to-sample tasks (Maki et al., 1977; Tranberg & Rilling, 1980). However, despite its demonstrated generality, retroactive interference in animal short-term memory has proved theoretically challenging.

Several effects of delay-interval illumination can be deduced from the simple rehearsal model (Equation 1) if we let changes in delay-interval stimuli attract attention and thus interrupt the rehearsal process. If the rehearsal parameter is temporarily reduced for the duration of the light, the model then predicts that brief episodes of distracting stimuli should have less effect than lights of longer duration, and that the temporal point of interpolation of a brief distractor should not matter. This simple hypothesis seemed capable, at the time, of explaining much of what was known about effects of delay-interval illumination on pigeons' delayed matching-to-sample performance. Both Grant and Roberts (1976) and Maki et al. (1977) showed that short-duration lights inserted within the delay had less effect than did light throughout the delay, and performance seemed equally affected by brief lights interpolated at the beginning and at the end of the delay. Grant and Roberts also showed that the performance decreased as the intensity of the light (and presumably its salience) increased. (See also D'Amato, 1973, and Herman, 1975, for similar data in monkeys and dolphins). Because the rehearsal process was viewed as only prolonging the residence of to-be-remembered events in short-term memory, the model allowed the development ("consolidation") of associations in long-term memory in spite of the presence of a post-trial distractor; Maki (1979a) showed that a post-trial houselight did not retard discrimination learning even though the light appeared to simultaneously disrupt short-term memory for trial outcomes.

The simple rehearsal hypothesis quickly ran into trouble however. Roberts and Grant (1978) convincingly showed that point of interpolation *did* matter; pigeons' delayed matching-to-sample choice accuracy was worse when light occurred at the end of a delay than when it occurred at the beginning. Without postulating some additional mechanism(s), the simple rehearsal hypothesis cannot cope with this "beginning-end" effect, as Roberts and Grant dubbed the effects of interfering light presented at the beginning versus the end of the delay. Another possible source of trouble came from reports that delay-interval illumination per se may not always be interfering. For example, Tranberg and Rilling (1980) trained pigeons to perform accurately when illumination occurred during delay intervals; the pigeons' performance was then disrupted by *darkness* during the delay, a result confirmed by Cook (1980) and in unpublished experiments in my laboratory. The possibility that pigeons might be disturbed by a change in stimuli customarily appearing in delay intervals is not so problematic as is the learning that is implied by the procedures in these studies. For example, Tranberg and Rilling exposed their birds to several cycles of training in which illumination conditions differed. The birds therefore learned and relearned to perform accurately under conditions that began by causing poor performance. I've called this relearning to perform delayed matching-to-sample-type tasks accurately in the presence of originally interfering delay-interval stimuli "recovery from retroactive interference." There is nothing in a simple rehearsal hypothesis (Equation 1) that predicts such recovery and, as far as I know, the possible mechanisms responsible for the recovery have not been discussed.

We were thus led to perform a series of experiments aimed at extending our knowledge of both beginning-end and recovery effects with the hope that we could reveal the memory and learning mechanisms underlying retroactive interference with delayed matching. We have now completed three systematic replications of a transfer experiment. The common design called for (1) training birds to perform a delayed conditional discrimination task at relatively long delays, (2) retraining different groups when light occurred early or late in the delay, and (3) finally transferring both groups to whole-delay illumination conditions. One goal was to determine whether a beginning-end effect would be found during recovery from interference produced by brief houselight presentations during the delay. Another goal was to determine how later recovery from interference produced by continuous delay-interval illumination would be influenced by prior recovery from illumination during part of the delay.

Naive White King pigeons (N = 24) were first shaped and trained on the following delayed condition discrimination task. Each trial began, after a 20-sec intertrial interval, with the presentation of a white disk on the center key of a three-key conditioning chamber. A single peck on the key darkened the key and resulted in 2-sec access to mixed grain (a "sample" of food) or 2-sec darkness (a "sample" of no food).   After a delay (initially O sec), the two side keys were lit with red and green disks.   A peck on either key darkened both keys and initiated a 2-sec period in which reinforcement could occur; if the red key was pecked following a sample of food or if the green key was pecked following a sample of no food, 2-sec access to grain reinforcement occurred.   Each session consisted of 64 such trials counterbalanced for samples and positions of correct side keys; unless otherwise specified, the sessions were conducted in darkness.   After each bird was stably performing at or above 87.5% correct (reinforced) choices, delay training began.   Delays were increased by 0.5 sec per day contingent on performance at or above the 87.5% criterial level.   After thirty days of delay training each bird reached a delay length that consistently supported criterial performance.   The birds were then divided into two groups so as to hold the average delay lengths in the two groups roughly the same.   The birds in Group LD (mean delay: 6.7 sec) were exposed to the houselight during the last 2 sec of each delay. The birds in Group DL (mean delay: 7.2 sec) were exposed to the houselight during the last 2 sec of each delay.   In the first replication, this training lasted 45 days.   In the second and third replications, this training continued until each bird had performed at the 87.5% level for three days followed by an additional seven overtraining days.   After this "part-delay" training was finished, all birds were trained under conditions of "whole delay" (LL) illumination; in each trial the houselight was on throughout the delay.   In the first replication, this stage continued for 45 days.   In the other two replications, this stage continued until each bird's performance had returned to criterial levels.

Despite differences between replications in length of training at each stage, the results were sufficiently similar to warrant combining the data from all three replications.   Our usual criterion was applied to the data from the first replication (87.5% correct during three successive days) as well as to the second and third replications so as to produce a comparable dependent measure (trials to criterion).   The data of main interest are presented in Table 7.2.   Median trials to criterion are presented for each stage and each group.   Also presented are ranges and the number of subjects achieving the criterion.[1]   First consider the results of the part-delay illumination stage.   Group LD recovered accurate performance significantly faster than did Group DL.   A beginning-end effect was thus shown in recovery from retroactive interference.   Second, consider the results of the whole-delay illumination stage.   The rate of recovery was determined by the groups' prior training; Group LD also

---

[1]Several birds failed to reach our criterion and the "mortality" problem was particularly severe during the whole-delay condition.   The birds that were eventually dropped typically showed near-chance levels of performance even after 3-5 months of training.   Statistical analyses were performed on available data from "survivors."   As best as can be determined, survivors' performances during earlier parts of the experiment were similar to those of the dropouts.   The dropout rates suggest that the estimated difficulty of recovery of accurate performance during the whole-delay illumination stage is underestimated by the trials to criterion measure and the difference between groups in both stages might be in fact larger than now estimated.

TABLE 7.2
Trials required for recovery
from retroactive interference as a function of
locus and amount of interfering light

		Stage	
Group		Part	Whole
LD	Median	720	672
	Range	320-1312	192-1728
	N	12	8
DL	Median	1664	2480
	Range	704-7074	1024-8256
	N	9	6

Note-- Group LD received light during the first 2 sec of the delay and Group DL received light during the last 2 sec of delay during the Part stage. Both groups received light during the entire delay during the Whole stage. Data are delayed matching trials to a criterion of 87.5% correct averaged over three 64-trial sessions.

recovered significantly faster in this stage than did Group DL.[2] Comparisons between stages within groups revealed no significant differences.

What is responsible for recovery from retroactive interference? Some simple habituation-like process (like learning to ignore originally distracting stimuli) is attractive from the standpoint of a rehearsal hypothesis but is inadequate to explain the data reported here. Explanations incorporating the notion of learned inattention can be made to accommodate the different rates of recovery in LD and DL conditions but stumble over the LL transfer results. Consider two examples. First, birds in the DL condition might find the occurrence of light more "surprising" than birds in the LD condition because of increased temporal uncertainty associated with long intervals. Given that surprising events are processed to a greater extent than are expected events (Wagner, 1978), the birds would be slower to adapt to the light in the DL condition than would be the birds in the LD condition. Second, attention to the light might be longer maintained in the DL condition because of its contiguity with end-of-trial events like comparison stimuli. According to conditioned attention theory (Lubow, Weiner, & Schnur, 1981), when two events E1 and E2 are arranged in a conditioning relationship E1-E2, habituation to E1 is retarded but does

---

[2]An extra group of four birds were included in the second replication. this group was trained with whole-delay illumination (LL) immediately after delay training. Two birds achieved criterial performances in 4224 and 4416 trials. Although tending to take longer than Group DL in both stages, the differences were not significant.

eventually occur. If we let the degree of habituation vary with the delay between E1 (light) and E2 (comparison stimuli), then habituation would occur faster in the LD condition (long E1–E2 interval) than in the DL condition. However, neither example of a learned inattention model anticipates the observation that recovery in the LL condition proceeds faster after recovery from LD than after recovery from DL.

An alternative explanation of retroactive interference effects would begin by arguing that introducing novel stimuli into a well-learned task could disrupt performance via generalization decrement. In its simplest form, the argument is that the original task consists of a sequence of sample–darkness–comparisons but that the transfer task consists of a different sequence, sample–light–comparisons. On this account, the bird must relearn the task under the altered stimulus conditions. This kind of explanation fares no better than a learned inattention explanation. Consider a specific example. The perception of comparison stimuli might be altered by a novel stimulus (light) during the delay (cf. Roberts & Grant, 1978).

The degree of perceptual alteration would be expected to increase with proximity of the light to the comparison stimuli. Consequently, the disparity between the original and transfer tasks would be greater for birds in the DL condition than for birds in the LD condition. While this explanation does accurately predict more rapid recovery in the LD condition, it predicts no differences in rates of recovery in the LL condition in which the stimuli are the same for both groups regardless of prior experience with the light (DL vs. LD).

Although neither the learned inattention nor the generalization decrement hypotheses adequately explain recovery from retroactive interference *by themselves*, some combination of the hypotheses might prove viable. Suppose that novel delay–interval stimuli have two effects. First, such stimuli are surprising, command attention, and diminish rehearsal of the to–be–remembered event. Second, such stimuli also act to distort the contents of working (or short–term) memory and thus render originally learned associations or rules ("if sample A, peck comparison B") inoperative. According to this two–process view, the bird must learn to respond to different memorial representations and to rehearse in the presence of distracting stimuli. In the LD condition, a dark delay intervenes between the light and the comparison stimuli. The bird thus has an opportunity to rehearse a memory consisting of the sample and the light; at the same time, the light becomes less distracting (for reasons suggested earlier). When faced with the LL condition, the bird then learns to respond to a memorial representation similar to that experienced in earlier training. In the other, DL condition, comparison stimuli immediately follow the light and there is no opportunity for rehearsal prior to the time of choice between comparison stimuli. Thus, the bird only rehearses, during the early part of the delay, a representation consisting of the sample plus darkness as it had originally been trained to do. Because of the contiguity of light and comparison stimuli, habituation to the light occurs slowly. When faced with the LL condition after recovery from DL, this bird learns to respond to memorial representations totally different from any so far experienced. The two–process explanation thus predicts faster recovery from LD than from DL (because of faster habituation to the light) and also predicts that subsequent recovery from LL will be faster following LD training than following DL training (because of differential opportunities to rehearse memorial representations containing light).

Given the present state of our knowledge, it is not clear that the two-process view provided above is the only or best explanation of retroactive interference effects. What is clear is that such effects are likely to be more complicated than implied by previous accounts (e.g., Maki *et al.*, 1977) based on the simple rehearsal model (e.g., Equation 1). What is also clear is that a complete theory of working memory in animals will need to incorporate some learning mechanism(s) capable of accounting for the recovery effects reported in this chapter.

B. SPATIAL MEMORY AND RESISTANCE TO RETROACTIVE INTERFERENCE

Studies of spatial memory in rats using radial–arm mazes lead to the impression of working memory for spatial locations as being exceptional. While memory for nonspatial events as studied in delayed conditional discriminations may last for minutes (D'Amato, 1973), spatial working memory appears to last for *hours* (Beatty & Shavalia, 1980a). Further, spatial memory does not appear susceptible to a broad range of potentially interfering (attentionally distracting) delay–interval stimuli; a variety of stimuli from different sensory modalities (Maki, Brokofsky, & Berg, 1979), performing in another maze (Beatty & Shavalia, 1980b; cf. Roberts, 1981), and even electroconvulsive shock early in the delay interval (Shavalia, Dodge, & Beatty, 1981) were all treatments shown to be without detectable influence. There was no such information available on pigeon spatial memory at the time we began our work described below.

Prompted by failures to demonstrate any sort of robust memory for spatial locations in birds using variants on the radial–arm maze procedures (e.g., Bond, Cook, & Lamb, 1981; Olson, unpublished data), we (Olson & Maki, 1983) simplified the task for the birds. In the usual 8–arm radial maze, rats appear to follow a "win–shift rule"; that is, once having found food in an arm, that arm is avoided. In our simplified procedure, we maintained the win–shift rule but reduced the number of locations to two. We constructed large T–mazes and performed a series of delayed alternation experiments. The T–mazes, enclosed by hardware cloth, were relatively open to the rich visual environment (rooms with many extramaze cues provided by stored racks of equipment, for example). The alleys, start boxes, and goal boxes were all 24 cm wide and 38 cm high. The start box was 53 cm long and the remainder of the maze (including alleys and goal boxes) was 252 cm in length. (For additional details, see Olson & Maki, 1983.)

The pigeons were given four delayed alternation trials per day. Each trial commenced with a forced choice. Food cups at the end of both right and left alleys were baited with two peas and four kernels of corn. One alley, chosen at random, was blocked and the pigeon was inserted into the start box. After the pigeon had walked to the end of the open alley and consumed the grain, it was removed from the maze. After a delay (initially 0 sec), both alleys were opened and the pigeon was returned to the start box for the free choice. If the pigeon chose the alley it had just visited, it encountered an empty food cup and was removed from the maze. If, however, the pigeon chose the other alley it was rewarded with the grain from the other food cup. Spatial locations of forced choice alleys were counterbalanced within days. ITIs were 30–45 min in the experiments reported here.

In the first experiment five White King pigeons were trained on the delayed alternation problem just described. We adopted a criterion of 87.5% correct (rewarded free choices) over four successive days (14/16 correct). The birds averaged 30 trials to criterion (not counting the four criterial days). We then began delay training. After each pair of days in which the criterion (7 correct out of 8 trials) had been attained, the delay was doubled (commencing with 0.5 min). During the delay, birds were detained in a holding cage. All birds progressed through the sequence to 8–min delays with little problem, averaging 98 total trials. After an additional 24 – 72 trials, two of the birds achieved 16–min delays. Clearly pigeons learn this problem with ease and can then tolerate delays many times longer than those obtained in more traditional tests of pigeon short–term memory using operant chambers (see Table 7.2). One correspondence with spatial memory characteristics of rats was thus established, i.e., relatively long persistence.

Encouraged by these results, we sought to determine whether other characteristics of rat spatial memory could be demonstrated in pigeons with our delayed alternation procedures. Resistance to retroactive interference (e.g., Beatty & Shavalia, 1980b; Maki *et al.*, 1979; Roberts, 1981) is of particular importance in the present context. Four of the birds continued training at 8–min delays. On some days, the lights in the room were turned off during delays; novel reductions in delay–interval illumination have been shown to interfere with pigeons' performance of delayed conditional discrimination tasks (Cook, 1980; Tranberg & Rilling, 1980). The birds averaged 84% correct during days in which the light was on during the delay (the normal illumination) and averaged 85% correct during days in which delays were spent in darkness. In another experiment four birds either spent the delay in the holding cage or else were transported through the laboratory back to their home cages in the colony room. The birds averaged 89% correct on holding–cage days and 92% correct on home–cage days. The memories displayed by pigeons in our delayed alternation situation thus seem resistant to sources of retroactive interference (changes in delay–interval stimuli) that have profound effects on performance of more conventional delayed conditional discrimination tasks.

The similarities of our results with results with pigeons to those previously obtained with rats in spatial tasks are marked -- ease of acquisition, accurate performance at relatively long delays, and resistance to retroactive interference. The congruences, however, do not in themselves show that our birds' performance was based on spatial cues. Therefore we asked whether the memories we were studying were actually spatial in nature. Accurate delayed alternation performance in our T–maze could be based on three sorts of cues. First, birds could use room (extramaze) cues to guide choices. Second, the birds could use more proximal (intramaze) cues arising from stimuli within the maze. Third, the birds could use their own behaviors (response cues) to guide choices.

Four birds were trained using the procedures described above. On control days, the birds performed the usual delayed alternation task with 2-min delays. On rotation days, the maze was rotated 180 degrees during the 2-min delay and the food cups were exchanged (so the reward remained in the same spatial location). If the bird's performance was based on extramaze cues it should choose the correct alley during the free choice; if it does so, however, it will turn in the same direction as it did during the forced choice and also travel through the same alley. The birds averaged 88% correct on control days and 94% correct on rotation days. A bird's accurate performance in our experiments was thus based on extramaze cues and not intramaze or response-produced cues.

The pigeon's rapid acquisition of the delayed alternation task in the T-maze may be due to their demonstrated reliance on the rich extramaze environment. However, it might also be due to the delayed alternation contingencies; learning to avoid places recently visited (a win-shift rule) may be very easy. To estimate the contribution of the rule we trained one group of pigeons with the win-shift rule (using the delayed alternation procedures already described) and another group with the win-stay rule. The win-stay rule defines a spatial delayed matching-to-sample task in that a repetition of forced choice during the free-choice part of a trial is rewarded. After 60 trials (15 days), the shift group averaged over 82% correct choices while the stay group remained at low levels of accuracy (50%). It is not clear whether pigeons might eventually learn the spatial matching (win-stay) task with extended training. It is clear, though, that pigeons come to our experiments equipped with a preexisting win-shift bias (as do other avian species in spatial tasks; see Kamil, Chapter 29, this volume).

We now have information on how working memory for spatial information compares to working memory for other stimuli in other tasks and we have that comparison within a single, well-studied species (pigeons). The dramatic difference between durations of spatial and nonspatial working memory is likely to be true of rats as well (see Wallace, Steinert, Scobie, & Spear, 1980, for a conditional discrimination benchmark). Acquiring this comparative data, however, has left us with yet another set of theoretical problems. Honig (1978) anticipated this state of affairs by acknowledging that the characteristics of working memory are likely to vary across procedures and species. We have already made some progress in identifying those procedural variations that might be associated with variations in working memory characteristics (Olson & Maki, 1983). Earlier in this chapter, when methods for studying working memory were outlined, two such procedural variations were noted. First, intertrial interval durations are typically much longer in studies of spatial memory than in experiments employing more traditional delayed conditional discrimination tasks. For example, whereas ITIs were 30 - 45 minutes in our delayed alternation, experiments just described, the intertrial intervals were only 20 *seconds* in the experiments on effects of delay-interval illumination reported earlier in this chapter. We know that intertrial interval duration determines the level of performance in both conventional delayed conditional discrimination tasks (e.g., Roberts & Grant, 1976) and spatial tasks (Roberts & Dale, 1981). Consequently, long intertrial intervals might be responsible for the accurate performance at long delays found with spatial tasks. A second task variable might be responsible for variations in resistance to retroactive interference from delay—interval stimulus changes. In conventional delayed conditional discrimination tasks, the to—be—remembered stimuli are rather simple (a colored pecking key, for example). In spatial tasks, the functional, extramaze cues are multidimensional (e.g., a window, a relay rack, the experimenter). In the spatial task, then, the memory trace is likely to be richer (composed of more attributes) than in the delayed conditional discrimination task. The resulting redundancy might then provide some measure of protection against interference from delay—interval stimuli in the spatial task. Now the challenge is to construct a theory of working memory from which the effects of task variables can be deduced. If it is to cope with these and other facts (e.g., Beatty & Shavalia, 1980a; Shavalia et al., 1981). such a theory is likely to be much more complex than the simplistic model of memory maintenance embodied in Equation 1.

IV. SUMMARY

Tasks like delayed matching—to—sample can be used to assess working memory in animals. Performance is accurate to the extent that the subject remembers a conditional stimulus (the sample) throughout a retention interval (delay). A common finding is that a change in delay—interval stimuli reduces matching accuracy (retroactive interference). One interpretation of retroactive interference earlier formulated holds that novel delay—interval stimuli capture some of the animal's limited attentional capacity thereby detracting from rehearsal of the sample memory and permitting trace decay. In its simple form, the rehearsal hypothesis fails to predict two sets of phenomena described in this chapter. First, with sufficient training, pigeons' delayed matching performance recovered from the initially detrimental effects of a change in delay—interval illumination. The rate of recovery was faster if the change occurred early than if it occurred late in the delay. Recovery from light interpolated early in the delay produced faster recovery from subsequent whole—delay illumination than did recovery from light interpolated late in the delay. Second, using a large T—maze, pigeons were found to perform delayed alternation well at long delays and to rely on extramaze cues. Memory for spatial locations exhibited in these experiments was resistant to potential sources of retroactive interference like changes in delay—interval illumination. At the present time, recovery from and resistance to retroactive interference remain problems for a theory of working memory.

ACKNOWLEDGMENTS

The author's research described in this chapter was supported by grant R01—MH31432 from the National Institute of Mental Health. I thank Becky Aaland, Curt Borchert, Laurel Knoell, Rebecca Nelson, Robin White, and particularly Kathryn Bengtson and Deborah Olson for their assistance with and contributions to many aspects of the work.

REFERENCES

Baddeley, A. D., & Hitch, G. Working memory. In G. H. Bower (Ed.), *The psychology of learning and motivation*. (Vol. 8.) New York: Academic Press, 1974.

Beatty, W. W., & Shavalia, D. A. Spatial memory in rats: Time course of working memory and effect of anesthetics. *Behavioral Biology*, 1980, *28*, 454–462. (a)

Beatty, W. W., & Shavalia, D. A. Rat spatial memory: Resistance to retroactive interference at long retention intervals. *Animal Learning and Behavior*, 1980, *8*, 550–552. (b)

Bond, A. B., Cook, R. G., & Lamb, M. R. Spatial memory and the performance of rats and pigeons in the radial-arm maze. *Animal Learning and Behavior*, 1981, *9*, 575–580.

Broadbent, D. E. Human perception and animal learning. In W. H. Thorpe & O. L. Zangwill (Eds.), *Current problems in animal behavior*. Cambridge, England: The University Press, 1961.

Cook, R. G. Retroactive interference in pigeon short-term memory by a reduction in ambient illumination. *Journal of Experimental Psychology: Animal Behavior Processes*, 1980, *6*, 326–338.

D'Amato, M. R. Delayed matching and short-term memory in monkeys. In G. H. Bower (Ed.), *The psychology of learning and motivation*. New York: Academic Press, 1973.

Grant, D. S., & Roberts, W. A. Sources of retroactive inhibition in pigeon short-term memory. *Journal of Experimental Psychology: Animal Behavior Processes*, 1976, *2*, 1–16.

Herman, L. M. Interference and auditory short-term memory in the bottle nosed dolphin. *Animal Learning and Behavior*, 1975, *3*, 43–48.

Honig, W. K. Studies of working memory in the pigeon. In S. H. Hulse, H. Fowler, & W. K. Honig (Eds.), *Cognitive processes in animal behavior*. Hillsdale, N.J.: Erlbaum, 1978.

Honig, W. K. Working memory and the temporal map. In N. E. Spear & R. R. Miller (Eds.), *Information processing in animals: Memory mechanisms*. Hillsdale, N.J.: Erlbaum, 1981.

Kamin, L. J. Selective association and conditioning. In N. J. Mackintosh & W. K. Honig (Eds.), *Fundamental issues in associative learning*. Halifax: Dalhousie University Press, 1969.

Lubow, R. E., Weiner, I., & Schnur, P. Conditioned attention theory. In G. Bower (Ed.), *The psychology of learning and motivation*. (Vol. 15.) New York: Academic Press, 1981.

Maki, W. S. Discrimination learning without short-term memory: Dissociation of memory processes in pigeons. *Science*, 1979, *204*, 83–85. (a)

Maki, W. S. Pigeon's short-term memories for surprising vs. expected reinforcement and nonreinforcement. *Animal Learning and Behavior*, 1979, *7*, 31–37. (b)

Maki, W. S. Directed forgetting in pigeons. In N. E. Spear & R. R. Miller (Eds.), *Information processing in animals: Memory mechanisms*. Hillsdale, N.J.: Erlbaum, 1981.

Maki, W. S., Brokofsky, S., & Berg, B. Spatial memory in rats: Resistance to retroactive interference. *Animal Learning and Behavior*, 1979, *7*, 25–30.

Maki, W. S., Moe, J. C., & Bierley, C. M. Short-term memory for stimuli, responses, and reinforcers. *Journal of Experimental Psychology: Animal Behavior Processes*, 1977, *3*, 156–177.

Miller, G. A., Galanter, E., & Pribram, K. H. *Plans and the structure of behavior*. New York: Holt, Rinehart and Winston, 1960.

Olson, D., & Maki, W. S. Characteristics of spatial memory in pigeons. *Journal of Experimental Psychology: Animal Behavior Processes*, 1983, *9*, 266–280.

Olton, D. S. Characteristics of spatial memory. In S. H. Hulse, H. Fowler, & W. K. Honig (Eds.), *Cognitive processes in animal behavior*. Hillsdale, N.J.: Erlbaum, 1978.

Olton, D., & Samuelson, R. J. Remembrance of places passed: Spatial memory in rats. *Journal of Experimental Psychology: Animal Behavior Processes*, 1976, *2*, 97–116.

Peterson, G. B., & Trapold, M. A. Effects of altering outcome expectancies on pigeons' delayed conditional discrimination performance. *Learning and Motivation*, 1980, *11*, 267–288.

Rescorla, R. A., & Wagner, A. R. A theory of Pavlovian conditioning: Variations in the effectiveness of reinforcement and nonreinforcement. In A. H. Black & W. F. Prokasy (Eds.), *Classical conditioning II: Current research and theory*. New York: Appleton–Century–Crofts, 1972.

Riley, D. A., & Leith, C. R. Multidimensional psychophysics and selective attention in animals. *Psychological Bulletin*, 1976, *83*, 138–160.

Roberts, W. A. Spatial memory in the rat on a hierarchical maze. *Learning and Motivation*, 1979, *10*, 117–140.

Roberts, W. A. Retroactive inhibition in rat spatial memory. *Animal Learning and Behavior*, 1981, *9*, 566–574.

Roberts, W. A., & Dale, R. H. I. Remembrance of places lasts: Proactive inhibition and patterns of choice in rat spatial memory. *Learning and Motivation*, 1981, *12*, 261–281.

Roberts, W. A., & Grant, D. S. Studies of short-term memory in the pigeon using the delayed matching-to-sample procedure. In D. L. Medin, W. A. Roberts, & R. T. Davis (Eds.), *Processes of animal memory*. Hillsdale, N.J.: Erlbaum, 1976.

Roberts, W. A., & Grant, D. S. An analysis of light-induced retroactive inhibition in pigeon short-term memory. *Journal of Experimental Psychology: Animal Behavior Processes*, 1978, *4*, 219–236.

Roitblat, H. L. Codes and coding processes in pigeon short-term memory. *Animal Learning and Behavior*, 1980, *8*, 341–351.

Santi, A., Tombaugh, J., & Tombaugh, C. N. Differential effects of omitting comparison stimuli on symbolic and identity matching-to-sample: Evidence for altered instructional processes in pigeon short-term memory. *Animal Learning and Behavior*, 1982, *10*, 152–158.

Shavalia, D. A., Dodge, A. M., & Beatty, W. W. Time-dependent effects of ECS on spatial memory in rats. *Behavioral and Neural Biology*, 1981, *31*, 261–273.

Tranberg, D. K., & Rilling, M. Delay-interval illumination changes interfere with pigeon short-term memory. *Journal of the Experimental Analysis of Behavior*, 1980, *33*, 39-49.

Wagner, A. R. Expectancies and the priming of STM. In S. H. Hulse, H. Fowler, & W. K. Honig (Eds.), *Cognitive processes in animal behavior*. Hillsdale, N.J.: Erlbaum, 1978.

Wagner, A. R. SOP: A model of automatic memory processing in animal behavior. In N. E. Spear & R. R. Miller (Eds.), *Information processing in animals: Memory mechanisms*. Hillsdale, N.J.: Erlbaum, 1981.

Wallace, J., Steinert, P. A., Scobie, S. R., & Spear, N. E. Stimulus modality and short-term memory in rats. *Animal Learning and Behavior*, 1980, *8*, 10-16.

Wright, A. A., & Sands, S. F. A model of detection and decision processes during matching to sample by pigeons: Performance with 88 different wavelengths in delayed and simultaneous matching tasks. *Journal of Experimental Psychology: Animal Behavior Processes*, 1981, *7*, 191-216.

8 HOW EXPECTANCIES GUIDE BEHAVIOR

Gail B. Peterson
University of Minnesota

I. INTRODUCTION

In a seminal paper, Trapold (1970) asked if animals' expectancies of different reinforcers are discriminably different. He reasoned that the answer to this question might be evident in performance on a two-choice conditional discrimination task in which the correct response (R_1) following one conditional cue (S_1) always produces a particular outcome (O_1), while the correct response (R_2) following the alternative cue (S_2) always produces another outcome (O_2). With experience on this procedure, the subject could correctly anticipate the outcome appropriate to each kind of trial at the moment the trial begins. In other words, S_1 could trigger an expectancy of $O_1(E_1)$, whereas S_2 could trigger an expectancy of $O_2(E_2)$. In keeping with the tradition of Hull–Spence r_g-s_g theory, Trapold assumed that the stimulus properties of expectancies become associated with whatever response is reinforced in their presence. To the extent that such expectancy stimulation is salient and varies with kind of outcome expected, the differential conditioning of expectancies to the conditional cues should augment the stimulus differential between the two kinds of trials and thus function to support higher levels of performance relative to animals for whom differential expectancy stimulation is not consistently available.

The operational relations and corresponding theoretical analyses of the Differential Outcomes procedure described above are shown in Figure 8.1 together with those of two control conditions, the Common Outcome procedure and the Nondifferential or Mixed Outcomes procedure. The Common Outcome procedure is the customary procedure for conditional discrimination experiments With this procedure, correct responses of both kinds yield the same reinforcer, typically food. With the Mixed Outcomes procedure, two kinds of outcomes (e.g., food or water) are given for correct responses, but in a randomly mixed, "unexpectable" fashion. As the diagrams in Figure 8.1 suggest, the theory assumes that expectancies are established by all of these procedures, but only the Differential Outcomes procedure conditions a distinct expectancy to each of the cues which, hence, is in a position to reliably mediate selection of the choice response.

Several experiments based on this rationale have yielded evidence corroborating the expectancy model. For example, Trapold (1970) went on to provide the original demonstration of a *differential outcomes effect* by finding that rats will learn to make R_1 to S_1 and R_2 to S_2 faster if correct S_1–R_1 occurrences are reinforced with food pellets and correct

135

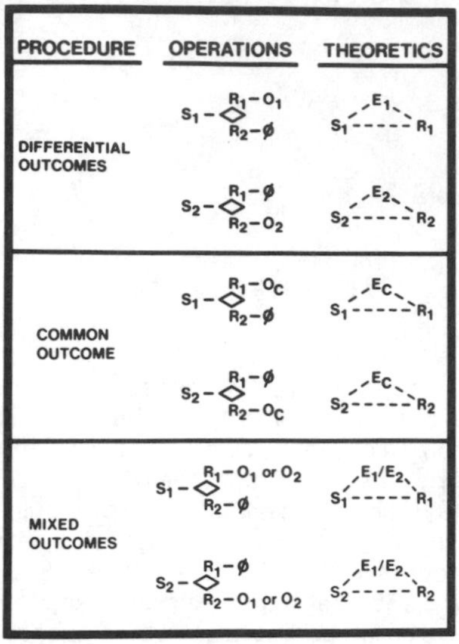

FIG. 8.1. Operational and theoretical schematics for the Differential Outcomes procedure and the Common Outcome and the Nondifferential or Mixed Outcomes control procedures. S_1 and S_2 are conditional cues, R_1 and R_2 are choice responses, O_1 and O_2 are differential outcomes, O_C is a single, common outcome, and $\emptyset$ is the consequence following incorrect responses. The diamond represents the choice point of the conditional discrimination. E_1, E_2, and E_C are expectancies of O_1, O_2, O_C, respectively. Solid lines mark procedural sequence relations while dashed lines represent hypothetical associative relationships.

S_2–R_2 occurrences with sucrose than if both S–R chains are reinforced with the same thing. Carlson and Wielkiewicz (1972) obtained a differential outcomes effect in rats when the outcomes were qualitatively and quantitatively identical but differed in the delay of presentation, O_1 being immediate reinforcement and O_2 being delayed reinforcement. In a subsequent experiment, Carlson and Wielkiewicz (1976) reported a similar finding in rats when the outcomes were large versus small food reward. Brodigan and Peterson (1976) and Peterson, Wheeler, and Armstrong (1978) demonstrated differential outcomes effects in pigeons using food and water as the different outcomes, as have Edwards, Jagielo, Zentall and Hogan (1982) using different kinds of grains (see Chapter 22, by Zentall, Hogan, & Edwards, this volume). A corresponding effect has also been demonstrated in pigeons when the outcome following correct occurrences of R_1 is food and that following correct occurrences of R_2 is a brief feedback tone (Peterson, Wheeler, & Trapold, 1980). Peterson and Trapold (1980, 1982) confirmed a number of theoretical predictions derived from the expectancy model using this food/tone Differential Outcomes procedure in a series of transfer experiments with pigeons. Many of these results have been replicated by DeLong and Wasserman (1981) and

Flynn (1981) using a slightly different procedure with pigeons.

A feature of several of these pigeon experiments has been manipulation of the time between offset of the conditional cue and onset of the choice stimuli. The typical finding is that expectancy stimuli are less subject to forgetting than are the nominal conditional cues. For example, Peterson et al. (1980) found that a 10-sec delay diminished the performance of a Common Outcome group to chance level, whereas food/tone Differential Outcomes groups were able to perform at a level of about 90% correct.

The present paper illustrates the utility of the outcome expectancy model further by applying it to the analysis of three related issues: 1) the learned equivalence of cues, 2) the learned distinctiveness of cues, and 3) the ambiguous-cue effect.

II. LEARNED EQUIVALENCE OF CUES

The theoretical analysis of the Differential Outcomes procedure implies that, to the extent that choice responding is controlled by the expectancy mediated process, it should not matter whether trials commence with S_1 or S_2 per se, as long as they commence with stimuli that evoke the appropriate expectancies. Thus, if a third stimulus (S_3) evokes E_1 while a fourth (S_4) evokes E_2, then S_3 and S_4 should be substitutable for S_1 and S_2, respectively, even though they have never before served as conditional cues for choice responding. That is, learned expectancies should function to mediate a learned equivalence of stimuli of the kind proposed by Dollard and Miller (1950). To test this, we trained pigeons on a conditional discrimination task in a standard three-key Skinner Box in which S_1 and S_2 were the colors red and green presented on the center key, and R_1 and R_2 were choices between a white vertical line on a black background or a white horizontal line on a black background presented on the two side keys. The left/right positions of R_1 and R_2 were counterbalanced across trials so that each choice stimulus appeared equally often in each location. Intermixed among these conditional color-to-line discrimination trials were center-key, single-stimulus (i.e., no choice) trials of either a white circle on a black background (S_3) or a white triangle on a black background (S_4). In a subsequent transfer test, S_1 and S_2 were removed from the proceedings entirely and replaced by S_3 and S_4.

Each trial began with the illumination of the center key with white light. A single peck to this white key initiated S_1, S_2, S_3, or S_4 for 4 sec. No responding was required to these stimuli; if any occurred, it had no programmed effect.[1] The only other responding required was a single peck choice-response directed to either the vertical or the horizontal line

---

[1]The first few days of training were an exception to this procedure. During that time trials began with immediate onset of S_1, S_2, S_3, or S_4, rather than with white light. The stimulus remained on until the subject pecked it once, when it went off and was followed by the next event in the sequence. This procedure was discontinued after 12 sessions, when the procedure described above was adopted.

as R_1 or R_2. R_1 was correct only following S_1 and R_2 only following S_2. Incorrect choices produced a 10-sec blackout followed by the usual 5-sec ITI and a repeat of the trial. This correction procedure continued until the trial was correctly completed.

The experimental design involved four groups of four pigeons each. For Group DD (Differential-Differential), the color-to-line problem was programmed according to a Differential Outcomes procedure in which O_1 was 3 sec access to grain while O_2 was a 0.75 sec, moderately loud 1-kHz tone. On the geometric form trials for this group, the circle was always followed by the food whereas the triangle was always followed by the tone. For Group NN (Nondifferential-Nondifferential), the color-to-line problem was programmed according to the Nondifferential Outcomes procedure so that each kind of trial was consequated equally often by both kinds of outcome. Similarly, for this group the circle and triangle were each followed by food on a random half of the trials and by tone on the other half. The procedures for the remaining two groups, Group DN and Group ND, involved combinations of these procedures such that the first letter in the group designation identifies the procedure on the color-to-line problem and the second letter the procedure on the geometric form trials.

For the transfer test, S_3 and S_4 replaced S_1 and S_2, and all subjects were trained on the Differential Outcomes procedure where O_1 occurred on S_3 trials and O_2 on S_4 trials. R_1 was correct only following S_3 and R_2 only following S_4.

The results of the experiment are presented in Figure 8.2. Groups DD and DN acquired the original conditional discrimination significantly faster than did Groups ND and NN. Moreover, the insertion of a 1-sec delay between offset of the conditional cues and onset of the choice stimuli impeded performance of the latter groups more than the former. When the delay was removed, Groups DD and DN exhibited essentially errorless levels of performance while Groups ND and NN performed at around the 85-90% correct level. All of these results correspond to previous observations from our laboratory (cf. Peterson et al., 1980).

The results of most interest to the present discussion are those of the transfer test. As expected from the theory, significantly more accurate choosing of the correct alternatives obtained for Group DD than for the other groups. This was due, presumably, to S_3 and S_4 evoking E_1 and E_2 (as had S_1 and S_2 in original training), and, in turn, E_1 and E_2 cuing R_1 and R_2, respectively. It should be noted that there was some decrement in Group DD's performance at the outset of the transfer test. This suggests that either the expectancies did not have exclusive control of R_1 and R_2 in original training (i.e., S_1 and S_2 exerted at least some direct stimulus control), or, alternatively, that if the expectancies did have exclusive control of the choice responses, then S_3 and S_4 did not invoke E_1 and E_2 as powerfully as S_1 and S_2 had. Further research is needed to resolve this issue. It is clear, however, that the superior transfer exhibited by Group DD relative to Group DN cannot be attributed simply to the former having learned to discriminate S_3 and S_4 during original training; since Group ND had the same S_3-O_1 and S_4-O_2 pairings during original training as Group DD, Group ND presumably would have learned to discriminate S_3 and S_4 too, and yet they did not show any particular benefit from this pre-

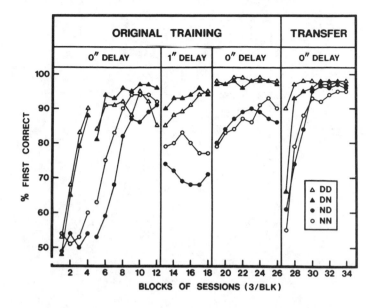

FIG. 8.2. Mean percent first correct during original training and during the transfer test. The first letter in the group designation identifies the procedure (Differential or Nondifferential Outcomes) used on the conditional choice task while the second letter identifies that used with the non-choice task. The curves are broken between blocks 4 and 5 because of a procedural change at that point (see Footnote 1).

transfer stimulus differentiation training. All in all, the overall pattern of results fits the predictions of model which attributes outcome-specific stimulus properties to expectancies.

III. LEARNED DISTINCTIVENESS OF CUES

For pigeons, line matching seems to be a more difficult problem than color matching (e.g., Carter & Eckerman, 1975; Maki, Riley, & Leith, 1976). The top-most panel of Figure 8.3, for example, presents data from our laboratory obtained from a group of pigeons trained concurrently on red/green color matching and vertical/horizontal line matching tasks. As in the preceding experiment, a trial began with white light on the center key and, following the trial-initializing peck, was followed by a 4-sec presentation of the sample. For this particular group, training was conducted according to the conventional procedure of reinforcing all correct matches with grain. Incorrect responses produced a 10-sec blackout followed by a repeat of the trial. Under the zero-delay condition, performance on the color matching task rose very quickly over the first 15 sessions to an asymptote of nearly 100% correct. Acquisition of the

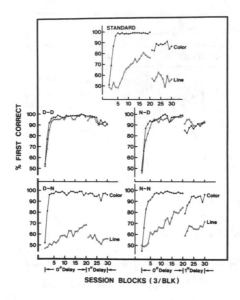

FIG. 8.3. Mean percent first correct on concurrent color matching and line matching tasks for the five groups of pigeons. For the Standard group, all correct responses yielded food reward. For the remaining groups, correct responses were reinforced according to either the Differential or Nondifferential Outcomes procedure. The first letter in the group designation identifies the procedure used on the color matching task while the second identifies that used on line matching.

line matching problem, however, proceeded much more slowly, requiring around 40 sessions to reach an asymptote of 75–80% correct. A 1–sec delay between the sample and choice stimuli reduced the level of correct performance to 90% on the color problem and 60% on the line problem.

Would conditioning distinctive expectancies to the sample stimuli in these tasks reduce the difference in difficulty? More specifically, would conditioning distinctive expectancies to the line orientations increase their functional distinctiveness as behavior controlling stimuli? The remaining panels of Figure 8.3 present data indicating that the answer is "yes." The first letter in the group designation identifies the procedure, Differential or Nondifferential Outcomes, used on the color matching problem while the second letter identifies the procedure used on the line matching problem. Except for the use of two outcomes (food and tone) instead of one, all aspects of the procedure employed with these four groups were the same as those employed with the Standard group. When line matching was trained via the Differential Outcomes procedure (D–D and N–D), the difficulty differential between the color and line tasks virtually disappeared. Moreover, the supplemental stimulus support the distinctive expectancies provided was sufficient to cope with the otherwise deleterious effect of the 1–sec delay. This improved performance on line matching cannot be attributed to the intermittent reinforcement schedule inherent to the Differential Outcomes procedure since the Nondifferential Outcomes procedure (D-N and N-N), which also provided 50% reinforcement of correct matching, did not improve performance. Except for Group D-N showing significantly better color matching under the 1-sec delay, the results for Groups D-N and N-N parallel closely those obtained under the Standard condition. Again, the overall pattern of results is consistent with the expectancy analysis outlined above.[2]

IV. AMBIGUOUS-CUE EFFECT

Basically, the ambiguous-cue problem consists of two simultaneous discriminations which share one stimulus in common. For example, on a random half of the trials the animal is confronted with a simultaneous discrimination between, say, red and yellow, while on the rest of the trials it is faced with a simultaneous discrimination between yellow and green. Thus, the yellow stimulus is common to both discriminations. Furthermore, this common stimulus serves as S- on one kind of trial (e.g., red+/yellow-) but as S+ on the other (e.g., yellow+/green-). This is sometimes referred to as the PAN procedure since, in the terms of this example, red is always positive (P), yellow is ambiguous (A), being sometimes positive, sometimes negative, and green is always negative (N). The alternative procedures for training the same nominal discriminations could be abbreviated accordingly as PNP (red+/yellow-; yellow-/green+) or NPN (red-/yellow+; yellow+/green-). As one might expect, the PNP and NPN procedures give rise rather quickly to very high levels of correct choice performance on both kinds of discrimination trials. The effect of the PAN procedure, however, is more interesting. The typical finding here is that, although the percent correct choice of the positive stimulus on NA trials is comparable to that on corresponding trials for the PNP and NPN control groups, performance on PA trials is significantly below the NA and control levels, (e.g., Boyer & Polidora, 1972; Fletcher & Garske, 1972; Leary, 1958; Richards & Marcattilio, 1975). Indeed, it is the NA > PA performance differential that defines the PAN ambiguous-cue effect. (However, see Hall, (1980) for a discussion of discrepant findings and a demonstration of stimulus salience factors contributing to the results of PAN experiments).

Why is responding so haphazard on the PA trials? One possibility (Leary, 1958) is that it is because the same response (approach and contact) has been conditioned to both P and A, and the resulting approach-approach conflict is resolved sometimes toward P, leading to a correct response, but sometimes toward A, leading to an incorrect response. Another possibility is that both stimuli evoke an expectancy of the same reinforcing outcome, and this common expectancy mediates a learned equivalence of cues which results in an overall reduction in the

---

[2]I thank Stephen Daniel for his assistance on this experiment.

discriminability of the alternatives so that the PA discrimination is, in effect, more difficult than the NA discrimination or its PN or NP counterparts. Since the standard PAN procedure perfectly confounds these possibilities, there is no way to confidently attribute the ambiguous–cue effect, either completely or partially, to one or the other.

The Differential Outcomes procedure offers a potential solution to this dilemma. Animals could be trained on a PAN procedure in which correct responses on PA trials would produce an outcome different from that given for correct responses on NA trials. This procedure unconfounds the response and expectancy factors discussed above in that, although the same approach and contact response would still be conditioned to both P and A, the stimuli would no longer evoke the same outcome expectancy. Instead, P would evoke an expectancy of the PA–trial outcome while A would evoke an expectancy of the NA–trial outcome. Thus, the response conflict would remain in place, but the (hypothetical) reduction in discriminability by overlapping expectancy stimulation would not. To the extent that the NA > PA difference is based on the response conflict inherent to PA trials, the effect should still be present even under the Differential Outcomes procedure. On the other hand, the effect should be offset under this procedure to the extent that it is based on commonality of expectancy stimulation.

We have done a preliminary experiment which suggests that the expectancy model may be relevant to an understanding of the PAN ambiguous–cue effect. Before we did that experiment, however, we felt it was important to replicate the PAN effect in our laboratory under the standard conditions. In addition, we wanted to examine the role played by the method of displaying the choice stimuli in the PAN problem. The conventional method of displaying the two stimuli to the pigeon by projecting one on a response key on one side of the intelligence panel and the other on another key on the other side of the panel probably results in the subject inadvertently seeing one stimulus before (or without) seeing the other. This circumstance would be of no great importance on the PNP or NPN procedures since, even if it were seen first, N would not solicit an approach response, but would probably provoke a withdrawal response instead and effectively direct the animal's attention to P anyway. Of course, if P were observed first, the subject would correctly respond to it immediately. This inadvertent stimulus selection effect could be important, however, on the PAN procedure. For instance, an important bit of learning on NA trials would take place if A were noticed first and responded to without making an observing response to the other key. On such occasions, "impulsive" responding to A would be reinforced, a behavior which would be maladaptive on PA trials since it would lead to errors. Indeed, this scenario suggests that the entire ambiguous–cue effect might conceivably be an artifact of stimulus display techniques which inadvertently promote the observing of only one of the stimuli in what is formally a two–stimulus display. To assess this possibility, we studied the PAN problem under two stimulus display conditions, Disjoint and Conjoint. For the Disjoint case, the stimuli were presented on the outside keys of a three–key panel in the conventional fashion. In the Conjoint display the two stimuli were projected on opposite left/right halves of the center key so that, for all practical purposes, the pigeon could not look at one stimulus without also seeing the other. This key was split so that responses to either half could be recorded separately. The display variable was manipulated within subjects so that each subject performed discriminations on the side keys and on the center key within the same session. As in the example, the stimuli were red, yellow, and green hues. Yellow served as the ambiguous stimulus for all subjects, but the roles of red and green as P or N were counterbalanced between subjects. In an effort to have the stimulus displays observed from approximately the same vantage point, each trial began with the onset of a keylight in the center of the *back* wall of the chamber. A single peck to this key extinguished it, and illuminated the display appropriate to that trial on the keys on the front wall. Thus, the subject's head was toward the rear and roughly in the center of the box when the display was presented. The left/right location of each stimulus was counterbalanced between trials so that, each stimulus occurred equally often in each location. Correct responses yielded immediate access to grain. Incorrect responses produced a blackout and a repeat of the trial.

In addition to training eight pigeons on the standard PAN procedure, eight others were trained on the control procedures, four on PNP and four on NPN. Since there were no significant differences in performance of PNP and NPN animals, their data were combined. The bar graph in the top part of Figure 8.4 presents the results in terms of the mean percent correct over four sessions of training. As can be seen, a significant PAN effect was obtained. Moreover, although the effect was significantly less pronounced under the Conjoint than under the Disjoint condition, the NA > PA difference was still significant with the Conjoint procedure.

Immediately following the completion of that experiment, the eight subjects of the PAN group were divided into two matched groups. For the next four sessions, one of these groups was trained using a Differential Outcomes routine whereby correct responses on PA trials yielded the grain reward as before while correct responses on NA trials now yielded only a feedback tone. The other group was trained on a Mixed Outcomes procedure which delivered the two kinds of outcomes with equal likelihood for all correct responses. The bar graph in the bottom of Figure 8.4 presents the mean percent correct over the four sessions under these conditions. The Differential Outcomes procedure appears to have disambiguated the ambiguous-cue problem. The right side of Figure 8.4 presents a more detailed look at the results of this experiment together with data from subsequent sessions (5-16) in which the Mixed PAN group was switched to the Differential Procedure and vice versa, with both groups finally being switched back to the standard (Common Outcome) procedure. The NA > PA effect disappears under the Differential procedure, and reappears under the control procedures, a pattern of results consistent with the expectancy analysis outlined above.

The results of the PAN experiments by Richards & Marcattilio (1975) are also amenable to an outcome-expectancy interpretation. They studied pigeons in situations where the probability of reinforcement on PA trials was always 1.0, but the probability of reinforcement on NA trials varied among 1.0, 0.50, 0.25, 0.125, and 0.0625. When the PA and NA reinforcement probabilities were equal, the NA > PA effect was observed. However, the greater the difference in the outcome probabilities, the less

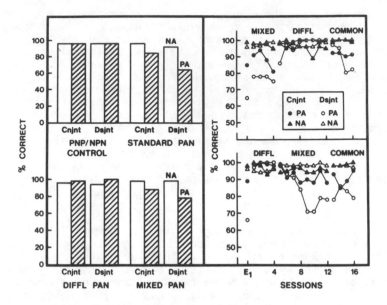

FIG. 8.4. The results of the two experiments on the PAN ambiguous–cue effect. The upper left panel presents the results of Experiment 1 which replicated the standard PAN effect. The lower left panel presents the results of the second experiment where the PAN problem was trained under either Differential or Mixed Outcomes procedures. The panels on the right present session–by–session means for the two matched groups of Experiment 2 in successive phases under the Differential, Mixed and Common Outcome procedures. Cnjnt = Conjoint Display; Dsjnt = Disjoint Display; E₁ = last session of Experiment 1.

marked this effect became, until it vanished (or reversed) at the greatest outcome probability differentials.

It must be noted that there is at least one major problem with the expectancy interpretation of these particular experiments. Specifically, with respect to the experiment reported here, it may be that the Differential Outcomes procedure improved performance on PA trials not because it eliminated common expectancy stimulation, but because it induced significant differences in the relative strengths of stimulus control the P and A stimuli had over the instrumental response. This possibility arises because of the large difference in the incentive values (reinforcer potencies?) of the food and tone outcomes. In other words, reinforcing the response to P with food and the response to A with tone may have strongly biased the PA response conflict in P's favor (cf. Hall, 1980).

However, a similar conflict should have been present with every presentation of the choice stimuli in our conditional discrimination studies employing the food/tone Differential Outcomes procedure (e.g., Peterson et al., 1980), and, furthermore, it should have led to reduced performance. We have consistently found enhanced performance under those conditions. Nevertheless, what is needed is a PAN experiment in which discriminably different outcomes of equal incentive value are employed.

V.   CONCLUDING REMARKS

Except for the use of the cognitive-sounding term expectancy, there is, in fact, very little that is necessarily cognitive about the model presented in this paper. Indeed, the formulation depicted in Figure 8.1 is pure S–R associationism, and is essentially identical to the mediational theories of 30 or more years ago (e.g., Hull, 1930; Hull, 1931; Osgood, 1953). Whether the expectancy is regarded as cognitive or noncognitive depends primarily upon where one believes it resides, i.e., whether one believes it is a central or peripheral event. This is, of course, a classic issue in animal learning, and one on which there has been little progress in the past half century. Peripheral responses identified as likely candidates for the position of expectancy have seldom been found to behave the way mediation theory says they should (cf. Rescorla & Solomon, 1967; Trapold & Overmier, 1972). Moreover, even if a situation were found in which a certain peripheral response correlated perfectly with the predictions of the theory, the possibility would remain that the response was merely the surface manifestation of some more primary central event. It should be noted, however, that the expectancy model does not require that the location of the expectancy be known. The only requirement is that the response to a stimulus is found to be predictable from a knowledge of 1) the particular outcome with which that stimulus has been associated and 2) the response the subject has most recently learned to make in the presence of stimuli associated with that outcome. This has been amply demonstrated to be the case. Thus, the abstract concept of the expectancy is useful independently of whether the "real" expectancy is ever identified.

As noted above, expectancy mediation of behavior is not a new idea. The Differential Outcomes procedure, however, is relatively new. Why did the development of a laboratory technique for studying expectancy mediation lag so far behind its conceptual development? In the main, I think this is attributable to a particular inhibitory effect which traditional reinforcement theory, especially the orthodox operant paradigm, has had on laboratory procedural variation. The idea of differentiation of behavior via differential consequences is, of course, the fundamental concept of operant conditioning. However, according to the traditional doctrine, the mechanism by which differential consequences control behavior is quite different from the expectancy mechanism. The emphasis in discussions of response differentiation within the operant paradigm has always been on differential reinforcement in the sense of reinforcing one class of behavior and not reinforcing another. No mention is made of the role of *anticipations* or *expectancies* of reinforcement or nonreinforcement in this process. Moreover, as far as the process of reinforcement is concerned, an implicit assumption of the traditional account seems to be that, to paraphrase Gertrude Stein, a reinforcer is a reinforcer is a reinforcer. On this view, qualitatively different reinforcing substances may differ quantitatively in their effectiveness as the reinforcer, but otherwise one reinforcer is the same as another. This belief, combined with technical convenience, has had the practical effect of limiting the range of reinforcers used in animal learning experiments to a select few. Indeed, food has principally emerged as the general representative of all reinforcing events. While the problems given animals have increased in complexity over the years with different stimuli cuing different responses at different times, etc., for the most part researchers have continued to apply *the* reinforcer at all junctures. The idea of using a variety of rewards in an experiment simply didn't occur to them because, presumably, using a single reward repeatedly did the same job as using a variety of rewards and, besides, it is technically more convenient to do the former than the latter. The belief in monolithic reinforcement theory has been so total that many were surprised a few years ago to learn that different kinds of reinforcers select topographically different patterns of behavior in instrumental conditioning situations (e.g., Jenkins & Moore, 1973; Moore, 1973; Peterson, 1975; Peterson, Ackil, Frommer, & Hearst, 1972). The relatively small amount of research completed thus far using the Differential Outcomes procedure strongly suggests that programming different outcomes in complex discrimination problems in a way that increases the formal structural differentiation of the overall pattern of events can have a dramatic effect on the way these problems are solved. Further study of situations involving such differential reinforcement and feedback may well reveal new insights into the problem solving strategies animals are capable of employing -- insights that have been obscured by the traditional single reinforcer methodology. However, whether the explication of these strategies will require the invocation of theoretical constructs which are significantly different from those of traditional S-R associationism remains to be seen.

REFERENCES

Boyer, W. N., & Polidora, V. J. An analysis of the solution of PAN ambiguous-cue problems by rhesus monkeys. *Learning and Motivation*, 1972, *3*, 325-333.

Brodigan, D. L., & Peterson, G. B. Two-choice conditional discrimination performance of pigeons as a function of reward expectancy, prechoice delay, and domesticity. *Animal Learning and Behavior*, 1976, *4*, 121-124.

Carlson, J. G., & Wielkiewicz, R. M. Delay of reinforcement in instrumental discrimination learning of rats. *Journal of Comparative and Physiological Psychology*, 1972, *81*, 365-370.

Carlson, J. G., & Wielkiewicz, R. M. Mediators of the effects of magnitude of reinforcement. *Learning and Motivation*, 1976, *7*, 184-196.

Carter, D. E. & Eckerman, D. A. Symbolic matching by pigeons: Rate of learning complex discriminations predicted from simple discriminations. *Science*, 1975, *187*, 662-664.

DeLong, R. E., & Wasserman, E. A. Effects of differential reinforcement expectancies on successive matching–to–sample performance in pigeons. *Journal of Experimental Psychology: Animal Behavior Processes*, 1981, *7*, 394–412.

Dollard, J., & Miller, N. E. *Personality and psychotherapy.* New York: McGraw–Hill, 1950.

Edwards, C. A., Jagielo, J. A., Zentall, T. R., & Hogan, D. E. Acquired equivalence and distinctiveness in matching to sample by pigeons: Mediation by reinforcer–specific expectancies. *Journal of Experimental Psychology: Animal Behavior Processes*, 1982, *8*, 244–259.

Fletcher, H. J., & Garske, J. P. Response competition in monkeys' solution of PAN ambiguous–cue problems. *Learning and Motivation*, 1972, *3*, 334–340.

Flynn, M. S. Differential outcomes in a successive delayed conditional discrimination procedure. Honors thesis, Dalhousie University, 1981.

Hall, G. An investigation of ambiguous–cue learning in pigeons. *Animal Learning and Behavior*, 1980, *8*, 282–296.

Hull, C. L. Knowledge and purpose as habit mechanisms. *Psychological Review*, 1930, *37*, 511–525.

Hull, C. L. Goal attraction and directing ideas conceived as habit phenomena. *Psychological Review*, 1931, *38*, 487–506.

Jenkins, H. M., & Moore, B. R. The form of the autoshaped response with food or water reinforcers. *Journal of the Experimental Analysis of Behavior*, 1973, *20*, 163–181.

Leary, R. W. The learning of ambiguous–cue problems by monkeys. *American Journal of Psychology*, 1958, *71*, 718–724.

Maki, W. S., Riley, D. A., & Leith, C. R. The role of test stimuli in matching to compound samples by pigeons. *Animal Learning and Behavior*, 1976, *4*, 13–21.

Moore, B. R. The role of directed Pavlovian reactions in simple instrumental learning in the pigeon. In R. A. Hinde & J. Stevenson–Hinde (Eds.), *Constraints on learning: Limitations and predispositions.* New York: Academic Press, 1973.

Osgood, C. E. *Method and theory in experimental psychology.* New York: Oxford University Press, 1953.

Peterson, G. B. Response selection properties of food and brain–stimulation reinforcers in rats. *Physiology and Behavior*, 1975, *14*, 681–688.

Peterson, G. B., & Trapold, M. A. Effects of altering outcome expectancies on pigeons' delayed conditional discrimination performance. *Learning and Motivation*, 1980, *11*, 267–288.

Peterson, G. B., & Trapold, M. A. Expectancy mediation of concurrent conditional discriminations. *American Journal of Psychology*, 1982, *95*, 571–580.

Peterson, G. B., Ackil, J., Frommer, G. P., & Hearst, E. Conditioned approach and contact behavior toward signals for food or brain–stimulation reinforcement. *Science*, 1972, *177*, 1009–1011.

Peterson, G. B., Wheeler, R. L., & Armstrong, G. D. Expectancies as mediators in the differential–reward conditional discrimination performance of pigeons. *Animal Learning and Behavior*, 1978, *6*, 279–285.

Peterson, G. B., Wheeler, R. L., & Trapold, M. A. Enhancement of pigeons' conditional discrimination performance by expectancies of reinforcement and nonreinforcement. *Animal Learning and Behavior*, 1980, *8*, 22–30.

Rescorla, R. A., & Solomon, R. L. Two–process learning theory: Relationships between Pavlovian conditioning and instrumental learning. *Psychological Review*, 1967, *74*, 151–182.

Richards, R. W., & Marcattilio, A. J. Intermittency of reinforcement during NA trials and performance on the ambiguous–cue problem. *Canadian Journal of Psychology*, 1975, *24*, 210–223.

Trapold, M. A. Are expectancies based upon different positive reinforcing events discriminably different? *Learning and Motivation*, 1970, *1*, 129–140.

Trapold, M. A., & Overmier, J. B. The second learning process in instrumental learning. In A. H. Black & W. F. Prokasy (Eds.), *Classical conditioning II: Current research and theory*. New York: Appleton–Century–Crofts, 1972.

9 COGNITIVE PROCESSES IN CEBUS MONKEYS

M. R. D'Amato
David P. Salmon
Rutgers University

I. INTRODUCTION

It is difficult for one working in animal cognition to avoid adopting a comparative view. When studying classical or instrumental conditioning, one is dealing with processes so pervasive and fundamental that, as history has shown, it is easy, if not natural, to focus on the processes themselves in relative isolation of their embodiment. In contrast, cognitive processes seem more directly an expression of the animal's particular endowments, immediately raising the issue of the relationship between the two. No one in their right mind would look for evidence of conservation of volume (Woodruff, Premack, & Kennel, 1978) in wasps or in rats, despite the propensity of some to stretch analogy to the breaking point (Epstein, Lanza, & Skinner, 1980). The simple observation, then, is that almost everyone interested in animal cognition worries about the relationship between specific cognitive manifestations and the phylogenetic standing of their bearer, and this has been true ever since the prospect of cognitive evolution emerged as a serious possibility.

Despite this long-standing interest, we have made relatively little progress in defining the cognitive capabilities of animals and specifying how these capabilities vary as a function of phylogenetic status. And no wonder, for in addition to the traditional problems of comparative psychology --those of attempting to equate for motivational, reinforcement, task variables and the like--cognitive processes, unlike classical and instrumental conditioning, are generally poorly defined and at the same time multitudinous. Cognitive maps, pattern learning, acquisition of natural concepts, working memory, directed forgetting, rehearsal, and a variety of other observed and inferred phenomena have been taken as expressions of cognitive processes in animals. It is perhaps inevitable that our desire to understand how cognitive capacities are related to phylogenetic status will require that we consider a vast array of cognitive competencies. But how much simpler the task would be if we could identify a relatively small number of kernel cognitive capabilities that would allow us, through their measurement, to make reasonable statements about the cognitive potentials and capacities of various species. Such good fortune is not likely to befall us, in the main, because of the dependence of cognitive capacities on the evolutionary history of the organism and the functions they serve in its survival. Thus a particular cognitive skill, such as the ability to form and maintain cognitive maps, could conceivably be elaborated to a greater degree at the level of the insect or bird (see Balda

& Turek, this volume, Chapter 28) than in some mammals. The more intricate task befalls us of delineating a wide variety of cognitive abilities and attempting to identify the pattern of their representation in the species that interest us. We discuss below a number of behavioral phenomena that might prove useful in this pursuit.

II.   PROCESSING OF THE IDENTITY RELATION IN MONKEYS

A variety of data suggest that there are important differences in the way primates and nonprimates (in particular the pigeon) process the identity relation. For example, if monkeys are trained to match to sample with one pair of stimuli, they frequently show very substantial generalization to a new pair of stimuli (e.g., D'Amato, 1971; Weinstein, 1941, 1945). Pigeons, on the other hand, often show little generalization under comparable procedures and the demonstration of any transfer at all may require special techniques (Urcuioli & Nevin, 1975; Zentall & Hogan, 1978).

Premack (1978) has commented in detail on this issue, and has pointed out that the difference observed between pigeons and primates might be due to the circumstance that monkeys most often were tested with stimulus objects, whereas pigeons were tested with two-dimensional stimuli; in addition, the amount of previous training in identity matching given pigeons and primates often differed. To this we might add that the pigeon subjects were most often experimentally naive, while the primate subjects were frequently test-wise.

In an attempt to evaluate the contribution of such factors to the differences observed in pigeons and monkeys, we trained three laboratory-born cebus monkeys (*Cebus apella*) to match to sample with two-dimensional samples (color and forms presented on IEEE stimulus projectors; cf. D'Amato, 1973). Two monkeys were experimentally naive, while the third (Atom) had previously been exposed to an auditory discrimination (D'Amato & Salmon, 1982). It must be emphasized that acquisition of identity matching by monkeys is by no means a rapid affair, whether two-dimensional cues or objects are employed as stimuli; more than 1000 training trials are often required before the animals reach a reasonably stringent criterion of accuracy. And although it is difficult to equate numbers of training trials in the procedures employed with monkeys and pigeons, there is no compelling evidence that monkeys generally learn identity matching faster than pigeons. In our own case, one monkey (Atom) learned to match the initial pair of stimuli extremely rapidly, in only 370 trials; Clea required 1570 trials and Kip, in spite of a variety of remedial techniques, did not learn the initial matching task until he had accumulated some 4500 trials. During acquisition the animals normally were given 48 trials a day; during testing for generalization of identity matching, they received 24 trials with the original sample set and 24 trials with the new pair of stimuli, trials with the new and old sample sets occurring in an unpredictable order.

The testing data for two of the animals appear in Figure 9.1. Kip, trained with a red disk and a white dot as the initial sample set, was extraordinarily slow in acquiring the original matching task and never

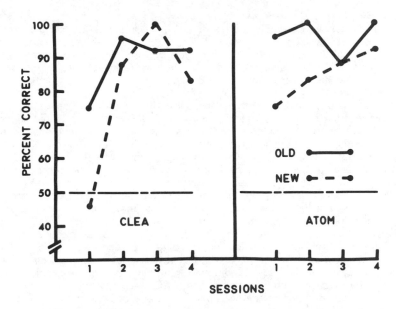

FIG. 9.1. Generalization of identity matching in monkeys that had no previous experimental experience with visual stimuli. Each session contained 48 trials, 24 with the "old" pair of sample stimuli and 24 with the new. Atom was trained and tested with simultaneous matching whereas 0-sec delay matching was employed with Clea.

reached the criterion of 90% correct in two successive sessions; he showed no evidence of transfer whatever (to an inverted triangle and a vertical line). Clea was trained with a sample set of white dot and a circle and was tested for generalization with the red disk and inverted triangle. Atom received initial training with the red disk and an upright triangle and was tested with the dot and the circle. Although Clea showed no evidence of transfer on the first 24 test trials with the new sample set, during the subsequent three test sessions her performance on the test set was indistinguishable from her performance on the old stimuli. Atom showed significant (p < .01) generalization to the new test pair during the very first 24 trials, his performance increasing thereafter. Such rapid generalization of identity matching is impressive in view of the fact that original training required hundreds of trials and taking into account the fact that, because the animals had not previously seen the test stimuli, some degree of stimulus learning had to occur before they could match successfully. It should also be pointed out that, unlike Atom, who was trained and tested with simultaneous matching, Clea had been trained with and tested with 0-sec delay matching. Consequently, her poor performance during the first 24 trials might be due in part to a memory decrement associated with the new sample set.

A reasonably comparable assessment of generalization of identity matching was made with pigeon subjects. The birds learned the initial identity matching task in 5 to 15 sessions of 96 trials each. They were tested with a new pair of samples, the old pair appearing on 32 of the daily trials, the new on 64. None of the birds showed the slightest trace of transfer to the new pair of stimuli during the initial test sessions; by the end of the 12th session one pigeon had reached the 90% correct level with the new stimuli whereas the other three averaged little better than 80% correct.[1]

These results indicate that the identity relation has a greater significance for primates than for pigeons. This is not to say that pigeons learn nothing about the identity relation in matching to sample, for Zentall and his associates have shown otherwise (e.g., Zentall, Edwards, Moore, & Hogan, 1981; Zentall, Hogan, & Edwards, this volume, Chapter 22). Pigeons apparently can generalize identity matching to some degree, but to a more limited extent and under a narrower set of conditions than is the case for primates. However this issue is ultimately resolved, there are additional considerations that support the view that primates and pigeons process the identity relation differently.

Pigeons trained in identity matching show considerable transfer to conditional discriminations that do not involve the identity relation (Holmes, 1979). In contrast, our experience has been that monkeys show little transfer from identity to conditional matching. For example, when subsequently trained on a conditional matching problem, Clea, who previously had demonstrated considerable transfer of identity matching, required about 1500 trials to reach the criterion of 90% correct in a single session. Moreover, when monkeys learn identity matching in the simultaneous condition, they normally generalize to 0-sec matching and longer delays very rapidly (D'Amato, 1971). With conditional matching, on the other hand, we have frequently found that performance deteriorates substantially upon shifting from simultaneous matching to 0-sec delay and from the latter to delays of 0.5 or 1.0 sec.

It seems fair to conclude from these results that monkeys process identity and conditional relations in quite different ways, and though pigeons may impose a similar distinction, it is not nearly as sharp. We recently obtained additional data that bear on this general point. Monkeys were trained on identity and conditional matching with two-sample sets until their reaction times to the comparison (choice) stimuli were comparable. There were no contingencies in place to encourage rapid responding to the comparison stimuli, but with extensive practice the animals nevertheless produced fast and relatively stable reaction times. A critical feature of the experiment was that the sample stimuli were the same for both the identity and conditional matching tasks. For example, if red and vertical line were the samples, then of course the animal would see red and vertical line as the comparison stimuli on an identity matching trial, but on a conditional matching trial it would see, say, dot and triangle. The experiment, which employed 0-sec matching throughout, consisted of a number of repetitions of the following sequence.

The animals received baseline training on identity matching and then, without warning, entered a test period during which 4 of the 28 trials of

---

[1] The data were collected by Robin Timmons in Dr. A. Tomie's laboratory.

a session, quasi-randomly distributed, were conditional matching trials. The assumption was that owing to the baseline training, the monkeys would adopt an identity matching set during the test sessions, and because the conditional matching probes conflicted with the identity matching set, response latencies would be elevated on such probe trials. With the reverse type of sequence (conditional matching baseline followed by test sessions with identity matching probes) we suspected that, if the identity relation has a special significance for monkeys, an increase in response latency on identity matching probe trials might not occur or might be smaller than that observed with conditional matching probes. Indeed, a comparable experiment in our laboratory with humans produced this asymmetry, which may reflect the operation of a kind of parallel processing with regard to identity matching. Apparently in humans, and possibly in monkeys, the comparator process that operates during identity matching can function equally efficiently whether or not the individual is preset for a different discriminative activity.

Figure 9.2 depicts representative results obtained with Fifi, who was trained with red disk and vertical line as the sample set, and dot vs. inverted triangle as the conditional comparison stimuli. It is clear that there was a difference in reaction times between the identity test trials and the conditional probes, but with identity matching probes embedded in conditional matching trials, reaction time was slightly faster on the identity matching trials. The typical pattern was that the first difference was statistically significant but not the second. This pattern was shown by Coco, who was trained with red and triangle as sample stimuli. Both monkeys were run through a second pass of identity and conditional matching probe trials. The same results were obtained, except that Coco produced a small, but significant, difference on the identity probes in favor of the conditional matching trials.

The third monkey, Dagwood, trained with the same arrangement of stimuli as Fifi, showed no difference in reaction times either with identity or conditional probes. Because reaction time is affected by the discriminability of the comparison stimuli, we retrained all three animals using the conditional comparison stimuli as samples, the former sample stimuli serving as the conditional comparison stimuli. With this new arrangement of stimuli, Fifi and Coco, on two replications, produced a significant difference in reaction times on conditional probes in favor of identity matching and no difference on identity probes. The same pattern was also obtained with Dagwood, indicating that the earlier failure was likely due to a small difference in the discriminability of the two sets of comparison stimuli in favor of conditional matching.

Our suspicion, bolstered by pilot data, is that pigeons would not show any evidence of this "parallel" processing of the identity relation. If this hunch is substantiated by future research, it would provide another point of difference in the manner that primates and pigeons process the identity relation, which may be indicative of their relative capacity for categorical perception (cf. Premack, 1978).

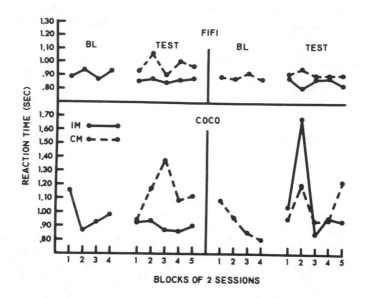

FIG. 9.2. Choice reaction times (RTs) on identity matching (IM) and conditional matching (CM) trials during baseline (BL) and testing in blocks of two sessions. Four of the 28 trials of each test session were probe trials, different from the baseline condition. Baseline RTs are based on the mean of the median RT of each of the two 24-trial sessions. The probe trial RTs are based on the mean of the median RT of the 4 probes of each test session. The RTs of the non-probe trials are based on the median RT of the last 20 of such trials (the first 4 trials of the test sessions never contained probe).

III.   ASSOCIATIVE SYMMETRY AND TRANSITIVITY IN CONDITIONAL RELATIONS

As another potential means of discriminating between the cognitive and cognitive-like mechanisms of monkeys and nonprimates on the one hand, and monkeys and higher primates on the other, we have been assessing the degree to which conditional associations develop bidirectionally. Stimulated by Asch and Ebenholtz's (1962) principle of associative symmetry in human learning, a number of investigators attempted to demonstrate associative symmetry or, at least, backward associations in pigeons that had learned conditional matching. The principle of associative symmetry as advanced by Asch and Ebenholtz proposes that forward and backward associations are of equal strength. Backward associations in animals might be of interest even if they fell short of this requirement. The strongest claim that can be made on the basis of the available pigeon studies, all employing conditional matching, is that if backward associations occur in pigeons, they are very weak and dissipate rapidly (Gray, 1966; Hogan & Zentall, 1977; Rodewald, 1974). In the best controlled of these studies (Hogan & Zentall), there was some evidence in one of three experiments for a small amount of backward associations, largely restricted to the first 15 test trials. On the other hand, Sidman et al. (1982) found no evidence whatever of backward associations in rhesus monkeys or in baboons with (conditional matching) procedures that produced virtual symmetry in children approximately 5 to 6 years of age. One possible complicating factor in their study is that all tests of backward associations were conducted with comparison stimuli consisting of vertical line vs. horizontal line, stimuli that monkeys frequently have considerable difficulty discriminating. Although the animals had learned, and were concurrently tested on, an identity matching problem involving these same stimuli, because a simultaneous matching procedure was employed, the sample stimulus on identity trials might have served to facilitate discrimination between the comparison line stimuli. Lacking this crutch on the conditional trials that tested for backward associations (colors served as samples), the animals' discrimination of the vertical and horizontal lines might have suffered. In any case, the conclusion that monkeys do not form backward associations of any appreciable strength would be fortified by verification with a wider range of test conditions.

In our own work, which is not yet completed, we trained monkeys on a conditional matching task to a high level of proficiency. To test for bidirectionality of the conditional relation we gave a single 24-trial test session in which we reversed the roles of the samples and the comparison stimuli, keeping intact the initial associations. Thus, if the original samples were red and vertical line, with triangle as the correct comparison stimuli when red was the sample and dot the correct stimulus when the vertical line was the sample, the "positive" backward-association test consisted of presenting the triangle as a sample with a response to red reinforced; when the dot appeared as the sample, a response to the vertical line was reinforced. To avoid stimulus configuration entering as a factor, all training and testing employed 0-sec conditional matching. There also was a "negative" backward-association assessment in which the correct response associated with the triangle sample was the vertical line, red being the correct comparison stimulus when the dot appeared as the sample. If the monkeys acquired backward as well as forward associations during the initial training, performance on the positive test ought to be superior to that observed on the negative backward-association test. Some animals received the positive test first followed by the negative and others got the reverse order. On occasion more than one negative and positive test was given on the same conditional discrimination, always with at least 8 baseline sessions separating each test.

After completion of the backward-association tests, in preparation for subsequent assessments of associative transitivity, the same monkeys were trained on a second conditional matching task in which the comparison stimuli of the first served as samples for the second and two new stimuli served as comparison stimuli. Positive and negative backward-association tests were also given with this second conditional discrimination.

Thus far, two of the six subjects have completed testing on the first and second conditional discriminations and one is partially finished. Table

9.1 summarizes the results obtained to date. In Coco's first discrimination the sample stimuli were an inverted triangle (T) for which the correct comparison stimulus was the red disk (R), and a white dot (D) which was conditionally paired with the vertical line (V). Thus in the positive backward–association test the samples were red and vertical line and the correct comparison stimuli were inverted triangle and dot, respectively. As may be seen in Table 9.1, a strong backward association effect, virtual symmetry, was obtained. In the second conditional discrimination, which involved the red disk and the vertical line as samples and a plus (+) and a circle (C) as comparison stimuli, there was only a very small superiority of positive over negative backward associations. Fifi, who had a different pairing of the six stimuli, showed a fairly strong backward–association effect on the first discrimination but an even stronger negative effect on the second conditional discrimination. The basis of the high performance on the negative backward–association test seemed to be a strong tendency to choose the triangle when the circle appeared as the sample, as though the animal's behavior was under the control of the similarity of samples and comparison stimuli to the exclusion of other factors. Dagwood has thus far progressed only to the first conditional discrimination, yielding a marked positive effect. Because of the possibility that stimulus similarity plays a strong role in this situation (all three instances of strong backward associations involved the same stimulus pairings), it is important that the discriminative stimuli be carefully controlled. Thus, we will have to await the results of the other animals in the experiment before we can have any confidence that the suggestion of backward associations in Table 9.1 has any generality.

TABLE 9.1
Percent Correct Responses on Positive and Negative Backward Association Tests

Subject	Conditional Discrimination	Training Associations	Tests Pos.	Neg.
Coco	First	T->R, D->V	92	29
	Second	R->+, V->C	47	44
Fifi	First	R->T, V->D	67	42
	Second	T->+, D->C	11	88
Dagwood	First	T->R, D->V	79	8

As already mentioned, in conditional matching tasks normal children 5 to 6 years of age demonstrate backward associations virtually to the point of associative symmetry, and the same result seems to hold for mentally retarded adolescents (Stromer & Osborne, 1982). Given that backward associations occur weakly, if at all, in pigeons and the suggestion from our data that they may be manifested somewhat more strongly in monkeys, this phenomenon may prove a useful tool for comparative cognition.

A. ASSOCIATIVE VERSUS INFERENTIAL TRANSITIVITY

As noted above, Coco and Fifi were trained on two conditional matching tasks in which the comparison stimuli of the first conditional discrimination served as the samples for the second. The purpose of this training was to determine whether monkeys would display associative transitivity when presented with a conditional discrimination task in which the samples were those of the first discrimination and the comparison stimuli were those of the second. Before testing for associative transitivity, the animals were alternated between the two conditional discriminations until they could shift from one to another while maintaining a high performance level. All transitivity tests were separated by at least 7 baseline sessions during which the subject was alternated between the two original conditional discriminations. The five transitivity tests, all consisting of a single 24-trial session in which the assigned correct responses were rewarded, were given in the order indicated in Table 9.2. The first and last were positive forward transitivity tests. Taking Coco as an example (see Table 9.1), associative transitivity would be displayed if this animal performed at a relatively high level when tested with inverted triangle and dot as samples and correct comparison stimuli of plus and circle, respectively. To provide a basis of comparison and to control for particular stimulus pairings, the second transitivity test maintained the pairings of the first but in the reverse order. The third and fourth pairings were negative forward and negative backward transitivity tests, respectively. If the monkeys performed at a high level on the positive forward transitivity tests, one would expect that they perform at a relatively low level on the negative forward test.

TABLE 9.2

Percent Correct Responses on the Various Associative Transitivity Tests

Transitivity Test	Stimulus Relations (for Coco)	Percent Correct	
		Coco	Fifi
Positive Forward	T->+, D->C	100	75
Positive Backward	+->T, C->D	33	54
Negative Forward	T->C, D->+	29	25
Negative Backward	C->T, +->D	46	71
Positive Forward	T->+, D->C	100	75

Table 9.2 shows that the transitivity results were stronger and more consistent than those obtained with backward associations. Both monkeys produced a very large difference in performance on the forward positive and negative transitivity tests, with performance on the backward transitivity tests falling in between. Coco's results were particularly dramatic. Fifi, who managed 75% correct responses on the positive forward tests, was correct on the first 7 trials on each of these tests before committing an error. Although the data are not all in, it appears from these two animals that associative transitivity is a strong phenomena in monkeys. Sidman et al. (1982), in the previously described study, ran a single test of associative transitivity on one monkey with negative results. However, as already noted, this test took place with vertical and horizontal lines as comparison stimuli, which may have been a source of difficulty.

Tests for transitivity have also been performed with chimpanzees (Gillan, 1981) and squirrel monkeys (McGonigle & Chalmers, 1977), although more was at stake in these studies than associative transitivity. The question that concerned these investigators was whether animals are capable of inferential transitivity of the sort that has been investigated in children (e.g., Bryant & Trabasso, 1971). The issues involved are somewhat complex but may be clarified by distinguishing between associative and inferential transitivity. By associative transitivity we refer to the behavior demonstrated by Coco and Fifi. There was nothing in the original conditional relations that demanded the experimenter pair samples and comparison stimuli as was done in the positive forward symmetry tests. The animals displayed transitivity presumably because, due to the prior training on the two conditional relations, a sample that appeared on the positive forward transitivity test elicited a representation of the correct comparison stimulus of the initial conditional discrimination. This served as the mediating (associative) "link" to draw the animal to choose the comparison stimulus that had been appropriate during the second conditional discrimination. While the associations are transitive, there is no transitive interference involved because there is nothing in the initial two conditional discriminations themselves to demand particular pairings of samples and comparison stimuli on the transitivity tests.

The situation is quite different in the the assessment of inferential transitivity in children. The relations involved in the initial discriminations (for example, A is longer than B, B is longer than C, etc.) logically demand transitivity if the original relations are understood (cf. Breslow, 1981). To be sure, associative transitivity may also be involved in these tests, which could mediate transitive behavior that might mistakenly be taken as inferential.

Breslow (1981) offered a "sequential-contiguity" model to explain transitive behavior in very young children (e.g., Bryant & Trabasso, 1971). The model accounts for transitive responses essentially in terms of a sequential association process, without granting transitive inference to the young children. Something like this model could also account for the results obtained with chimpanzees (Gillan, 1981) and squirrel monkeys (McGonigle & Chalmers, 1977). Interestingly, the latter authors also advanced a noninferential interpretation of the transitive responses of their subjects.

This possibility takes on added credence when one considers the procedures used in the Gillan, and McGonigle and Chalmers studies. In the former, there was no differential quantitative relationship established for different pairs of stimuli. The subjects were exposed to an A/B, pair, with B always reinforced and A never reinforced. They also received B/C and C/D, etc., pairs in which the second stimulus was always reinforced and the first, never. All "interior" stimuli were therefore associated with the same overall percentage of reinforcement, and it is therefore difficult to see how an ordering based on quantitative variation could have been established during training that would have generated inferential transitivity. McGonigle and Chalmers (1977) used a similar procedure, with a weight cue correlated with the reinforcement cue. Although Gillan (1981) showed that he could eliminate transitive behavior by presenting the extreme stimuli together during training in a way that prevented linear ordering of the stimuli, this result, too, could be explained by Breslow's sequential-contiguity model.

Apparently, then, it remains for future research to determine whether animals are capable of inferential transitivity. On the other hand, there is less question about the occurrence of associative transitivity, and it might prove fruitful to assess this capacity in animals of different phylogenetic status.

IV. ABSTRACTING THE COMPONENTS OF A COMPOUND STIMULUS

An area in which the performance of pigeons and monkeys on relatively simple laboratory tasks falls far short of their cognitive capacities as inferred from complex discriminative tasks either in the laboratory or in the field, lies in the task of abstracting the components of a compound stimulus, or in different terms, decomposing a compound stimulus into its constituent components (see Riley, Chapter 19, this volume). Several years ago Maki and his associates investigated shared attention in pigeons in the context of matching to sample with element and compound stimuli. The compound stimuli were formed by superimposition of two elements, such as vertical lines superimposed on a red disk (Maki & Leith, 1973; Maki & Leuin, 1972; Maki, Riley, & Leith, 1976). A major result of these studies was that when presented with a compound sample stimulus followed by element comparison stimuli (e.g., vertical lines + red as sample and red vs. green as comparison stimuli), matching behavior was poorer than when an element served as the sample stimulus, and this performance difference persisted throughout extensive experience with element and compound samples. Maki and his associates interpreted this result in terms of shared attention. When presented with a compound sample for a relatively brief duration, the pigeon was obliged to share its attention between the two elements of the compound, and because of a limited information-processing system neither element was processed to the same degree as when they occurred alone as samples. This interpretation thus implies that the pigeon decomposes the compound sample into its components, rather than processing it as an integrated, unitary stimulus. In different terms, the pigeon abstracts the elements from the compound sample.

We investigated this phenomenon extensively using cebus monkeys as subjects, thinking that if pigeons were capable of decomposing compounds into their components, monkeys should have little trouble performing the same feat. We ran several experiments employing different kinds of compound stimuli, all formed by superimposition, and extending the range of testing conditions from simultaneous matching to delayed matching with retention intervals ranging to 128 sec. We approached the problem with the view that if the sample stimuli were presented until the monkey terminated them, rather than for a very brief period, the monkey should be able to process fully both elements of the compound samples. As a consequence, with nominal retention intervals, matching on compound sample trials ought to be as good as matching on element sample trials. Despite an extraordinary amount of experience with a variety of compound samples and their components, there was no evidence that the monkeys decomposed the compound samples. Rather, they were persistent in treating the compounds as unitary stimuli.

Figure 9.3 presents the results of one study (Cox & D'Amato, 1982, experiment 4) that are representative of the major results. The figure shows that with simultaneous matching (SIM), performance on compound sample trials in which the comparison stimuli were elements (CSS) was substantially poorer than performance on trials in which both the sample and the comparison stimuli were elements (SSS). (The letters of the triad refer to the sample, correct and incorrect comparison stimuli, respectively.) This result seems particularly damaging to the shared-attention view. It is difficult to see why, if the monkeys abstracted the elements of the compound sample, they would not match the relevant component to the correct comparison stimulus to the same degree as on element sample trials, given that sample and comparison stimuli were available concurrently. On the other hand, assuming that the animals organized the compounds as unitary stimuli, a stimulus generalization-decrement interpretation would predict poorer performance on compound sample trials, inasmuch as an exact match to the sample was not available on such trials. Consistent with this interpretation, Figure 9.3 shows that when the compound sample appeared as a comparison stimulus (CCS), performance was no different from that observed on element sample trials (SSS). The CSC condition, in which a compound appeared as sample and a different compound as the incorrect comparison stimulus, was included to equate the number of times that a compound comparison stimulus was the correct and the incorrect comparison stimulus.

Because the shared-attention view grants animals the capacity to decompose compounds into their component elements, the memorial burden on compound-sample trials will be greater than on element sample trials (two elements vs. one). Treating compound samples as essentially element samples, albeit rather complex ones, the generalization-decrement hypothesis assumes an equal memory load on compound and on element sample trials. We had reason to believe that if the former view was correct, the SSS and CSS conditions should produce divergent retention gradients, whereas parallel or convergent gradients ought to be observed if the generalization-decrement view was valid. All four experiments produced essentially parallel retention gradients (Cox & D'Amato, 1982).

On the basis of our results with monkeys and other results with pigeons (Farthing, Wagner, Gilmour, & Waxman, 1977; Lamb & Riley, 1981;

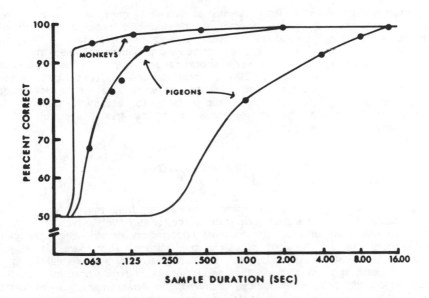

FIG. 9.3. Retention gradients generated by element (S) and compound (C) samples on 4 types of trials: element sample and comparison stimuli (SSS), compound sample with element comparison stimuli (CSS), compound sample with compund correct and element incorrect comparison stimuli (CCS) and compound sample with element correct and compound incorrect comparison stimuli (CSC). (From Cox & D'Amato, 1982.)

Roberts & Grant, 1978), it appears that neither monkeys nor pigeons show much disposition to decompose the elements of superimposed or "unified" (Riley, Chapter 19, this volume) compound stimuli. The generality of this conclusion is extended by the work of Rescorla and his associates on within-event learning. Rescorla and Durlach (1981) reported a number of studies employing rats and pigeons in which compounds based on taste elements or visual elements were presented in sensory preconditioning paradigms. As an illustration, rats were exposed to a taste compound consisting of sucrose plus quinine, subsequently injected with lithium chloride after ingesting only one component of the compound, and finally tested for a taste aversion to the other component. The uniform result was that the reaction that had been conditioned to the one element of the compound generalized substantially to the other. Detailed analysis of the phenomenon suggested to Rescorla and Durlach that the basis of the observed "within-event" learning was not that the rats, during their initial exposure to the compound taste, decomposed it into its elements and formed an association between the two elements. Rather, they favored the view that a unitary representation of the compound stimulus was formed, and that the conditioned reaction produced by the nontrained element was essentially an instance of stimulus generalization.

The apparent reluctance of pigeons and monkeys to decompose

compound stimuli into their elements seems inconsistent with other observations, for example, reports of the pigeon's rapid acquisition of concepts based on natural and artificial categories (Herrnstein, 1979, and Chapter 14, this volume). Perhaps compounds in which the elements bear a fixed relationship to each other discourage decomposition, particularly in animals (Cox & D'Amato, 1982). Or perhaps the ability to decompose compounds composed of tightly coupled elements is a highly developed cognitive skill not essential to such tasks as assigning objects to categories. Whatever the answer turns out to be, this is a problem that deserves further investigation,

V.   OTHER PHENOMENA

We wish to mention three other phenomena that might have some significance for comparative cognition. The speed with which an animal can process a stimulus to the point of recognition should have significant implications for a variety of cognitive mechanisms. In humans, as we all know, a well-learned stimulus can be recognized in a few milliseconds. Several years ago we showed that cebus monkeys, highly practiced on the task and the relevant stimuli, showed no decrement in matching performance as sample presentation time was reduced to very short durations, about 60 msec (D'Amato & Worsham, 1972). Moreover, retention of the briefly presented samples over delay intervals as long as 4 min was equal to that observed with much longer sample durations. In short, the monkey's perceptual/storage processing mechanisms appear to act as digital rather than as analog devices. Unless practice is prolonged, however, some dependence of matching performance on sample duration is likely to be observed (Herzog, Grant, & Roberts, 1977). With the pigeon, on the other hand, it appears that the dependence of matching performance on sample duration is much stronger, even after extreme amounts of practice with brief sample durations (Figure 9.4). This might prove a fruitful research area for comparative cognition. How do different species compare with regard to the rapidity with which they learn to recognize briefly presented stimuli? When the function relating recognition to stimulus duration is steepened by practice with the same stimuli, does this improvement transfer to new stimuli or does the animal have to start from scratch all over again? What are the limits that practice can confer on speed of recognition? Finally, what role does stimulus modality play?

Recently we have become interested in how animals process structured auditory stimuli, which we refer to as "tunes." Researchers employing monkeys as subjects have been perplexed for many years by the inordinate difficulty that these animals frequently encounter when learning discriminations based on auditory stimuli (compared with comparable visual discrimination tasks). We thought this difficulty might reside, in part, in the general use of rather simple auditory stimuli. Although we were successful in training monkeys on a tune-based operant discrimination, we were surprised, to put it mildly, that rats learned almost the identical discrimination much faster than the monkeys (D'Amato & Salmon, 1982).

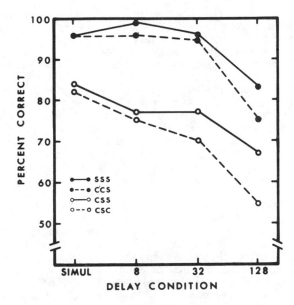

FIG. 9.4. Delayed matching-to-sample performance as a function of sample duration for highly-practiced monkeys (based on data from D'Amato & Worsham, 1972), for very-highly-practiced pigeons (middle curve, based on data from Maki et al., 1976), and for moderately-practiced pigeons (based on data from Grant, 1976). The identity matching task incorporated a 0-sec delay for the pigeons, 2-sec delay for the monkeys.

They also generalized the tune discrimination across intensity and frequency transformations to the same degree as the monkeys.

In an effort to determine the degree to which rats make use of the structure available in tunes, we are currently running a study in which one group of rats is trained with tunes as S+ and S-, each having a pattern that should facilitate their recognition. The mean frequencies of the tunes are separated by about 1.5 octaves, so that the discrimination can be formed on the basis of mean frequency as well as pattern. For a second group of rats, the same tunes provide the discriminative stimuli; however, on each trial the notes of a given tune are presented in random order, eliminating the pattern cues. If the first group of rats makes use of the patterns available in the tune, it should acquire the discrimination considerably faster than the second. Thus far, our expectations have not been confirmed. Although there exists a difference between the groups, it is quite small, and transfer tests do not suggest that the patterns exert a strong controlling influence. We plan to run the same study with monkeys. Possibly the monkeys will show a much larger difference between the two conditions, even though they may generally require more training than the rats to learn the discriminations.

The ability of animals to perceive structure or pattern in sequences of stimuli is a topic that has attracted interest in recent years (e.g., Hulse, 1978). Because patterned auditory stimuli are easy to generate with currently available electronic equipment and because such stimuli are independent of reinforcement modalities, they may prove useful in the pursuit of this interesting area (see also Chapter 11, this volume, by Hulse, Cynx, & Humpal).

Although it was once thought that animals had only a limited appreciation of past events, it is now clear that many species excel in what has come to be called "episodic" memory, memory for specific past events. And it may not be outlandish to suggest that "semantic" memory might also be considerable in some species (D'Amato, 1977). But although animals' appreciation of the past can be formidable, their contemplation of the future seems rudimentary at best. Using performance on delay of reward and delay of punishment tasks as indices, we have argued that monkeys and other animals are poorly disposed toward developing foresightful behavior, meaning by the latter the capacity to act in terms of the future rather than the present utility of objects or events (Cox & D'Amato, 1977; D'Amato & Cox, 1976). Where an animal's relationship to its environment requires that it act in terms of the future utility of objects or events, the appropriate mechanisms apparently have been built into the organism, e.g., hoarding, deposition of fat prior to hibernation, etc. Acting in terms of the future rather than the present utility of objects or events is something that does not come easy even for humans. Such foresightful behavior seems to depend upon our ability to imagine future scenarios, and it would be interesting to see whether, for example, we could train apes to "save for a rainy day."

VI.   SUMMARY AND CONCLUSIONS

There is a natural, and quite understandable, tendency of investigators to stress the cognitive accomplishments of the animal subjects with which they are most intimately connected. So much so, that one must be careful that reported complex cognitive processes are not "in the eyes of the beholder." (Anthropomorphism used to be the accusation.) During our 17 years of research experience with cebus monkeys, which Premack (1978) has referred to as the poor man's chimpanzee, we have alternated between marvelling at their cognitive accomplishments and being plunged to the depths of despair over their inability or extraordinary reluctance to learn a variety of apparently simple tasks. Over and over again, a monkey exposed to a task somewhat different from that on which it previously exhibited expert performance would perform miserably, and only after weeks or months of intensive training would the animal acquire the new task. Even when the transitions from one task to another were aided by shaping techniques, this problem often occurred. Thus, in addition to the apparent cognitive limitations mentioned above, it must be added that cebus monkeys (and other monkeys as well) are not spontaneously flexible creatures. It is true that they can be taught flexibility by means of a learning-set approach, but this, too, becomes rather specific to the task at hand. Related to, and possibly a cause of, this inflexibility is the strong control exerted by contextual cues. One could cite innumerable illustrations, but we will mention only one.

We trained cebus monkeys on a simultaneous conditional discrimination with auditory samples (tunes) and visual comparison stimuli. Needless to say, the required training was extensive, throughout which the houselight was off during the beginning of the sample presentation period. A trial began with the playing of the tune and the extinction of the houselight (placing the monkey in near-total darkness). After 3 or 4 playings of the tune, the visual stimuli were presented along with the illumination of the houselight. Because the tune continued playing until a response was made, the monkeys had considerable experience with the tunes and the illuminated houselight in conjunction. We then made what for humans, at least, would be considered a small procedural change; the houselight remained on throughout the sample presentation period. Previous experience naturally elicited the expectation that some decrement in performance would occur upon this change in context, but it provided no guide to the degree and persistence of the subsequent decline. For example, one animal that had been responding at better than 90% correct, declined to approximately 38% in the houselight-on condition. It took more than 400 trials with various remedial procedures before the performance of this animal on the conditional discrimination was disassociated from the houselight condition. Such findings raise the interesting question of how the degree of control by nonrelevant contextual stimuli depends on phylogenetic status (cf. Tranberg & Rilling, 1980 vs. Salmon & D'Amato, 1981).

Another interesting issue is whether there is any connection between an animal's ability to decompose a compound sample formed of static, superimposed elements and its susceptibility to control by contextual stimuli. If we view the conjunction of contextual stimuli and relevant discriminative cues as forming a complex compound stimulus, then control of discriminative behavior by the former might be attributed to the animal's failure to decompose the compound into its (rather complex) components. If this speculation has merit, young children, whose matching behavior seems much less controlled by contextual cues than that of monkeys (Weinstein, 1941), ought to perform better than monkeys on the task of abstracting the components of a compound stimulus.

Some 50 years ago Kluver (1933), in his detailed monograph devoted to behavior mechanisms in monkeys, decried dependence on comparing animals' performance on specific tasks as a means of determining their "relative intelligence"--or what we today would call cognitive capacities. Although he conceded that there were certain benefits to such an approach, he emphasized the importance of searching for "mechanisms" or "properties reacted to" as more germane goals. His message seems relevant even today. Because of "similar" behaviors on "similar" tasks, we are sometimes too quick to see homologies in the behavior of animals and humans where only superficial analogies exist. If we knew more about the underlying mechanisms, we would be less likely to fall into such errors. On the other hand, knowledge of mechanisms can only come about through refinements and clever modifications of the tasks themselves. Although progress has not been notable in the intervening half century, given the increased attention that talented investigators have devoted to comparative cognition in recent years, the chances of discerning and defining such mechanisms seem far better today than they were in Kluver's time.

ACKNOWLEDGMENTS

The research from our laboratory described in the paper was supported by grants from the National Science Foundation and a grant from NIMH.

REFERENCES

Asch, S. E., & Ebenholtz, S. M. The principle of associative symmetry. *Proceedings of the American Philosophical Society*, 1962, *106*, 135–163.

Breslow, L. Reevaluation of the literature on the development of transitive inferences. *Psychological Bulletin*, 1981, *89*, 325–351.

Bryant, P. B., & Trabasso, T. Transitive inferences and memory in young children. *Nature*, 1971, *232*, 456–458.

Cox, J. K., & D'Amato, M. R. Disruption of overlearned discriminative behavior in monkeys *(Cebus apella)* by delay of reward. *Animal Learning and Behavior*, 1977, *5*, 93–98.

Cox, J. K., & D'Amato, M. R. Matching-to-compound samples by monkeys *(Cebus apella)*: Shared attention or generalization decrement? *Journal of Experimental Psychology: Animal Behavior Processes*, 1982, *8*, 209–225.

D'Amato, M. R. Sample familiarity and delayed matching in monkeys. *Psychonomic Science*, 1971, *25*, 179–180.

D'Amato, M. R. Delayed matching and short-term memory in monkeys. In G. H. Bower (Ed.), *The psychology of learning and motivation*. New York: Academic Press, 1973.

D'Amato, M. R. Memory: Animal studies. In B. B. Wolman & L. R. Pomeroy (Eds.), *International encyclopedia of psychiatry, psychology, psychoanalysis, and neurology*. (Vol. VII.) New York: Aesculapius, 1977.

D'Amato, M. R., & Cox, J. K. Delay of consequences and short-term memory in monkeys. In D. L. Medin, W. A. Roberts, & R. T. Davis (Eds.), *Processes of animal memory*. Hillsdale, N.J.: Erlbaum, 1976.

D'Amato, M. R., & Salmon, D. P. Tune discrimination in monkeys *(Cebus apella)* and in rats. *Animal Learning and Behavior*, 1982, *10*, 126–134.

D'Amato, M. R., & Worsham, R. W. Delayed matching in the capuchin monkey with brief sample durations. *Learning and Motivation*, 1972, *3*, 304–312.

Epstein, R., Lanza, R. P., & Skinner, B. F. Symbolic communication between two pigeons *(Columba livia domestica)*. *Science*, 1980, *207*, 543–545. (a)

Farthing, G. W., Wagner, J. M., Gilmour, S., & Waxman, H. M. Short-term memory and information processing in pigeons. *Learning and Motivation*, 1977, *8*, 520-532.

Gillan, D. J. Reasoning in the chimpanzee: II. Transitive inference. *Journal of Experimental Psychology: Animal Behavior Processes*, 1981, *7*, 150-164.

Grant, D. S. Effect of sample presentation time on long-delay matching in the pigeon. *Learning and Motivation*, 1976, *7*, 580-590.

Gray, L. Backward association in pigeons. *Psychonomic Science*, 1966, *4*, 333-334.

Herrnstein, R. J. Acquisition, generalization, and discrimination reversal of a natural concept. *Journal of Experimental Psychology: Animal Behavior Processes*, 1979, *5*, 116-129.

Herzog, H. L., Grant, D. S., & Roberts, W. A. Effects of sample duration and spaced repetition upon delayed matching-to-sample in monkeys *Macaca arctoides* and *Saimiri sciureus*. *Animal Learning and Behavior*, 1977, *5*, 347-354.

Hogan, D. E., & Zentall, T. R. Backward associations in the pigeon. *American Journal of Psychology*, 1977, *90*, 3-15.

Holmes, P. W. Transfer of matching performance in pigeons. *Journal of the Experimental Analysis of Behavior*, 1979, *31*, 103-114.

Hulse, S. H. Cognitive structure and serial pattern learning by animals. In S. H. Hulse, H. Fowler, & W. K. Honig (Eds.), *Cognitive processes in animal behavior*. Hillsdale, N.J.: Erlbaum, 1978.

Kluver, H. *Behavior mechanisms in monkeys*. Chicago: University of Chicago Press, 1933.

Lamb, M. R., & Riley, D. A. Effects of element arrangement on the processing of compound stimuli in pigeons *(Columba livia)*. *Journal of Experimental Psychology: Animal Behavior Processes*, 1981, *7*, 45-58.

Maki, W. S., & Leith, C. R. Shared attention in pigeons. *Journal of the Experimental Analysis of Behavior*, 1973, *19*, 345-349.

Maki, W. S., & Leuin, T. C. Information processing by pigeons. *Science*, 1972, *176*, 535-536.

Maki, W. S., Riley, D. A., & Leith, C. R. The role of test stimuli in matching to compound samples by pigeons. *Animal Learning and Behavior*, 1976, *4*, 13-21.

McGonigle, B. O., & Chalmers, M. Are monkeys logical? *Nature*, 1977, *267*, 694-696.

Premack, D. On the abstractness of human concepts: Why it would be difficult to talk to a pigeon. In S. H. Hulse, H. Fowler, & W. K. Honig (Eds.), *Cognitive processes in animal behavior*. Hillsdale, N.J.: Erlbaum, 1978.

Rescorla, R. A., & Durlach, P. J. Within-event learning and Pavlovian conditioning. In N. E. Spear & R. R. Miller (Eds.), *Information processing in animals: Memory mechanisms*. Hillsdale, N.J.: Erlbaum, 1981.

Roberts, W. A., & Grant, D. S. An analysis of light-induced retroactive inhibition in pigeon short-term memory. *Journal of Experimental Psychology: Animal Behavior Processes*, 1978, *4*, 219-236.

Rodewald, H. K. Symbolic matching-to-sample by pigeons. *Psychological Reports*, 1974, *34*, 987-990.

Salmon, D. P., & D'Amato, M. R. Note on delay-interval illumination effects on retention in monkeys *(Cebus apella)*. *Journal of the Experimental Analysis of Behavior*, 1981, *36*, 381-385.

Sidman, M., Rauzin, R., Lazar, R., Cunningham, S., Tailby, W., & Carrigan, P. A search for symmetry in the conditional discriminations of rhesus monkeys, baboons, and children. *Journal of the Experimental Analysis of Behavior*, 1982, *37*, 23-44.

Stromer, R., & Osborne, J. G. Control of adolescent's arbitrary matching-to-sample by positive and negative stimulus relations. *Journal of the Experimental Analysis of Behavior*, 1982, *37*, 329-348.

Tranberg, D. K., & Rilling, M. Delay-interval illumination changes interfere with pigeon short-term memory. *Journal of the Experimental Analysis of Behavior*, 1980, *33*, 39-49.

Urcuioli, P. J., & Nevin, J. A. Transfer of hue matching in pigeons. *Journal of the Experimental Analysis of Behavior*, 1975, *24*, 149-155.

Weinstein, B. Matching-from-sample by rhesus monkeys and by children. *Journal of Comparative Physiology*, 1941, *31*, 195-213.

Weinstein, B. The evaluation of intelligent behavior in rhesus monkeys. *Genetic Psychology Monographs*, 1945, *31*, 3-48.

Woodruff, G., Premack, D., & Kennel, K. Conservation of liquid and solid quantity by the chimpanzee. *Science*, 1978, *202*, 991-994.

Zentall, T. R., & Hogan, D. E. Same/different concept learning in the pigeon: The effect of negative instances and prior adaptation to the transfer stimuli. *Journal of the Experimental Analysis of Behavior*, 1978, *30*, 177-186.

Zentall, T. R., Edwards, C. A., Moore, B. S., & Hogan, D. E. Identity: The basis for both matching and oddity learning in pigeons. *Journal of Experimental Psychology: Animal Behavior Processes*, 1981, *7*, 70-86.

III.   SEQUENCE MEMORY

David S. Olton
Matthew L. Shapiro
Stewart H. Hulse
The Johns Hopkins University

I. INTRODUCTION

Memories differ in the extent to which the information to be remembered is associated with the context in which it was learned. Some information is strongly linked with this temporal/personal context, while other information is not. This distinction has proved useful in many different fields: predators searching for prey in natural habitats (Olton, Handelmann, & Walker, 1981), animals learning discrimination tasks in the laboratory (Honig, 1978), memory processing in humans (Tulving, 1972), and the amnesic syndrome following brain damage in both animals (Olton, Becker, & Handelmann, 1979; Olton, Becker, & Handelmann, 1980) and humans (Rozin, 1976; Schacter & Tulving, 1983; Squire & Cohen, 1983; Winocur, 1982).

Several different theoretical frameworks and sets of terminology have been introduced to investigate the importance of temporal associations in memory. The one most relevant to the procedures used here is that of Honig (1978); *working memory* associates an event with its temporal/personal context, while *reference memory* processes information independently of its temporal/personal context.

The Delayed Matching–To–Sample task is one example of a task requiring working memory. At the beginning of each DMTS trial, a sample stimulus is presented and then removed. After a delay interval, this sample stimulus is presented together with other stimuli, and the animal must choose the original sample to get a reward. Because the sample stimulus varies from trial to trial, the animal must remember the specific sample presented at the beginning of each particular trial. Working memory is the process responsible for associating the sample stimulus with the trial in which it is presented.

In contrast, reference memory is the process responsible for general rules and procedures which are useful for all trials of a task. Every procedure has some reference memory components, including one like delayed matching–to–sample, which also has a working memory component. The reference components of delayed matching–to–sample include the rules of the task that remain the same from trial to trial such as: don't respond during the time out, make an initiating response at the beginning of the trial to turn on the sample stimulus, wait during the delay, get food from the food hopper when the feeder operates, etc. Thus, information in reference memory can effectively guide behavior which is not dependent on any particular trial, but is based upon rules, procedures, and contingencies that specify relations among classes of events that are consistent for many trials.

The components of many tasks can be described easily in terms of working memory and reference memory processes. The components of other tasks pose more difficulty. Consider, for example, the differences between discrete trial and continuous trial procedures. Both have been considered examples of tasks requiring working memory (Honig, 1978, pp. 214–218), yet a fundamental distinction exists between the two. In a discrete trial procedure, such as DMTS, the sample stimulus changes randomly from trial to trial. Thus, the experimental procedure requires the animal to associate the sample stimulus of each trial with only that trial in order to choose correctly when the comparison stimuli are presented. In a continuous trial procedure, such as delayed alternation, the correct response changes from trial to trial but in a *predictable* sequence. Thus the animal might determine the correct response for any given trial on the basis of information from the immediately preceding trial, but in principle could use information from the function describing the entire sequence.

The continuous trial procedure is an example of serial pattern learning, which includes many tasks having components that may be described in terms of either working or reference memory. Consider a task in which four responses (A, B, C, D) are to be made in a serial order: A then B then C then D (A->B->C->D. Reference memory is clearly used to remember the overall sequence. But what type of memory does the animal use to remember which response in this sequence is correct at any given time? Reference memory may be used because the correct pattern of responses remains the same for every trial. Furthermore, serial patterns are often described in terms of the rules that connect the elements of a pattern (Hulse, 1978), and rules are processed by reference memory. However, the correct response changes as the animal moves through the pattern. Thus, working memory might be used to remember the immediately preceding response, in the same way that it is used to remember the sample stimulus in DMTS.

The hippocampal system is required for working memory as defined by the discrete trial procedure. Both animals (see Olton, in press, for a review) and humans (Sidman, Stoddard, & Mohr, 1968) with damage to the hippocampal system are unable to choose accurately in the components of tasks that require working memory, such as DMTS. These same individuals are capable of learning other tasks, however, which are dependent on the types of rules processed by reference memory (Squire & Cohen, 1983).

The present experiment was designed to determine the extent to which serial pattern learning is similar to a discrete trial procedure that requires working memory, or represents a form of a continuous trial procedure with rules which may be solved on the basis of reference memory. Rats were trained in a serial pattern discrimination, given lesions of the fimbria-fornix, which should eliminate working memory, and returned to the task. If serial pattern learning requires working memory and rats determine the correct response on the basis of the immediately preceding response, then rats with fimbria–fornix lesions should be unable to perform the task correctly. If, on the other hand, a serial pattern has emergent properties that are processed independently of its discrete components so that it can be learned on the basis of only reference memory, then rats with these lesions ought to be able to perform correctly.

II. METHODS AND PROCEDURES

A. SUBJECTS AND APPARATUS

The subjects were 12 albino rats from Charles River weighing about 350 grams when they arrived at the laboratory. An elevated radial maze with four arms, constructed of wood, was similar to that described in detail elsewhere (Hulse & O'Leary, 1982). To confine rats between runs, a guillotine door was placed at the entrance to each arm. The maze was surrounded by many extramaze cues.

B. PROCEDURE

1. General. Rats were handled and shaped to run to the ends of the arms for food pellets (45 mg sucrose pellets). The rats were then trained preoperatively with one of four food quantities (18, 6, 1, or 0 sucrose pellets) placed in the food cup at the end of each arm. One food quantity was place in each cup at the beginning of a trial and was not replaced during that trial. For each rat, a given food quantity was always associated with the same arm. Thus the optimal strategy was to go to each of the three arms with food once and only once, and go to these arms in the order of the amount of food at the end of the arm: 18, 6, then 1. Each rat was assigned one of the 24 possible arrangements of the four food quantities on the four arms for the entire experiment. After reaching a predetermined criterion of choice accuracy, each rat was given either a control operation, or a lesion of the fimbria–fornix (FFx) or the amygdala (Am). Postoperatively the rats were given the same tests they had been given preoperatively, as well as a series of transfer tests.

At the beginning of each *trial*, the appropriate food quantity was placed in each food cup. The rat was placed in the center of the maze with all of the guillotine doors shut. At the start of each *run*, one or more of the doors were opened, and the rat was allowed to go the the end of an arm and eat whatever food was there. The doors to the other arms were closed. When the rat returned to the central platform, the door to the arm he had just chosen was also closed, and he was confined to the central platform for five seconds. This procedure continued until the end of a trial, when the rat was returned to his home cage in the testing room for an intertrial interval of about five minutes. Each test session was composed of six trials. At least one test session was given each day, five days a week, and rats received as many as seven test sessions during a week.

Each run was either a choice run or a forced run. For a *Choice Run*, all guillotine doors were opened simultaneously, allowing the rat access to all four arms. For a *Forced Run*, only one guillotine door was opened, giving the rat access to only one arm.

Training involved two general classes of trials: *No-Force Trials* and

Force Trials (see Table 10.1).   In No-Force Trials, all of the guillotine doors were opened at the start of each run so that all runs were choice runs.   In Force Trials, only one guillotine door was opened for each of the first one or two runs, forcing the rat to go to the available arm.   Two types of Force Trials were given.   In *Single Force Trials*, the first run was a Forced Run.   All subsequent runs were Choice Runs.   Each Force Trial was named after the number of food pellets on the arm to which the forced run was made.   For example, in a Force-6 trial, only the door to the arm with 6 pellets was open.   In *Double Force Trials*, the first two runs were Forced Runs, each to a different arm.   All subsequent runs were Choice Runs.   All combinations of Double Force Trials to baited arms were given, and are defined on the basis of the arms for the Force Runs.

TABLE 10.1
Forced Runs and Optimizing Choice Patterns for Each Type of Trial

Trial	Forced Runs to Arms	Optimizing Response Pattern for the Subsequent Choice Runs
No Force		18-6-1
Single Force		
Force-18	18	6-1
Force-6	6	18-1
Force-1	1	18-6
Double Force		
Force 18-6	18-6	1
Force 18-1	18-1	6
Force 6-18	6-18	1
Force 6-1	6-1	18
Force 1-18	1-18	6
Force 1-6	1-6	18

An *Optimizing Choice Pattern* was defined as going once to each available arm with food in order of descending food quantity.   Hence this pattern depended on the type of trial (see Table 10.1).

C. PREOPERATIVE TRAINING

1. Acquisition.   Rats were given No-Force, Force-1, and Force-6 Trials during acquisition.   Each test session was composed of three No-Force

Trials and three Force Trials. No-Force Trials were the 1st, 3rd, and 5th trials of a session. Force-6 and Force-1 Trials were the 2nd, 4th, and 6th Trials of each session. The Forced Trials alternated so that during one day the rats received two Force-6 Trials and one Force-1 Trial, and during the next day received two Force-1 Trials and one Force-6 Trial. Each rat was tested until he reached a criterion in which at least 75% of the choice patterns of a test session were optimizing for five consecutive test sessions.

2. *Surgery.* Surgical procedures have been described in detail elsewhere (Becker, Walker, & Olton, 1980).

D. POSTOPERATIVE TESTING

1. *Reacquisition.* All rats were given No-Force, Force-6, and Force-1 trials as for preoperative testing described above. This testing continued until all rats had reached at least the same criterion of performance achieved preoperatively during the No-Force Trials.

2. *Transfer tests.* These tests were composed of Force Trials that were not given during preoperative training. Each test session had three Force Trials and three No-Force Trials. The transfer tests included all of the Single Force and Double Force Trials that had not been given previously.

A variety of other tests were given to examine the extent to which choice behavior was controlled by: response chains; the amount of time elapsed since the previous response; intramaze cues such as those coming from the food, the maze, or the behavior of the rat; and the amount of food eaten at the end of an arm (see Olton & Samuelson, 1976; Olton, 1979; Walker & Olton, 1979 for a discussion of the appropriate designs).

III. RESULTS

A. GENERAL

The experiment is still in progress, so we don't have histology for all the rats. The particular surgical procedures used here have produced highly consistent results in the past, however, (Becker, Walker, & Olton, 1980), and the lesions we have examined here are similar.

Because of the preliminary nature of this report, only the major differences in the behavior of rats with fimbria-fornix lesions and the rats in the other two groups are presented, and statistical analyses have often been omitted. Nonetheless, the differences in behavior were substantial and stable, are being replicated now with additional rats, and lead to a consistent interpretation of the effects of the fimbria-fornix lesion.

B. PREOPERATIVE ACQUISITION

The acquisition of the task was similar to that reported by Hulse and O'Leary (1982). The rats developed optimizing patterns after 5 weeks of testing. During the criterion test sessions, at least 75% of the responses for No-Force Trials were optimizing. After being forced to Arm-6, the rats went next to Arm-18 and then to Arm-1 (87% of the Force-6 Trials); after being forced to Arm-1, they went next to Arm-18 and then to Arm-6 (95% of the Force-1 Trials).

C. POSTOPERATIVE TESTING

1. Rats with control operations. These rats showed optimizing response sequences for both the tests that they had received preoperatively, and all the transfer tests that they received postoperatively. These optimizing patterns usually occurred the first time the test was presented, and continued for all subsequent trials. There was no evidence that response chains, the amount of time since the previous response, intramaze cues, or the amount of food eaten at the end of an arm had a significant influence on choice accuracy.

2. Rats with amygdaloid lesions. For the first week of testing, these rats had a slight decrease in the percentage of trials in which they made optimizing choice sequences. Subsequently, their choice patterns were essentially the same as those exhibited by rats with control operations. Hence, the data from the two groups were combined in Table 10.2, which presents the modal response patterns seen in each type of trial.

3. Rats with fimbria-fornix lesions.
No-Force Trials. During the first week of behavioral testing, rats with fimbria-fornix lesions never entered the arms of the maze in an optimizing sequence. In many instances, they repeated choices to Arm-18 and Arm-6, and returned to arms entered in the immediately preceding run, a behavior rarely seen preoperatively. The pattern of choices gradually improved so that after six weeks, these rats had optimizing sequences in 81% of the trials.
Force Trials. In contrast, the pattern of choices in Force-6 trials was always different from that which occurred preoperatively. After the Forced Run to Arm-6, the rats responded appropriately by going next to Arm-18, but then responded inappropriately by returning to Arm-6, after which they responded appropriately by going to Arm-1. The pattern of responses following a force to Arm-1 was also different than that which occurred preoperatively. Following the Forced Run, the rats responded appropriately by going first to Arm-18 and then to Arm-6, but then almost inevitably returned to Arm-1, a response almost never seen preoperatively.
Choice patterns during some of the postoperative transfer tests were also different from those of control rats, and are summarized in Table 10.2. They show the following characteristics. (1) When the rats went to

TABLE 10.2
The Modal Choice Pattern for Each Type of Trial

Trial	Rats with Control Operations	Rats with FFx Lesions
No Force	18-6-1 (O,NC)	18-6-1
Single Force		
18	6-1 (O,NC)	6-1
6	18-1 (O,NC)	18-6-1
1	18-6 (O,NC)	18-6-1
Double Force		
18-6	1 (O,NC)	1
18-1	6 (O,NC)	6 (1)
6-18	1 (O,NC)	6 (1)
6-1	18 (O,NC)	18 (6-1)
1-18	6 (O,NC)	6 (1)
1-6	18 (O,NC)	18 (6-1)

Note-- The numbers indicate the arm that was chosen: 18 is Arm 18, 6 is Arm 6, 1 is Arm 1, 0 is the arm without the food. "NC" means that no choice was made. Data in parentheses indicate responses when the rat had been to each of the three arms with food and removed all of it. If no data are presented in parentheses, the rats tended to distribute their choices among the arms.

an arm in the correct serial order (Arm-18, then Arm-6, then Arm-1), they continued the serial pattern in the correct order and did not return to an arm until they had obtained the food from all of the arms on the maze. (2) When the rats went to an arm out of the correct serial order (Arm-1 or Arm-6 before Arm-18, Arm-1 before Arm-6), they subsequently went to the arm earliest in the sequence that had not already been chosen (as did control rats), but then returned to every arm in the sequence after that arm, even if they had already entered the arm and removed all the food. This description is applicable to the results of both the single Force Trials and the Double Force Trials.

If the rats entered Arm-18 first during either the No-Force Trials or the Single Force Trials, they subsequently went to Arm-6 and remembered that choice as indicated by the fact that they then went to Arm-1. If, however, rats went to Arm-6 first (as during Force-6 Trials), they did not remember this response (although they did remember the subsequent response to Arm-18). Following the force to Arm-6, they went to Arm-18 and then returned to Arm-6. Thus, after a response to Arm-18, the response to Arm-6 was remembered when it was made in the correct serial order (Arm-18 (then Arm-6), but it was not remembered when it was made out of order (Arm-6 then Arm-18).

A similar result was found in the Double Force Trials. When these involved the correct serial order (Arm-18 then Arm-6), the rats did not return to Arm-6 and made the next response to Arm-1. When these forces involved the incorrect serial order (Arm-1 then Arm-6, or Arm-1 then Arm-1), the rats returned to Arm-6 after going to Arm-18.

The same pattern occurred for the response to Arm-1; if Arm-1 was chosen in the correct serial order (i.e., after Arm-18 and then Arm-6, the rats rarely returned to the arm for the next run. If, however, Arm-1 was chosen out of sequence, the rats inevitably returned to the arm. A similar case was found for the Double Force Trials. if Arm-1 was entered before both Arm-18 and Arm-6 (1-18, 1-6, 18-1, 6-1), the rats inevitably went back to it after responding to Arm-6.

IV. DISCUSSION

The results of the present experiment demonstrate that the memory of items presented in a serial pattern can be independent of the memory of those same items when not presented in that pattern. A model of these choice patterns can be provided by assuming that (1) the rat has within his memory a bin to record a response to Arm-18, another bin for a response to Arm-6, and a third for a response to Arm-1; and (2), the rat chooses the arms represented by these bins in order, unless information is already in the bin, indicating that the arm has been chosen. A normal rat has these bins available in parallel, and can access information in all of them throughout a trial. A rat with fimbria-fornix lesion either has these bins available only in series, or cannot access information in them throughout a trial. Each of these possibilities will be discussed in turn.

The *serial model* suggests that as a result of fimbria-fornix lesion, a rat is able to store information in each of the memory bins only in serial order. Only after the bin for Arm-18 is filled can information be loaded into the bin for Arm-6, and only after the bins for both Arm-18 and Arm-6 are filled, can information be loaded into the bin for Arm-1. Thus both normal rats and rats with lesions had information about the serial order of the bins (as indicated by the fact that both chose appropriately during No-Force Trials), but normal rats were able to address these bins flexibly (indicated by their choosing correctly in all of the transfer tests and showing complete transitivity of choices), while rats with fimbria-fornix lesions were able to address these bins only in a particular order: 18 then 6 then 1.

The *access model* suggests that rats with fimbria-fornix lesions are able to store information in each bin as the appropriate response is made (and thus the bins are in parallel as in control rats), but that when a response is made to an arm, the memory for all arms later in the sequence is reset (Olton & Samuelson, 1976). Thus the memory of previous responses to these arms is no longer available, and the rat returns to them.

At the conference, D. A. Riley suggested an experimental test to distinguish between these two alternatives.   In this test, rats with lesions are forced to Arm–6, and then given a choice between Arm–6 and Arm–1 (the doors to the other two arms remaining closed).   If the rats fail to store the information about the force to Arm–6 (as predicted by the serial model), they should go back to Arm–6 for the subsequent choice run.   If, however, the rats store this information as well as normals but no longer have access to it when they make a response to Arm–18 (as predicted by the access model), then they should not return to Arm–6 but go to Arm–1. The appropriate tests are now in progress.

The interpretation offered above assumes that the rats have a representation of the serial pattern provided by the different quantities of food on the ends of the arms, and make their choices on the basis of this representation.   Many lines of evidence suggest that this assumption is correct.   In the present experiment, transfer tests demonstrated that the rats with control operations did not use response chains, intramaze cues, the amount of food eaten, or the time since the last choice to determine which arm they should enter.   Rather, they identified each arm on the basis of its location in space, and determined whether an arm has been chosen by remembering whether they had been there.   These conclusions are consistent with those from a long series of experiments carried out in mazes (see Olton, 1979, for a review).   The ability of animals to learn the structure of serial patterns and respond on the basis of these patterns has also been well documented, using procedures both similar to and different from the ones used here (Hulse & Dorsky, 1977; Straub & Terrace, 1981; Straub, Seidenberg, Bever, & Terrace, 1979; see also a review by Hulse, Cynx, & Humpal, and Chapter 11, this volume).

Although the components of many tasks can be conveniently identified as requiring either working memory or reference memory, such is not the case for the elements of a serial pattern.   On the one hand, across trials the pattern is the same for every trial, and can be thought of as an example of the type of information processed by reference memory.   On the other hand, within each trial the element of the pattern that is correct changes after each response, and may be thought of as an example of the type of information processed by working memory.

The same difficulty of interpretation occurs with any task in which events during trial n are not independent of those during trial n–1. Consider, for example, a continuous trial delayed alternation task in which the correct series of responses is left–right–left–right etc.   This is a form of serial pattern learning in which the pattern is repeated continuously. Because the correct response during any given series is a predictable element in a sequence, the same issues raised in the preceding paragraph are relevant here also.

A basic assumption of the present series of experiments is that rats with fimbria–fornix lesions can be considered to be functioning without working memory.   If that is true, then the dissociations exhibited by these rats clearly demonstrate that remembering information presented in a serial pattern does not require working memory.   Consequently, discrete trial procedures such as delayed matching–to–sample, and continuous trial procedures that involve serial patterns, cannot be considered as equivalent tests of working memory.   In addition, elements remembered in the context of a serial pattern are removed from the domain of working memory and placed in the domain of reference memory.

As illustrated here, and summarized elsewhere (Olton, Becker, & Handelmann, 1979), animals with damage to the hippocampal system are able to learn tasks that involve serial order.   Human amnesics can also learn sequential tasks that are based upon rules or procedures mediated by reference memory.   One of the most dramatic illustrations of this ability has recently been reported for the Tower of Hanoi problem (Cohen & Corkin, 1981; Squire & Cohen, 1983).   The successful solution of this problem requires five disks to be moved from one peg to another in a certain order.   The optimal solution requires 31 moves.   Patient HM, whose severe amnesia is thought to be due to damage to the hippocampus, was able to solve the problem in only 32 moves.   In both animals and humans, then, damage to the hippocampal system prevents the association of information with its temporal/personal context, while allowing the processing of information that is based on rules, procedures and reference memory.

However, this result does not indicate how an animal with damage to the hippocampal system can track choices within serial patterns, nor how human amnesics can learn procedural tasks without personal/contextual memory.   One possible explanation of the rats' behavior might rest on a combination of short-term memory and reference memory.   Amnesics with damage to the hippocampal system can perform accurately on tests of short-term memory (e.g., digit span tests and the Seashore Test of tonal memory), but perform poorly on tests of working memory such as delayed matching-to-sample tasks (Milner, Corkin, & Teuber, 1968; Sidman, Stoddard, & Mohr, 1968; Wickelgren, 1968).   Hence distinguishing short-term memory and working memory has both conceptual and empirical support.   For both rats and human amnesics with hippocampal damage, the rules underlying serial patterns may provide a reference memory context for correct responses, while short-term memory may hold information about the position within the pattern.   Thus a combination of intact reference memory for the rules defining a serial pattern along with an intact short-term memory for the position within the pattern may be sufficient to perform correctly on serial pattern tasks.

ACKNOWLEDGMENTS

This research was supported in part by research grant MH24213 from the National Institute of Health to D.S.O., and by a research grant NSF BNS 8014137 from the National Science Foundation to S.H.H.   The authors thank Steve Salzar, Philip Kwait, Carl Pronsky, Sarah Raskin, David Abramson, Rosy Benarroch, Mark Drews, Suzanne Malveaux, Suzette Malveaux, Carol Prescott, Mark Raccasi, Amy Muffolett, Eric Sipos, Judy Rudnick, Kathy Heim, Barbara Knowlton, Bob Findling, Joe Schonfeld, and others for helping to test the rats, Jeff Cynx, Cathy Cramer, R. Church, Chris Haig, John Humpal, Tim Moran, Priscilla Kehoe, and E. Tulving for comments on the ideas expressed here, Carmen Tirado for histology, and J. Krach and O. Rossman for typing.

REFERENCES

Becker, J. T., Walker, J. A., & Olton, D. S. The neuroanatomical bases of spatial memory. *Brain Research*, 1980, *200*, 307–320.

Cohen, N. J., & Corkin, S. The amnesic patient: Learning and retention of a cognitive skill. *Society for Neuroscience, 11th Annual Meeting*, 1981, *7*, 235. (Abstract.)

Honig, W. K. Studies of working memory in the pigeon. In S. H. Hulse, H. Fowler, & W. K. Honig (Eds.), *Cognitive processes in animal behavior*. Hillsdale, N.J.: Erlbaum, 1978.

Hulse, S. H. Cognitive structure and serial pattern learning by animals. In S. H. Hulse, H. Fowler, & W. K. Honig (Eds.), *Cognitive processes in animal behavior*. Hillsdale, N.J.: Erlbaum, 1978.

Hulse, S. H., & Dorsky, N. P. Structural complexity as a determinant of serial pattern learning. *Learning and Motivation*, 1977, *8*, 488–506.

Hulse, S. H., & O'Leary, D. K. Serial pattern learning: Teaching an alphabet to rats. *Journal of Experimental Psychology: Animal Behavior Processes*, 1982, *8*, 260–273.

Milner, B., Corkin, S., & Teuber, H. L. Further analysis of hippocampal amnesic syndrome: 14 year follow-up study of H. M. *Neuropsychologia*, 1968, *6*, 215–234.

Olton, D. S. Mazes, maps, and memory. *American Psychologist*, 1979, *34*, 583–596.

Olton, D. S. Memory functions and the hippocampus. In W. Seifert (Ed.), *Neurobiology of the hippocampus*. New York: Academic Press, 1983. In press.

Olton, D., & Samuelson, R. J. Remembrance of places passed: Spatial memory in rats. *Journal of Experimental Psychology: Animal Behavior Processes*, 1976, *2*, 97–116.

Olton, D. S., Becker, J. T., & Handelmann, G. E. Hippocampus, space, and memory. *The Behavioral and Brain Sciences*, 1979, *2*, 313–322.

Olton, D. S., Becker, J. T., & Handelmann, G. E. Hippocampal function: Working memory or cognitive mapping? *Physiological Psychology*, 1980, *8*, 239–246.

Olton, D. S., Handelmann, G. E., & Walker, J. A. Spatial memory and food searching strategies. In A. C. Kamil & T. D. Sargent (Eds.), *Foraging behavior: Ecological, ethological, and psychological approaches*. New York: Garland STPM Press, 1981.

Rozin, P. The psychobiological approach to human memory. In M. R. Rosensweig & E. L. Bennett (Eds.), *Neural mechanisms of learning and memory*. New York: Academic Press, 1976.

Schacter, D. L., & Tulving, E. Memory, amnesia, and the episodic/semantic distinction. In R. L. Isaacson & N. E. Spear (Eds.), *Expression of knowledge*. New York: Plenum, 1983.

Sidman, M., Stoddard, L. T., & Mohr, J. P. Some additional quantitative observations of immediate memory in a patient with bilateral hippocampal lesions. *Neuropsychologia*, 1968, *6*, 245–254.

Squire, L. R., & Cohen, N. R. Human memory and amnesia. In J. L. McGaugh, G. Lynch, & N. M. Weinberg (Eds.), *Conference on the neurobiology of learning and memory*. Guilford Press, 1983.

Straub, R. O., & Terrace, H. S. Generalization of serial learning in the pigeon. *Animal Learning and Behavior*, 1981, *9*, 454–468.

Straub, R. O., Seidenberg, M. S., Bever, T. G., & Terrace, H. S. Serial learning in the pigeon. *Journal of the Experimental Analysis of Behavior*, 1979, *32*, 137–148.

Tulving, E. Episodic and semantic memory. In E. Tulving & W. D. Donaldson (Eds.), *Organization of memory*. New York: Academic Press, 1972.

Walker, J. A., & Olton, D. S. The role of response and reward in spatial memory. *Learning and Motivation*, 1979, *10*, 73–84.

Wickelgren, W. A. Sparing of short-term memory in an amnesic patient: Implications for a strength theory of memory. *Neuropsychologia*, 1968, *6*, 235–244.

Winocur, G. The amnesic syndrome: A deficit in cue utilization. In L. S. Cermak (Ed.), *Human memory and amnesia*. Hillsdale, N.J.: Erlbaum, 1982.

11 COGNITIVE PROCESSING OF PITCH AND RHYTHM STRUCTURES BY BIRDS

Stewart H. Hulse
Jeffrey Cynx
John Humpal
The Johns Hopkins University

I. INTRODUCTION

Recently, there has been a rekindling of interest at both an empirical and theoretical level in animals' capacity to manage abstract forms of serially-organized stimulus information (Hulse, 1978; Straub, Seidenberg, Bever, & Terrace, 1979; Straub & Terrace, 1981). bindex<Seidenberg, M. S.> While the theoretical underpinnings of serial behavior in animals have been cast traditionally in the chains of stimulus and response (Hull, 1931; Skinner, 1934), there has been an expanding exploration of newer, cognitively based models as theoretical accounts for serially organized behavior (Hulse, 1978; Hulse & O'Leary, 1982; Roitblat, 1983).

It is to some of this empirical work and theoretical analysis of serial learning in animals that this article is addressed. In particular, we have been studying discrimination and generalization among acoustic patterns organized according to rules of human rhythm and pitch perception. Our research identifies some of the principles of serial information processing that animals and humans share, and some they do not. Also, the work presumably illuminates basic processes in acoustic communication among animals, and sheds additional light on possible analogies between animal communication and human language (Marler, 1970).

We first outline some principles basic to serial pattern learning, then describe some pertinent features of rhythm and pitch perception. We then turn to some experiments designed to test birds' capacity to process such acoustic information.

II. ALPHABETS, RULES, AND PATTERNS

An alphabet, in serial pattern learning, is a set of stimuli in which the members of the set are discriminable from one another and conform, at a minimum, to the properties of an ordinal scale (Jones, 1974; Hulse & O'Leary, 1982). Alphabets may be of two types, *direct* or *derived*. A direct alphabet is one in which the stimuli are ordered because of some inherent property of the dimension from which the stimuli are drawn; generally such alphabets are selected from physical dimensions such as light wavelength, sound intensity, or sound frequency. Derived alphabets, on the other hand, depend for their ordering properties on learning. Thus, the letters A, B, and C are without order as symbols -- until the student learns in kindergarten or the first grade the utterly arbitrary order in which they occur in the Roman alphabet. Similarly, the set of digit names and the names of the set of notes comprising the musical scale of C major acquire their ordering properties through learning. In the latter case, learning is based on a direct correspondence between a derived alphabet (the names for the notes of the C major scale) and a direct alphabet (the set of frequencies for the pitches of the C major scale).

A serial pattern is formed by rules operating on either a direct or derived alphabet. Thus, given an initializing stimulus, B, on the derived alphabet of Roman letters, the pattern BCDE can be generated by three successive applications of a NEXT rule to the alphabet. The pattern could be extended, BCDEBCDE, giving it a hierarchical structure, by applying a REPEAT rule to the subset BCDE (Simon & Kotovsky, 1963; Restle, 1970; Jones, 1974). In another domain, such as music, the sequence beginning with middle C, C-D-E-F#, defines a four-tone sequence on a scale, the whole-tone scale in particular, in which an octave is divided into six equal intervals on a log scale of frequency.

Thus, serial patterns are described in terms of explicitly defined formal structure. The behavioral consequences of variations in that formal structure can then be explored. For example, in studies with human subjects, it has been found that people (a) find patterns with formally complex as compared with formally simple rule structures more difficult to learn; (b) have a predilection to extrapolate patterns, inducing a rule from a sample of a pattern and applying the rule to supply the next stimulus in a pattern sequence; (c) tend to make errors in anticipating the next stimulus in a pattern at those points where structural rules change in the pattern; and (d) find patterns easier to learn if they are "chunked" or grouped into subunits by temporal pauses or other cues located, similarly, at those points where structural rules change in the pattern (Bower & Winzenz, 1969; Deutsch & Feroe, 1981; Jones, 1978; Kotovsky & Simon, 1973; Restle, 1970; Restle, 1972; Restle & Brown, 1970; Vitz & Todd, 1969).

In a recent series of experiments, we have been able to show that rats learn to anticipate the elements of direct and derived alphabets based on food quantities according to many of the same principles demonstrated by humans (Hulse, 1978; Hulse & Dorsky, 1977; Hulse & Dorsky, 1979; Fountain & Hulse, 1981; Hulse & O'Leary, 1982). Thus, rats find formally complex patterns harder to learn than formally simple patterns, extrapolate patterns, and so on. Terrace and his associates (e.g., Straub et al., 1979; Straub & Terrace, 1981) have been pursuing a similar line of research with pigeons with notable success. Thus, there appear to be many functional parallels between the way in which people and animals process serial information.

III. PITCH AND RHYTHM PERCEPTION

Sounds can vary in frequency, intensity, and temporal structure. The sounds we studied in the research reported here were, on the one hand, pure tones of constant frequency that varied in temporal structure. On the other, they were sequences of pure tones arranged at equal intervals on the direct alphabet defined by a log scale of acoustic frequency. To set equal intervals, we chose frequencies that divided octaves into 6 equal log units. Expressed in musical terminology, our stimuli were drawn from the derived alphabet of the whole tone scale.

A major property of acoustic alphabets based on log scales of frequency is that a series of tones drawn from the alphabet to form a serial pattern remains perceptually constant within broad limits as long as the *ratios* between neighboring frequencies remain constant. That is, humans typically respond to *relative* rather than absolute frequencies of a series of tones as a constant percept, and perceptual constancy is maintained over substantial changes in the general frequency range of the tones involved (Deutsch, 1969, 1978a; Shepard, 1965). An example of this phenomenon is *octave generalization*. Here, tones whose frequencies stand in the ratio of 2:1 (which defines an octave) tend to sound alike. The phenomenon of octave generalization has been studied rather extensively with human subjects (e.g., Attneave & Olson, 1971), but hardly at all with animals (Blackwell & Schlosberg, 1943; D'Amato & Salmon, 1982).

Perceptual constancy along the log scale of acoustic frequency need not be limited to single tones and octaves. If a familiar *series* of pitches -- a tune or melody -- is transposed from one octave to another, or -- in the case of music -- from one key to another, perceptual constancy obtains and the tune remains readily identifiable. As a matter of fact, it may take a fairly sophisticated ear to recognize that a change in key has taken place if a moderately long time interval elapses between key changes. The upshot of all this, once again, is that people respond to changes in melodic contour in a *relative* way. Pitch structures remain perceptually constant as long as the note-to-note *ratio* of frequencies in a pattern remain the same.

Rhythm is based upon a serial pattern of (a) tones and intertone intervals, (b) temporally organized accents upon tones, or (c) a combination of the two. *Tempo* is defined by the average rate with which a given rhythmic structure is produced. A given rhythmic pattern will maintain perceptual constancy over a broad range of tempos. As we have seen, tone patterns drawn from a log scale of frequencies maintain their perceptual constancy on the basis of constant ratios among pitch frequencies in a melody. In parallel fashion, rhythmic structures maintain perceptual constancy on the basis of constant ratios among the durations of their temporal units. Just as the melody of a familiar tune sounds the same across changes in key, so does the tune tend to maintain its rhythmic structure as its tempo increases or decreases (Deutsch, 1978a).

IV.   EXPERIMENTS

A.   BASIC CONSIDERATIONS

Our first concern was a choice of species that would be most suitable for the work.   Clearly, an animal was called for that indicated in some direct fashion that it made significant use of acoustic information.   Ideally, because of the nature of our stimuli, we wanted *a priori* evidence that the species could process arbitrary acoustic patterns as opposed to a restricted set of species-specific songs or calls.   For this reason, a mimicking species of bird seemed ideal, specifically, the European starling (*Sturnis vulgaris*).

Our second concern was a choice of task that would most successfully tap any putative sensitivity to rhythmic and pitch structure. The possibilities sorted themselves into tasks which would require the animal to produce sound patterns in some way (the *production* method), and those that would require the animal to listen to two or more sound patterns, indicating discrimination among them through an operant procedure (the *reception* method).   While the production method held some distinct advantages, the procedural problems and the potential species constraints involved in developing a production procedure appeared formidable.   Furthermore, we could gain just as much information about basic perceptual capacity if we were to ask animals to make a series of simple discriminations and transfers among sound patterns that were carefully designed to distinguish among a restricted set of possible perceptual outcomes.   We chose the reception method.

Finally, there were some important considerations arising from the nature of our stimuli.   In many acoustic experiments with animals (see D'Amato & Salmon, 1982, for a review), it is typical for stimuli to be simplified in the interests of making a discrimination or some other perceptual process easy to master.   Sometimes, however, that which is simple from the experimenter's point of view may not be simple from the animal's vantage point.   Thus, a simple two- or three-tone sequence may take even a rhesus monkey thousands of trials to learn with a production procedure (Cowey & Weiskrantz, 1976).   Sometimes, auditory discriminations involving relatively complex multi-element acoustic stimuli are substantially easier to learn than discriminations involving "simpler," one- or two-tone stimuli (D'Amato & Salmon, 1982).   Because pitch and rhythmic structures -- viewed as aspects of serial patterns -- generally require extended stimulus samples for their very definition, we determined to present our subjects with stimulus patterns that would be (a) lengthy, and (b) structured to highlight perceptually the pattern features of interest. Instead of impoverishing the stimulus pattern -- in the interests of "simplifying" it -- by reducing it to some small set of stimulus events, we chose to simplify it by organizing it, using many exemplars highlighting those pattern features the bird was to learn.

B. RHYTHM DISCRIMINATION AND GENERALIZATION

Our first experiments were designed to assess the capacity of starlings to make the most fundamental of discriminations among temporally-based serial acoustic structures: a discrimination between a rhythmic and an arrhythmic pattern. Given the discrimination, we then changed tempos to see if the patterns would retain their perceptual constancy, that is, if the discrimination would maintain itself as the tempos were increased or decreased.

Four birds learned to discriminate between a hierarchical rhythmic pattern and an arrhythmic pattern. Two birds learned to discriminate between a linear rhythmic pattern and the same arrhythmic pattern. All patterns consisted of 2000 Hz tones and silent intertone intervals generated on line by a computer. The hierarchical rhythmic pattern was composed of two subpatterns, one containing 4 100-msec tones separated by 100-msec intertone intervals, and the other consisting of an intertone interval of 800-msec. The two subpatterns were related in turn by a higher-level ALTERNATE rule, so that the overall pattern was a hierarchical structure in which the two subpatterns alternated indefinitely. The linear rhythmic pattern was constructed, similarly, from tone and intertone intervals each equal to 100-msec that were related by a single ALTERNATE rule. The computer generated the arrhythmic pattern on line by alternately selecting at random from a list of 11 tone and intertone durations ranging from 30 to 300 msec. The lists were chosen to have the same mean duration as the tone and intertone durations of the rhythmic patterns averaged over 4 sec. It is important to recognize that the arrhythmic pattern changed constantly as it was produced by the computer, so it contained no consistent temporal cues whatsoever.

The birds learned a go-right/go-left conditional discrimination in a wire mesh test cage suspended from the ceiling of an IAC acoustic chamber. Sound patterns were delivered through a speaker located directly over the test cage. A trial began with the illumination of the center of three keys located in a horizontal row on one wall of the test chamber. A peck on this key -- an observing response -- darkened the key and turned on one of the two sound patterns with p = 0.5. A sound pattern continued for a minimum of 4 sec during which time pecks on any key had no programmed consequences. At the end of the 4-sec interval, the two side keys were illuminated, the sound continuing. A peck on one, for example, the right key, was correct for one pattern, say the rhythmic pattern. If the computer was producing the rhythmic pattern on that trial, a peck on the right key turned off the sound pattern, darkened both side keys, generated 1.5-sec access to food from a magazine, and instituted a 5-sec intertrial interval. The next trial then began with the illumination of the center key and a fresh random selection of a pattern. A peck on the left key (associated with the arrhythmic pattern) -- an error -- turned off the sound pattern, all key lights, and introduced a 10-sec time-out in which lights at the top of the test cage were turned off. The time-out was followed by the 5-sec intertrial interval and the illumination of the center key for the beginning of the next trial, a correction trial which repeated the same pattern. Correction trials continued until a correct response occurred. The birds were trained in 1-hr daily sessions for 117–119 days at which time they had reached asymptotes, for different birds, of 81 to 96% correct. Our first important result, then, is that starlings can learn discriminations between rhythmic and arrhythmic sound patterns.

Following initial discrimination training, transfer tests began in which the birds were tested for their ability to maintain the discrimination when the patterns were increased or decreased in tempo. The computer produced the tempo changes by multiplying the durations of the original tone and intertone intervals by one of 7 constants arranged at equal intervals from 0.5 to 2.0 on a log scale. The 0.5 and 2.0 multipliers doubled and halved the tempos, respectively. A transfer test consisted of a daily session of 200 trials or 1 hr, whichever occurred first, on one of the new tempos. Transfer tests were separated by daily sessions incorporating a return to the baseline tempo. Two tests were conducted at each transfer tempo.

If, like humans, the birds transposed the rhythmic structures on the basis of constant *relative* time between successive temporal intervals in the rhythmic patterns, then we anticipated the discrimination would be maintained with great accuracy across tempo changes -- the generalization gradient would be flat. If, however, the birds solved the original discriminations on the basis of some absolute temporal feature or features of the original patterns, then the rhythmic structures should rapidly change perceptually, and discrimination should fail. The result should be a generalization gradient with discrimination accuracy (between the rhythmic and arrhythmic patterns) falling sharply with tempo change.

Figure 11.1 shows that for both linear and hierarchical rhythmic structures, the generalization gradients across tempo changes are quite flat. There is an indication that discrimination performance fell somewhat at the extreme tempos ($p < 0.05$), but discrimination performance was still well above chance, with the exception of that for the linear pattern for one bird at the 2.0 tempo. An examination of the data for the first 10 trials of the very first transfer test showed discrimination accuracies ranging from 100% (three birds) to 70% (one bird), so the birds generalized their accurate performance to the new tempos virtually immediately. The simplest summary of these data is that, like humans, starlings generalize a rhythmic discrimination across tempos with great facility.

These data are important for at least two reasons. First, they are to be contrasted with the typical generalization gradient for a sensory dimension which ordinarily shows a sharp decline in performance as stimuli become farther removed from the original training stimulus. While most gradients show, in other words, that stimuli become increasingly dissimilar the farther they are removed from an original stimulus (Honig, 1965) our flat gradients show the birds were perceiving the changed stimulus patterns as remarkably similar to their respective original training stimuli over a range encompassing 2 log (to the base 2) units.

Second, the birds had to be transferring the discrimination on the basis of relational properties between successive tone and intertone intervals in the patterns, because those were the only properties that remained constant across the transfers. That is to say, all absolute features (i.e., the absolute duration of all tone and intertone intervals) changed from one tempo to the next; the only constant feature was the ratio of the duration

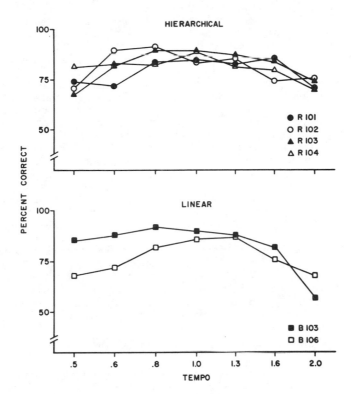

FIG. 11.1. Generalization gradients for individual birds for transfer of the hierarchical and linear rhythmic discriminations across tempos. The baseline tempo was taken as 1.0, and the computer generated the transfer tempos by multiplying the baseline tempo by the numbers indicated on the abscissa. Each data point represents mean percent correct averaged over two 1-hr test sessions.

of successive temporal intervals within a rhythmic structure. This statement must be hedged to the extent that birds in the hierarchical rhythmic discrimination might have generalized on the basis of the absolute length of one subpattern within the rhythmic structure. Thus, across all transfers, the 800-msec interval of the hierarchical rhythm in the baseline discrimination remained longer than any interval in the arrhythmic pattern, and the birds could have maintained the discrimination on the basis of that interval. While such may indeed have been the case for the hierarchical structure, the performance of the birds in the discrimination involving the linear rhythmic pattern shows the hierarchical case would be at most an isolated exception not characteristic of the general case.

Most important, a control test in which we transferred birds trained on the hierarchical pattern to the linear pattern showed momentary losses in discrimination accuracy of 4%, 8%, 8%, and 18%, for the four birds, respectively.   While these are statistically significant decreases for the birds as a whole (p < 0.01), discrimination accuracy remained well above chance for all birds, ranging from 76% to 85% correct responses. Apparently, the birds did indeed abstract some *general* concept of rhythm which they transferred across the rhythmic conditions tested.

One important transfer test remains to be done. We have yet to test for generalization across tempos when the test incorporates a change in the rhythmic patterns that *distorts* a constant ratio among tones and intertone intervals. If generalization is taking place relationally, as the data so clearly suggest, failure of generalization should result if some nonrelational property -- such as a *constant* tone duration -- were introduced into the rhythmic patterns across tests at different tempos. Thus, for example, we might keep the duration of tones constant at their baseline value of 100 msec while varying the duration of intertone intervals from session to session while holding them constant within a session.   In this case we would use the same intertone durations incorporated in the constant-ratio transfers.   This "counterexperiment" is currently under way.

We have shown that humans and at least one species of bird share the capacity to respond in a highly organized, relational way to rhythmic patterns of acoustic stimuli.   The evolutionary continuity-noncontinuity controversy aside, good relational learning is relatively rare outside the primates (Premack, 1978) yet starlings perform quite well in learning and generalizing at least one class of relational discrimination.

C. PITCH DISCRIMINATION AND GENERALIZATION

Our second class of experiments was designed, first, to assess the capacity of starlings to make the most fundamental discrimination in pitch: whether a sequence of tones is rising or falling. Given the discrimination, we then asked if the starlings could generalize the discrimination across changes in frequency. If starlings perceive pitch structures like humans, a change in frequency -- such as a shift of a pitch pattern from one octave to another -- should do little to discrimination accuracy because, as we have seen, perceptual constancy holds for humans as long as *ratio* relations among neighboring acoustic frequencies remain constant.

The stimuli we used consisted of sequences of 4 tones that ascended or descended at equal whole-tone intervals on the alphabet of the whole-tone scale.   The 4-tone sequences were configured temporally just like the hierarchically-rhythmic pattern described earlier: the duration of each of the four tones and their intertone intervals was 100 msec, and each subpattern of four tones was separated by an 800 msec pause.   Four exemplars each of rising and falling patterns were used, each exemplar beginning on a different tone with the restriction that all the tones of all the sequences had to be contained within one octave.   The set of tones from which exemplars were chosen were sine waves with frequencies of 2093, 2349, 2637, 2960, 3322, 3729, and 4186 Hz. These frequencies are well within the normal range of frequency sensitivity of starlings (Trainer, 1946). The set of exemplars is shown in Table 11.1.

Several features of the stimulus sequences are noteworthy. First, each

TABLE 11.1

Patterns Used in Rising and Falling Pitch Discrimination

ORDINAL LOCATION OF TONE			
First	Second	Third	Fourth
Rising exemplars			
2093	2349	2637	2960
2349	2637	2960	3322
2637	2960	3322	3729
Falling exemplars			
2960	3322	3729	4186
2960	2637	2349	2093
3322	2960	2637	2349
3729	3322	2960	2637
4186	3729	3322	2960

Note-- Tabled values are sinewave frequencies in Hertz.

four-tone sequence contained not just 1 or 2, but 3 exemplars of the pattern feature the birds were to learn: a rising or falling shift in pitch from one tone to the next. Formally, that is to say, a sequence was generated by initializing it on one of four pitches, and then applying a NEXT or "+1" rule three times to the alphabet of ascending or descending whole tones. Second, each class of rising or falling sequences contained not just 1 or 2, but 4 exemplars of the class. In effect, varying the stimulus sequences from trial to trial (only one exemplar was used within a trial) trained the birds to generalize across sequences within a given class. Third, the sequences were constructed with an eye to making any given pitch a component of as many different serial locations in as many different sequences as possible. This was done to minimize the possibility that the birds would learn the discrimination on the basis of pitches unique to any of the stimulus sequences. Obviously, within the other constraints imposed (e.g., that all pitches had to occur within a single octave and all sequences had to be 4 tones in length) this desideratum could only be approached. The single instance where a given pitch occurred in all locations in all rising and falling sequences was 2960 Hz. Other frequencies (e.g., 2349 Hz, 2673 Hz) were common to two or three locations, and so on. Only two frequencies were unique to a given serial location in a rising and falling sequence; thus, when they were used in an exemplar pattern, 4186 Hz was always first in a falling sequence (and last in a rising sequence), while 2093 Hz was always first in a rising sequence (and last in a falling sequence). Finally, the temporal configuration of the pattern presented to the birds was designed to group the 4-tone sequences according to the perceptual feature of rising and falling pitch we wanted the birds to learn. We could not be certain without direct analysis, of course, which feature or features of the stimuli the bird would ultimately process. Once again, our main initial goal was to enrich and configure the stimuli perceptually so as to simplify them in terms of the features of interest, and thus force the desired discrimination.

Four birds were tested in the same apparatus that was used for the rhythm discriminations, but a go/no-go conditional discrimination procedure was used in place of the go-right/go-left procedure because birds failed to achieve a reliable discrimination with the latter (a phenomenon we are pursuing for its own sake). With the new procedure, just two keys were used, the center key and one side key. A trial began with the illumination of the center key. A peck on the center key turned on, with p = 0.5, an exemplar of the rising or falling pitch patterns. The pattern continued for 4-sec; pecks on any key during this interval had no programmed consequences. At the end of the 4-sec listening period, the center key went out and the side key was illuminated, the pattern continuing. If the pattern was a falling pattern, a peck on the side key turned off the sound pattern and the key light, and produced 1.5-sec access to the food magazine. After a 5-sec intertrial interval, the center key was illuminated for the beginning of the next trial. If the sound pattern was a rising pattern, the bird had to withhold a peck for 4 sec to enter the intrial interval. If the bird pecked within this 4-sec interval, a 10-sec timeout ensued to be followed by the intertrial interval. Daily sessions lasted 1 hr. Training continued for 50 days.

Following initial training, a transfer test was undertaken in which all the sequences of original training were shifted down one octave in pitch. That is, the frequencies of all the pitches were divided by 2. As before, a daily session consisted of 1 hr of testing, but birds continued on the pitch-shifted patterns for various numbers of days as described below.

1. Discrimination learning. Data for the last 4 days of initial training showed that the discrimination between the rising and falling pitch patterns was well learned. The mean go/no-go latency differences averaged over these 4 days were 3.8 sec for all four (ps < 0.01). The birds were pecking on trials for which a go response was correct within 0.2 sec or so, and withholding pecks on no-go trials for the full 4-sec no-go interval.

2. Frequency transfer. When the pitch range of the patterns was shifted down one octave at transfer, go/no-go latency differences -- and the pattern discrimination -- *disappeared altogether*. All birds responded with a short latency, roughly 0.2-0.4 sec, to both rising and falling patterns. In other words, there was *absolutely no sign* of immediate transfer of the discrimination from one octave to the other. This result stands in utter contrast to human pitch perception.

The birds recovered the discrimination in the new pitch range at varying rates. Using an arbitrary criterion of relearning in which the no-go latency was three times the go latency, one bird, B105 met the criterion on the second day of transfer. At that time, the mean go/no-go latency difference was 2.0 sec; the no-go latency was 2.5 sec. A second bird, B101, required 12 days to reach the criterion. For this bird, the mean go/no-go latency difference was 0.9 sec; the no-go latency was 1.4 sec. For the third bird, B104, 16 days were required to reach the criterion, and at that time the mean go/no-go latency difference was 0.9 sec; the no-go latency was 1.3 sec. The fourth bird was not tested.

In an effort to produce performance on the transfer comparable to that obtained on the original discrimination, the birds continued on the transfer test for varying numbers of days. Bird B105 showed a go/no-go latency difference of 3.4 sec, 90% of the baseline difference, on the 10th day of transfer. Bird B101, however, achieved a mean latency difference of only 2.9 sec (76% of baseline) after 40 days on the transfer, while B104, similarly, had a mean latency difference of 3.0 sec (80% of baseline) after 40 days. While the latter two birds thus learned the transferred discrimination, they failed to duplicate baseline performance levels even after substantial training.

At this time, one bird, B105, was returned to the baseline discrimination to see if the discrimination with the original pitches could be recovered following the transfer. Recovery was immediate and virtually perfect. On the first day of the change, the mean go/no-go latency difference was 3.7 sec; the mean no-go latency was 3.9 sec. Comparable performance was obtained for a second day.

Next, bird B105 was transferred to pitch patterns in the octave *above* the original training octave, that is, to frequencies ranging from 4186 Hz to 8272 Hz. Procedures were identical to those used during the first transfer test. Once again, the discrimination was completely lost, to be recovered over 14 days of testing in the new pitch range. Figure 11.2 summarizes the data for B105, not only for this transfer, but from the beginning of the experiment.

To further analyze transfer processes, bird B105 was next exposed to a series of rising and falling four-tone patterns in which the component pitches were shifted, in a series of probe trials, one semitone from the pitches of the original patterns. The frequencies of the new pitches remained within the octave range of the original discrimination, but were located half-way between the pitches used originally. Thus, rising and falling patterns were constructed from a list of tones, 2218, 2484, 2793, 3136, 3520, and 3951 Hz. The testing procedure was just like that in original training, except that on a random 20% of the trials one of three different probe sequences was introduced in place of the tone sequences used during original discrimination training. On the first day of testing two falling probes and one rising probe were used, while on the second day, two rising and one falling probe were used.

The outcome of this test was clear. The bird maintained the rising-falling discrimination perfectly. On trials involving patterns of the original discrimination, mean go and no-go latencies were 0.5 and 3.5 sec, respectively. On probe trials, the comparable figures were 0.4 and 3.7 sec. While data for only one bird are available at this time, it is evident that the bird could generalize rising and falling pattern structures to new

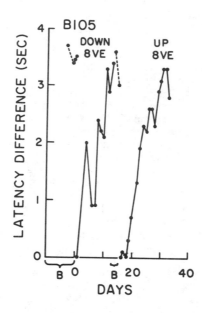

FIG. 11.2. Baseline and transfer performance for Bird B105. Baseline data are shown prior to the first transfer. The first transfer was a downward shift in frequency of one octave. Retraining in the new octave range was followed by a return to testing with the baseline frequencies. A second transfer, an upward shift in frequency of one octave relative to baseline, then ensued. Transfers to new frequency ranges caused a complete loss of the discrimination which was followed by relearning. Returns to baseline produced immediate recovery of the discrimination.

frequencies *within the original octave range.*

Two features of the data bear comment. First, the birds learned the serial pitch discrimination and they learned it well. While we do not yet know the exact features of the rising and falling sequences the birds were using to form the discrimination, the birds were able to respond accurately –– in spite of considerable variation in the rising and falling sequences from trial to trial. After the fact, this may seem straightforward, but the data stand in contrast to a number of experiments showing that animals learn a discrimination based on "simpler" stimuli involving two or three serial elements with relatively great difficulty (see D'Amato & Salmon, 1982, for further discussion of this literature). Our efforts to configure and otherwise simplify the stimulus patterns for discrimination training were, quite evidently, successful.

The most remarkable observation, however, was the birds' utter failure to show rapid transfer of the discrimination across an octave change in the frequency range of the pitches. The fact that the birds were specifically trained by the structure of the original stimulus sequences to generalize from one rising (or falling) sequence to several others makes this result even more remarkable. While the birds appeared to show good transfer within the original pitch range as they learned the original discrimination, they were unable -- without considerable training -- to transfer the structural information so as to perform well in the new range.

3. *Conclusion.* At an empirical level, the data say that starlings, like humans, can learn to generalize well on the basis of frequency patterns within the frequency range in which they were trained, but unlike humans, starlings fail to generalize pitch information across relatively large changes in pitch like that involved in a frequency shift of an octave (Attneave & Olson, 1971). Attneave and Olson found, however, that musically trained subjects, or subjects familiar with the transferred melody performed much better than musically naive subjects. Perhaps, like humans, starlings require experience with acoustic structures before successful transfers across large shifts in frequency will occur. Yet, in the final analysis, it may be that starlings are truly incapable of such transfers.

Another intriguing possibility, however, is that the birds learned the original discrimination not on the basis of any relational features among the pitches in the various sequences after all, but rather on the basis of some *absolute* feature or features of the patterns. There is some evidence that at least one species of bird, the barn owl, has the capacity to remember sound patterns on the basis of their absolute pitch (Konishi & Kenuk, 1975; Quine & Konishi, 1974). Perhaps the starlings were mostly memorizing the absolute pitches involved in the eight pitch sequences used in the original discrimination. That is, given the sequence of tones, 2349 - 2637 - 2960 - 3322 Hz, for example, the birds simply memorized the pitch of the four tones unique to that sequence -- or, as Bower suggested during Conference discussion -- the last two tones of each sequence. Stimulus ordering within the pattern would have to enter the picture at some level, otherwise the birds would fail to distinguish falling patterns from their rising counterparts -- the rising and falling sequences beginning with 2960 Hz in particular. Also, an absolute pitch hypothesis would have to be reconciled with the fact that the birds *could* generalize well to new pitch sequences within the original training range of frequencies. Finally, for a precise statement of such an hypothesis, we need to know just how much information (i.e., how many tones of a sequence) a bird needs to have before a correct discrimination can be made between a rising and falling pattern. (Preliminary data suggest very clearly that the bird can respond correctly given the first two tones of a sequence.) The further exploration of the generalization gradient for pitch patterns that these data demand is currently underway.

V. SOME SPECULATIONS

Rule-based serial patterns formed with principles of human rhythm and pitch perception in mind afford a very useful tool for the comparative study of serial pattern learning. Clearly, our research is a wedge into a rich and pithy field. The observations that starlings generalize rhythmic pattern discriminations across tempos very well, but generalize pitch pattern discriminations across range changes in frequency with far less facility, are important. They invite speculation on several counts.

First of all, it now appears that the capacity for the perception of at least primitive rhythm and pitch structures is not restricted to humans -- though the capacity must be limited by species constraints with a range yet to be determined. Second, because the starling generalizes so well across tempo changes in rhythmic patterns, but fails to do so across large frequency changes in pitch patterns, could it be that temporal structure is subordinate to changes in pitch or other acoustic dimensions for generation of the adequate stimulus for communication? Or do temporal and pitch structures interact in some fashion to produce the adequate stimulus? Perhaps pitch changes occur in bird song primarily to delineate the *temporal* structure of the song -- assuring that rhythmic generalization which might otherwise occur will *not* take place! We know of no work to examine this idea, but it would be interesting to see how starlings or other birds would respond to natural rhythm in birdsong if birdsong were reduced to a monotone. The relevant experiments are obvious and not difficult to do.

Finally, the generality across species of serial pattern learning as exemplified by rhythm and pitch sensitivity is important not only at the behavioral level, but also for those interested in the processing of complex acoustic patterns at the neural level -- given, among other things, that there is evidence for hemisphere-specific mediation of music (and, of course, language) perception in humans (Bever & Chiarello, 1974; Deutsch, 1978b) and birdsong in birds (Nottebohm, 1981). The use of patterned rhythmic and pitch stimuli would provide a useful tool to study homologous and other relationships in the neural substrate for acoustic communication across species.

Our report underscores the utility of a cognitive, information-processing approach to the study of animal learning and behavior. It is doubtful if our work would have been undertaken without the implicit assumption that sensitivity to the formal properties of serial patterns, expressed through pitch and rhythm perception would generalize in some way to other species. The limits on that generalization are far from realized, and we should learn a great deal about animal -- and human -- communication as the data accrue.

ACKNOWLEDGMENTS

Many people contributed to the research, and we thank them all: Deborah Kohl, Chris Crowder, Spencer Fisher, Paul Garlinghouse, Steven Gucker, Stephen Hulse, and Tom Park. W. T. Green, Richard Wurster, and John Yingling provided invaluable technical help and computer and programming expertise. The research was supported by National Science Foundation Research Grant BNS-8014137.

REFERENCES

Attneave, F., & Olson, R. K. Pitch as a medium: A new approach to psychophysical scaling. *American Journal of Psychology*, 1971, *84*, 147–166.
Bever, T. G., & Chiarello, R. Cerebral dominance in musicians and nonmusicians. *Science*, 1974, *185*, 537–539.
Blackwell, H. R., & Schlosberg, H. Octave generalization, pitch discrimination, and loudness thresholds in the white rat. *Journal of Experimental Psychology*, 1943, *33*, 407–419.
Bower, G. H., & Winzenz, D. Group structure, coding, and memory for serial digits. *Journal of Experimental Psychology: Monographs*, 1969, *80*, 1–17. Part 2.
Cowey, A., & Weiskrantz, L. Auditory sequence discrimination in *Macaca mulatta*:   The role of the superior temporal cortex. *Neuropsychologia*, 1976, *14*, 1–10.
D'Amato, M. R., & Salmon, D. P. Tune discrimination in monkeys *(Cebus apella)* and in rats. *Animal Learning and Behavior*, 1982, *10*, 126–134.
Deutsch, D. Music recognition. *Psychological Review*, 1969, *76*, 300–307.
Deutsch, D. The psychology of music. In E. C. Carterette & M. P. Friedman (Eds.), *Handbook of perception*. (Vol. 10.) New York: Academic Press, 1978. (a)
Deutsch, D. Pitch memory: An advantage for the left handed. *Science*, 1978, *199*, 559–560. (b)
Deutsch, D., & Feroe, J. The internal representation of pitch sequences in tonal music. *Psychological Review*, 1981, *88*, 503–522.
Fountain, S. B., & Hulse, S. H. Extrapolation of serial stimulus patterns by rats. *Animal Learning and Behavior*, 1981, *9*, 381–384.
Honig, W. K. Discrimination, generalization, and transfer on the basis of stimulus differences. In D. I. Mostofsky (Ed.), *Stimulus generalization*. Stanford: Stanford University Press, 1965.
Hull, C. L. Goal attraction and directing ideas conceived as habit phenomena. *Psychological Review*, 1931, *38*, 487–506.
Hulse, S. H. Cognitive structure and serial pattern learning by animals. In S. H. Hulse, H. Fowler, & W. K. Honig (Eds.), *Cognitive processes in animal behavior*. Hillsdale, N.J.: Erlbaum, 1978.
Hulse, S. H., & Dorsky, N. P. Structural complexity as a determinant of serial pattern learning. *Learning and Motivation*, 1977, *8*, 488–506.
Hulse, S. H., & Dorsky, N. P. Serial pattern learning by rats: Transfer of a formally defined stimulus relationship and the significance of nonreinforcement. *Animal Learning and Behavior*, 1979, *7*, 211–220.
Hulse, S. H., & O'Leary, D. K. Serial pattern learning: Teaching an alphabet to rats. *Journal of Experimental Psychology: Animal Behavior Processes*, 1982, *8*, 260–273.
Jones, M. R. Cognitive representations of serial patterns. In B. Kantowitz (Ed.), *Human information processing: Tutorials in performance and cognition*. Hillsdale, N.J.: Erlbaum, 1974.

Jones, M. R. Auditory patterns. The perceiving organism. In E. C. Carterette & M. P. Friedman (Eds.), *Handbook of perception*. (Vol. 8.) New York: Academic Press, 1978.

Konishi, M., & Kenuk, A. S. Discrimination of noise spectra by memory in the barn owl. *Journal of Comparative Physiology*, 1975, *97*, 55–58.

Kotovsky, K., & Simon, H. A. Empirical tests of a theory of human acquisition of concepts for sequential patterns. *Cognitive Psychology*, 1973, *4*, 399–424.

Marler, P. Birdsong and speech development: Could there be parallels? *American Scientist*, 1970, *58*, 669–674.

Nottebohm, F. A brain for all seasons: Cyclical anatomical changes in song control nuclei of the canary brain. *Science*, 1981, *214*, 1368–1370.

Premack, D. On the abstractness of human concepts: Why it would be difficult to talk to a pigeon. In S. H. Hulse, H. Fowler, & W. K. Honig (Eds.), *Cognitive processes in animal behavior*. Hillsdale, N.J.: Erlbaum, 1978.

Quine, D. B., & Konishi, M. Absolute frequency discrimination in the barn owl. *Journal of Comparative Physiology*, 1974, *93*, 347–360.

Restle, F. Theory of serial pattern learning: Structural trees. *Psychological Review*, 1970, *77*, 481–495.

Restle, F. Serial patterns: The role of phrasing. *Journal of Experimental Psychology*, 1972, *92*, 385–390.

Restle, F., & Brown, E. R. Serial pattern learning. *Journal of Experimental Psychology*, 1970, *83*, 120–125.

Roitblat, H. L. The meaning of representation in animal memory. *The Behavioral and Brain Sciences*, 1982, *5*, 353–406.

Shepard, R. N. Approximations to uniform gradients of generalization by monotone transformations of scale. In D. I. Mostofsky (Ed.), *Stimulus generalization*. Stanford: Stanford University Press, 1965.

Simon, H. A., & Kotovsky, K. Human acquisition of concepts for sequential patterns. *Psychological Review*, 1963, *70*, 534–546.

Skinner, B. F. The extinction of chained reflexes. *Proceedings of the National Academy of Sciences*, 1934, *20*, 234–237.

Straub, R. O., & Terrace, H. S. Generalization of serial learning in the pigeon. *Animal Learning and Behavior*, 1981, *9*, 454–468.

Straub, R. O., Seidenberg, M. S., Bever, T. G., & Terrace, H. S. Serial learning in the pigeon. *Journal of the Experimental Analysis of Behavior*, 1979, *32*, 137–148.

Trainer, J. E. *The auditory acuity of certain birds*. PhD Thesis, Cornell University, 1946.

Vitz, P. C., & Todd, R. C. A coded element model of the perceptual processing of stimuli. *Psychological Review*, 1969, *76*, 433–449.

12 ORDER COMPETENCIES IN ANIMALS: MODELS FOR THE DELAYED SEQUENCE DISCRIMINATION TASK

R. G. Weisman
R. von Konigslow
Queen's University

I. INTRODUCTION

The hypothesis that humans and other animal species share common cognitive abilities is probably the oldest and most interesting idea in comparative psychology (Darwin, 1920; Morgan, 1896; Romanes, 1898).

Yet the evolution of intellect has not been an active area of research over the past fifty tears. Psychologists have, instead, chosen to study two quite primitive intellectual mechanisms, i.e., classical conditioning and simple instrumental learning. This concern for "elementary" learning processes has produced some good science, but it has left us with a sadly impoverished view of the animal mind. We simply do not know very much about the mental lives of animals, and until we do, we will not be able to speak to the similarities and differences between the cognitions of humans and other animals.

In an effort to reduce the obsession of our discipline with simple learning principles, a number of investigators have begun to seek exquisite complexity, rather than elegant simplicity, in the study of animal intellect. The most important aspects of this cognitive research strategy are a concern for the "classical" mental competencies (Bindra, 1976), such as spatial and temporal representation and an interest in testing information processing models of the mental processes involved in these competencies (Wasserman, 1981; Roitblat, 1982).

Mapping of the temporal relationships between events is one of the most important ways in which we and other animals come to know about the world. The problem of how temporal succession is represented in the human mind was posed early in the history of Psychology (Volkmann, as cited by James, 1890). Order processing, because it is so central to human thought and language, continues to be a topic of much excellent research and theorizing (e.g., Kotovsky & Simon, 1973; Restle, 1970; Sternberg, 1975; Warren & Ackroff, 1976).

Recently, there has been a renewal of interest in the representation of stimulus order in animals, a research problem pioneered by Woodbury (1943). Several investigators (Cowey & Weiskrantz, 1976; Devine, Burke, & Rohack, 1979; Hulse & Dorsky, 1977; Straub, Seidenberg, Bever, & Terrace, 1979; Wasserman, Nelson, & Larew, 1980; Weisman, Wasserman, Dodd, & Larew, 1980) using different paradigms have demonstrated the discrimination of stimulus order in animals.   These experiments have been important in opening the problem of order processing in animals to investigation.   Of course, these studies of sequence discrimination are not all replications of the same task, nor does it seem likely that order representation in these various tasks will be explained by only one model of order competence.   This chapter explores possible explanations of order processing during one kind of sequence discrimination task.   We shall start by describing the sequence discrimination task, and then proceed to describe the models investigated.   This research aims, at one level, to provide successful models for cognition during the task, and at another level, to gain insight into the nature of the cognitive system performing the task.

II.   THE DELAYED SEQUENCE DISCRIMINATION TASK

Delayed sequence discrimination tasks (DSD) require animals to discriminate between distinct temporal orders of events.   That is, the same events are used, but the temporal order differs from one sequence to another. Several sequences of events are presented during a delayed sequence discrimination task.   Following one particular sequence, responding during a subsequent test period is reinforced (usually with food); following all other arrangements of the same events, responding is not reinforced. Discrimination of the order of two events has been studied using both dogs (Woodbury, 1943) and rats (Kosiba & Logan, 1978).   Our prior research (Weisman et al., 1980, Experiments 1 & 2; Weisman & DiFranco, 1981) has successfully demonstrated two-event DSD and tested various information processing models of order representation in pigeons.   These tasks are useful for testing models of order competence because they are representative of an important form of order search, one in which a single sequence is sought among many.   There are other forms of order search, for example two sequences may be sought among many, as in Weisman et al. (1981, Experiment 3).   This latter research is likely to require different models than are presented here.

   We shall describe our version of the two-event delayed sequence discrimination task and some typical results from this form of discrimination learning in order to acquaint the reader with the competencies we wish to model.   The task is presented in discrete trials, each trial consists of a series of events and interevent intervals.   A sample sequence of colored lights (S1 and S2) signals reinforcement or nonreinforcement dependent on pecking during a subsequent test stimulus. The intertrial interval continues from the offset of the preceding trial until the onset of S1.   The interstimulus interval continues from the offset of S1 until the onset of S2, the retention interval continues from the offset of S2 until the onset of the test stimulus.   The S1-S2 sequence presented on positive, reinforced trials is designated AB, because the actual colors and their locations vary between counterbalancing groups and experiments.   Other sequential arrangements of A and B, e.g., AA, BB and BA serve as negative sequence types correlated with nonreinforcement during the test stimulus.

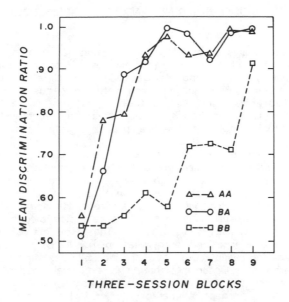

FIG. 12.1. The mean discrimination ratio for six pigeons trained in a two-event delayed sequence discrimination task for nine blocks of three sessions each.

Figure 12.1 presents the mean discrimination ratio for six pigeons during the acquisition of a two-event task. The pigeons learned a two-event discrimination in which **AB** was the positive sequence type and **AA**, **BB** and **BA** were negative sequence types. The discrimination ratio shown in Figure 12.1 was calculated separately for each negative sequence during each session by dividing test-stimulus key pecks per sec for the positive sequence by test-stimulus key pecks per sec for both the negative and positive sequence. This ratio approaches 0.50 when, as in early sessions of training, the negative sequences are poorly differentiated from **AB**, and approaches 1.00 when, as in latter sessions, the differentiation of negative sequences is further advanced. These results, and those of our prior published experiments (Weisman *et al.*, 1980; Weisman & DiFranco, 1981) leave little doubt that pigeons can represent and remember the positive sequence, **AB**, and discriminate it from the negative sequences, **AA, BA,** and **BB**. Negative sequences that include the same final event as the reinforced sequence (i.e., **B**) are more slowly discriminated from **AB**; in most instances, **BA** is discriminated more slowly than **AA**.

III.   MODEL BUILDING AND ITS TERMINOLOGY

In a cognitive analysis, after an interesting competence has been demonstrated, the next step, the important step, is to ask how the animal does it. What model or models describe how pigeons represent and utilize order information in distinguishing between sequences of events in the delayed sequence discrimination task? Our models use terminology common in the study of cognition, but it is still worthwhile to review some of the major concepts before proceeding to models and model testing.

The three concepts reviewed are recognition, perceptual memory and instructional memory. The recognition processes are the comparison of working memories for recent events with reference memories for past events and decision making concerning similarities and differences between these representations. Recognition occurs throughout a trial. Pigeons must recognize the sample stimuli as part of the sequence discrimination task if further order processing is to occur. The products of this stimulus recognition task are organized *perceptual memories* of the sample stimuli. In our models, representations of S1 and S2 as stimuli are cast as perceptual memories. Order information is a normal product the stimulus recognition process and, thus part of the perceptual memory of a stimulus. For example, the perceptual memory of S1 has the order tag "ITI preceded this event" and the perceptual memory of S2 has the order tag "S1 preceded this event." This kind of order tag is implicit in the difference between a perceptual memory of S1 and a perception in the presence of S2. Especially important to the delayed sequence discrimination task, however, is the process of *temporal pattern recognition*. This is our term for the process whereby animals compare the order of the sample sequence to that of the reinforced sequence and make decisions based on that comparison. These decisions are the representational products of temporal pattern recognition during a trial. They are, if you will, expectations concerning trial outcome. Honig (1978) has suggested that in delayed discriminations animals often encode and remember information about probable trial outcome, specifically, *instructions* concerning what to do later in the trial. We wish to modify Honig's instructional concept of outcome expectations in two ways. First, animals make decisions concerning what is likely to occur, e.g., food or no food, as well as what to do, e.g., peck or do not peck. The former are instructions in the sense of knowledge concerning a particular fact or circumstance, the latter are instructions in the sense of directions for action. Second, Honig conceived of instructions as more or less binary, e.g., respond or don't respond, but decisions can just as easily be thought of in terms of varying confidence that responding will or will not lead to some outcome. We will use the term instruction to mean a quantitative expectancy, concerning an event, an action, or both, resultant from a pattern recognition process.

Our models identify stages in the processing of trial information. One can think of perceptual and instructional memories as occurring at different "levels" (Craik & Lockhart, 1972) in a cognitive system. Recognition processes, such as stimulus recognition and temporal pattern recognition, move the focus of working memory from one level, or representational system, to another in the animal's mind. Generative models of the delayed sequence discrimination task combine models of the temporal pattern recognition process and of the outcome instruction rule in order to predict the detail of performance. Models of temporal pattern recognition tell us when perceptions of the sample stimuli are translated into decisions concerning those stimuli; models of the outcome decision rule tell us how the latter decisions are translated into separate overall instructions for each sequence type. In this chapter we shall describe these models and their experimental tests. Three models for the temporal pattern recognition process are presented, then four models for the outcome instruction rule are discussed. In both instances experimental evidence is used to reduce the number of alternatives by falsifying otherwise plausible models.

IV. MODELS FOR TEMPORAL PATTERN RECOGNITION

We shall present three kinds of temporal pattern recognition models. All three require pigeons to recognize the sample stimuli as events relevant to the task and to retain perceptual information concerning these stimulus and their order for greater or lesser amounts of time. But the models differ markedly in when and how during during a DSD trial they allow pigeons to recognize the pattern of events to be that of a reinforced or nonreinforced trial type. The models may be thought of as being on a continuum from comparing the sample with the reinforced stimuli one at a time as they are presented to comparing the sample with the reinforced stimuli only once after the whole of the sample sequence has been presented.

In prior work, we (Weisman & DiFranco, 1981) labelled our models with reference as to where in a two-event trial, temporal pattern recognition occurred, i.e., during the test stimulus, during S2, or during both S1 and S2. The models have been modified both in name and in function to apply to temporal pattern recognition in n-event delayed sequence discrimination tasks.

Model 1, the perceptual memory trace recognition model, requires pigeons to retain perceptual memories of all of the sample stimuli and their order until the test stimulus, whereupon this information is compared to a referenced memory of the reinforced sequence. While the model can take slightly different forms in predicting delayed sequence discrimination performance (Weisman & DiFranco, 1981) it will suffice to describe only a "one buffer" model for the two-event task here. According to this version of the perceptual trace model perceptual memories of S1, S2, and their order tags form a holistic pattern that is compared with reference memory for the reinforced sequence only during the test stimulus. We term this perceptual pattern holistic because it can only be compared with the reinforced sequence as a whole, not event by event.

Model 2, the temporal position recognition model, requires pigeons to compare perceptual representations of the current sample stimulus with reference memory representations of the reinforced sequence, one temporal position at a time, and to carry forward to the test stimulus an expectancy, i.e., the outcome instruction, resultant from successive decisions based on those comparisons. Successively, S1 is compared with the item in the first position, and then S2 with the item in the second position in the pigeon's reference memory of the reinforced sequence.

Model 3, the part-sequence order recognition model, states that pigeons construct perceptual units from the current sample sequence that are smaller than the complete reinforced sequence. In a version of the model appropriate for two-event delayed sequence discrimination the units are pairs of stimuli; the S1-S2 sample pair is compared with reference memory for the reinforced sequence during S2. This is the same as saying that the model compares events in the reinforced and the sample sequence in pairs rather than one at a time.

Each of the three models just described is competent, in the sense that it could be used to generate successful discrimination between sequences of events. That is, one could write a computer program based on each model that would indeed sort out sequences according to their order as required by the task. Some, perhaps all three, of the models are unsuitable because they fail to predict the observed course of sequence discrimination, and ultimately because they fail to predict the way in which animal's minds work. The models should be easy to falsify; remember that how much pigeons know about the relationship between the sample sequence and the reinforced sequence at various points in a trial is specified by each of the temporal pattern recognition models. The perceptual trace model states that pigeons do not know whether the sequence they are viewing is positive or negative until the test stimulus is presented. The temporal position model permits pigeons to know whether the stimuli in the sample sequence are, or not, part of the reinforced sequence, one-by-one, as they are presented. The part-sequence order model allows pigeons to recognize negative sequences after some of the sample stimuli, in this instance a pair, have been presented.

These distinctions are useful, not only because they make experimental tests of the models easier, but more importantly, because they suggest different sorts of animal minds. The temporal position and part-sequence order models imply that animals can shift from remembering retrospective perceptual information to remembering prospective instructional information during a delayed sequence discrimination trial, while the perceptual memory model implies that animals process all sample stimuli alike, i.e., perceptually, quite independent of the task. Thus, the former models suggest a mind active in rapid analytic decision making, while the latter model suggests a mind capable of impressive perceptual organization but rather sluggish at decision making.

We (Weisman, Gibson, & Rochford, 1981) have conducted an experimental test designed to distinguish between the temporal pattern recognition models. The procedure allows pigeons to tell us when, in the trial, they recognize that the sample sequence is negative, i.e., precedes a test stimulus that ends without the reinforcer. The paradigm is an adaptation of the Honig (1978) advance key procedure. In general, advance key procedures allow animals to terminate the current trial and institute a new trial in a successive "go/no go" discrimination. Honig (1978) used the procedure to allow pigeons to terminate negative trials in a delayed discrimination experiment. We used the procedure to allow pigeons to terminate negative trials (Weisman, Gibson, & Rochford, 1981). In our procedure two keys are used. After a warning signal, the advance key, colored green, and the continue key, colored blue, and of course, S1, an overhead light, are presented simultaneously. If the pigeon pecks the continue key, then, after a short ISI, S1 is replaced by S2, overhead. If the pigeon pecks the advance key, the trial is terminated, then, after a brief presentation of the warning signal, a new trial begins. Provided that the pigeon pecks the continue key, the choice between the continue and advance keys is repeated again in S2, and then in the test stimulus. During the test stimulus, the continue key is white, rather than blue, and the pigeon must peck to receive food at its offset on positive trials.

The preceding may be termed a forced choice advance key procedure, because it requires the animal to actively respond in order to either continue in the same trial or advance to a new trial. In the Honig *et al.* (1978) procedure, a pigeon's failure to advance from a negative trial can mean either a lapse in attention or a failure to discriminate. In the present adaptation, if the pigeon chooses to advance, we infer that the bird knew, i.e., recognized, that the sample stimulus, say S1, presented on that trial was not part of the reinforced sequence. If the bird chooses to continue, however, it is not necessarily because it knows that the sample stimulus is in the reinforced sequence. Instead, the bird may peck the continue key in order to produce additional perceptual information needed later in the trial.

Figure 12.2 presents the results for three pigeons trained in the two-event task with the advance-continue procedure. The positive sequence was **AB** and the negative sequences were **AA, BB, BA,** and **XX.** The **XX** sequence was a "catch" trial during which neither overhead light was turned on, but the test stimulus was presented. Plotted in Figure 12.2 is the percentage of trails of each sequence type on which the pigeons chose to advance to a new trial as function of stimulus position in the trial sequence: S1, S2, T (test) or N (no advance response). Each bird chose to use the advance key (a) during S1 on at least 90% of the **BB, BA,** and **XX** trials, (b) during S2 on approximately 80% of the AA trials, but (c) on virtually none of the AB trials. Results for other pigeons, not shown here, are in good agreement with those shown in Figure 12.2.

These results place important constraints on the viability of our temporal pattern recognition models. That pigeons can recognize **B** in the S1 position, i.e., **BB** and **BA,** and **A** in the S2 position, i.e., **AA,** as instances of negative sequences clearly falsifies the perceptual trace recognition model. Recall that this model postponed comparison of the sample sequence with the reinforced sequence until the test stimulus. These same results cast considerable doubt upon the part-sequence order model as well. This model would require temporal pattern recognition after presentation of the first pair of events, i.e., S1 and S2. So, unless one can successfully redefine a pair of events, it would seem that the part-sequence order model has been falsified, too. The temporal position model, in contrast, is able to predict the results of the advance key experiment quite effortlessly. This model, recall, would require pigeons to recognize that **B** was not S1 in the reinforced sequence during S1, and that **A** was not S2 in the reinforced sequence during S2, which is certainly a nice description of our results.

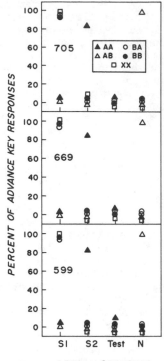

FIG. 12.2. Mean percentage of trials terminated by an advance key response summed over eight sessions for three pigeons.

V.   MODELS FOR THE OUTCOME INSTRUCTION RULE

Prediction of DSD performance does not follow simply from the knowledge that pigeons compare events in the sample with those in the reinforced sequence one at a time. For example, one might predict that the **BB** and **BA** sequences are easily discriminated from **AB**, because these sequences are recognized as negative early in a trial. The facts of delayed sequence discrimination performance are quite different; discrimination is less accurate during **BA** and especially **BB** trials than during **AA** trails. The relationship between the pigeon's performance and its knowledge of the individual events in the sequence is not obvious and requires formal analysis. We turn now to this analysis of how decisions concerning the temporal pattern presented during a trial are translated into an expectancy concerning trial outcome. We shall present four models for the outcome instruction rule. Each model is capable of generating quantitative outcome instructions from the sequences of decisions produced by the temporal position recognition model. One of the models (the binary outcome instruction rule) is noncumulative, i.e., does not require summation of successive decisions. The remaining models (the additive, weighted additive and multiplicative instruction rules) are cumulative, i.e., do require summation of successive decisions. These models are discussed in detail after an introduction to their shared logic.

TABLE 12.1
Model 2
Decisions Output from the Temporal
Position Recognition Model in Successive Comparisons
Between the Sample and the Reinforced Sequence

Obtained Rank Order	Sequence Type	Qualitative Decisions		Sequence Independent		Sequence Dependent	
		S1	S2	S1	S2	S1	S2
(low) 1	AA	yes	no	X	Y	X	W
2	BA	no	no	Y	Y	Y	W
3	BB	no	yes	Y	X	Y	Z
(high) 4	AB	yes	yes	X	X	X	Z

Note-- Where according to the sequence independent rule, X = value (yes), Y = value (no), and according to the sequence dependent rule, for S1, X = value (yes), Y = value (no), and for S2, Z = value (yes), W = value (no).

Since the temporal position recognition model provides the basis for all four instruction rules, it is useful to review the workings of the model in some detail. Table 12.1 shows a summary of the decision taken by model 2, event-by-event, with respect to S1 and S2 for each sequence type. The sequence types are ranked from low responding (i.e., high discriminability), to high responding (i.e., low discriminability). The reinforced sequence, AB, generated the most responding and the other sequences were discriminated from it when they generated less responding. Next, to the right in Table 12.1 are the yes and no decisions generated by matches and mismatches betwen the sample sequence and the reinforced sequence. Sequence independent decision models convert these matches and mismatches to only two alternative quantitative values, say, X = value (yes) and Y = value (no), independent of whether they are decisions about S1 or S2. The resultant decisions for each sequence are shown next to the right in Table 12.1. Sequence dependent decision models convert decisions concerning S1 and S2 to two separate pairs of alternative quantitative values, i.e., for S1, X = value (yes) and Y = value (no), and for S2, Z = value (yes) and W = value (no). The resultant values are shown last on the right in Table 12.1. The two versions of the temporal pattern recognition decision model differ importantly. The sequence dependent model requires a separate pair of values for each successive decision, while the sequence independent model requires only one pair of values for any number of successive decisions.   Implicit, so far, in this discussion have been certain inequalities between the quantitative values denoted in Table 12.1.   It is necessary that X > Y and Z > W, if matches are to have higher decision values than mismatches in the overall outcome instruction rule.   These inequalities will prove useful in subsequent evaluation of the outcome instruction models.

Each model generates a quantitative instruction, here termed a value, equivalent to the model's confidence that the ongoing trial type is positive. The higher this overall value the greater the confidence in a positive outcome for that trial type.   Explicit in our comparisons between predictions generated by the outcome models and patterns of discriminability generated by real pigeons are the assumptions that (a) some quantitative instruction is an important measure of pigeons' confidence in a positive trial outcome, and (b) pigeons' key pecking rates during the test stimulus are at least ordinally related to their confidence.   According to this logic, the adequacy of the models can be tested by comparing for each sample sequence the ranks of values obtained from the four outcome instruction rules with the ranks obtained from pigeons performing the DSD task.   As in Table 12.1, Tables 12.2, 12.3, 12.4, and 12.5 list the sequence types in order from least to most responding during discrimination training, i.e., the rank order obtained from pigeons is **AA** = 1, **BA** = 2, **BB** =3, **AB** = 4.   Shown in Tables 12.2, 12.3, 12.4, and 12.5 are the ranks for each sequence type as predicted by one of the outcome instruction rules.   Keep in mind, during the ensuing analysis, that higher predicted instruction values and predicted ranks are equivalent to higher predicted key pecking rates and less accurate predicted discrimination from **AB**.   Remember, too, that if our evaluation of a particular outcome model uncovers an inconsistency between obtained and predicted ordinal ranks, then the model is falsified.

Models 5 and 6, the noncumulative, or binary rules were proposed by Weisman and DiFranco (1981).   They state that after a pigeon recognizes an initial departure in the sample sequence from the reinforced sequence, it then retains reduced expectancy of a positive outcome until trial offset, uninfluenced by subsequent sample sequence information.   The sequence independent version, Model 4, assigns a predicted instruction value of Y to all of the negative sequences, which then have tied ranks of 2.   The sequence dependent version, Model 5, assigns an instruction value of Y to both BA and BB, which have tied ranks of 2.5.   Thus, while both versions of the noncumulative outcome rule are able to predict a higher instruction for the positive than the negative sequence types, neither version is able to predict important differences in discriminability among the negative types.

Models 5 and 6 accumulate decisions concerning the individual sample stimuli by simple addition.   Again we evaluate the sequence independent model shown at the left in Table 12.3, and the sequence dependent model shown at the right in Table 12.3 separately.   Both models predict a higher response rate on **AB** trials than on any negative trial and higher response rates on **AA** than **BA** trials.   The evaluation of these predictions is straightforward: for the sequence independent model, if X > Y, then (a) **AB** (value) = X + X > **BB** (value) = Y + X, (b) **AA** (value) = X + Y >

TABLE 12.2
Models 4 and 5
Evaluation of the Sequence Independent
and Dependent Noncumulative Instruction Rules

Sequence Types in Obtained Order	Sequence Independent Model (4) Predicted				Sequence Dependent Model (5) Predicted			
	S1	S2	Value	Rank	S1	S2	Value	Rank
AA	X	Y	Y	2	X	W	W	1
BA	Y	Y	Y	2	Y	W	Y	2.5
BB	Y	X	Y	2	Y	Z	Y	2.5
AB	X	X	X	4	X	Z	Z	4

Note-- Implies X > Y, Z > W, Y > W.

BA (value) = Y + Y, and for the sequence dependent model, if X > Y and Z > W, then (c) AB (value) = X + Z > BB (value) = Y + Z, (d) AA (value) = X + W > BA (value) = Y + W. In fact, in every application of the two-event delayed sequence discrimination experiment, we have observed just the reverse order in the discriminability of AA and BA. Thus, the additive models fail to predict the obtained order of discriminability among the negative sequences. Although both models are falsified and therefore probably of little further interest, the sequence dependent rule, which (given Y > W) predicts the rank order discriminability of BB, fares better than the sequence independent rule, which does not. It is worth noting that weighted additive models are a subset of the sequence dependent additive model, and therefore fare as well as the latter.

Models 8 and 9 sum decisions concerning the individual sample stimuli multiplicatively. according to these models, the decision value for S1 is multiplied by the decision value for S2. The outcome instruction value is simply the product of the two decisions. Evaluation of the sequence independent model is quite straightforward: if X > Y, then XX > XY > YY. Model 8, thus, predicts greatest responding on AB trials, least responding on BA trials and intermediate responding on AA and BB trials. The sequence independent model is easily falsified; pigeons discriminate BB with greater difficulty than AA, and, as stated above, discriminate AA with less difficulty than BA. Evaluation of the sequence independent model is only slightly less straightforward. If X > Y, Z > W, and W < O, then AB (value) = XZ > BB (value) = YZ > BA (value) = YW > AA (value) = XW. That W is negative follows from YW > XW. The sequence dependent multiplicative model generates the same rank ordering as obtained from the results of two-event delayed sequence discrimination training experiments conducted in our laboratory.

Finally, we present model 10, a hybrid multiplicative instruction rule. The model uses a sequence independent rule, but requires a single sequence dependent parameter, C, which is added to the S1 decision value

TABLE 12.3
Models 6 and 7
Evaluation of the Sequence Independent
(Unweighted) and Dependent (Weighted)
Additive Instruction Rules

Sequence Types in Obtained Order	Sequence Independent Model (6) Predicted		Sequence Dependent Model (7) Predicted	
	Sum	Rank	Sum	Rank
AA	X + Y	2.5	X + W	2
BA	Y + Y	1	Y + W	1
BB	Y + X	2.5	Y + Z	3
AB	X + X	4	X + Z	4

Note-- Implies X > Y, and Z > W.

TABLE 12.4
Models 8 and 9
Evaluation of the Sequence Independent
and Dependent Multiplicative Instruction Rules

Sequence Types in Obtained Order	Sequence Independent Model (8) Predicted		Sequence Dependent Model (9) Predicted	
	Product	Rank	Product	Rank
AA	XY	2.5	XW	1
BA	YY	1	YW	2
BB	YX	2.5	YZ	3
AB	XX	4	XZ	4

Note-- Implies X > Y, Z > W, W < 0.

prior to multiplication by the S2 decision value, as shown to the left in Table 12.5. In evaluating the model we shall require $X > Y$, $Y < 0$, and $C > |Y|$. That Y is negative follows from $YY + CY > XY + CY$ and that $C > |Y|$ follows (given $X > Y$ and $Y < 0$) from $Y + C > 0$ as in $(Y + C)X > (Y + C)Y$, (see also Table 12.5). Then, we can rank the sequences as follows: **AB** (value) = $(X + C)X > $ **BB** (value) = $(Y + C)X > $ **BA** (value) + $(Y + C)Y > $ **AA** (value) = $(X + C)Y$. It should be clear from this evaluation that the model predicts the obtained rank order discriminability of the sequence types perfectly.

The hybrid rule combines the mathematical and psychological simplicity of a sequence independent model with the predictive competence of a sequence dependent rule. The latter requires four decision parameters for

TABLE 12.5
Model 10
Evaluation of a Sequence Independent
Multiplicative Instruction Model with One
Sequence Dependent Parameter

Sequence Types in Obtained Order	Predicted Product	Rank	Illustrative Numerical Evaluation Product	Value
AA	(X+C)Y = XY + CY	1	(1+2)(-1) =	-3
BA	(Y+C)Y = YY + CY	2	(-1+2)(-1) =	-1
BB	(Y+C)X = YX + CX	3	(-1+2)(1) =	1
AB	(X+C)X = XX + CX	4	(1+2)(1) =	3

Note-- Implies $X > Y$, $Y < 0$, $C > |Y|$.
Numerical Evaluation $X = 1$, $Y = -1$, $C = 2$.

two-event delayed sequence discrimination, quite a few for the pigeon to keep track of and for us to evaluate. The former requires two decision parameters (X and Y) and one sequence parameter (C) for two-event delayed sequence discrimination. The pigeon needs only two decision values while accumulating an outcome instruction and we need solve for only three parameters.

For those who prefer numerical analysis to the algebra of inequalities, we have included an example of the model to the right in Table 12.5. The sample values are $X = 1$, $Y = -1$ and $C = 2$, which are chosen only because they are the smallest whole numbers available to satisfy the inequalities implied in the previous algebraic evaluation. the predicted and obtained ranks of the sequence types are, of course, in agreement with the previous algebraic evaluation.

The outcome instruction rules just reviewed can be organized into two classes: cumulative and noncumulative. It is probably more important to know that pigeons cumulate their sequential decisions, than to know precisely how they cumulate them. Initially, we supposed that once a pigeon had evidence that a sample sequence was negative, it could ignore further evidence, however tempting, and continue to remember not to peck. The results of the preceding analysis of the two-event DSD task made it quite clear that pigeons do cumulate their successive decisions to form an outcome instruction. And, it would appear that they cumulate their temporal pattern recognition decisions multiplicatively in this task.

VI. BROADER ISSUES IN ANIMAL COGNITION

The study of order processing during the delayed sequence discrimination task is important in its own right, but it is important, also, to the more general question of the nature of animal intellect. That is, successful modeling of task may tell us something about animals' cognitive systems.

The temporal position model differs, in interesting, nontrivial ways from the perceptual trace model of the delayed sequence discrimination task. The temporal position model suggests that animals use rule driven, limited resource, utility based evaluation strategies. The perceptual trace model suggests that animals use holistic, perceptually based strategies. The temporal position model requires some form of event-by-event pattern recognition, i.e., a set of rules govern the step-by-step comparison of sample sequence stimuli with representations in reference memory. The perceptual trace model requires one rather global instance of temporal pattern recognition. The temporal position model places a limited load on the resource of working memory by only requiring retention of the results of successive instances of pattern recognition over the trial. In contrast, the perceptual memory model places a large load on working memory by requiring retention of the entire stimulus sequence until the test. The temporal position model makes utility in predicting trial outcome a task characteristic important for the organization of cognition, while the perceptual trace model implies that the retention of a sequence is independent of its later utilization in a task.

As a result of his study of auditory sequence discrimination in young adult humans, Warren (1974; Warren & Ackroff, 1976) proposed two types of sequence processing, holistic pattern recognition and direct identification of the sequence components and their order. In the context of the present discussion these terms would appear to apply to the perceptual trace, and temporal position, models, respectively. The present analysis certainly suggests that in delayed sequence discrimination and perhaps other tasks as well animals are capable of direct identification of sequence components and their order and can use rule driven, limited resource, utility based strategies to solve their problems. It is obvious that this does not eliminate the possibility that under other circumstances animals may use holistic, perceptually based strategies to solve their problems.

An animal employing a perceptual trace recognition process to integrate order information from a sequence of events makes only one decision concerning trial outcome, and that decision comes rather close in time to the outcome itself. This would appear to reduce the complexity of the possible instruction rules for a perceptual trace process considerably. On the other hand, an animal employing a temporal position process to recognize individual stimuli in a sequence faces a rather more difficult task, integrating information from those separate decisions into a single outcome instruction. Thus, the complexity of the possible rules necessary to generate an instruction from a temporal position recognition process is correspondingly greater.

In the evaluation of models of the pigeons' outcome instruction for the DSD task, we found it easy to falsify a noncumulative rule. Also, we were unable to find support for any additive linear model of the outcome instruction, despite their ubiquity in theories of human decision making (Dawes, 1979), Of course, it would be foolish to suppose that pigeons always employ multiplicative sequential decision rules and never employ noncumulative and linear decision rules.

The problem of sequential decision making has been the subject of extensive theoretical analyses, not in the context of animal cognition, but in the literatures of machine pattern recognition (Rosenfeld & Pfaltz, 1966) and human decision making (Dawes, 1979). Thus, cognitive tasks as diverse as digital picture processing and medical diagnosis seem to require rules for the integration of information from independent sequential decisions. Pigeon cognition, too, appears to use the logic of sequential decision making.

ACKNOWLEDGMENTS

This research was supported by Grant No. A01382 from the Natural Science and Engineering Council of Canada. Rainer von Konigslow is a Senior Consultant at AES Data, Ltd., Montreal, Quebec. We thank Gordon Bower, Susan Brown, and Kathy Martin for their thoughtful criticism of earlier versions of this chapter.

REFERENCES

Bindra, D. *A theory of intelligent behavior.* New York: Wiley, 1976.

Cowey, A., & Weiskrantz, L. Auditory sequence discrimination in *Macaca mulatta*: The role of the superior temporal cortex. *Neuropsychologia,* 1976, *14*, 1-10.

Craik, F. I. M., & Lockhart, R. S. Levels of processing: A framework for memory research. *Journal of Verbal Learning and Verbal Behavior,* 1972, *11*, 671-684.

Darwin, C. *The descent of man: And selection in relation to sex.* New York: Appleton, 1920. Originally published, 1871.

Dawes, R. M. The robust beauty of improper linear models in decision making. *American Psychologist,* 1979, *54*, 571-582.

Devine, J. D., Burke, M. W., & Rohack, J. J. Stimulus similarity and order as factors in visual short-term memory in nonhuman primates. *Journal of Experimental Psychology: Animal Behavior Processes,* 1979, *5*, 335-354.

Honig, W. K. Studies of working memory in the pigeon. In S. H. Hulse, H. Fowler, & W. K. Honig (Eds.), *Cognitive processes in animal behavior.* Hillsdale, N.J.: Erlbaum, 1978.

Hulse, S. H., & Dorsky, N. P. Structural complexity as a determinant of serial pattern learning. *Learning and Motivation,* 1977, *8*, 488-506.

James, W. *The principles of psychology.* New York: Holt, 1899.

Kosiba, R., & Logan, F. A. Differential trace conditioning to temporal compounds. *Animal Learning and Behavior,* 1978, *6*, 205-208.

Kotovsky, K., & Simon, H. A. Empirical tests of a theory of human acquisition of concepts for sequential patterns. *Cognitive Psychology,* 1973, *4*, 399-424.

Morgan, C. L. *An introduction to comparative psychology.* London: Walter Scott, 1896. Originally published in 1894.

Restle, F. Theory of serial pattern learning: Structural trees. *Psychological Review,* 1970, *77*, 481-495.

Roitblat, H. L. The meaning of representation in animal memory. *The Behavioral and Brain Sciences*, 1982, *5*, 353–406.

Romanes, G. J. *Mental evolution in animals*. New York: Appleton, 1898. Originally published in 1883.

Rosenfeld, A., & Pfaltz, J. L. Sequential operations in digital picture processing. *Journal of the Association of Computer Machinery*, 1966, *13*, 471–494.

Sternberg, S. Memory scanning: New findings and current controversies. *Quarterly Journal of Experimental Psychology*, 1975, *27*, 1–32.

Straub, R. O., Seidenberg, M. S., Bever, T. G., & Terrace, H. S. Serial learning in the pigeon. *Journal of the Experimental Analysis of Behavior*, 1979, *32*, 137–148.

Warren, R. M. Auditory pattern recognition by untrained listeners. *Perception & Psychophysics*, 1974, *15*, 495–500.

Warren, R. M., & Ackroff, J. Two types of auditory sequence perception. *Perception & Psychophysics*, 1976, *20*, 387–394.

Wasserman, E. A. Response evocation in autoshaping: Contributions of cognitive and comparative–evolutionary analyses to an understanding of directed action. In C. M. Locurto, H. S. Terrace, & J. Gibbon (Eds.), *Autoshaping and conditioning theory*. (Vol. 35.) New York: Academic Press, 1981.

Wasserman, E. A., Nelson, K. R., & Larew, M. B. Memory for sequences of stimuli and responses. *Journal of the Experimental Analysis of Behavior*, 1980, *34*, 49–59.

Weisman, R. G., & DiFranco, M. P. Testing models of delayed sequence discrimination in pigeons: Delay intervals and stimulus durations. *Journal of Experimental Psychology*, 1981, *7*, 413–424.

Weisman, R. G., Gibson, M., & Rochford, J. Advances in delayed sequence discrimination. Paper presented at the annual meeting of the Psychonomic Society, Philadelphia Society. November, 1981.

Weisman, R. G., Wasserman, E. A., Dodd, P. W. D., & Larew, M. B. Representation and retention of two–event sequences in pigeons. *Journal of Experimental Psychology: Animal Behavior Processes*, 1980, *6*, 312–325.

Woodbury, C. B. The learning of stimulus patterns by dogs. *Journal of Comparative Physiology*, 1943, *35*, 29–40.

13   SELF REPORTS BY RATS OF THE TEMPORAL PATTERNING OF THEIR BEHAVIOR: A DISSOCIATION BETWEEN TACIT KNOWLEDGE AND KNOWLEDGE

Charles P. Shimp
University of Utah

> *There is a story going the rounds of a man who was driving a horse uphill with a heavy load, his dog running along beside the wagon.*
> *Suddenly the horse stopped short in his tracks and spoke: 'Listen! I'm sick of this!' he said. 'I don't pull this load another step.'*
> *'Well, I'll be darned!' said the man, taken aback. 'I never heard a horse talk before.'*
> *'Neither did I,' said the dog.*
> *Now, this story may seem a little fantastic to the layman, but, from the data that I have been able to gather, the thing is not beyond the realm of possibility. At any rate, not the dog's end of it. The man's remark seems a little far-fetched.*
> *From My Ten Years in a Quandary by Robert Benchley.*

I. INTRODUCTION

This chapter describes an experiment that addresses two problems. First, it explores the generality of earlier empirical research on the way reinforcement can impose serial patterning on laboratory animal behavior. Second, it explores a theoretical conjecture about a possible mechanism for the way reinforcement might establish serial patterning. This conjecture holds that reinforcement acts on remembered behaviors; behaviors an organism thinks it has recently emitted (Shimp, 1975, 1976a, 1978). This experiment, then, investigates first the generality of empirical descriptions of serial patterning of behavior and second, a possible theoretical mechanism for this patterning.

Over the past 10 to 15 years, a body of literature has been slowly accumulating on the way reinforcement imposes various sorts of temporal patterning on animal behavior (Fetterman & Stubbs, 1982; Gibbon, 1977; Hinson & Staddon, 1981; Platt, 1979; Schwartz, 1980; Shimp, 1969, 1976a, 1979; Silberberg, Hamilton, Ziriax, & Casey, 1978; Straub, Seidenberg, Bever, & Terrace, 1979). This growing literature parallels in certain of its features the literature on schemata and scripts in humans. Parallel features include a concern for the identification of behavioral patterns that have become unitized or chunked and a concern for the criteria by means of which unitization can be recognized.   In the animal literature on temporal patterning, especially that dealing with interresponse times and similar temporal patterns, unitized behavioral patterns seem to have some of the properties of automated motor programs: it is as though once a memory representation of a particular program is activated, the corresponding temporal pattern of behavior runs off automatically (Shimp, 1969, 1975, 1976a).

Most of this research with animals has employed pigeon subjects. Indeed, some of the results that seem more difficult to assimilate theoretically, and therefore that seem more theoretically diagnostic, have been obtained only with pigeons.   For example, when two simple temporal patterns, such as interresponse times, are reinforced such that one pattern is always, say, three times longer than the other, one finds that preference for the shorter pattern increases as the absolute durations of both are increased (Hawkes & Shimp, 1974; Shimp, 1983).   This result has proven to be diagnostic of weaknesses in several theories (Gibbon, 1977; Shimp, 1978), but some evidence suggests that it might have limited generality, indeed, that it might be sharply restricted to key pecking of pigeons.   One such line of evidence derives from recent work by Donahoe, Palmer, and Strickney (1981).   They found evidence to implicate a role for autoshaped key pecking: They suggested that this increasing preference depends on an increasing frequency of autoshaped key pecks.   If this interpretation were generally applicable, it would cast serious doubt on the interpretation of interresponse times as chunked and unitized higher-order units.   Instead, the behavioral units in these experiments might be only the elementary key pecks themselves.   Furthermore, since autoshaping seems to depend heavily on the particular species and response under consideration (Terrace, 1981; Wasserman, 1981), it would seem unlikely that the detailed results obtained heretofore only with pigeons' key pecking would have any important generality.

There is a second reason for some concern about the generality of results obtained so far with pigeons' key pecking.   As noted above, one possible mechanism for the origin of higher-order unitized patterns of behavior is an organism's memory for its own recent behavior: perhaps a pattern becomes unitized to the extent to which it repeatedly is remembered as having occurred just before the delivery of a reinforcer. The repeated activation of the same memory representation of the pattern, at the moment of reinforcement, might contribute to the gradual unitization of a corresponding motor program (Shimp, 1975).   However, it seems clear there are enormous species and response differences in the degree to which an organism can remember what it recently has done.   Pigeons, for example, can sometimes remember only for a few seconds whether they recently have pecked a key (Kramer, 1982) or which key they have recently pecked (Shimp, 1976b), but at least in some cases rats can remember for relatively long durations where they recently have been (see Chapter 24, by Roberts, this volume; Olton, 1979).   Thus, if memory for recent behavior contributes to the unitization of behavioral patterns, one might expect corresponding differences across species in the extent to which various behaviors can be unitized.

These two reasons lead one to expect limited generality of the results on the local organization of behavior which to date rest mostly on pigeons' key pecking. The present experiment explores this possibility of limited generality. Here the temporal patterns are based on rats' lever pressing on spatially distinct levers. Rats and pigeons may be noticeably different with respect to spatial memory, various spatial strategies, and autoshaping, so it was expected that the experiment might provide results on the local organization of behavior distinctly different from those obtained previously. On the other hand, should the results resemble those obtained earlier with another species and a different response, it would seem as though a significant empirical generalization might have been achieved.

One last issue needs to be discussed before the experiment proper can be described. Consider in greater detail the proposed theoretical mechanism for the development of temporal patterning, specifically the idea that temporal patterns are established to the extent to which they are remembered. A straightforward way to evaluate this idea is to use the behavioral patterns as stimuli in a short-term memory procedure. According to the conjectured mechanism, one ought to find that the behavioral patterns that can be shaped and unitized can also be remembered (Shimp, 1978). It would be difficult to maintain the conjecture should there prove to be no particular correlation between patterns that are established and unitized, and patterns that can be remembered. On the other hand, such a lack of correlation would be an example of a particularly intriguing phenomenon in its own right. It would exemplify a type of dissociation between what may be called tacit knowledge and knowledge (Shimp, 1982a, 1983).

Consider how this is so. The production of temporal patterns of behavior as a result of their reinforcement may be viewed as a learned skill. Such an acquired skill is by definition a kind of knowledge, but the organism that displays it need have no awareness of it. The actual performance of the acquired skill of pressing levers in an adaptive, temporally patterned way can be thought of as a rat's tacit knowledge of the reinforcement contingency. If the organism is a human, there may be no ability verbally or otherwise to describe or otherwise acknowledge the possession of such a skill. Accordingly, it may be called tacit knowledge: The organism does not know that it possesses it. In the present experiment we propose not only to arrange to have a rat display such a skill, or tacit knowledge, but also to report if it remembers what behavioral patterns it produces. That is, we propose in a sense to ask a rat to describe its own behavior. A self report may be said to tell us something about what the rat knows about what it is doing. To the extent to which the rat can report to us what it is doing, we could say not only that its skill displays tacit knowledge, but that its self reports of its skill display something like knowledge. To the extent to which a rat's self report of its behavior faithfully mirrors that behavior, we can say that the rat knows what it is doing. The mechanism proposed above for the development of unitized behavioral patterns suggests just this kind of self awareness.

The preceding may be rephrased as follows. I propose a method for the study of the relation between behavior and a kind of metabehavior, or behavior that is about other behavior. This method is based on the assumption that the relation between behavior and metabehavior resembles the relation between tacit knowledge and knowledge (or alternatively, between knowledge and metaknowledge, see Shimp, 1982a). Further, I propose to use this method to investigate the conjecture that unitized temporal patterns, such as interresponse times, develop only when a subject knows it has emitted a pattern before reinforcement, so that the subject may be said to attribute reinforcement to the pattern of behavior it knows it just emitted. On the other hand, if a subject displays various behavioral patterns but can not report that it is doing so, we would have a display of tacit knowledge without corresponding knowledge: the subject would not in a sense know what it was doing. Or, we might find that a subject displays only a poor skill but can accurately describe what it is doing. These various possibilities describe possible kinds of dissociations between tacit knowledge and knowledge. These dissociations would seem to falsify the proposed mechanism for the development of unitized behavioral patterns.

A striking dissociation between tacit knowledge and knowledge has been identified in pigeons when the absolute durations of reinforced temporal patterns are varied as in the present experiment (Shimp, 1983).

In such a case, what a pigeon does and what it "says" it does may be quite different. First, the accuracy of the skill is not correlated with the accuracy of self report. Second, increasing preference accompanies decreasing accuracy of self report: it is as though the difference between two unitized temporal patterns becomes more and more important while an organism can only more and more poorly tell the difference between them. The present experiment was designed to discover if there is a similar dissociation between tacit knowledge and knowledge in the case of the rats' lever pressing.

II. METHOD

A. SUBJECTS

Two Long Evans rats were five months old at the start of the experiment. They were maintained at 70% of their free-feeding weight except that one gram was added each week in an attempt to compensate for growth. Water was continuously available in the home cages and an approximately 14 hr/10 hr light/dark cycle was maintained in the colony room.

B. APPARATUS

A Terak computer arranged all experimental contingencies and recorded the data. This computer was interfaced to two standard, two-lever rat chambers manufactured by Campden Instruments. Noyes food pellets were dispensed into a trough situated midway between the left and right levers. Reinforcement consisted of the delivery of two such pellets in the space of approximately half a second. Each chamber was housed in its own experimental room and sounds from outside the rooms were masked by fans and either white noise or classical music and speech from a local FM radio station.

C. PROCEDURE

The basic procedure was similar to that of earlier experiments on the differential reinforcement of two classes of discriminated interresponse times (e.g., Hawkes & Shimp, 1974). Two temporal patterns, a shorter and a longer, were concurrently reinforced. A variable-interval schedule arranged a distribution of minimum interreinforcement intervals and a pattern-selection procedure randomly assigned each reinforcer arranged by the variable-interval schedule to one of the two patterns.

Consider the details of the variable-interval schedule. A single timer queried a random number table every one sec to determine whether to arrange a reinforcer. The probability of arranging a reinforcer every one sec was 0.050, so that the schedule was a random interval 20 sec schedule. Once a reinforcer was arranged, the 1-sec timer stopped until the reinforcer was collected, or until the reinforcer was cancelled by an excessively long pause. If a rat paused for longer than the upper bound of the longer reinforced pattern, an arranged reinforcer was cancelled. This was to discourage long pauses.

Now consider the pattern-selection procedure. The durations of the reinforced pairs of patterns, shorter and longer, are shown for each condition in Table 13.1. Throughout the experiment the longer pattern either was exactly or approximately three times longer, and also three times wider, than the shorter pattern. Thus, as the absolute duration was varied, relative duration remained fixed. The assignment of each reinforcer arranged by the variable-interval schedule to a pattern was random in the sense that the patterns were equally likely to be selected and the pattern chosen for one reinforcer was independent of the pattern chosen for the preceding reinforcer.

TABLE 13.1
Experimental Conditions

Condition number	Reinforced classes of interresponse times (lower and upper bounds in sec)	
	Shorter	Longer
1	1.5 - 2.5	4.5 - 7.5
2	2.5 - 4.0	7.5 - 12.0
3	4.0 - 7.0	12.0 - 18.0
4	1.5 - 2.5	4.5 - 7.5
5	1.0 - 2.0	3.0 - 6.0
6	0.75 - 1.5	2.0 - 3.5
7	0.5 - 1.0	1.5 - 3.0

The patterns in Table 13.1 are described only in terms of their durations. Also relevant to the collection of a reinforcer was which lever was pressed. If a reinforcer had been arranged and assigned to the shorter pattern, only a left–lever press terminating the shorter pattern would be reinforced. A right–lever press would terminate the pattern but would not be reinforced. Similarly, if a reinforcer had been arranged and assigned to the longer pattern, only a right–lever press terminating the longer pattern would be reinforced, a left lever press would terminate the pattern but would not be reinforced. As a result of this arrangement either pattern could be initiated or terminated by a press to either lever, but only those shorter (longer) patterns terminated by a left (right) lever press could be reinforced.

The times when a lever press could initiate reinforcement were signalled by the onset of two white lights, one directly over each lever, and by the offset of the houselight.

Each pattern was timed from the release of a lever and there was a minimum interval of 0.1 sec during which the computer did not look for a response. That is, a lever had to be released and then a tenth of a sec had to elapse before any subsequent lever press would be recognized. A lever press occurring before 0.1 sec had elapsed reset the 0.1–sec timer. The houselight was off while this timer ran.

Sessions were conducted six days each week, and conditions were changed every 15 days.

To review the procedure, imagine the following protocol from condition 1. A rat might press the left lever. A tenth of a second later the houselight comes on. After another 1.4 sec both lever lights come on and the houselight goes off. Now a press of either lever terminates the shorter pattern, but a press of the left lever is reinforced if the variable–interval schedule has arranged a reinforcer and if the reinforcer has been assigned to the shorter pattern. Let us suppose the left lever is pressed but no reinforcer is delivered. As after every lever press, for the first 0.1 sec the chamber is dark and then after another 1.4 sec both lever lights come on and the houselight goes off. This time let us suppose the rat does not press either lever while the lever lights are on, i.e., during the time a lever press would terminate an interresponse time in the shorter category. Accordingly, after another second elapses the houselight comes on and the lever lights go off since the width of the shorter pattern is one sec (as shown in Table 13.1 for condition 1). After another 2 sec the lever lights come on again and a press of either lever terminates the longer pattern. A press of the right lever delivers a reinforcer if the variable–interval schedule has arranged a reinforcer and if it has been assigned to the longer pattern. Shorter patterns terminated by right–lever presses or longer patterns terminated by left–lever presses are never reinforced. All responses when the lever lights are off, including those terminating interresponse times longer than the longer reinforced pattern, simply restart the interresponse time clock.

Finally, let us review what behavior corresponds to a learned skill or tacit knowledge and what behavior corresponds to knowledge. The claim is that the sheer temporal distribution of lever pressing, regardless of which levers are pressed, represents adaptive behavior similar to a human skill, which a human might or might not be able to describe verbally. Such a skill is a form of tacit knowledge. Now, if a human can provide verbal self reports of such a skill, we say he has knowledge of the skill. In the present case, the analogy is to say that a rat has knowledge of its temporally-patterned behavior if its non-verbal self reports accurately describe its own temporally-patterned behavior. *If* a rat *knows* how long it has been since the previous lever press, it should be able to press not only some lever but the correct lever. If on the other hand, it can emit lever presses that are correctly spaced but can not terminate interresponse times on correct levers, we would say it had a skill, but did not know what it was doing.

We ask, to what extent is a rat's knowledge of its skill dependent on the nature of that skill? When does a rat know what it is doing and when does it not?

III. RESULTS

The chief results fall into three categories. First, one can plot relative-frequency distributions of interresponse times for each lever. Such distributions indicate the extent to which responses are temporally distributed in an adaptive manner. Figure 13.1 shows individual and average distributions for the last day of each of two conditions, 4 and 7. These conditions provide an opportunity to see patterning in rats' behavior for a contingency (condition 4) corresponding to that for which pigeons' behavior is clearly bimodal and one (condition 7) for which pigeons' behavior is less clearly bimodal or perhaps unimodal (Shimp, 1981, 1983).

As Figure 13.1 shows, the same pattern emerges here if one looks at the total responding on both levers. In condition 4, modes appeared at locations in each lever's distribution appropriate to the reinforced class on that lever. While a similar tendency can be seen for each separate lever for condition 7, it is equally clear that, had all the presses been on the same lever, the overall distribution certainly would not have been strikingly bimodal. In terms of the qualitative trends in the distributions, the curves in Figure 13.1 are difficult to distinguish from those produced by pigeons (Shimp, 1968, 1981, 1983).

In passing, note that Figure 13.1 shows that lever pressing is by no means restricted entirely to those times, signalled by the lever lights, when reinforcement is potentially available. However, Figure 13.1 shows responding notably more accurately clustered near times of reinforcer availability than in pilot data not presented here, obtained under conditions where reinforcement availability was not signalled. A similar effect occurs with pigeon subjects, in which case signalling reinforcer availability improves accuracy but leaves a fair amount of responding remaining at times when reinforcement is unavailable.

Second, one can look at preference for one reinforced pattern, say the shorter, as a function of the absolute durations of the reinforced patterns. To look at preference, we count the number of interresponse times falling in each reinforced class, for each of the last five days of a condition and take the fraction equal to the number of shorter patterns divided by the total of shorter and longer patterns. The resulting average

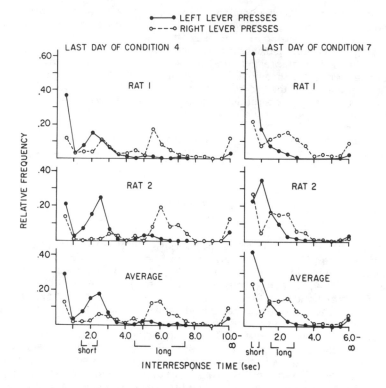

FIG. 13.1. Relative frequencies of interresponse times on left (solid figures) and right (open figures) levers in 0.5-sec categories, for the last days of condition 4 and of condition 7.

and standard error for each condition is shown in Figure 13.2. In the present experiment these numbers can be computed in diverse ways. Figure 13.2 shows the results of two of these ways. The solid figures are based on a definition of the shorter (longer) pattern as the number of interresponse times in the shorter (longer) class terminated exclusively by presses on the left (right) lever. The open figures are based on a definition of the shorter (longer) pattern as the number of interresponse times in the shorter (longer) class terminated by presses on either lever. As the figure shows, both definitions lead to the same qualitative result: preference for the shorter pattern increases as the absolute durations of both patterns increased. If rats' performance resembles that of pigeons (Hawkes & Shimp, 1974), preference for the shorter pattern should approximate indifference (50–50) when the lower bound of the shorter pattern is about half a sec and should rise beyond time–allocation matching to 80 percent or more by the time this lower bound has been extended to 4 sec. As can be seen in Figure 13.2, the performance of the present two rats was strikingly similar to these expectations. The dashed horizontal line represents time-allocation matching, the locus of points such that equal amounts of time would be spent engaging in the two equally reinforced patterns, when one is three times longer than the other. If one imagines smooth curves drawn through the points in Figure 13.2, the curves would cross the time-allocation matching line at about 1.5 sec for Rat 1 and perhaps between 2.5 to 3 sec for Rat 2. The average curve crosses the matching line at about 2.5 sec. These results also are remarkably similar to those for pigeons as reported by Hawkes and Shimp (1974).

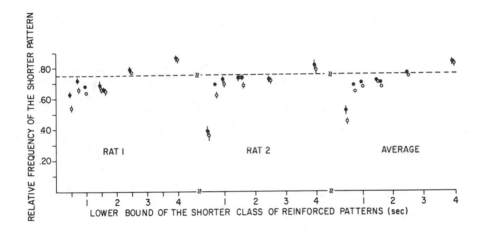

FIG. 13.2. The relative frequency of the shorter reinforced pattern averaged over the last five days of a condition, as a function of the length of the reinforced patterns defined in terms of the lower bound of the shorter reinforced pattern. The longer pattern was three times as long as the shorter. Vertical lines represent plus and minus one standard error. Solid figures represent relative frequencies based exclusively on reinforced categories (left lever presses terminating shorter patterns and right lever presses terminating longer patterns) while open figures represent relative frequencies based on presses on either lever terminating either pattern. Of the two pairs of figures at 1.5 sec, the right-hand pair represents the replication. The dashed horizontal line represents time-allocation matching.

Third, one can compare the extent to which preference for a pattern covaried with the extent to which a rat was able to report which pattern it just produced. Figure 13.3 shows the probability of a correct response, that is the fraction of all lever presses terminating shorter or longer patterns that were to the correct lever, as a function of the degree of preference for the shorter pattern.

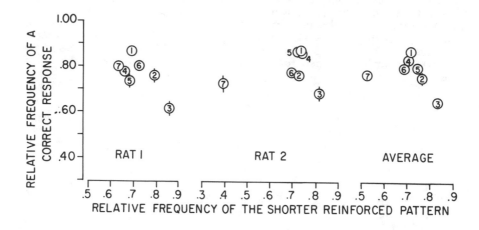

FIG. 13.3. The relative frequency of a correct response (a response to the left lever after a shorter pattern or a response to the right lever after a longer pattern) as a function of the relative frequency of the shorter reinforced pattern. Numbers within symbols refer to condition numbers. Vertical lines represent plus and minus one standard error. No line appears when it is less than the diameter of the symbol, or, for Rat 2, when symbols cluster closely together.

Recall the present claim that the lever a rat presses to terminate a pattern may be viewed as a measure of the rat's knowledge about what it just did. If it knows it can terminate a shorter pattern, it should press the left lever, if it knows it can terminate a longer pattern, it should press the right lever. There is a sense, therefore, in which the probability of terminating a pattern with a press on the correct lever may be said to represent the extent to which a rat knows what it is doing. Figure 13.3 shows that as preference for the shorter pattern increased, the accuracy of the self reports of the most recent pattern decreased. Thus, when preference indicated the difference between the two patterns mattered more and more, a rat knew less and less about which pattern it was terminating.

IV. DISCUSSION

The results are strikingly similar to those obtained previously with pigeons' key pecking. The shapes of the interresponse time distributions resemble those obtained with pigeons (Shimp, 1968, 1981, 1983), preference for one pattern over the other depended on pattern duration in a manner closely resembling that obtained with pigeons (Hawkes & Shimp, 1974), and finally, a dissociation was obtained between the preference for one pattern over the other and the extent to which a rat was able successfully to report what it was doing (Shimp, 1983).

These similarities suggest a greater generality than one previously might have cared to attribute to earlier work on the local organization of pigeons' key pecking. The concern about generality raised by Donahoe *et al.*'s (1981) work, if not entirely laid to rest, at least is diminished, since now we have generality over rats' lever pressing as well as over discriminated (Hawkes & Shimp, 1974) and non-discriminated (Shimp, 1981, 1983) key pecking by pigeons. Furthermore, it should be added that several phenomena related to the local organization of behavior seem to generalize over a variety of different kinds of temporal patterns of key pecking, such as interchangeover times (Shimp, 1979) and lengths of runs in discrete-trials settings (Shimp, 1982b), as well as interresponse times. These forms of generality encourage the view that reinforcement shapes the local temporal structure of behavior in rather robust ways and that these temporal shaping phenomena are the kinds of quantitative effects that seem to justify theoretical attempts to explain them (Gibbon, 1977; Shimp, 1978). In addition, the robustness of these temporal shaping phenomena also encourages the continued search for the kinds of parallels between animal and human behavior that were mentioned in the introduction. Thus, it would now seem even more justified to ask about general similarities between cognitive concepts such as a script or schema (e.g., Bower, Black, & Turner, 1979), in which an hierarchically organized structure determines the temporal patterning of human behavior, and the kinds of hierarchical structures of unitized temporal patterns that have been proposed to explain behavior of the sort reported in the present experiment (Shimp, 1978).

In addition to generalizing previous results on the way reinforcement shapes the local organization of behavior, the experiment also sheds light on a proposed mechanism for the development of unitized behavior patterns, and in general on the relation between tacit knowledge and knowledge. The relation between the performance of a learned skill and self reports of that performance has been studied in two ways with infrahumans. One way involves symbolic matching-to-sample probes where correct responses depend on previous behavior (Shimp, 1981, 1983) and a second way involves an autoshaping procedure in which a reinforcer is delivered automatically at the end of one stimulus if the previous pattern was a shorter one, but not if it was a longer one, and a reinforcer is delivered at the end of a second stimulus if the previous pattern was a longer one, but not if it was a shorter one. An animal's knowledge of the pattern it just emitted is interpreted in terms of the relative rate of responding in the presence of stimuli that signal reinforcement (Shimp, 1982a). The procedure of the present experiment is much simpler than the symbolic matching–to–sample procedure and seems as convenient and easy to arrange as the autoshaping procedure. Thus, we now have three ways by means of which we can ask a non-verbal organism to provide self reports of its own temporally patterned behavior.

An important next step would seem to be to ask what kind of integrative theory can assimilate the present dissociation. That is, why, as preference becomes more extreme, does the animal appear to know less and less about what it is doing. This result seems counter–intuitive. It would seem natural for an animal, as two response alternatives become more difficult for it to discriminate, to care less about the difference between them: for example, Baum (1974) has suggested that undermatching in concurrent performance is attributable to a failure to discriminate perfectly between response alternatives. Here, however, the opposite result was obtained: as the subject knew less and less about the two response alternatives, it cared more and more about the difference between them.

The present dissociation, and related dissociations between behavior and metabehavior in pigeons (Shimp, 1983) may be related to various other dissociations, such as those obtained with split brain preparations and amnesiacs. Similarly it might be related to dissociations between perceptual–motor skills in general and verbal self reports of those skills (Nisbett & Wilson, 1977; Jacoby & Dallas, 1981) and perhaps also to dissociations between various levels of intentionality as described by Dennett (Dennett, 1983).

V. SUMMARY

An experiment was conducted on the local organization of behavior in rats in order to examine the generality of previous work on pigeons' key pecking. Lever pressing by two rats was reinforced according to a concurrent schedule of reinforcement for shorter and longer classes of interresponse times. Shorter and longer reinforced classes were terminated by presses on left and right levers, respectively, and were cued by illumination changes within the chamber: the two reinforced behavioral patterns differed not only in temporal duration but also in the spatial location of their terminal lever presses. The interresponse–time distributions showed reasonably good control by the durations of the reinforced classes. As in previous experiments with pigeons, preference between patterns depended not only on relative but also on absolute durations of the patterns: preference for the shorter pattern increased as the absolute durations of both were increased, even though the longer was always three times as long as the shorter. This result has proven to be theoretically diagnostic and apparently has considerable empirical generality. Finally, the procedure permitted a kind of self-description by a rat of its own performance: the temporal patterning of behavior may be viewed as a skill, and the relative frequency with which a given temporal pattern is terminated by a response to the correct lever may be viewed as the accuracy of a rat's self characterization of overall performance of that skill. The relation between performance of that skill and a rat's self report of its performance was similar to that obtained previously in pigeons, to the extent that a dissociation was obtained between degree of preference for a pattern and accuracy of self report: the greater the preference for one pattern over the other, the less accurate was a rat's discrimination between the two patterns. This dissociation resembles a variety of other dissociations between tacit knowledge and knowledge in pigeons, and suggests that animals resemble people in that only sometimes do they know what they are doing.

ACKNOWLEDGMENTS

This research was supported in part by NIMH Grant 16928 and by a Bio-Medical Sciences Support Grant to the University of Utah. The author would like to express his gratitude to Stephen Hutton and David Bodily for their help in conducting the experiment and in analyzing the data. The author also would like to thank his colleagues at the University of Utah, Larry Jacoby, William Johnston, and Raymond Kesner for conversations about the general topic of dissociations between tacit knowledge and knowledge. The author is grateful also to Charles Catania, Peter Harzem, Philip Hineline, and John Wearden for conversations on the same topic.

REFERENCES

Baum, W. M. On two types of deviation from the matching law: Bias and undermatching. *Journal of the Experimental Analysis of Behavior,* 1974, *22,* 231–242.

Benchley, R. C. *Talking dogs. In My ten years in a quandary.* Garden City, N.Y.: Blue Ribbon Books, 1940.

Bower, G. H., Black, J. B., & Turner, T. J. Scripts in memory for text. *Cognitive Psychology,* 1979, *11,* 177–220.

Dennett, D. C. Intentional systems in cognitive ethology: The "Panglossian Paradigm" defended. *The Behavioral and Brain Sciences,* 1983, *6,* in press.

Donahoe, J. W., Palmer, D. C., & Strickney, K. J. *Unified reinforcement principle: Some related experiments.* Paper delivered at the Twenty-second Annual Meeting of the Psychonomic Society, 1981.

Fetterman, J. G., & Stubbs, D. A. Matching, maximizing, and the behavioral unit: Concurrent reinforcement of response sequences. *Journal of the Experimental Analysis of Behavior,* 1982, *37,* 97–114.

Gibbon, J. Scalar expectancy theory and Weber's law in animal timing. *Psychological Review,* 1977, *84,* 279–325.

Hawkes, L., & Shimp, C. P. Choice between response rates. *Journal of the Experimental Analysis of Behavior,* 1974, *21,* 109–115.

Hinson, J. M., & Staddon, J. E. R. Maximizing on interval schedules. In C. M. Bradshaw, E. Szabadi, & C. F. Lowe (Eds.), *Quantifications of steady-state operant behavior.* Amsterdam: Elsevier, 1981.

Jacoby, L. L., & Dallas, M. On the relationship between autobiographical memory and perceptual learning. *Journal of Experimental Psychology: General*, 1981, *110*, 306–340.

Kramer, S. P. Memory for recent behavior in the pigeon. *Journal of the Experimental Analysis of Behavior*, 1982, *38*, 71–85.

Nisbett, R. W., & Wilson, T. D. Telling more than we can know: Verbal reports on mental processes. *Psychological Review*, 1977, *84*, 231–259.

Olton, D. S. Mazes, maps, and memory. *American Psychologist*, 1979, *34*, 583–596.

Platt, J. R. Interresponse–time shaping by variable–interval–like interresponse–time reinforcement contingencies. *Journal of the Experimental Analysis of Behavior*, 1979, *31*, 3–14.

Schwartz, B. Development of complex, stereotyped behavior in pigeons. *Journal of the Experimental Analysis of Behavior*, 1980, *33*, 153–166.

Shimp, C. P. Magnitude and frequency of reinforcement and frequencies of interresponses times. *Journal of the Experimental Analysis of Behavior*, 1968, *11*, 525–535.

Shimp, C. P. Optimal behavior in free–operant experiments. *Psychological Review*, 1969, *76*, 97–112.

Shimp, C. P. Perspectives on the behavioral unit: Choice behavior in animals. In W. K. Estes (Ed.), *Handbook of learning and cognitive processes*. (Vol. 2.) Hillsdale, N.J.: Erlbaum, 1975.

Shimp, C. P. Organization in memory and behavior. *Journal of the Experimental Analysis of Behavior*, 1976, *26*, 113–130. (a)

Shimp, C. P. Short–term memory in the pigeon: The previously reinforced response. *Journal of the Experimental Analysis of Behavior*, 1976, *26*, 487–493. (b)

Shimp, C. P. Memory, temporal discrimination, and learned structure in behavior. In G. H. Bower (Ed.), *The psychology of learning and motivation*. (Vol. 12.) New York: Academic Press, 1978.

Shimp, C. P. The local organization of behavior: Method and theory. In M. D. Zeiler & P. Harzam (Eds.), *Advances in analysis of behavior, Vol. 1: Reinforcement and the organization of behavior*. Chichester, England: Wiley, 1979.

Shimp, C. P. The local organization of behavior: Discrimination of and memory for simple behavioral patterns. *Journal of the Experimental Analysis of Behavior*, 1981, *36*, 303–315.

Shimp, C. P. Metaknowledge in the pigeon: An organism's knowledge about its own adaptive behavior. *Animal Learning and Behavior*, 1982, *10*, 358–364. (a)

Shimp, C. P. Choice and behavioral patterning. *Journal of the Experimental Analysis of Behavior*, 1982, *37*, 157–169. (b)

Shimp, C. P. The local organization of behavior: A dissociation between a pigeon's behavior and a self–report of that behavior. *Journal of the Experimental Analysis of Behavior*, 1983, *39*, 61–68.

Silberberg, A., Hamilton, B., Ziriax, J. M., & Casey, J. The structure of choice. *Journal of Experimental Psychology: Animal Behavior Processes*, 1978, *4*, 368–398.

Straub, R. O., Seidenberg, M. S., Bever, T. G., & Terrace, H. S. Serial learning in the pigeon. *Journal of the Experimental Analysis of Behavior*, 1979, *32*, 137–148.

Terrace, H. S. Introduction: Autoshaping and two–factor learning theory. In C. M. Locurto, H. S. Terrace, & J. Gibbon (Eds.), *Autoshaping and conditioning theory*. New York: Academic Press, 1981.

Wasserman, E. A. Response evocation in autoshaping: Contributions of cognitive and comparative–evolutionary analyses to an understanding of directed action. In C. M. Locurto, H. S. Terrace, & J. Gibbon (Eds.), *Autoshaping and conditioning theory*. (Vol. 35.) New York: Academic Press, 1981.

IV.   CONCEPT FORMATION AND PROCESSING OF COMPLEX STIMULI

14 OBJECTS, CATEGORIES, AND DISCRIMINATIVE STIMULI

R. J. Herrnstein
Harvard University

I. INTRODUCTION

A science-writer popularizing cognitive psychology in the N. Y. Times *Sunday Magazine* recently said, "We human beings. . .are concept-making creatures: Unlike any other animal, we have a natural ability to group objects or events together into categories (Hunt, 1982a, p. 48)."

Elsewhere, he drove the point home even harder: "One of the really astounding things the mind does, apparently effortlessly, is group similar things and make a concept of them. This is something that, as far as we know, no animal can do (Hunt, 1982b, p. 63)." A widely-used textbook (Glass, Holyoak, & Santa, 1979) tells beginning students of cognition that, "One of the most pervasive aspects of human thought is the tendency to divide the world into categories (p. 326)." In an engineering journal that publishes on pattern recognition, particularly on simulation by computer, two scientists wrote "the human visual is the only effective pattern classification system known (Howard & Campion, 1978, p. 32)."

Several recent books surveying and analyzing categorization as a subject in cognitive psychology, review only data on human subjects (Rosch & Lloyd, 1978; Smith & Medin, 1981). The numerous journals that have lately featured research on categorization rarely publish or cite anything but human data. In these discussions, as in the present one, categorization is taken to mean simply a manifestation of rules for sorting instances with some degree of parsimony. No attempt will be made here to distinguish categorization from such allied notions as classification, discrimination, or conceptualization, although it is conceivable that various distinctions could profitably be drawn.

While some students of categorization, particularly those with a perceptual orientation, take a broader view, the topic as a whole has a clear tendency toward what can be called an anthropocentric bias — the assumption without proof that a phenomenon found in people is found only in people. Classifying in humans is so intimately bound with language that it is easy to suppose that anyone or anything that classifies does it in some sense linguistically. Language as we know it could not have arisen in a non-classifying creature. No function of language is more pervasive than the attaching of names to classes of objects or events. Naturally, then, many who study categorization in people incidentally assume they are studying an aspect, or at least a cognitive precursor, of language. In almost all of the recent research on human classification, the primary data come from subjects using words to express underlying classifications. The results are often taken as bearing especially on the psychology of language. And, when language ability is tested in a near relation like a chimp, the test often involves categorization. As Premack (1976) put it: "Sorting or preverbal classification. . .is an immediate precursor of verbal classification. Probably it should be distinguished from discrimination, which would seem to be a more distant precursor (p. 215)." Among the differences between "classification" and "discrimination," said Premack, is that discrimination is more "stimulus-specific." Chimpanzees and the other higher primates at or over the threshold of language, he suggested, classify; lower creatures discriminate.

Language clearly involves far more than mere categorization (see Ristau & Robbins, 1982 for an excellent discussion), but if categorization is neither as rare, as high, or as linguistic a cognitive capacity as is often supposed, then the anthropocentric bias can lead to two kinds of distortion. It can lead to underestimating the distinctiveness of human cognition and overestimating the distinctiveness of classification. First, it can divert attention from whatever really distinguishes human cognition. If categorization itself is ruled out as a defining feature of human psychology, more specialized hypotheses spring to mind. Is it language with grammar and predication, a large repertoire of categories in a characteristic hierarchical structure, an ability to apply logical structures, etc.? When students of human categorization fail to question how or if animals differ, as they usually fail to do, they risk underestimating the human specializations. On the other side of the coin, the anthropocentric bias leads to overly specialized explanations of categorization. When attention focuses exclusively on what human beings do, theories of classification probably fail to capture its most general features.

This paper is prompted mainly by a concern with the second issue and will therefore attempt to describe categorization in animals, hence in its more general dimensions. But a more general account of categorization should also help forestall the first kind of distortion by setting a baseline for categorization in the absence of language. Many of the issues that have arisen in the study of human categorization arise with animals too, but their implications change when the subjects are mute like pigeons or monkeys. As should become clear, the most challenging puzzles surrounding categorization have more to do with stimulus generalization and discrimination than with the specialized capacities of the human mind.

II. A REPRESENTATIVE CATEGORY

The first experiment I will describe (Herrnstein, 1979) resembles a conventional stimulus discrimination experiment except for the stimuli being discriminated. Its procedure and results are approximately representative of a number of experiments on categorization by animals. Pigeons were shown the same 80, 35 mm slides once per daily session from the beginning of the experiment. Before this, the pigeons had never seen slides. The slides were projected successively in a different order every session. Half of them, the "positive instances," contained fully or partially obstructed views of trees.

The 40 other slides, the "negative instances," contained no trees.   The 80 slides had been picked from a large collection assembled for several category-learning experiments, not necessarily involving trees.   Positive instances sometimes contained trees only incidentally or in the distance. An attempt was made to span a range of difficulty in the visibility of trees, and also to make the positive and negative instances comparable except for the presence or absence of trees, at least to a human observer.   The pigeons saw the slides projected for about 45 sec on a small screen next to the standard pigeon response-key.   Brief food reinforcement came intermittently for pecking if the slide on the screen was positive and not at all if it was negative.

Since the pigeons saw the same 80 slides daily, mere discrimination may not seem like convincing evidence of an over-arching "tree" category. The pigeons could have been learning 80 distinct patterns, 40 of which happen to be positive and 40 negative.   We would like to know if this is what the pigeon, with its relatively small nervous system, is limited to.   An alternative is that the pigeon, like a human observer, sees each slide as representing objects in a three-dimensional space.   The patterns of light, shade, and color would then be seen as aspects of a finite collection of objects, and the pigeon's task is to respond if there is a tree among them. The speed of discrimination suggests the latter interpretation.   The slowest of the four pigeons in the experiment was discriminating significantly by the fifth session, after having seen the 80 slides only four times.   The other three pigeons were discriminating significantly by the second session, having seen the slides only once before.   Occasionally, in experiments like this, pigeons seem to get the idea before the first session is over and start discriminating significantly during the first rotation of the carousel slide tray.

To say the pigeon gets an idea is to say it has an over-arching principle of classification.   Once a category forms, the pigeon knows, as it were, that certain slides are positive, even if they have never been associated with reinforcement.   In this sense, categorization transcends mere induction or, for that matter, mere conditioning.   But, to be useful, a category has to be at least nearly right.   Were the pigeons really looking for trees, or is there some other category or perhaps a set of categories that overlaps with the experimenter's notion of trees?   This proves to be a hard (perhaps unanswerable) question, but we can chip away at it.   We can test to see if the pigeons' principle of classification applies to new instances the way our category of tree does.

After the pigeons were stably discriminating between the original 40 positive and 40 negative instances, they were tested with new slides.   Five positives and five negatives were replaced with five new slides containing trees and five new slides without them, interspersed among the 70 remaining original slides.   Then, a few sessions later, another ten new slides were substituted for ten of the original set.   Would the pigeons correctly classify new positives and negatives on first presentation?   All the pigeons generalized, which means they were sorting slides according to criteria that at least approximated trees for a large proportion of the instances examined.   In terms of rates of pecking, the new positives were ranked as high as or higher than the training stimuli, as if they were generally better exemplars of the pigeons' category, which they might well have been. The new negatives were also ranked slightly higher than the old slides, as if they were less decisively negative. One interpretation of these deviations from perfect generalization is that new slides evoked over-responding, hence higher ranks for both positive and negative instances, but another interpretation, favored by results in experiments to be described later, is that the pigeon's principle of inclusion is a bit more tolerant than ours. It is more likely to misclassify a negative than a positive.

Successful generalization indicates a degree of correspondence between pigeon and human categorization, but it does not prove perfect correspondence. For example, it does not enable us to conclude that pigeons saw trees as a single category or as several categories whose union approximates trees. The pigeons may have been sorting by multiple rules that were redundantly exemplified in our slides.

III.   RISKS AND BENEFITS OF CATEGORIZATION

A benefit of categorization is to define a class speedily, without having to be conditioned to one instance after another. But categorization has risks. Suppose the experimenter had in mind, not trees in general, but some other principle distinguishing the 40 positive and 40 negative instances. Other principles are clearly possible, even if outlandish or unreasonable. A finite sample of positive and negative slides always lends itself to many alternative rules of discrimination. It could have been only the kinds of trees shown in the sample chosen -- oaks, maples, ashes, black spruce, white pine, etc. It could have been certain kinds of trees when the sun was high; different, perhaps overlapping, kinds when the sun was low. It could have taken location into account, or the presence of other objects or coordinations of objects. It could have depended on sequences of stimuli -- oaks at a distance only after no trees at all for two slides. If the experimenter was using rules like these, and the pigeon inferred, simply, trees, then many of the generalization tests would have been failed. Generalization would also suffer if the pigeon extracted those complex rules when all that was needed was to look for trees. Given any finite collection of instances, the "right" principle of categorization is just one of a set of equivalent principles that would divide the initial collection the same way. The right principle is the one that generalizes beyond the initial collection. Since it depends on instances not yet experienced, the right principle can only arise from dispositions already present in the perceiver. In the experiment just described, what the experimenter had in mind converged fairly well with what the pigeon evidently used as a principle of categorizing.

When there is such a convergence, categorization should speed discrimination. This implication was tested directly by Herrnstein and de Villiers (1980) with pigeons sorting underwater photographs taken by a recreational scuba diver for purposes other than to be used in an experiment. One group of pigeons was reinforced for pecking when a slide contained fish and not otherwise. Non-fish photos often contained other underwater creatures -- for example, turtles, shrimp, starfish, and an occasional scuba diver. The pigeons saw the same 80 slides daily in different orders, until discrimination reached a specific criterion of accuracy. A second group of pigeons saw the same slides in the same orders, but the positive category included 20 fish pictures and 20 non-fish pictures, and so did the negative category. For the first group, a fish category would help; for the second group, it would hurt because it would tend to confuse positive and negative instances. Moreover, for the second group there was no apparent correct principle of sorting except to learn whether each stimulus was positive or negative. The first group learned to sort at the criterion level more than twice as rapidly as the second group. The clear instances of fish were sorted easily and well by the first group but with difficulty by the second group, presumably because the pigeons had trouble distinguishing the fish in the positive set from the fish in the negative set. The benefit of forming a generalizable category depends on its being correlated with reinforcement, as fish was for the first group but not the second.

In both of these experiments, pigeons displayed concepts after training with 80 instances, half pro and half con, but smaller numbers may often be sufficient. Research has not yet uncovered how, or even if, subjects are sensitive to the trade-off between the speed in applying a principle of categorization and the risk of applying the wrong one. Malott and Siddall (1972) showed that pigeons required as few as six and no more than three dozen positive and negative instances to discriminate and generalize photos of human faces, torsos, or people in groups.

Cerella's (1979) results on oak leaf discrimination hold the record for speed in generalizing a category. In the first study, pigeons daily saw 40 individual silhouettes of white oak leaves and 40 individual silhouettes of other broad leaves, as illustrated in Figure 14.1. White oak leaves, shown at the top, vary in the number, size, and placement of lobes, among other features. Sycamore, sassafras, and elm leaves, shown at the bottom, also have varying shapes and lobes. Nevertheless, after about two dozen daily sessions with the 80 stimuli, the pigeons were discriminating virtually perfectly and generalized without decrement to 40 new instances of white oak leaves.

Next, Cerella trained another group of pigeons with only a repeated presentation of a single white oak leaf -- the upper middle one in Figure 14.1 -- versus 40 non-oak broad leaves. Learning to the same criterion was about twice as fast, and generalization to 40 new instances of white oak leaves was just as good. Prompted by this remarkable finding, Cerella tested the possibility that the pigeons were not really learning the category of white oak leaves at all, but simply to respond to any leaf except those specifically excluded by having been negative instances. Two new groups were trained, each with the 40 non-oak leaves but one with 40 repetitions of a single positive oak leaf and the other with 40 different positive instances. As before, the former group learned quicker. Both groups generalized without decrement to 40 new negative instances, not responding to them. From this, we can conclude that pigeons extract something like the general shape of white oak leaves from even a single instance when it is contrasted with 40 non-instances.

In the next study, a new group of pigeons started with sessions in

FIG. 14.1. Representative white oak leaves (top) and non−oak leaves (bottom) in Cerella (1979).

which they saw only repetitions of a single white oak leaf but no non−oak leaves, then were tested with that one leaf, with 20 other white oak leaves, and with 20 non−oak leaves. If the pigeons had pecked the screen on which these slides were projected, presumably insuring that they saw the leaves,[1] then they generalized to the new oak leaves but not to the non−oak leaves. The category of white oak leaves was evidently induced from a single instance without any contrasting leaves to compare it with.

Finally, Cerella tried to train four new pigeons to discriminate between repetitions of one particular white oak leaf and 40 other white oak leaves. After more than 12 times as many sessions as it took to learn to discriminate a white oak leaf from non−oak leaves, two pigeons were still responding at chance levels and two were apparently stuck at sub−criterial discriminations. Even though white oak leaves would probably be easily discriminable when seen together, when seen successively they apparently fall in a single category for pigeons. A strong prior disposition to form

---

[1]In the first version of the experiment, the slides were projected on a screen next to the response key; in the test, the pigeons pecked about equally to the three categories, as if they never noticed the slides on the screen. Then, they were retrained pecking at the screen itself while they were shown only repetitions of the single oak leaf. In the retest, pecking to white oak leaves in general unambiguously separated from that to non−oak leaves. It is possible that these results were contaminated by the first, unsuccessful test, even though reinforcement was never associated with anything but the single white oak leaf.

generic classes, like white oak leaf, may be adaptive in nature, where generic rather than individual recognition is needed, but not here.

IV. PATTERNS VERSUS OBJECTS

In another experiment, Cerella (1977) tested the pigeon's ability to see simple figures as representations of solid forms when doing so would be beneficial.   First, the pigeons learned to discriminate between line drawings of squares and random quadrilaterals, an easy task.   Then, squares and random quadrilaterals were replaced by computer-generated projections of cubes in random orientation as the positive class and distortions of the projections as the negative class.   The bottom row in Figure 14.2 illustrates a cube and its distortion.   To distort a cube, each vertex was moved a small, random distance on the plane of the page, resulting in a figure with the same number of lines, vertices, and regions as a cube but lacking its projective geometry.   Shifting from square to cube abolished the discrimination and 150 more sessions of training with the full range of orientations of cube and non-cube failed to restore it.

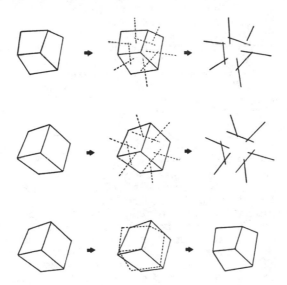

FIG. 14.2. The bottom row shows, on the left, an example of a cube and, on the right, a non-cube, obtained by shifting each vertex of a cube a small, random distance.   The two upper rows show the positive (middle) and negative (top) line tangles, obtained by rotating each line segment in a cube and a non-cube by 90° around its midpoint. (From Cerella, 1977.)

With four new pigeons, Cerella attempted a more gradual exposure to the cube -- non-cube discrimination. First, only repetitions of a single orientation of the cube were used as the positive class while the negative class sampled over all orientations of non-cubes. When discrimination reached a certain level, the set of positive stimuli was enlarged to include additional orientations of the cube, immediate neighbors of the original cube in angular rotation. If discrimination again reached the criterion level, a still wider range of orientations was sampled. The full set of orientations is contained within ±45 degrees of rotation along each axis, which Cerella divided into 12 neighborhoods including the original fixed orientation, totalling over 9000 different views of a cube.

The question was how far the pigeons could progress toward the full set of orientations for the positive class. All four pigeons met the criterion out to ±24° of rotation along each axis, somewhat more than half the set, but none made it to ±45°, the full set. Levels of accuracy and learning speed varied little at each stage until ±24°, then both measures declined even for the decreasing sample of pigeons that discriminated with the larger sampling sets.

Cerella's experiment has not been exactly duplicated with human subjects, but he commented in passing that human subjects mastered his cube problem "immediately and completely." Perkins has shown (Perkins, 1972; Perkins & Deregowski, unpub.) that, by the age of three or so, human beings can reliably discriminate between line drawings of rectilinear and non-rectilinear boxes. The capacity to sort correctly, he found, does not depend on practice with specific stimuli. The occasional misclassifications of the non-rectilinear boxes - false alarms, in other words -- were largely confined to almost rectilinear boxes, while false dismissals -- boxes misclassified -- are evidently spread more or less irregularly over the full range of views. Perkins' experiments confirm the casual impression that we can pretty well recognize a line drawing of a rectilinear parallelopiped such as a cube, without needing specific instruction for each view.

In contrast, Cerella's data argued against a comparable capacity in pigeons. With a final procedure, he should have been able to detect any trace of an over-arching three-dimensional category. The positive and negative stimuli in the preceding experiment were transformed so as to preserve two-dimensional complexity in some sense while disrupting the relation to the third dimension. The upper rows of Figure 14.2 show how each line segment of a negative (above) or positive (middle) stimulus was rotated 90° around its midpoint, producing a tangle of lines conveying no three-dimensionality to human observers.[2] Figure 14.3 shows a subset of orientations of both cubes and line tangles. Except for the change in stimuli, the preceding experiment was exactly repeated with four new pigeons. For human observers, said Cerella, the line tangles constituted a much more difficult, perhaps unlearnable, discrimination, compared to the

---

[2]Note that it is merely a fact, not a necessity, that we fail to see the three-dimensionality of these stimuli. One can conceive of a perceptual system that would detect both the underlying cubical projection and its transformation. To the owner of such a system, our inability to see the cube might look just as pathetic as the pigeon's inability to see the cubical invariance does to us.

cubes and non-cubes they come from.

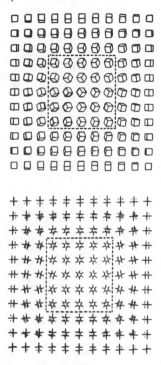

FIG. 14.3. Upper panel shows some of the projections of a cube; lower panel, the corresponding line patterns.   The central projection in each panel was the starting orientation, and the dotted lines enclose approximately the range of orientations successfully learned by the pigeons in each experiment. (From Cerella, 1977.)

Figure 14.4 compares the four pigeons in each experiment, to see if pigeons also find cubes easier than lines.   The upper graph shows how many pigeons mastered the discrimination with the increasing samples of stimuli.   In both experiments all pigeons learned to discriminate positive and negative instances for orientations covering ±20° around the original orientation.   Three of the four pigeons failed to discriminate the line patterns beyond ±20°, but one did so all the way out to ±46°, the full set.   For cubes, no subject learned the full set, but the decline after ±24° was gradual.   No clear difference in discriminating lines or cubes can be read into this measure.

The two other measures in Figure 14.4 also suggest no difference in difficulty.   Approximately the same number of sessions to criterion were taken in both experiments, as the middle panel shows, until the groups started dwindling because the criterion of learning was not being met.   The bottom panel plots a measure of accuracy -- relative frequency of responding to positive stimuli -- which also corresponded quite well in

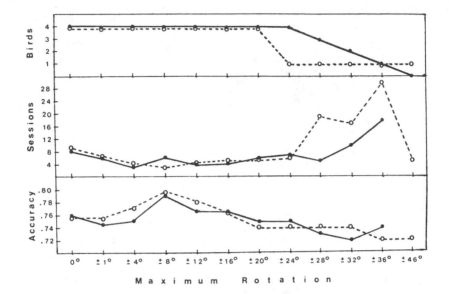

FIG. 14.4. Three measures of performance for pigeons discriminating cubes from non-cubes (solid lines) or the line tangles derived from them (dashed lines). Upper panel shows the number of pigeons out of four that reached criterion of discrimination at successively larger angles of rotation. Middle panel shows the average number of sessions taken to meet criterion for pigeons meeting the criterion. Bottom panel shows the accuracy of discrimination at criterion, as measured by the relative frequency of responding to positive stimuli. (From Cerella, 1977.)

the two experiments. Both the middle and bottom panels indicate a slight improvement with practice until the countervailing effect of the larger sampling set reversed the trend. Peak performance came at about ±8° of rotation in both experiments, which further suggests equivalent difficulty in the two discriminations. Since, by all measures examined, the cubes were just as hard to discriminate as the line patterns, the parsimonious inference is that pigeons see line projections of cubes as just so many lines in a plane.

In other experiments, too, patterns that we see three-dimensionally, pigeons apparently see two-dimensionally. Cerella (1980) trained pigeons to discriminate between equilateral triangles and various other geometrical forms -- e.g., circles, diamonds, and stars -- on a field divided horizontally into a white and black region. The discrimination was learned rapidly and well. Then, a series of probe trials presented either obstructed or partial triangles. Figure 14.5 illustrates a full triangle, a couple of partial "triangles" (which are really four-sided figures) and an obstructed triangle, the difference between partial and obstructed being whether the truncated triangle abutted the black region. Amount of truncation varied, and, as would be expected, pigeons responded more to more nearly complete triangles. However, for every degree of truncation, partial triangles evoked more responding than obstructed. The reverse may be expected for a human observer, who would probably see an obstructed triangle as, in fact, a whole triangle obstructed in an inferred third dimension. For a two-dimensional perceiver, partial triangles probably evoke more responding because their location on the slide more nearly matches that of the complete triangle; obstructed triangles are further to the right.

FIG. 14.5. A complete triangle, on the left; two partial triangles, in the center, and an obstructed triangle, on the right; all examples of the type of stimuli used in Cerella (1980).

Pigeons trained to discriminate pictures of Charlie Brown from other characters in Peanuts -- Snoopy, Linus, Lucy, etc. -- also responded as if the patterns were seen two-dimensionally. After learning with considerable difficulty, the pigeons generalized roughly equally to partial or obstructed views of Charlie. They generalized to scrambled or inverted views, as illustrated on the right in Figure 14.6 (Cerella, 1980). No scrambling more extensive than moving thirds of Charlie up or down was attempted, so we cannot guess about the limiting resolution of the pigeon's category, but it was evidently at least as fine as a third of Charlie Brown. Less than a whole Charlie, as illustrated by the partial figure on the left, was also responded to as if it were a bona fide positive instance. Even though only complete, unscrambled Charlie's were associated with reinforcement, the category extended to various sorts of contextually impossible instances, meaning that they will not turn up in Peanuts nor in the three-dimensional space we infer for it. We see Charlie Brown as an integrated object, but from Cerella's data, it seems that pigeons see Charlie as just a pattern that can be subdivided or inverted without losing any essence.

FIG. 14.6. From left to right, incomplete, scrambled, and inverted positive stimuli, as used in Cerella (1980). (From Cerella, 1982.)

From his findings with oak leaves, cubes, pseudo-cubes, triangles, and Charlie Brown, Cerella (1982) concluded that pigeons are controlled by local features of the two-dimensional patterns, rather than by either more global two-dimensional or by three-dimensional invariances conveyed by nouns like cube or Charlie Brown. The lobes of the oak leaf, Charlie's torso, the angles of a cubical projection or a triangle exemplify the controlling features that Cerella deduced from his results, rather than objects like a whole person or a Gestalt like a triangle or a cube. A somewhat different version of a feature theory will be proposed later, after more of the empirical literature has been reviewed.

At this point, we may wonder whether animals ever see objects in two-dimensional representations, such as photographs, rather than just flat patterns of lines, light, shade, and color. The notion of an "object" is somewhat opaque philosophically, but we can finesse it for our purposes by assuming that objects are what an animal sees when it looks at the three-dimensional world. Does it see anything of the sort in a picture? Despite occasional claims that not even pre-literate people see objects in pictures, the evidence is clear that at least primates do. Viki, the home-reared chimp, spontaneously recognized familiar objects in pictures, according to her foster parents (Hayes & Hayes, 1953). She imitated actions caught by still pictures -- clapping, sticking out her tongue; she put her ear to a picture of a wristwatch; made food barks to a picture of candy, and so on. A discrimination between realistic pictures was learned faster than one between nonsense pictures.

Object-picture, or picture-object, transfers are hard to interpret as anything but evidence that pictures sometimes represent the three-dimensional world. A series of experiments on chimps and orangutans by Davenport and his associates (Davenport & Rogers, 1970, 1971; Davenport, Rogers, & Russell, 1973, 1975) showed not only object-picture transfers, but transfer from touch to object or picture and vice versa, from object or picture to touch. In one experiment (Davenport et al., 1975), chimps successfully matched the feel of an object to a photograph or a drawing of it, after being trained with touch-sight matches for other kinds of objects. A 20-sec delay between the felt and seen stimuli did not abolish discrimination. In a non-verbal species like the chimp, a delayed, cross-modal match like this argues for imagistic, rather than propositional, internal representations of stimuli. It also suggests strongly that the animal sees objects in drawings and photos.

Monkeys, too, apparently see object in pictures. Butler and Woolpy (1963) assessed rhesus monkeys' preferences for photos that were color or black and white, changing or unchanging, in-focus, upright movies or out-of-focus, inverted movies, and so on. Their preferences were consistent with seeing objects, rather than just flat patterns. Butler and Woolpy concluded that the monkeys "are responding to the projected images as being representative of real images (p. 328)." A variety of other ingenious experiments with monkeys (e.g., Humphrey, 1974; Redican, Kellicutt, & Mitchell, 1971; Sackett, 1965; Sands, Lincoln, & Wright, 1982) leave no doubt that Butler and Woolpy drew the right conclusion.

Finally, how about pigeons and other non-primates? It may seem plausible that a generalized concept of tree or fish, and so on, in photographs must be based on some sort of object invariance, but can we prove it? Besides the experiments on categorizing photographic slides that raised the question in the first place, there is far less evidence one way or the other about non-primates because fewer relevant experiments have been done. In one study (Looney & Cohen, 1974), pigeons pecked at the head region of a color photo of a pigeon. The experiment used an aggression-arousing schedule known to induce peck attacks by one pigeon at another, especially at its head. An attack at the head in a photo suggests that the photo looked like a pigeon to the aroused pigeon. In another experiment, Lumsden (1977) showed that, after discrimination training between objects, the generalization gradient for various orientations of the positive object was essentially duplicated for photos or line tracings of the object in various orientations. This, too, suggests that the photos and tracings looked like the object, but there was only one pigeon used in the study. In an experiment in progress in our laboratory, pigeons appear to have learned to sort photos on the basis of being upside-down or right-side-up, a principle that seems to require seeing the photos as representations of objects, not just as so many local features. Photos being seen for the first time are correctly classified as upside-down or right-side up, but probably only if they contain certain relevant cues to orientation which we have not identified with any certainty. Cabe (1976,

1980) concluded that pigeons transfer a discrimination learned with objects to one with photos of the objects, much as Davenport's apes did, but his evidence was not as direct, based as it was on time to reverse a discrimination. But in another experiment (Cabe & Healey, 1979), pigeons failed to generalize from objects to line drawings of them. All told, it is hard to say whether pigeons are poorer at object–picture transfers than primates or just that the evidence is weaker because it is scarcer.

If pigeons can see objects in pictures, as most of the data suggest, then some of Cerella's apparently contradictory results need explaining, particularly the experiment on cubes. It is possible that the pigeon's problem was not so much cubes as line projections. They may have been less stymied by photographs of rectangular blocks in various orientations, which would have provided more cues to the third dimension. In the Charlie Brown experiment, successful learning required that backgrounds be blacked out early in training. Only after Charlie was being discriminated from Lucy, Snoopy, etc. were backgrounds reintroduced. Before that, the pigeons appeared to be having trouble seeing the figure–ground separation needed to isolate the Peanuts characters from the cartoon strip backgrounds. Yet, in an experiment using actual photographs (Herrnstein, Loveland, & Cable, 1976), pigeons learned to categorize instances of a particular woman, as distinguished from other people, without needing any special treatment of backgrounds or anything else about the stimuli. Unposed photographs of a person in any orientation or context, at any distance, wearing a variety of outfits, at all seasons of the year, constitute a much larger set of stimuli than Charlie Brown, and that may be their advantage. A three–dimensional invariant may be induced only when the two–dimensional patterns are sufficiently varied. The problem with Cerella's cubes might therefore have been solved by using stimuli that were more diverse at the two–dimensional level, like photos or realistic drawings.

V. NATURAL CATEGORIES

Another approach to the pigeon's success with photos of trees or people and its difficulty with cubes and Charlie Brown is to contrast natural and artificial categories. Natural categories have been much in the air since it turned out that organisms are not indifferent to where the dividing lines between classes are placed. Human beings, for example, seem to cut the color continuum (Berlin & Kay, 1969) into particular bands, whatever the color names of their language. Not just color divides into natural categories, but other sorts of stimuli too (Herrnstein, 1982). Organisms impose a structure on experience that cannot be derived a priori from an analysis of just the stimuli alone. The imposed structure often allows organisms to categorize easily instances not experienced before, according to rules that have so far eluded physical specification.

By now, this line of thought has become so widely accepted that it may be due for revision itself. Usually we rely on intuition or common sense when we invoke the naturalness of categories. Charlie Brown seems artificial; trees seem natural. But the concept of a natural category resists definition, perhaps because it is a natural category itself. Let us consider several interpretations of naturalness.

Are categories natural because they classify objects in a creature's natural environment, such as trees for pigeons?   It may seem so, but the pigeons used in these experiments vary widely in their personal experiences with the categories being reinforced, and no effect of personal history has been encountered.   In our experiments on the tree category, for example, some of the pigeons came from local breeders, raised in colonies of active homing pigeons and doubtless with plenty of experience with trees.   But other pigeons had been caged all their lives and had little or no tree experience.   Pigeons from both backgrounds categorized trees easily, equally easily as far as we could tell.   The pigeons in Cerella's experiment had never seen oak leaves up close, yet they showed a strong propensity to classify them.

If individual experience does not account for naturalness, could it instead be collective experience registered in the genes?   An individual pigeon may not experience trees, but pigeons in general do.   Like a migration pathway or a mating ritual or other species-specific behavior, could natural categories be inborn principles of categorization?   Our results with pigeons looking for fish in underwater photographs suggests no. Neither individually nor collectively have pigeons experienced today's fish, for it has been at least 50 million years since birds and fish shared an environment, and both kinds of animals have changed considerably in the interim, at least on the average.   Fish come in a variety of shapes and colors comparable to that of trees, but the pigeons learned to classify them anyway and to generalize to new instances.   The discrimination level for the category of fish was slightly lower than for trees, but slightly higher than for bodies of water or an individual person (Herrnstein, Loveland, & Cable, 1976).   If any of these categories counts as natural, all of them do.   In a generalization test for the fish category, fish (positive instances) generalized better than non-fish (negative instances), again suggesting a more permissive criterion by the pigeons than the experimenters, just as occurred with generalization of the tree category. Trees, in fact, generalized better than non-trees even when they were the negative class, not correlated with reinforcement (Herrnstein, 1979).   We can learn something from this recurrent permissiveness (see Herrnstein, Loveland, & Cable, 1976 for additional examples).   If pigeons were categorizing by an internal template based on positive instances, they should be less, not more permissive, for new instances should deviate more on the average from the template than the instances used in the training.   In that case (and assuming no template for negatives), negative instances should generalize better than positive.   Given an internal template mirroring exemplars, generalization should yield more false dismissals than false alarms; the data typically show the reverse.   Natural categories are evidently not templates based in any simple way on the exemplars the subjects see in the experiments.

Figure 14.7 reproduces the three "best" fist and three "best" non-fish slides, as measured by the pigeons' average ranks.   The best positive instances illustrate what to a human observer would be the canonical view of fish, a single fish seen from the side.   Yet, in our slides, canonical views were more the exception than the rule.   Almost half the positive

instances contained more than one fish; fish were often truncated by the
edges of the photograph or by obstructing objects. Usually the fish were
oblique to the line of sight. The photographer was apparently more
interested in coral than fish most of the time, as most positive instances
contained fish only incidentally. It is apparently not just past experience
that makes canonical views special (see Herrnstein & de Villiers, 1980, for
discussion).

FIG. 14.7. Based on pooled rankings by seven pigeons in Herrnstein &
de Villiers (1980), the three best fish (left) and non-fish (right) slides.
(From Herrnstein & de Villiers, 1980.)

Underwater pictures are natural for pigeons neither individually nor
collectively. Yet fish are natural in some sense that may seem, at this
point, to be relevant. They are natural in a way that telephones or
bicycles or city streets are not. Fish, like trees or bodies of water or
human beings, inhabit a natural environment that human intervention has not
re-arranged. Perhaps this is the kind of naturalness that creatures, barring

human, find congenial for classifying.

The evidence reviewed so far may seem to support this interpretation of naturalness. Trees, fish, leaves, persons, and water are learned readily, while line drawings of cubes and Charlie Brown, both man-made, give the pigeons trouble. And, in unpublished work, pigeons had difficulty categorizing photos of chairs, food cups, bottles, or wheeled vehicles. At best, limited and particular instances of those categories were responded to, and, at worst, responding remained close to random for many session. But if we look more broadly at the literature, counter-examples can be found, and if we look more critically at the distinction between natural and "man-made" itself, it tends to dissolve.

Let us take an empirical example first. Morgan, Fitch, Holman, and Lea (1976) trained pigeons to discriminate the printed capital letter A from the printed numeral 2. In training, instances were repeatedly drawn from 18 different typefaces, shown in Figure 14.8, along with positive and negative rates of responding for three pigeons. In every case, responding to the A was higher than to the 2; no particular typeface was consistently hard or easy to discriminate. Generalization to 22 additional, and widely varying, typefaces was virtually complete, and again, no particular typeface consistently generalized better or worse than the others. One pigeon was also tested with handprinted A's and 2's from 10 people. In every case but one, the A evoked more responding than the 2; in the one exception, the person by mistake provided a lower-case cursive a instead of a hand-printed capital. In its learnability and open-endedness, A acts like a natural category, yet it is man-made if anything is.

Though open-ended enough to generalize to novel instances, the A-category is circumscribed enough to have allowed Morgan and his associates to find at least some of its distinguishing features. Such a possibility is not readily available with natural categories, like fish or trees. In a test using mutilated A's and 2's, a fairly clear order in the rates of responding emerged. Such A-like features as an angle with its apex at the top or two projections at the bottom seemed to favor responding. 2-like features, such as curves or loops, favored non-responding. Another test interspersed without reinforcement other letters of the alphabet. Figure 14.9 shows the 25 letters besides A, not in the Helvetica medium used but a similar LeRoy typeface. The letters are arrayed top-down and left-right in decreasing order of responding, with ties shown together. The highest four: R, H, X, and K, have the two bottom projections, and the only other letter that does, M, is not far below. At the other end, six letters were tied with zero responding: C, G, I, J, L, and Z. None has the bottom projections nor the angle at the top characteristic of A's. Also, to casual inspection, these six letters include the most 2-like features, assuming generalization to mirror reversals.

The authors concluded "that no single feature could be called necessary and sufficient (p. 65)" to account for the pigeon's rule for responding. The graded levels of responding yielded by this method suggest that, for the pigeon, being an A is not an all-or-none matter, but something quantitative - a matter of more or less. A more complete study (Blough, 1982) confirmed these results and added that the errors pigeons make in discriminating between letters are consistent with the similarities and difference people report. The pigeon's alphanumeric

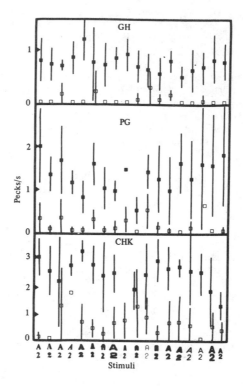

FIG. 14.8. Eighteen typefaces used to train pigeons to discriminate A from 2, along with average rates of pecking and the standard deviations for each of three subjects. Filled squares are for A's; open squares, for 2's. (From Morgan et al., 1976.)

categories seem to be, Morgan *et al.* pointed out, polymorphous, a term (Ryle, 1951) for concepts unified by degrees of overlap rather than by necessary and sufficient elements. They quoted Ludwig Wittgenstein on polymorphous groupings: "if you look at them you will see something that is common to *all*, but similarities, relationships, and a whole series of them at that. . .sometimes overall similarities, sometimes similarities of detail (Wittgenstein, 1968)."

The results and the account of them both fit the pattern of natural categories -- open-ended classes lacking necessary or sufficient criteria for membership. Yet, letters and numbers for pigeons are not natural in the ordinary sense. As a man-made natural category, letters are an empirical counter-example to the importance of the distinction between being natural and being man-made.

Conceptually as well, the distinction has problems. In an unpublished study, pigeons learned to discriminate fairly well between dogs and non-dogs. It seems likely that they could do as well with unicorns and non-

A = positive

R   N   T   P

HX   M   F   Q

K   BU   DV   ES

W   Y   O   CGIJLZ

2 = negative

FIG. 14.9. After learning to discriminate between A's and 2's, two pigeons were tested for generalization with the other 25 letters of the alphabet in a typeface that closely resembles that shown here. The order of average rates of pecking to the other letters are shown with the highest at the upper left and the lowest (i.e., zero) at the lower right. Ties are at the same level. (Derived from Morgan et al., 1976.)

unicorns, given enough exemplars. If so, what would it say about natural categories that they include fictitious objects? Even for real objects, some groupings may be learnable as a category and others, not. In no ordinary sense are unicorns and letters natural and chairs and vehicles not natural.

Morgan et al. (1976) found that pigeons were unable to learn to classify computer-generated random-dot patterns on the basis of symmetry around the vertical axis. A few individual patterns were learned, but not the general category. But using stimuli in which symmetry is much more apparent to a human observer, Delius and Habers (1978) succeeded in training pigeons to discriminate bilateral symmetry from asymmetry and to generalize to new examples. In addition to simpler stimuli, Delius and Habers used simultaneous presentations of a positive and a negative instance while Morgan et al. used the more standard, and more difficult, successive discrimination procedure. Bilateral symmetry, an aspect of many animals found in nature, is apparently harder for pigeons to learn than, say, the man-made letter A, which, note, happens to be bilaterally symmetrical also.

Poole and Lander (1971) successfully trained pigeons to discriminate portrait photos of various breeds of pigeons from those of other birds, non-bird animals, and inanimate objects. In generalization tests, the pigeons responded more to pictures of breeds of pigeons not shown before than to pictures of other bird species or other animals or objects.   Most pigeon breeds are man—made, not found in nature except for human intervention.   Yet, after learning to classify a few breeds, pigeons generalized the principle to other breeds.   More generally, all the taxonomic categories of biology are man—made.   Categories like trees, fish, people, and so on, are human inventions, one system of classification out of any number of possible ones.   Other systems may yield categories pigeons cannot learn, yet they would be no less natural than ours.

None of the obvious senses of naturalness hold up.   Natural categories may not be common in a creature's experience, have no special status in its species' history, and they can be man—made.   The one thing about them we can be sure of, oddly enough, is that they are classifications that interest psychologists and others because they defy simple specification. They seem to be easy to use but hard to characterize.   It may not seem too helpful to attack a riddle by noting how tough it is to solve, but enumerating the reasons for the difficulty may be a start.   Here are three reasons:

A. FLEXIBLE FEATURES

In Morgan *et al.* (1976) the critical features for responding to A's were the dual bottom projections and an acute angle at the top.   Other elements may also have contributed, such as the cross bar in A's, H's, and, in certain typefaces, R's.   These were not the features found by Lubow (1974), who trained pigeons to distinguish, in black and white aerial photographs, between man—made objects such as buildings or streets, and natural settings such as woods or fields.   The training was at least partially successful and generalized with decrement to new instances.   The controlling features appeared to be straight lines and right angles and, to a lesser extent, high contrast regions.   The manifest features of each categorization are evidently tailored for the task at hand.

Using Barbary doves as subjects, Hrycenko and Harwood (1980) established a discrimination between the two forms shown in Figure 14.10.   Generalization tests presented seven different forms, also shown in the figure.   Various plausible stimulus dimensions were examined to see which controlled responding to the generalization forms.   The main controlling dimension was a measure of compactness, the length of the perimeter divided by the square root of the area.   It can hardly be an accident that the index of compactness was minimal for the O and maximal for the I—like forms.   In generalization, responding was most fully accounted for by a stimulus dimension that maximized the distance between the two forms discriminated in training.   Another important dimension was a measure of complexity, the sum of the length of the perimeter and the number of independent line segments, which was also minimal for O and maximal for I. Dimensions like area, which did not separate the O and the I to any great extent, did not much control responding in generalization.

If area alone differentiated between positive and negative instances in a discrimination, it would doubtless become a decisive dimension.   Pigeons

TRAINING SHAPES

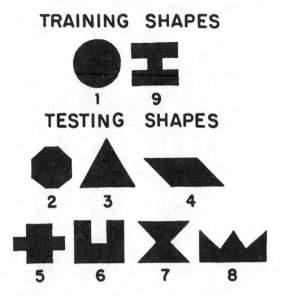

TESTING SHAPES

FIG. 14.10. Training and testing shapes used for Barbary doves. (From Hrycenko & Harwood, 1980.)

would, for example, use area (or some close correlate) as a dimension in discriminating between large and small green circles. A significant dimension in one setting may fail to be a controlling dimension in another, as the results of these few experiments demonstrate. Consequently, the features that identify an object's membership in one class may come from dimensions that have little to do with the dimension involved in some other class. Common sense leads to the same conclusion. The features that identify a bird for us probably consist of feathers, beaks, wings and so on. But feathers, beaks and the like play no role in deciding if a tree is an elm or if a car is a Buick. Features can be small, as in the shape of a fish crow's tail feathers or the subtle cues that allow us to discriminate between identical twins; they can be large -- a thundercloud in the sky, a bus coming around a corner. They can also be relatively simple physically, like the yellow bill that helps identify a black bird as a starling. Or they can themselves defy precise physical specification, like the leaves on a tree. If leaves are features of trees, then what are the features of leaves?

Theories of categorization often assume, in effect, something like an alphabet of features (Tversky, 1977), out of which a given discrimination problem spells, as it were, the relevant categories. This conception neglects an essential step in the process, one which greatly increases its analytic complexity. The conventional feature .theory presupposes a repertoire of features, but the evidence suggests an earlier stage in the categorizing process. Categories are composed, not of fixed features, but of features dependent on the category being composed. A better metaphor than an alphabet is to picture the categorizing process as a kind of perceptual carpentry, shaping the features the way a carpenter shapes the elements of a chest of drawers.

The carpenter metaphor is chosen to focus on the idea of shaping. Reinforcement is known to "shape" responses out of a stream of movement. The response class may be narrow or broad, highly specified in topography or only loosely so, depending on the contingencies of reinforcement. There may be constraints on operant conditioning, but they are secondary compared to the plasticity of movement under the control of reinforcement. But besides shaping response topographies, reinforcement appears also to shape perceptual features. From the available stimulus dimensions, differential reinforcement selects those that differentiate positive and negative instances. The resulting features are then likely to be woven into a polymorphous rule for categorizing -- the next topic.

B. POLYMORPHISM

Figure 14.11 (adapted from Dennis, Hampton, & Lea, 1973) shows two groups of clusters, the Y's and the N's, separated according to a principle expressible in ten words. The principle involves only the dimensions of shape, shade, and symmetry, each of which comes in only two values -- black or white, round or triangle, and asymmetrical or symmetrical. Even with the clues, and in spite of its logical simplicity, this remains an extraordinarily difficult rule for people to formulate, apparently because no feature is necessary or sufficient.

The rule is: Y's contain at least two of symmetry, blackness, or circularity. Because no feature is necessary or sufficient, the rule is polymorphous. In experiments on Cambridge University undergraduates, Dennis and his associates found polymorphous rules far more difficult to extract explicitly or to generalize to new instances than conjunctions (and relations) or disjunctions (or relations). Indeed, if subjects solved the problem at all, they usually bypassed the simple polymorphism and came up instead with an unwieldy disjunction of conjunctions. The classification in Figure 14.11 could, for example, be formulated as: Y's are black and circular, or black and symmetrical, or circular and symmetrical, or black, circular, and symmetrical.

For conjunctive rules, features are necessary but not sufficient; for disjunctive rules, features are sufficient but not necessary. For polymorphous rules, no feature is necessary or sufficient, but each contributes incrementally to membership. Membership thus becomes like having a quantitative value greater than whatever happens to be the prevailing criterion. In this example, circularity, symmetry, and blackness could each be assigned a value of, say, 1.0 while other attributes earned 0.0. The criterion for selection would then be a value of at least 2.0. As simple as it seems, such rules are difficult to spot and the difficulty in extracting them helps explain the inscrutability of natural categories, which are evidently polymorphous, as philosophers in the tradition of Wittgenstein have convincingly argued.

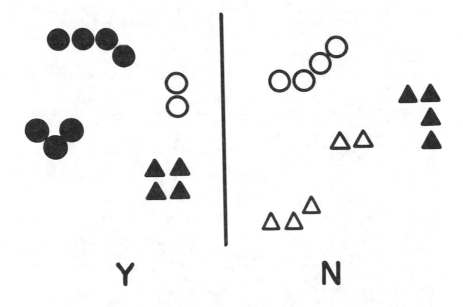

FIG. 14.11. A polymorphous rule separates the Y's from the N's. See text. (Adapted from Dennis et al., 1973.)

The everyday rules of classification are polymorphous because reinforcement in nature is correlated with objects, rather than stimuli. From an invariant object, the stimuli reaching the organism on different occasions are likely to vary polymorphously. Presumably, if polymorphous rules define a set of exemplars that resemble each other perceptually, then they are mastered without having been analyzed into their elements. We can assume that this explains how the pigeons in Lea and Harrison's experiment (1978) mastered a classification problem logically equivalent to the one exceeding the analytic powers of the Cambridge undergraduates in Dennis *et al.* (1973). The biological solution to the complexities of polymorphism appears to be dimensions of similarity or generalization.

C. GENERALIZATION

Jenkins and Harrison (1960, 1962) trained pigeons to peck in the presence of a 1000 Hz tone, then tested for generalization to other frequencies and also to silence. They found no generalization decrement, not even to silence; pecking occurred at more or less the same rate whatever was or was not sounding. The pigeons were apparently not listening. A second group of pigeons received initial training that differed in one detail -- pecking was still reinforced in the presence of a 1000 Hz tone, but it was unreinforced in silence. The pigeons learned to discriminate between a tone, which happened to be 1000 Hz, and silence; they must have been listening. In generalizing afterwards, the gradients were approximately symmetrical against the logarithm of frequency. For each relative distance above or below 1000 Hz, there was about the same decrease in responding. Finally, pigeons were required to discriminate between a 1000 Hz tone and a 950 Hz tone, which they learned to do. Once again, they were tested for generalization to a wide range of frequencies. After reinforcement for listening to a small change in frequency, the gradient fell off much more rapidly than before. Now small changes in frequency resulted in large changes in responding.

These three procedures illustrate how the stimulus dimension in control is selected and shaped by contingencies of reinforcement, from not listening at all to listening to quite fine shifts in frequency. That much is obvious, but two other points may be less so, and these additional considerations help explain why natural categories resist analysis. First, the particular shapes of generalization gradients depend not just on reinforcement, but also on the inherent characteristics of the sensory system. Discriminating between 1000 Hz and silence may activate a frequency gradient of some sort, but not necessarily that it be logarithmic. For colors (Blough, 1961), the gradients tend to be asymmetrical against log frequency. The difference may be due to a difference between color perception, with its relatively sharp transitions between primary hues and pitch perception, which lacks comparably sharp transitions (Herrnstein, 1982). Reinforcement contingencies select among perceptual dimensions and also select a level of resolution on a dimension, but the dimensions probably have built-in textures that reinforcement does not control. The interaction between reinforcement and generalization is where innate factors contribute to the categorical structure. On the response side, the comparable constraint is the inherent organization of topographies -- such as the relative integrity of a pigeon's peck or the tendency to generalize movements across bilateral structures, such as the two human hands. In both cases, the contribution of the organism sets limits on how much can be accounted for by reinforcement alone.

The second point is that perceptual dimensions do not always correspond to a physical dimensions, as pitch almost does to frequency. A neat correspondence between a perceptual and a physical dimension is ideal for experiments on generalization, but it is probably a rarity, relatively speaking. The psychological dimensions of sound, for example, are more than two-dimensional, more than just a matter of pitch and loudness, with their physical dimensions of frequency and intensity. Sounds vary in perceived density, volume, noisiness, consonance, and so on, each of which is multidimensional in physical terms (Guirao & Stevens, 1964; Stevens, 1960, 1971; Terrace & Stevens, 1962). Yet each perceptual dimension could in principle be differentially associated with contingencies of reinforcement. When subjects generalize from, say, one white oak leaf to another, or from one picture of a body of water to another, they are showing proximity along perceptual dimensions whose physical description may be obscure, complex, or both. Because animals possess so many dimensions[3] for generalizing along, categorization cannot be accounted for by the few dimensions with simple physical correlates most often studied in the psychological laboratory.

VI. OVERVIEW

Human language may depend on categorization, but the evidence shows clearly that categorization does not depend on language. Generalization of discriminative stimuli shades into categorization as the contingency of reinforcement is associated with actual objects (or representations of them), rather than the usual abstracted lights or tones of psychological research. The complexity and variability of natural stimuli uncovers a capacity to see through stimulus variation even in relatively simple organisms. Aside from the flexibility of features, categorization may, in fact, not even depend on an ability to learn, for, in principle, instinctive reactions may also be under the control of open-ended polymorphous classes of stimuli. To categorize, which is to detect recurrences in the environment despite variations in local stimulus energies, must be so enormous an evolutionary advantage that it may well be universal among living organisms. Seen in this light, categorization is just object constancy, which is perhaps the fundamental constancy toward which all other perceptual constancies converge. Rather than psycholinguistics, it is the psychology of perception and stimulus discrimination that impinge most directly on categorization.[4]

Since categorization is widely dispersed among species, with no obvious variation in the ease with which it is accomplished, it follows that the underlying physiological structures must either be reasonably general and elementary or that there have been many physiological solutions to the object constancy problem. The two alternatives are not mutually exclusive. Categories that are fairly easy to describe physically do not seem to be significantly easier for animals, and may even be harder, than categories that are hard to describe in those terms. For example, pigeons appear to

---

[3]There is an extreme version of this point, holding that any differential reinforcement procedure establishes an appropriate dimension for categorization, which is to say that organisms are limitlessly malleable by reinforcement on the stimulus side. The fact that subjects, in fact, fail to learn in certain categorization tasks, or even that some tasks are easier to master than others, refutes this extreme hypothesis. The evidence against it is at least as strong as the now-discredited equipotentiality hypothesis for the reinforcement of responses (Seligman, 1970). There may be many potential dimensions for categorization in any given modality, but probably not a limitless number of them.

[4]A fuller treatment would have to deal with subsidiary issues. Nothing in the analysis requires the subject to use only a single psychological dimension or a single prototypical representation. Some fish, for example, look like snakes, others look like bats. A subject may induce multiple prototypes, and treat approximations to each as positive. Also, perception should not be taken narrowly, as if it were entirely localized in sense receptors. What is intended is the entire system dealing with categorization, including inferential stages that may be more relevant to higher organisms than to pigeons. (Herrnstein & de Villiers, 1980, discuss inferential factors affecting the category of fish for human beings but not for pigeons.)

find patches of colored light a harder category to form than photographs of trees (Herrnstein, 1979). That no man-made machine behaves this way suggests that existing machines differ in some basic way from living organisms, as far as categorization is concerned. The likeliest difference is in the perceptual dimensions themselves, which are hard to unearth and probably even harder to simulate with machines.[5] The standard dimensions of physics, taken one at a time, do not begin to exhaust the dimensions entering into categorization by animals or humans.

None of the issues arising from the question of categorization has been resolved. At this point, the task is simply to see what the issues are: when and how differential reinforcement shapes the features entering into a polymorphous rule; what the perceptual dimensions are and how they can be physically represented; what the physiological structures are, and how they might have evolved.

ACKNOWLEDGMENTS

Grant No. IST-8100404, from NSF to Harvard University, supported the preparation of this paper and the new research reported in it. A shorter version was presented at the 1982 meeting of the Eastern Psychological Association. Comments by John Cerella, Herbert S. Terrace, and William Vaughn were most helpful and are much appreciated.

REFERENCES

Berlin, B., & Kay, P. *Basic color terms: Their universality and evolution.* Berkeley: University of California Press, 1969.

Blough, D. S. The shape of some wavelength generalization gradients. *Journal of the Experimental Analysis of Behavior*, 1961, *4*, 31-40.

Blough, D. S. Pigeon perception of letters of the alphabet. *Science*, 1982, *218*, 397-398.

Butler, R. A., & Woolpy, J. H. Visual attention in the rhesus monkey. *Journal of Comparative and Physiological Psychology*, 1963, *56*, 324-328.

Cabe, P. A. Discrimination of stereometric and planometric displays by pigeons. *Perceptual and Motor Skills*, 1976, *42*, 1243-1250.

Cabe, P. A. Picture perception in nonhuman subjects. In M. A. Hagen (Ed.), *The perception of pictures.* (Vol. 2.) New York: Academic Press, 1980.

Cabe, P. A., & Healey, M. L. Figure-background color differences and transfer of discrimination from objects to line drawings with pigeons. *Bulletin of the Psychonomic Society*, 1979, *13*, 124-126.

---

[5]A comparable gulf separates natural and machine locomotion. Birds inspired man-made flight, but airplanes with flapping wings are not the outcome. The most common machine mode of ground travel is wheels; few, if any, animals move on wheels. Underwater locomotion also evolved differently for animals and machines. Categorizing machines may be useful, just as locomoting machines are, but it is probably futile, if not misleading, to think of one as a model of the other.

Cerella, J. Absence of perspective processing in the pigeon. *Pattern Recognition*, 1977, *9*, 65-68.

Cerella, J. Visual classes and natural categories in the pigeon. *Journal of Experimental Psychology: Human Perception and Performance*, 1979, *5*, 68-77.

Cerella, J. The pigeon's analysis of pictures. *Pattern Recognition*, 1980, *12*, 1-6.

Cerella, J. Mechanisms of concept formation in the pigeon. In D. J. Ingle, M. A. Goodale, & R. J. W. Mansfield (Eds.), *Analysis of visual behavior.* Cambridge, Mass.: MIT Press, 1982.

Davenport, R. K., & Rogers, C. M. Intermodal equivalence of stimuli in apes. *Science*, 1970, *168*, 279-280.

Davenport, R. K., & Rogers, C. M. Perception of photographs by apes. *Behaviour*, 1971, *39*, 318-320.

Davenport, R. K., Rogers, C. M., & Russell, I. S. Cross modal perception in apes. *Neuropsychologia*, 1973, *11*, 21-28.

Davenport, R. K., Rogers, C. M., & Russell, I. S. Cross-modal perception in apes: Altered visual cues and delay. *Neuropsychologia*, 1975, *13*, 229-235.

Delius, J. D., & Habers, G. Symmetry: Can pigeons conceptualize it? *Behavioral Biology*, 1978, *22*, 336-342.

Dennis, I., Hampton, J. A., & Lea, S. E. G. New problem in concept formation. *Nature*, 1973, *243*, 101-102.

Glass, A. L., Holyoak, K. J., & Santa, J. L. *Cognition*. Reading, Mass.: Addison-Wesley, 1979.

Guirao, M., & Stevens, S. S. Measurement of auditory density. *Journal of the Acoustical Society of America*, 1964, *36*, 1176-1182.

Hayes, K. J., & Hayes, C. Picture perception in a home-raised chimpanzee. *Journal of Comparative and Physiological Psychology*, 1953, *46*, 470-474.

Herrnstein, R. J. Acquisition, generalization, and discrimination reversal of a natural concept. *Journal of Experimental Psychology: Animal Behavior Processes*, 1979, *5*, 116-129.

Herrnstein, R. J. Stimuli and the texture of experience. *Neuroscience and Biobehavioral Reviews*, 1982, *6*, 105-117.

Herrnstein, R. J., & DeVilliers, P. A. Fish as a natural category for people and pigeons. In G. H. Bower (Ed.), *The psychology of learning and motivation.* (Vol. 14.) New York: Academic Press, 1980.

Herrnstein, R. J., Loveland, D. H., & Cable, C. Natural concepts in pigeons. *Journal of Experimental Psychology: Animal Behavior Processes*, 1976, *2*, 285-302.

Howard, J. M., & Campion, R. C. A metric for pattern discrimination performance. *IEEE Transactions on Systems, Man, and Cybernetics*, 1978, *8*, 32-37.

Hrycenko, O., & Harwood, D. W. Judgments of shape similarity in the Barbary dove (*Streptopelia risoria*). *Animal Behavior*, 1980, *28*, 586-592.

Humphrey, N. K. Species and individuals in the perceptual world of monkeys. *Perception*, 1974, *3*, 105-114.

Hunt, M. How the mind works. *New York Times Magazine*, 24 January 1982, 30ff. (a)

Hunt, M. It's smart to have crazy ideas. *Self Magazine*, March, 1982, 63–64. (b)

Jenkins, H. M., & Harrison, R. H. Effect of discrimination training on auditory generalization. *Journal of Experimental Psychology*, 1960, *59*, 246–253.

Jenkins, H. M., & Harrison, R. H. Generalization gradients of inhibition following auditory discrimination learning. *Journal of the Experimental Analysis of Behavior*, 1962, *5*, 435–441.

Lea, S. E. G., & Harrison, S. N. Discrimination of polymorphous stimulus sets by pigeons. *Quarterly Journal of Experimental Psychology*, 1978, *30*, 521–537.

Looney, T. A., & Cohen, P. S. Pictorial target control of schedule–induced attack in White Carneaux pigeons. *Journal of the Experimental Analysis of Behavior*, 1974, *21*, 571–584.

Lubow, R. E. High–order concept formation in the pigeon. *Journal of the Experimental Analysis of Behavior*, 1974, *21*, 475–483.

Lumsden, E. A. Generalization of an operant response to photographs and drawings/silhouettes of a three–dimensional object at various orientations. *Bulletin of the Psychonomic Society*, 1977, *10*, 405–407.

Malott, R. W., & Siddall, J. W. Acquisition of the people concept in pigeons. *Psychological Reports*, 1972, *31*, 3–13.

Morgan, M. J., Fitch, M. D., Holman, J. G., & Lea, S. E. G. Pigeons learn the concept of an "A". *Perception*, 1976, *5*, 57–66.

Perkins, D. N. Visual discrimination between rectangular and nonrectangular parallelopipeds. *Perception & Psychophysics*, 1972, *12*, 396–400.

Poole, J., & Lander, D. G. The pigeon's concept of pigeon. *Psychonomic Science*, 1971, *25*, 157–158.

Premack, D. *Intelligence in ape and man.* Hillsdale, N.J.: Erlbaum, 1976.

Redican, W. K., Kellicutt, M. H., & Mitchell, G. Preferences for facial expressions in juvenile rhesus monkeys (*Macaca mulatta*). *Developmental Psychology*, 1971, *5*, 539.

Ristau, C. A., & Robbins, D. Language in the great apes: A critical review. In J. S. Rosenblatt, P. A. Hinde, C. Beer, & M.–C. Busnel (Eds.), *Advances in the study of behavior.* (Vol. 12.) New York: Academic Press, 1982.

Rosch, E., & Lloyd, B. B. (Eds.) *Cognition and categorization.* Hillsdale, N.J.: Erlbaum, 1978.

Ryle, G. Thinking and language. *Proceedings of the Aristotelian Society, Supplement*, 1951, *25*, 65–82.

Sackett, G. P. Response of rhesus monkeys to social stimulation presented by means of colored slides. *Perceptual and Motor Skills*, 1965, *20*, 1027–1028.

Sands, S. F., Lincoln, C. E., & Wright, A. A. Pictorial similarity judgments and the organization of visual memory in the rhesus monkey. *Journal of Experimental Psychology: General*, 1982, *111*, 369–389.

Seligman, M. E. P. On the generality of the laws of learning. *Psychological Review*, 1970, *77*, 406–418.

Smith, E. E., & Medin, D. L. *Categories and concepts.* Cambridge, Mass.: Harvard University Press, 1981.

Stevens, S. S. The psychophysics of sensory functions. *American Scientist*, 1960, *48*, 226–253.

Stevens, S. S. Perceived level of noise by Mark VII and decibels (E). *Journal of the Acoustical Society of America*, 1971, *51*, 575–601.

Terrace, H. S., & Stevens, S. S. The quantification of tonal volume. *American Journal of Psychology*, 1962, *75*, 596–604.

Tversky, A. Features of similarity. *Psychological Review*, 1977, *84*, 327–352.

Wittgenstein, L. *Philosophical investigations*. Translated from the German by G. E. M. Anscombe. Oxford: Blackwell, 1968. Third edition.

15   IN WHAT SENSE DO PIGEONS LEARN CONCEPTS?

S. E. G. Lea
University of Exeter

I. INTRODUCTION

Much recent work on complex discrimination learning in animals makes reference to "concepts." But what do we mean when we talk about a non-human animal using, learning or acquiring a concept? The present paper attempts to give a coherent answer to this question, and to indicate what sort of behavioral data we might collect in order to see whether non-humans do possess concepts. In other words, I am going to attempt to give behavioral meaning to the concept of "concept."

Since we mostly think of concept formation as a characteristically human activity, it would be natural to start an investigation of animal concept formation with the animals that seem to be nearest to us in phylogenetic terms, the great apes. The arguments for starting with them are the stronger in that conceptualization is, as we shall see, closely linked to language, and everyone at this meeting will have firmly in mind the studies of language acquisition, or purported language acquisition, by great apes. Somewhat perversely perhaps, I am not going to talk about apes, but about pigeons instead. The reason for this is simple: I know a little about pigeons and next to nothing about apes. . . but more seriously, the comparative psychologist's original good reasons for working with "simple animals" do apply with some force here. The manifest "intelligence" of great apes offers some traps for the unwary: it is all too easy to attribute to them human-like mental processes just because they look like humans. The kind of investigation I intend demands a rigorous adherence to Lloyd Morgan's canon. In these circumstances there is something to be said for working with an animal whose intelligence is unlikely to be overrated by anyone. The laboratory pigeon fits this bill rather precisely.

II.   CONCEPT DISCRIMINATIONS

A.   CURRENT RESEARCH ON ANIMALS' CONCEPTS

1.   Verbal concepts versus logical concepts.   Two kinds of investigation of non-human animals' learning have been said to involve the use of concepts.   One kind concerns abstract concepts like number, symmetry, or, the one that has been studied in most detail, sameness (see, for example, Chapter 22, by Zentall, Hogan, & Edwards, this volume; also Delius & Habers, 1978; Gillan, Premack, & Woodruff, 1981).   The other kind involves complex perceptual discriminations.   The archetype of these latter experiments is the demonstration by Herrnstein and Loveland (1964) that pigeons could be trained to discriminate pictures containing people or parts of people from pictures that did not.   Most such experiments have involved visual discrimination by pigeons, although there is some closely related work on auditory discrimination of phonemes by chinchillas (e.g., Burdick & Miller, 1975).   As is now well known, pigeons have been taught to discriminate pictures according to the presence or absence in them of such objects as pigeons, oak leaves, artifacts, fish, the letter "A," and a variety of other visual patterns.

2.   Characteristics of visual concepts used in current research.   It is with this second type of "concept formation" problem that I am concerned here.   The discriminanda in these experiments generally have two properties.   First, all the positive stimuli would be labelled by a human subject (and, more specifically, are labelled by the experimenter) with a single name, e.g., "person," "artifact," etc.   Secondly, there is no obvious perceptual property that all positive stimuli (and/or all negative stimuli) have in common.   For example, it is not the case that all the positive slides in Herrnstein and Loveland's experiment had a lower average brightness than any of the negative slides.   So in order to explain pigeons' capacity to solve such discrimination problems, we have to postulate some learning capacity beyond what is needed to solve, say, the simple red/green or horizontal/vertical discriminations that are the stock-in-trade of the operant laboratory.   In passing we might note there is no obvious a priori reason why a red/green discrimination should be simpler than a person/non-person discrimination.   We are accustomed to think it so, but that says more about our habits of thought than about animal -- or human -- powers of discrimination.   Around 8% of the male population suffers from aberrancies of discrimination between red and green lights, after all, yet this is credibly claimed to have gone unnoticed until the time of Dalton.   If a comparable proportion had suffered from comparable aberrancies of discriminating people from non-people, I feel the fact might have become widely known with somewhat less delay.   Nonetheless, for the time being I am going to stay within the prejudice that discriminating persons from non-persons, for example, is a more difficult task than discriminating red from green, and that it therefore requires a more advanced discrimination process.

B. CONCEPT DISCRIMINATION: A DEFINITION

But what is the extra (or alternative) discrimination process? When we refer to these complex discrimination experiments as involving "concept formation" or "higher order concept formation," we beg the question of what is to be discriminated, and the question of how the subject organism achieves the discrimination. Now, the question of the discriminanda is relatively simple. They certainly are defined in terms of a concept -- a human concept, existing in the mind of the experimenter. (Strict behaviorists are please not to object to my giving the experimenter a mind. I mean this only in the sense in which we talk of the "mind" in the language of common discourse, and my meaning is as much or as little susceptible to behaviorist explication as any other lay use of the word "mind.") But if the discriminanda clearly involve a concept, the mechanism of discrimination just as clearly might not. Most of this paper will be concerned with the alternative, non-conceptual accounts of that mechanism. To keep matters clear, therefore, I propose that we should refer to studies of the type pioneered by Herrnstein and Loveland as "concept discrimination experiments." This terminology deliberately leaves open the question of whether the subject uses a concept to make the discrimination between the stimuli concerned, while emphasizing that the stimuli really are defined in terms of a concept -- the one held by the experimenter. Note that I shall try to avoid the phrase "concept formation" altogether: it seems to me that it is used too loosely. Some concept discrimination tasks no doubt can be said to involve "concept formation," in that if the subject uses a concept at all, it must be one that it did not have prior to the experiment. An example would be Herrnstein and De Villiers' (1980) demonstration that pigeons could discriminate fish from non-fish. But in other cases the most obvious guess is that, if the pigeon is capable of having concepts, it will possess the concept concerned before we ever start the experiment: all we do is to teach it to attach certain responses to a category it can already recognize. This seems very likely to be the case for the "person" concept, and even more so for the "pigeon" concept studied by Poole and Lander (1971). The term "concept discrimination" is neutral between these two interpretations.

III. MECHANISMS OF CONCEPT DISCRIMINATION

A. RECOGNITION OF INDIVIDUAL STIMULI

With the terminology, let us hope, cleared up, let us turn from the question of what is to be discriminated to the question of how it is discriminated. There are two obvious ways in which an organism might come to perform accurately on a concept discrimination. The first is to learn how to respond to every one of the stimuli to discriminated.

   1. Practicality of individual stimulus recognition. To recognize every instance of a stimulus class defined in terms of a human concept may sound like an impossible feat, but in fact the possibility needs to be taken very seriously (see Chapter 20, this volume, by Wright, Santiago, Sands, & Urcuioli). Although the number of possible pictures of "a person" or "not a person" is indeed infinite, the number actually used in any particular experiment is likely to be finite and rather small. Most experimenters use automatic slide projectors to present their stimuli, and these typically hold only a limited number of slides, 80 or 100. Unless a fresh set of slides is used for each session, we have to consider the possibility that the subjects simply learn, separately, how to respond to each one of the 80 or 100 stimuli they see. Bearing in mind that the response to be learned is very simple (peck or don't peck), this seems perfectly plausible. Recent results obtained in Herrnstein's laboratory (Greene, in press) have shown that pigeons are capable of learning to respond differentially to slides that differ only in the tiny differences of detail introduced by taking two pictures of the same scene, with the same camera position, in rapid succession. In comparison the task of coding 80 quite distinct pictures as positive or negative seems almost trivial. Of the existing experiments on concept discrimination, probably only that of Herrnstein, Loveland, and Cable (1976) comes close to avoiding this interpretation: they changed the slides they used each day, using a total pool of about 800 slides each of which was apparently only seen by the pigeon on 10–15 trials in the entire experiment.

2. The generalization test. Traditionally, experimenters have not felt it proper to describe their results as "concept formation" unless concept discrimination generalizes to new instances of the (experimenter-defined) concept. Doesn't this rule out any possibility that subjects have simply learned to respond to individual training stimuli? Unfortunately, it does not. The "new" instances will, inevitably, be somewhat similar to the training stimuli, and simple stimulus generalization will therefore be enough to ensure that they will be correctly discriminated. Of course this account begs certain questions about the origin of such generalization. In the language of pattern recognition theory, this kind of learning corresponds to a template matching account of categorization, and such accounts notoriously have difficulty in explaining generalization to new instances. These difficulties are not necessarily insuperable, but the means needed to overcome them have an important bearing on the general questions at issue here: I will return to this point below.

B. FEATURE THEORY

The alternative account of concept discrimination is phrased in terms of common features of the positive and/or negative feature sets. As was stated above, it is of the essence of the concept discrimination task that there should be no single, simple, perceptual feature common to all members of the stimulus set, and a feature-learning account must deal with this in one of two ways. It follows that there are two different versions of feature theory.

1. Single features. The first possibility is to suggest some thoroughly abstruse single common feature. A popular suggestion is some property of the spatial frequency spectrum of the stimuli, following the proposal of Campbell and Robson (1968) that Fourier analysis in the spatial domain may play some part in visual pattern recognition. For example, in the analogous problem of human face recognition, it has been suggested that the low—spatial frequency portion of the spectrum may have some special role (Frisby, 1979, gives this as an explanation of the way blurring makes Harmon's 1973, Abraham Lincoln figure recognizable). This kind of proposal has all the usual attractions of the mathematically difficult, but that should not be allowed to hide the fact that it is empirically empty. If anyone wants to believe that there is some unique property of the spatial frequencies of pictures of, say, pigeons, the onus is on him or her to produce it. Until that is done, no one else is obliged to take the suggestion seriously.

2. Multiple features. But there is a much more serious version of the common—feature account. This was adopted explicitly by Herrnstein *et al.* (1976), and by Morgan, Fitch, Holman and Lea (1976). It postulates that, though no single feature can account for discrimination, the subject can solve the problem by using some conjunctive, disjunctive or additive combination of features. Discriminations that are deliberately set up to be soluble in these terms are fairly readily tackled by both human and pigeon subjects (Bruner, Goodnow, & Austin, 1956; Lea & Harrison, 1978). Probably most people working on pigeons' concept discriminations believe that this is how the pigeons do solve these problems, or at least we did until Greene's recent data, mentioned above, shook our faith in anything but the merest pattern learning (see also Vaughan & Greene, in press).

Unlike the hypothesis that animals learn about the individual stimuli in a concept discrimination task, the hypothesis that what is learned is a set of features can immediately explain the fact that new instances are correctly discriminated; provided the new instances contain the right combination of features, they should be responded to just as vigorously as the stimuli used in training.

3. Features and similarity. I said earlier that, if we suppose that individual stimuli are what is learned, we must go on to specify a means of stimulus generalization. For without some such mechanism, an individual—stimulus theory gives no account of transfer to new instances of the concept. Now, stimulus generalization has to be restricted, or maximal, to stimuli that are "similar" to the training stimuli. But what defines similarity between two stimuli that happen both to be instances of a particular concept in the experimenter's mind? The simplest possibility, by far, is that two stimuli are similar (whether or not they are both instances of the same concept) in so far as they have features in common. Thus, the idea of feature analysis (see Chapter 19, this volume, by Riley) immediately crops up even if we suppose that concept discriminations are solved by the subject learning to respond to every individual stimulus.

4. Feature analysis versus feature learning. Note, though, that what is required here could be feature analysis without feature learning. That is, at least with discriminands defined in terms of human natural—language concepts (like "person" or "fish"), it is conceivable that a naive pigeon approaching the problem with a fixed set of ways of analyzing the perceptual world, and fixed saliences for different features of it, might see new instances of experimental concept as being similar to training stimuli unconditionally.   Artificially constructed stimulus sets are likely to be needed to investigate whether "acquired distinctiveness" of features is a necessary part of concept discrimination.   (Note the echoes of the disputes of previous decades in animal discrimination learning here.)   For example, Blough's experiments on letter confusions, reported in Chapter 16 of this present volume, suggest that saliences may be fixed, in that he was able to obtain coherent confusion space and clustering data from a situation where numerous different discriminations were trained one after another, so that with an "acquired distinctiveness" argument one would have expected cue saliences, and hence the basis for confusibility, to vary from subtask to subtask.

If a concept discrimination is learned in terms of a feature analysis, however, that necessarily implies that feature saliencies have changed as a function of experience: for it is usually easy to see some other task, which the subjects could perfectly well have been set instead, which would require a quite different set of features to be treated as salient.

C. SUMMARY OF CONCEPT DISCRIMINATION MECHANISMS

To summarize what I have said so far, therefore, we have that:

1. There are two possible mechanisms by which animals might learn concept discriminations of the type pioneered by Herrnstein and Loveland.   These are learning each individual stimulus, and learning about features of the stimuli that are correlated (not necessarily perfectly) with membership in the experimenter-defined concept classes.

2. If individual stimuli are learned, we have to specify a stimulus generalization mechanism to account for response to new instances of the concept.   The clearest candidate is a measure of features in common.   This might or might not involve the saliencies of individual features changing in order to facilitate discrimination.

3. If features of stimuli are used to make the concept discrimination, then feature saliences must be adjustable in the light of experience: I am calling this "feature learning" to contrast it with "feature analysis" which could involve fixed saliences.

IV. BEYOND CONCEPT DISCRIMINATION

A. FEATURES AND CONCEPTS

Many psychologists, particularly those with a broadly behavioral or experimental orientation, have been inclined to take feature learning as a sufficient criterion for saying that an animal has acquired a concept. It is this tradition that lies behind the use of the term "higher order concept formation" to describe what I have called concept discrimination. The idea is that since even simple discriminations involve feature learning, they also involve concept formation; complex discriminations must clearly therefore involve some more elaborate kind of concept. Explicit statements of the equivalence of feature analysis and concept formation are not easy to find, but methodologies that assume it are common, even in human experimental psychology; for example, see the mathematical study of concept formation (Levine, 1975), or the experiments of Rosch and her colleagues (Rosch & Mervis, 1975).

B. THE MENTALISTIC APPROACH TO CONCEPTS

However, there is also a persistent tradition that something more than feature analysis is needed if we are to talk in terms of concepts.

1. Verbal expression of a rule. Many investigators working with human subjects wait for a verbal expression of the rule defining a concept, even when behavior clearly indicates concept discrimination (Bruner *et al.*, 1956, Chapter 4; Dennis, Hampton, & Lea, 1973; Smoke, 1932). But though this provides some evidence for a general feeling that concept formation is something more than feature learning, it is no help in the present context; pigeons cannot talk, which is precisely the reason for being interested in their concept formation.

2. Mental manipulation. Other investigators have proposed other criteria for saying that subjects have concepts. Piaget and Inhelder (1969, p. 44) oppose a "constructive structuration" to "mere abstraction and generalizations from experience": the latter is obviously another way of describing feature analysis, while it is the former that Piaget and Inhelder see as the essential feature of conceptualization. Geach (1957, p. 12 ff.) flatly refuses to use the word "concept" unless the organism concerned can discuss its own concepts. Note that this goes beyond the common requirement for verbal expression of the concept: what Geach wants is that the subject should be able to manipulate the concept mentally.

3. Mental structure. Obviously, opinions of this sort tend to come from psychologists whom we should class, probably with their agreement, as mentalists. Is there any way in which we can give a behavioral sense to their assertions? I do not pose this question out of any personal distaste for mentalism as such, though Piaget and Inhelder's criterion does strike me as obscure, while Geach is plainly obscurantist. Yet I too feel that concept formation involves something more than feature learning, and I should like to be able to express that feeling in anyone's theoretical language. Clearly the traditional test of concept discrimination, transfer to new instances, is inadequate for the task of giving a behavioral sense to this "something else," since such transfer can be accounted for in terms of feature analysis of stimuli, possibly without even specifying that subjects learn what cues are relevant.   But is it possible to improve upon the transfer test?

To me, what the mentalists seem to be saying is that to say of a person that he or she has a certain concept is to say that he or she has some unique mental structure which is active when and only when an instance of that concept is presented in the external, physical environment, or when associated concepts are active in the mental environment. It is this idea of a special mental structure that needs to be given a behavioral sense.

C. A POSSIBLE PHYSIOLOGICAL IMPLEMENTATION

It may be easier to see what is meant by a "unique mental structure" if we consider the physiological substrate which may, perhaps, underlie it. Recall the phrase "higher order concept," often used to describe natural concepts like "person" or "fish."   In terms of physiology, what this calls to mind is the punningly-titled "grandmother cell" hypothesis current in visual physiology (for one of the earliest and most authoritative versions of this, see Barlow, 1972; for a more recent and perhaps more readable version, see Frisby, 1979, Chapter 5). According to this notion, simple perceptual features are extracted at low levels of the visual system -- in frogs, in the eye itself; in cats, in the first cortical cells reached by incoming visual information.   At this stage, the visual machinery can "recognize" orientations, line lengths, perhaps corners. . . the stuff of simple discrimination learning tasks, in fact.   But recognition of more elaborate objects requires the putting together, through a hierarchy of conjunctively and disjunctively triggered neuronal "gates," of many such simple features. At the top of this hierarchy (hence, in the position of a grandmother in a family tree of neurones), a level of complexity is reached at which a cell is triggered only on the presentation of some unique (but varying) object, defined by a disjunction of conjunctions of features -- for example, one's grandmother.

Clearly this grandmother cell gives one possible physical implementation to the mentalists' idea of a unique mental structure corresponding to a unique concept. If we want to produce a behavioral implementation of comparable attractiveness, we need to ask what properties of the grandmother cell idea make it a satisfying implementation of our intuitive mentalistic notions.   Its chief merit seems to be that it provides for an autonomous representation of Granny. We can see that such a cell would have, as it were, a life of its own: learning to find one's way to Grandmother's house, for example, might involve setting up neuronal connections between the Grandmother cell and similar higher order cells for, say, woods and wolves.   Such connections would not necessarily imply any connection between individual features of grandmother and individual features of woods or wolves. "Knowing" that Grandmother lives in the wood would not imply any general association between green leaves and grey hair, for example -- indeed, a system that forced such an association would clearly be inefficient.

D. INSTANCE-TO-CATEGORY GENERALIZATION AS A TEST FOR THE EXISTENCE OF A CONCEPT

A behavioral description of concepts must also capture this autonomy. Before I am prepared to say that a pigeon has a concept of "person," I want to know whether what I teach it about one instance of "person," say a tall man wearing a white coat, it will generalize to another instance of the same concept, say a short woman wearing a maroon sweater -- and that it will generalize in this particular way rather than to all white objects (e.g., refrigerators or laboratory doors), or all objects 179 cm tall. This looks like the traditional transfer test used in most concept discrimination experiments, but it is not. The essence of what I propose is that the concept should be more than the sum of its component features, or component instances. As we have seen, simple generalization to new instances can be guaranteed by feature learning, or by feature-mediated stimulus generalization following learning of individual instances. What I am suggesting is that we should look at generalization in the reverse direction, from single instances to the concept as a whole.

1. A general experimental design. the kind of experiment I propose as a test for the existence of concepts in this sense has three stages.

1. We teach the subject a concept discrimination in the usual way, and establish what the features are that make this discrimination possible (presumably there will be several of them).

2. EITHER   We establish some new response for which a single instance of the concept is a discriminative stimulus.

    OR   We establish some new response for which a single feature, relevant to the original concept but not sufficient to define it, is a sufficient discriminative stimulus.

3. We examine the subject's response on the new task to other instances of the concept or to other features relevant to it (i.e., to instances or features not used in the second stage).

Both versions of this experiment are needed if we are to be sure that any apparent concept learning is not the result simply of learning how to respond to many different instances or to many different features.

2. *Generalization from a single instance.*   It is probably easiest to see the point of this with the help of an idealized example.   Consider first the case where we give new training on a single instance of the original stimulus set.   Suppose that we start with 26 stimuli, which we can label A. . .Z, and that there is some human concept of which A. . .M are instances while N. . .Z are not.   By prior experimental work, however, we establish that for a naive pigeon, the amount of stimulus generalization from A to any one of B. . .M is, on average, the same as the amount of generalization from A to any one of N. . .Z, and that this is true for all 26 stimuli (this is, after all, an idealized example!).   We now train a pigeon to the point where it is discriminating the two groups of stimuli adequately, with (let us say) A. . .M as positive stimuli and N. . .Z as negative stimuli.   This completes Part 1 of the experimental design described above.   Now we need to take some new task, and train the pigeon so that just one of the previously positive stimuli (A, say) has a different significance.   For example, we might train a partial reversal, such that A now became a negative stimulus, and Z a positive stimulus (this version of the design obviously owes something to Kendler and Kendler, 1968).   This completes Part 2.   Finally, we retest the original discrimination.   In the case where Part 2 involves a reversal, we would ask: does the reversed significance of the particular instances of the concept used in Part 2 generalize to the other instances?

Let us suppose that such generalization of reversal does occur.   What does that tell us?   *Ex hypothesi,* it cannot be due to simple stimulus generalization: we established (in this world of ideal experiments) that stimulus generalization from A to other instances of the concept (B. . .M) was no greater than to non-instances (N. . .Z), and the same is supposed to be true of Z.   So if there is now preferential generalization from A to B. . .M, and from Z to N. . .Y, it seems that this generalization must have been mediated by some structure that is peculiar to the pigeon in question, by virtue of its prior training in concept discrimination, in Part 1 of the experiment.   This is the autonomous mental structure which we have been seeking as the essence of possessing a concept.

3. *Generalization from subsets of relevant features.*   Unfortunately, that conclusion is not quite watertight.   There is a simpler possibility for something that might change "within" the pigeon as a result of concept discrimination training.   If the pigeon acquires the concept discrimination through a process of feature learning, that will change the pattern of stimulus generalization between the instances and non-instances.   If features that tend to discriminate between the two groups of stimuli "acquire distinctiveness," that will mean that even if stimulus generalization from A to B. . .M was the same as to N. . .Y before discrimination training, this need not be true afterwards.   Therefore we also need the second version of the experiment, which is, unfortunately, not quite so easy to describe, although the principle is essentially the same.

Consider one of the "artificial polymorphous concepts" used by Dennis *et al.* (1973) and Lea and Harrison (1978).   In these experiments, stimulus sets were constructed by using a number of bipolar features.   For example, we can construct 32 stimuli from 5 features, and these can be denoted by the binary numbers 00000 to 11111, where each binary digit represents the value in that particular stimulus of one of the features. Now let us designate as positive any stimulus in which three or more features are at the "1" level, and designate as negative any stimulus of which this is not true.   This will give us 16 positive stimuli and 16 negatives. We now teach some pigeons to discriminate these positive and negative stimuli; this is an artificial concept discrimination task, and it constitutes Part 1 of our experimental design.   Lea and Harrison showed that pigeons could learn two out of three discriminations of this general class, and in unpublished work my colleague Catriona Ryan and I have started teaching pigeons the three out of five case, though in the unideal world it proves to be quite difficult to construct suitable stimuli.

In Part 2 of this variant of the design, we want to give one of the features (not an individual instance) of this concept a new significance. Again it will be convenient to do this by partial reversal. Let us take the following stimuli, all of which were positive, and make them negative:

01110 01101 01011 00111 11110 11101 11011 10111

and the following stimuli, all of which were negative, and make them positive:

10001 10010 10100 11000 00001 00010 00100 01000

In this partial reversal, the correlation of all features except the first with reinforcement is reversed.   The correlation of the first feature with reward is zero, so there is no information in the stimulus set which would enable the subject to know whether or not it has changed its significance. This completes Part 2 of the design.

Now we can proceed to test for generalization to the critical feature. To do this we can use the following stimuli:

10011 10101 10110 11001 11010 11100

00011 00101 00110 01001 01010 01100

whose status as positive or negative depends upon the significance of the first feature.   If we find that a pigeon trained in this way now treats the first six of these stimuli as negative, and the last six as positive, we can claim that the reversal learning given in Part 2 has generalized to the crucial feature.   With appropriate counterbalancing, we can ensure that there is no unconditional basis for such interfeature generalization.   If it occurs, therefore, it must be due to something that lies within the pigeon; in a word, to a concept.

This is much more nearly a watertight demonstration of a concept. There is one problem however, which was brought out by Sanders (1971) in connection with a much simpler partial reversal design, of the Kendler and Kendler type.   There is one thing that the five features in the artificial example just given have in common other than a possible concept, and that is their correlation with reward.   Could not our pigeon simply learn, in the partial reversal phase, that "whatever was good is now bad; whatever was bad is now good?" It is possible to think of experiments that might test this account, but it is not possible to make them simple, or to describe them simply. At this stage I suggest that we need to look for preliminary evidence in favor of concepts, using the kind of design I have outlined here, rather than try to solve a further set of problems which will only be relevant if the present type of experiments give positive results.

   4. An experiment in progress. I do not intend to talk about data in this paper. But to give you an idea of how this works out in practice, I might mention briefly some experiments that Catriona Ryan and I are carrying out. In these experiments, pigeons were taught wholly artificial "concepts," of which the "instances" were letters of the alphabet -- ten letters forming the positive set, and ten the negative set. The acquisition data from some of these experiments, and the method of getting at the underlying features, are described briefly in Lea and Ryan (in press). The second stage of the experiments, training on a new task, consists simply of giving reversal training on a subset of the training letters. We can them look to see whether this reversal generalizes to the non-reversed letters, and to features of the letters that were irrelevant to the reversed discrimination, or retained their pre-reversal correlation with reinforcement during it. So far the results show unambiguously that generalization to non-reversed letters occurs, but we have yet to obtain clear evidence for generalization to non-reversed features. But nor do we have evidence against it. We have a technique for testing transfer to non-reversed features using our letter stimuli, but it depends heavily on elaborate statistical analyses, and, frankly, it needs to be a lot more convincing than the data are so far. For unambiguous data on feature relevance, we now think that we must use artificial stimuli whose feature content can be controlled, as in the idealized experiment described above. Experiments with these new stimuli are still at the stage of training in the original concept discriminations.

E. ARE THERE ANY "SIMPLE" DISCRIMINATIONS?

This brings me to a final point. I have suggested that we only refer to "concepts" when there are no simple single perceptual features on which a discrimination could be based. That is consistent with the philosophical discussions of the polymorphous nature of everyday concepts (e.g., Wittgenstein, 1968) from which many of us who work in this area draw inspiration. But it is not to say that simpler discriminations may not also depend upon concepts. I believe that I have a concept of, say "green," and I am quite prepared to believe that a pigeon has such a concept too. It is just that at present I see no way to demonstrate that when a pigeon (or an earthworm: cf. Gea, 1957, p. 12) discriminates green from another color, it does so by means of a concept. Perhaps the only way of demonstrating that would be to show that complex discriminations are always mediated by such mental autonomous entities. Any apparent parsimony in treating single-feature discriminations differently might then disappear, and we should be free to consider the pigeon's discrimination even of something so simple as color or line orientation as being mediated by a concept.

REFERENCES

Barlow, H. B. Single units and perception: A neuron doctrine for perceptual psychology? *Perception*, 1972, *1*, 371–495.
Bruner, J. S., Goodnow, J. J., & Austin, G. A. *A study of thinking*. New York: Wiley, 1956.
Burdick, C. K., & Miller, J. D. Speech perception by the chinchilla: Discrimination of sustained /a/ and /i/. *Journal of the Acoustical Society of America*, 1975, *58*, 415–427.
Campbell, F. W., & Robson, J. G. Application of Fourier analysis to the visibility of gratings. *Journal of Physiology*, 1968, *197*, 551–556.
Delius, J. D., & Habers, G. Symmetry: Can pigeons conceptualize it? *Behavioral Biology*, 1978, *22*, 336–342.
Dennis, I., Hampton, J. A., & Lea, S. E. G. New problem in concept formation. *Nature*, 1973, *243*, 101–102.
Frisby, J. P. *Seeing*. Oxford: Oxford University Press, 1979.
Geach, P. *Mental acts*. London: Routledge and Kegan Paul, 1957.
Gillan, D. J., Premack, D., & Woodruff, G. Reasoning in the chimpanzee: I. Analogical reasoning. *Journal of Experimental Psychology: Animal Behavior Processes*, 1981, *7*, 1–17.
Greene, S. L. Figure–ground relations in concept formation. In M. L. Commons, R. J. Herrnstein, & A. R. Wagner (Eds.), *Quantitative analyses of behavior, Vol. IV: Discriminative processes*. Cambridge, Mass.: Ballinger, 1984. In press.
Harmon, L. D. The recognition of faces. *Scientific American*, 1973, *229*, 71–82.
Herrnstein, R. J., & DeVilliers, P. A. Fish as a natural category for people and pigeons. In G. H. Bower (Ed.), *The psychology of learning and motivation*. (Vol. 14.) New York: Academic Press, 1980.
Herrnstein, R. J., & Loveland, D. H. Complex visual concept in the pigeon. *Science*, 1964, *146*, 549–551.
Herrnstein, R. J., Loveland, D. H., & Cable, C. Natural concepts in pigeons. *Journal of Experimental Psychology: Animal Behavior Processes*, 1976, *2*, 285–302.
Kendler, H. H., & Kendler, T. S. Mediation and conceptual behavior. In K. W. Spence & J. T. Spence (Eds.), *The psychology of learning and motivation*. (Vol. 2.) New York: Academic Press, 1968.
Lea, S. E. G., & Harrison, S. N. Discrimination of polymorphous stimulus sets by pigeons. *Quarterly Journal of Experimental Psychology*, 1978, *30*, 521–537.
Lea, S. E. G., & Ryan, C. M. E. Feature analysis of pigeons' acquisition of discrimination between letters. In M. L. Commons, R. J. Herrnstein, & A. R. Wagner (Eds.), *Quantitative analyses of behavior, Vol. IV: Discriminative processes*. Cambridge, Mass.: Ballinger, 1984. In press.
Levine, M. A. *A Cognitive theory of learning*. Hillsdale, N.J.: Erlbaum, 1975.

Morgan, M. J., Fitch, M. D., Holman, J. G., & Lea, S. E. G. Pigeons learn the concept of an "A". *Perception*, 1976, *5*, 57–66.

Piaget, J., & Inhelder, B. *The psychology of the child*. London: Routledge and Kegan Paul, 1969. Translated by H. Weaver.

Poole, J., & Lander, D. G. The pigeon's concept of pigeon. *Psychonomic Science*, 1971, *25*, 157–158.

Rosch, E., & Mervis, C. B. Family resemblances: Studies in the internal structure of categories. *Cognitive Psychology*, 1975, *7*, 573–605.

Sanders, B. Factors affecting reversal and nonreversal shifts in rats and children. *Journal of Comparative and Physiological Psychology*, 1971, *74*, 192–202.

Smoke, K. L. An objective study of concept formation. *Psychological Monographs*, 1932, *42(191)*, 1–46.

Vaughan, W., & Greene, S. L. Acquisition of absolute discriminations in pigeons. In M. L. Commons, R. J. Herrnstein, & A. R. Wagner (Eds.), *Quantitative analyses of behavior, Vol. IV: Discriminative processes*. Cambridge, Mass.: Ballinger, 1984. In press.

Wittgenstein, L. *Philosophical investigations*. Translated from the German by G. E. M. Anscombe. Oxford: Blackwell, 1968. Third edition.

16   FORM RECOGNITION IN PIGEONS

Donald S. Blough
Brown University

> *In a sense, there is only one problem of psychophysics,*
> *namely, the definition of the stimulus.   In this same*
> *sense, there is only one problem in all of psychology --*
> *and it is the same problem. . .   The complete definition*
> *of the stimulus. . . involves the specification of all the*
> *transformations of the environment, both internal and*
> *external, that leave the response invariant (Stevens, 1951,*
> *p. 31).*

I. INTRODUCTION

When we study concept formation or perception, we seek to define the
stimulus.   The typical experiment on concept formation confronts the
subject with transformations of the environment, marks some as "correct,"
and records the emergence of an invariant response.   The typical
experiment on perception confronts the subject with transformations of
the environment and attempts to classify those that leave the response
invariant.   The main difference in these experiments is the degree of
control over the definition of the stimulus that is attempted by the
experimenter -- relatively much for concept formation, relatively little for
perception.   If your subject largely dictates the classification of stimuli (as,
say, "near" or "bright" or "tree"), you are studying perception, and you try
to find out what stimulus variables control the classification that your
subject generates: the "cues to depth," the stimuli affecting brightness, or
the variables that define "tree."

It has turned out to be very hard to define visual forms in objective
terms.   This is a classic problem for the study of form perception; it is
exemplified in this book by the research of Herrnstein (see Chapter 14),
but it is by no means restricted to "natural concepts."   In the past 30
years, steps toward stimulus specification have been attempted through
application of information theory, more recently though feature analysis,
and more recently still through spatial frequency analysis.   Despite these
attempts, it is still not possible, in general, to state rules that enable us to
predict which physical entities will be classified by an observer as a
particular form or object.   Until it is, we shall probably continue the
questionable practice of defining form stimuli by our own classificatory
responses: "the stimulus was a triangle centered on the response key"
(meaning, "what looks like a triangle to me was projected on the response
key").

If we are largely stuck with defining forms by our own invariant responses to ill—defined conjunctions of stimulus events, we may at least try to understand the degree to which our classifications agree with those of other species. Similarities and differences in such classifications may help us to understand the processes underlying form recognition; a comparative approach to perception has a number of other advantages as well (see Cabe, 1980, for a review). For the pigeon, the basis for cross-species comparison is already well established in visual psychophysics, for we know a lot about the pigeon's basic visual functions -- color vision, acuity, flicker sensitivity, and the like. Though distinctive in certain respects (e.g., pigeons see ultraviolet light) these functions are in many ways similar to those of humans (cf. Donovan , 1978). Thanks to Herrnstein and his colleagues (see, for example, Chapter 14, this volume, by Herrnstein) we know that pigeons easily learn to recognize people, trees, water, and even fish. This paper suggests how the gap between simple psychophysics and picture perception might be bridged by applying analytical procedures to the perception of relatively simple patterns.

II. LETTER RECOGNITION

To teach pigeons the letters of the alphabet seems an odd pastime, calculated only to enrage ethologists. However, it does have some merit. If any set of stimuli uncovers human—pigeon differences, this one might, since the forms were devised by humans and are overlearned by them, while for pigeons they are novel and seem to have no ecological validity whatever. Further, a number of sets of human data on letter discrimination are available. Most important, letters are relatively simple forms, so it seems feasible to describe them in objective terms, using, for example, features or spatial frequencies. Letters are also diverse enough that animals might classify them in many different ways; the classification that an animal actually generates may thus be relatively informative, and may potentially be correlated with an objective description.

Three pigeons were set the task of learning to discriminate each letter of the alphabet from all other letters. I cannot recount the full method here (see Blough, 1982); briefly, the study went like this. The letters were learned one at a time. A target letter (for example, a "B") appeared on the face of a small TV monitor on each of 675 test trials per day for four days. On a given trial, this letter was behind one of three glass keys, while another "distractor" letter (say, a "T") appeared behind both of the other keys. The letters were small, only 2mm high. The pigeon was rewarded for about 1/10 of its "correct" pecks to the target letter. A single peck terminated the trial and the target appeared with a new distractor pair about 1.5 sec later.

After four days of training with a given target (e.g., "B") the bird went on to a new one (e.g., "W"). Of course, it did poorly on the new task at first, especially when the old target appeared as a distractor -- for example, "W B B." Hence, the final data analysis used only the last two of the four days for a single letter and entirely excluded trials with the previous target (B, in this example). The three birds learned the letters in different orders.

The primary datum was the percent correct for each two-letter comparison (reaction times turned out to be uninformative). For each letter pair there were two percents -- one with each member of the pair as "target." It is important to note that these percents were collected during acquisition. That is, after four days the birds were still far from asymptote on many of the pairs. This tended to spread out the percentage scores, which averaged just above chance (33%) for the most similar letters and above 90% for the most different letters. The experiment thus says something about recognition during formation of the "letter concept": this could be different than, say, similarity judgements involving fully formed concepts.

Each bird yielded a data matrix showing percent correct for each letter paired with all other letters. These matrices were folded, combining the two measures for each letter pair, and averaged. The resulting mass of numbers was rendered meaningful by feeding it to two scaling programs (cf. Shepard, 1980). One of these, ALSCAL, placed the letters into a multidimensional space where similarity is represented by relative proximity. The two-dimensional result appears in Figure 16.1.

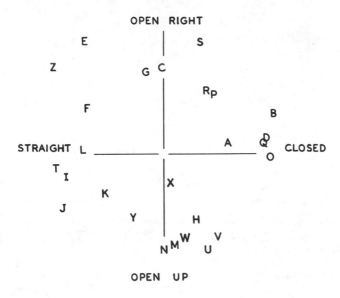

FIG. 16.1. Two-dimensional representation of letter similarity produced from error data by the ALSCAL procedure. Generally, where two letters are far apart the pigeon made few errors when discriminating them and where close together, few errors. Ideally, the closest pair would have the most errors, the next closest the next most, and so on. (From Blough, 1982.)

Figure 16.1 seems to make some sense, though "dimensions" are not clearly evident; the labels on Figure 16.1 only roughly characterize the forms nearby. Nonetheless, this picture looks much like similar pictures based on human data (e.g., Podgorny & Garner, 1979). Of course, three, four, or more dimensions provided a better fit to the data, but the results are very hard to depict.

However, neither pigeon nor human letter recognition data fits very comfortably in a space with visualizable dimensions. CLUSTER, the second procedure employed, is less constraining. It groups together letters that show similar patterns of confusion with other letters. The result is pictured in Figure 16.2.

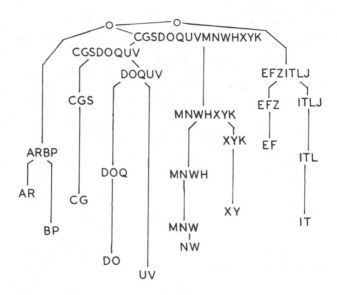

FIG. 16.2. Hierarchical clusters of letters produced from error data by the CLUSTER procedure. The distance up the ordinate is related to the similarity of letters in clusters at that level. For example, since the UV cluster is the lowest, U and V are the most similar letters, according to this algorithm. The letters were formed of dots, and are not exactly represented here. (From Blough, 1982.)

CLUSTER groups letters according to computed "distances," but it is roughly true that the lower the position of the cluster, the more confused are its letters. Thus, U and V were the most often confused with each other, and they form the lowest cluster. Going up the figure, as the distance allowed for joining a group gradually increases, more letters join and the clusters become larger. Though the letters used in the experiment were made of dots and had a stiff computerized look, human observers generally judge letters in the same group to be similar. Features common to groups suggest themselves; for example, "small closed loop" helps characterize the leftmost group.

Quantitative comparisons support such intuitive judgements.   Podgorny and Garner (1979) asked human subjects to rate the similarity of letter pairs, using letters of the same dot-matrix format that I used with the pigeons.   The correlation between the (folded) table of similarities and the pigeon error matrix was 0.68.   This correlation is actually higher than any I have found between data sets from human subjects.   For example, Podgorny and Garner also obtained reaction times while their subjects discriminated letters from each other; the correlation between the judged similarities between pairs of letters and the time required to discriminate those letters was 0.59.

All of this suggests that some common processes determine letter recognition in pigeons and people.   Just what these processes might be remains uncertain.   Possibly they are quite "low level"; they may have to do with common features, for example, or spatial frequency analysers.   Some attempts have been made to describe human letter discrimination in terms of features (Gibson, 1969), but there is some chance that data from studies of the present sort may be more amenable to such analysis than human data currently available.   For example, human similarity judgements may be confused by a mixture of strategies or past learning effects, as when human subjects note that a "B" "contains" a "P."   Discriminability data from humans may be distorted because, in order to get enough errors, the experimenter often must degrade the forms with visual noise or high-speed presentation.

III.   CONTEXT EFFECTS: "PATTERN SUPERIORITY"

The letter experiment just described exemplifies one approach to the analysis of form perception: classification of related forms.   A second approach explores the interaction between parts of stimulus patterns.   The Gestalt psychologists stressed that perception is inherently relational; they provided many demonstrations that local recognition is affected by events nearby in space and time (Kohler, 1947).   Cognitive psychologists have recently rediscovered this pervasive reality and have provided a new list of demonstrations, now often classified as "context effects."   Some of the new demonstrations seem more amenable to systematic study than the classic examples, and the "pattern superiority effect" may be one of these. A study performed by Mai-Ran Cheng and me investigated instances of this effect in pigeons.

The pattern superiority effect is said to occur if simple stimuli, such as lines of differing orientation, are more easily discriminated when they are embedded in a form than when they appear alone.   The key to the demonstration is that the embedding form by itself adds no discriminative information.   Cheng and I used the set of forms shown in Figure 16.3. The only difference between the forms in the upper and lower rows is a right-angle element common to the forms in the lower row.

Our pigeons discriminated between two of the lines in the upper row of Figure 16.3 and also between the corresponding forms in the lower row.   For example, a bird might have to discriminate the first form in the upper row from the third form; this performance would be compared with

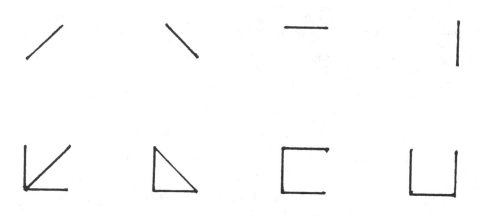

FIG. 16.3. Forms used in the pattern superiority experiment. The lower row differs from the upper only in that the same right-angle form is added to each line. Despite the lack of information provided by the added form, the forms in the lower row are easier to discriminate than those in the upper row.

the discrimination of the first and third forms in the lower row.

We used the same subjects and apparatus that were used in the letter recognition task just described. The procedure was similar, in that each trial provided one S+ target and two S− nontargets. However, the pigeon worked on two discriminations each day, one a "line" discrimination and one a "pattern" discrimination. Two "line" forms were chosen for each bird from those depicted in the upper row of Figure 16.3, together with the two corresponding forms from the lower "pattern" row. One line and its corresponding pattern were designated targets (S+), the others nontargets (S−). As in the letter experiment, the target appeared behind one key and the non-targets appeared behind both of the other keys. There were 600 trials per day, 300 "line" trials and 300 "pattern" trials, in randomized blocks of 24 trials. The birds worked for two days on a given set of forms; two birds completed four sets of forms and one did five.

This was an easy discrimination for the birds; they typically responded well above chance on both discriminations on the first day of training. The data combined over the two training days show a highly significant difference in favor of the "pattern" forms. The only two instances in which "line" gave fewer errors seemed clearly due to negative transfer; the positive stimuli from the previous set had become the negative stimuli for the cases in question. Since the pattern was better learned before the stimulus shift, it was associated with greater negative transfer after the shift. Interestingly, there was no sign of negative transfer when previous non-targets were made targets. This is consistent with other research indicating that pigeons often learn little about negative stimuli in a discrimination (e.g., Carter & Eckerman, 1975; Zentall, Edwards, Moore, & Hogan, 1981; also see Chapter 22, this volume, by Zentall, Hogan, & Edwards). All the birds also showed a highly significant difference in reaction times for the two types of targets; they responded more quickly to the patterns than to the lines.

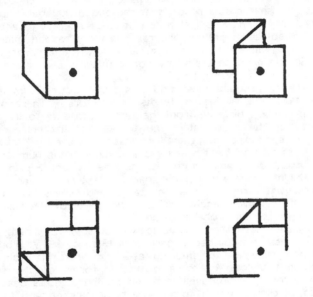

FIG. 16.4. As in Figure 16.3, the forms in each row here differ only in the position and angle of a single line. Human subjects find the "objects" in the upper row easier to discriminate than the forms in the lower row. Pigeons find them equally hard.

We also tried to see if pigeons display an "object superiority effect." Here again, forms differ by the position of a single line. Some studies with humans have shown that discrimination is facilitated if the forms can be seen as simple three-dimensional objects. We used stimuli that had provided this effect in a study by Weisstein and Harris (1974). The pairs of stimulus sets used are shown in Figure 16.4. After from 6 to 8 days of training on a single set, the birds barely showed significant learning of the discrimination, and only one bird showed a significant difference between the two tasks -- in favor of the "flat" forms! This outcome suggests a failure to see the simple line drawings in depth, a finding consistent with other results (Cerella, 1977; Cabe, 1980). Experiments in progress in our lab strongly suggest that forms whose surfaces are indicated by solid colors, rather than simple outlines, do yield an object superiority effect in pigeons.

IV.   CONTEXT EFFECTS: VISUAL SEARCH

The pattern superiority effect has attracted interest partly because it seems surprising that adding "irrelevant" information to a display enhances discriminability.   Of course, the circumstances of this enhancement are quite special: human subjects report that the added material is incorporated into a new unitary form.   As just noted, it is of interest to see when and how such perceptual integration occurs.   However, added material almost always interferes with recognition in humans.   Like patterns of integration patterns of interference generated by added material can provide important clues to processes of form recognition.

Suppose that a subject must find a specific target form in a display that also contains other, irrelevant forms ("distractors").   These other forms might vary in number, distance from the target, similarity to the target, and so on.   The time taken to detect the target, and the number of errors made, can tell something about the processes involved.   For example, if the time it takes to find the target rises linearly with the number of forms, the subject might be examining the forms one by one -- that is, "serial processing" might underlie the performance.   If time and errors are high when the irrelevant forms are near the target and diminish when the forms are spread out, this might imply that different parts of the visual field are associated with different processing "channels."

Thinking along these lines, I have looked at some parameters of visual search in pigeons.   In published experiments (Blough, 1977; Blough, 1979), pigeons were rewarded for pecking at a simple form, such as an "o," displayed along with one or more distractor forms on the face of an oscilloscope screen.   A photocell mounted on the pigeon's beak enabled a computer to determine which form was being pecked and the time the bird needed to find the form.   The forms were composed of dots; three examples are shown at the top of Figure 16.5.

These experiments showed clear interference effects: increasing the number of distractor forms or their similarity to the target increased the time required to peck the target, as shown in Figure 16.5.   Errors increased as well, though this is not shown in the Figure.   Though there is no doubt about the effects of the number and similarity variables, it is hard to infer the nature of the processes involved.   It may be luck, for example, that for birds 1 and 3 the reaction time functions for distractors similar to the target look much as they ought if the pigeons were doing a serial search: noting the log scale on the abscissa, you can see that reaction time for these birds rises nearly linearly with number of forms. One among several other suggestive findings was that a mixture of several different distractors was not especially disruptive.   This finding agrees with other results from pigeons (noted above) and humans which suggest that non-targets are not subjected to extensive processing.

In these initial experiments, the presence of non-target context material clearly disrupted recognition, despite the fact that the various

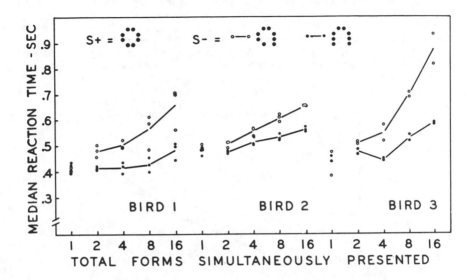

FIG. 16.5. Reaction times of birds searching for a target "o" amid various numbers of relatively similar or dissimilar distractors. The points represent medians of individual sessions; the lines connect means of these medians. (From Blough, 1979.)

forms were rather widely separated on the display screen. More recent studies (unpublished) have looked explicitly at the spatial separation between targets and irrelevant forms. The method was relatively conventional. Pigeons looked through two glass keys at the face of a TV monitor; behind one of the keys an ATARI home computer placed a black target "X" on a white ground.

In the first condition in this series, three letters appeared behind each key arranged in a row and spaced 0.8, 4.0, 7.2, or 13.6 mm apart. The center letter on one side was the target "X"; the other letters were randomly drawn from the letters of the alphabet A through N. All the letters were capitals formed within the standard 4x5 dot matrix used by the Atari computer for letter display. A single peck at one of the keys terminated a trial; 500 trials constituted a daily session (with additional trials imposed by a standard correction procedure).

Figure 16.6 shows the mean data for five days for the 4 different interletter distances. Clearly, the target was harder to find if the distracting letters were placed very close to it. Without going into the various processing models that attempt to handle this situation (e.g., Estes, 1972; Wolford, 1975), we can speculate on some of the factors that might contribute to interference here. For example, a relatively simple masking effect might make the letters harder to see when they are close to each other, regardless of their shape. Or, if two forms are close together elementary figural processes might make them appear as a single form that conceals its components -- as presumably happened in the pattern superiority experiments described above.   Both of these hypotheses, which could be much more precisely specified, apply only to very small interletter distances.   An alternative approach suggests wider processing "channels" which become overloaded when adjacent forms are relatively similar.

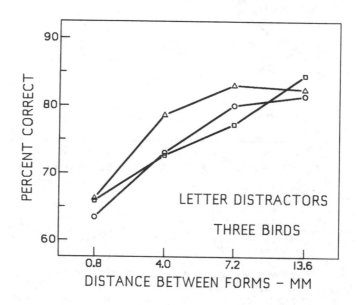

FIG.   16.6.   Accuracy of pecking at an "X" flanked by random letters at the distances shown.   Three random letters appeared on one key; the "X" and two other letters appeared on the other key.

To look at this matter, a second condition was run, which differed from the first only in that a single very distinctive form (a black spot) was displayed along with the "X."   The results of this are shown in Figure 16.7. As you can see, when the spot was right next to the target, it interfered with recognition only slightly less than did the random letters.   (Note that Figures 16.6 and 16.7 have different ordinate scales.)   However, when the spots were 4 mm or more from the letters, they interfered rather little and (more importantly) the interference did not vary with distance.   Similar interactions of target–distractor similarity with interform distance have been found in human subjects.   The data suggest, then, that effects due to very close proximity (masking, form distortion) may be separable from effects over a wider field.

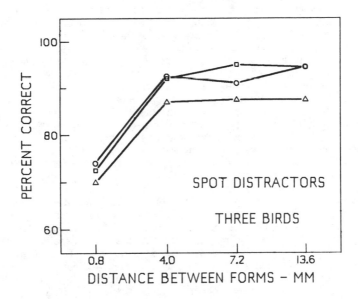

FIG. 16.7. Accuracy of pecking at an "X" flanked by black spots at the distances shown. Compare with Figure 16.6.

V. DISCUSSION

This paper has suggested three approaches to the study of form perception in pigeons. Each was a "recognition" task: the bird was required simply to peck at particular target forms. In the letter experiment, errors were taken as a measure of form similarity. In the pattern superiority experiment, errors decreased when target lines were integrated into larger target forms. In the search experiments, errors reflected variables that interfere with recognition. These results suggest that our current sketchy understanding of animal form perception may be expanded by relatively simple methods. I should emphasize, though, that these experiments are far from the first to study animal reactions to simple forms; I have been able neither to give adequate credit to previous researchers (cf. Zusne, 1970; Cabe, 1980) nor to provide the detailed connection with existing human data that would give the reported results somewhat more meaning.

As I have indicated, the results overall look very much like those produced by humans in related experiments. To some observers, this makes the results uninteresting. I have yet to hear a cogent argument for this view. The most seriously misguided source for this reaction is the stimulus error, the deeply embedded assumption that perceptions mirror the physical world -- that we see what is "out there." In the case of letter discrimination, for example, this assumption could lead to the statement that "A and R *are* similar, so it is no wonder that pigeons and humans both see them so." Persons who think this way altogether miss the point of perceptual and psychophysical research, yet in subtle guises such thinking is astonishingly widespread. Consider, for example, the assertion that "experiments with human subjects have already shown us the similarity relations between letters, so an animal experiment on this subject is just an uninteresting replication." Such remarks once more reflect the stimulus error. This one conceals the assumption that similarity relations exist independently of the perceiving organism, for only on this assumption does replication across species become predictable and thus uninteresting.

Once one clearly recognizes "seeing" as an organismic process rather than a window on the world, it becomes clear that comparative study of this process must be informative. If creatures as different as pigeons and people see letters in very similar ways, we know a great deal that we did not know before. Alternative hypotheses can be eliminated in huge blocks: letter similarity is not a function of human verbal learning, of symbol manipulation, of partially uncrossed visual tracts or a convoluted cortex. Since letters are arbitrary forms for pigeons, the results suggest that perception of this sort of form may be a fundamental function of the way that higher nervous systems are constructed. Against this background, species differences, where found, can be equally suggestive. Why do pigeons and people seem to react differently to line drawings (cf. Cabe, 1980)? Does this result from learning experiences unique to humans? Is it due to particular structural differences? Beyond the insight directly gained from comparative psychophysical results, there opens out the whole spectrum of research made possible by the use of non-human subjects: motivational manipulations, physiological interventions, control of experiential histories, to name a few. The potential of such methods in the study of cognition and perception is largely untapped, but surely exciting.

ACKNOWLEDGMENTS

The preparation of this paper and the research reported were supported in part by NSF grant BNS-8025515. Figures 16.1 and 16.2 are reprinted with permission of **Science**, Copyright 1982 by the American Association for the Advancement of Science.

REFERENCES

Blough, D. S. Visual search in the pigeons: Hunt and peck method. *Science*, 1977, *196*, 1013–1014.

Blough, D. S. Effect of the number and form of stimuli on visual search in the pigeon. *Journal of Experimental Psychology: Animal Behavior Processes*, 1979, *5*, 211–223.

Blough, D. S. Pigeon perception of letters of the alphabet. *Science*, 1982, *218*, 397–398.

Cabe, P. A. Picture perception in nonhuman subjects. In M. A. Hagen (Ed.), *The perception of pictures*. (Vol. 2.) New York: Academic Press, 1980.

Carter, D. E. & Eckerman, D. A. Symbolic matching by pigeons: Rate of learning complex discriminations predicted from simple discriminations. *Science*, 1975, *187*, 662–664.

Cerella, J. Absence of perspective processing in the pigeon. *Pattern Recognition*, 1977, *9*, 65–68.

Donovan, W. J. Structure and function of the pigeon visual system. *Physiological Psychology*, 1978, *6*, 403–437.

Estes, W. K. Interactions of signal and background variables in visual processing. *Perception & Psychophysics*, 1972, *12*, 278–286.

Gibson, E. J. *Principles of perceptual learning and development*. New York: Appleton–Century–Crofts, 1969.

Kohler, W. *Gestalt psychology*. New York: Liveright, 1947.

Podgorny, P., & Garner, W. R. Reaction time as a measure of inter– and intraobject visual similarity: Letters of the alphabet. *Perception & Psychophysics*, 1979, *26*, 37–52.

Shepard, R. N. Multidimensional scaling, tree–fitting and clustering. *Science*, 1980, *24*, 390–398.

Stevens, S. S. Mathematics, measurement and psychophysics. In S. S. Stevens (Ed.), *Handbook of experimental psychology*. New York: Wiley, 1951.

Weisstein, N., & Harris, C. S. Visual perception of line segments: An object–superiority effect. *Science*, 1974, *186*, 752–755.

Wolford, G. Perturbation model for letter identification. *Psychological Review*, 1975, *82*, 184–199.

Zentall, T. R., Edwards, C. A., Moore, B. S., & Hogan, D. E. Identity: The basis for both matching and oddity learning in pigeons. *Journal of Experimental Psychology: Animal Behavior Processes*, 1981, *7*, 70–86.

Zusne, L. *Visual perception of form*. New York: Academic Press, 1970.

17 ACQUISITION OF FUNCTIONAL SYMBOL USAGE IN APES AND CHILDREN

E. Sue Savage-Rumbaugh
Emory University

One of the more striking differences in the symbolic behavior reported for chimpanzees and young children (at the one word stage) is the apparent absence of any sort of indicative or declarative symbol usage on the part of the chimpanzees (Terrace, Petitto, Sanders, & Bever, 1979)

Chimpanzees do not seem to use symbols to declare or indicate what they are about to do or to draw the attention of others to objects of interest. For example, after working with Nim, Terrace observed that

> The function of the symbols of an ape's vocabulary appears to be not so much to identify things or to convey information as, for example, [Skinner's concept of "tacts"] as it is to satisfy a demand that it use that symbol in order to obtain some reward [Skinner's concept of "mands"] (Terrace *et al.*, 1979, p. 900).

Terrace *et al.* (1979) are not alone in this view. The overriding theme of the reports which describe symbol usage in apes (Gardner & Gardner, 1975; Miles, 1976; Patterson, 1978; Premack, 1976; Rumbaugh & Gill, 1976) suggest that the two basic contexts of chimpanzee-produced symbols are requesting desired objects and associatively labeling objects displayed by the experimenter.

Why should the circumstances in which chimpanzees typically employ symbols seem to center on their immediate physical desires and their responses to questions by their teachers (Gardner & Gardner, 1975; Sanders, 1980) while symbols are employed by children to describe what they do, to draw attention to what they are looking at, as well as to request things which they need? Is this difference merely a function of the fact that human beings are a more open and sharing species while chimpanzees are very self-centered as those who have worked with apes suggest (Premack, 1976; Rumbaugh & Gill, 1976; Terrace *et al.*, 1979)?

The suggestion offered here is that this difference is not a function of species specific tendencies alone. Rather, it obtains because of how symbols are taught to apes. Chimpanzees do not acquire symbols on their own, as do human children and the molding-modeling method used by most ape-language researchers encourages associative-imitative symbol usage. It is by no means clear that apes are able to make the transition from the associative-imitative symbol production, which sign training stresses, to the representational symbol usage exhibited by human children.

A symbol is not to be deemed a representation, simply because an ape can produce it when shown an object. In fact, the quality of "representation" is not something which a symbol possesses in and of itself, it is a quality which can only emerge in functional use between individuals. This makes representational symbol usage an inter-individual phenomenon, as opposed to an individual capacity. A symbol can only be said to function in a representational fashion when individuals are capable of using symbols to convey or specify actions and goals that are not known to all parties in advance of the use of the symbols. Representational symbolic requests then come to be ways of coordinating behaviors between individuals because each can make clear his desires and intentions to the other. Representational symbolic requests function as effective communicators to the degree that they encode something which is not already presupposed (or given by virtue of the context) by both parties.

Let us suppose, for example, that a caretaker approaches a child (or chimpanzee) carrying three new toys; a doll, a ball, and a music box. The child, upon seeing these, extends his hand, smiles, and says 'ba.' If the caretaker knows, based upon past experience with this child, that the child uses the names of these toys, or similar toys reliably, then it will be presumed that the child wants the ball and the caretaker will give it to him. If in the caretaker's past experience, however, the child calls everything 'ba' then it will be presumed that the child wants a toy but the caretaker will be uncertain as to which toy. In that case the caretaker will probably approach the child and watch to see which toy the child reaches for or looks at (Figures 17.1).

Although 'ba' as expressed by the child, can encode the specific object which the child wants, the desire that an object be given is expressed by the child's extended hand. If the caretaker were to enter the room carrying only the ball, the expression of the word 'ba' would be redundant. If the child simply looked toward the single object in the caretaker's hands and extended his hand, the caretaker would not wonder, "Which toy does the child want?" The desired object, by virtue of its singular presence, would be presupposed. Only when the caretaker enters the room carrying a number of toys do we find that the expression of a specific referent such as "ball" adds to the information already expressed by the extended arm.

Thus we see that for a symbol to convey information that is not apparent to both parties in advance, it must in some way delimit a number of alternative aspects of a situation. Another way in which a symbol can accomplish the same thing is when one party uses it to convey information to another about an object that is absent and hence cannot be referred in the present context except through the use of a symbol.

Symbols, as taught to chimpanzees by the molding-modeling methods, serve neither of the purposes described above. These methods simply teach the ape to make the sign in the presence of the object. Terrace (1981) has argued that the difference between chimpanzees and children is that apes only produce signs in the presence of objects when they obtain a reward for so doing, while children utter the names of objects for the

FIG. 17.1. Situation A. Adult enters room carrying one toy. Object transfer possibilities from child's perspective: (1) Adult gives toy. (2) Adult does not give toy. Situation B: Adult enters room carrying five toys. Object transfer possibilities from child's perspective: (1) Adult gives all toys. (2) Adult gives toy a. (3) Adult gives toy b. (4) Adult gives toy c. Situation C: When more than one possibility for transfer exists, the adult does not have to display each alternative if the child has a clear referent such as ball.

sheer joy of so doing, or for drawing the attention of others to objects of interest.

It is suggested here that this is an overly simplified characterization of the situation. While it is correct to say that children learn language spontaneously, with no apparent reinforcement, while chimpanzees do not -- before statements regarding *usage* differences are considered -- we must ascertain that indeed chimpanzees have learned symbols as defined above.

Are chimpanzees really learning symbols as names to stand for or replace indicative pointing gestures as do children? Or are apes, in fact, learning something else, and if they are, why?

In order to answer this question we must look more carefully at the type of gestural communication which the ape spontaneously develops prior to symbol training and compare the ape's skills in this domain with the type of gestural communicative capacities which the young child displays prior to the onset of one-word speech.

Both the human child (Gray, 1978) and the young chimpanzee (Savage, 1975) learn to use the extended hand gesture as a means of requesting object transference. Both come to do so without specific intentional training on the part of older conspecifics. That learning is probably involved, however, is suggested by the recent work of Gray (1978) which shows that human mothers are part of an interactive developmental process in which the mother's constantly changing response to the baby's changing grasping behaviors serves to promote increasingly sophisticated and structured patterns of exchange which culminate in the emergence of gestures.

The human child, after developing gestures of requesting goes on to develop indicative pointing gestures which are used for both near and distant objects. These pointing gestures appear prior to the onset of one word speech and are used to coordinate the child's attention with others. That is, the child points at an object, often while uttering a nonsense vocalization, in order to get others to look at the object to which he is pointing. The child is, at a gesturally indicative level, already engaging in the behavior that will be called "naming" with the onset of one-word speech. Chimpanzees do not spontaneously point at objects and vocalize to draw the attention of others to those objects. Pointing, itself, has not been observed in wild populations (van Lawick-Goodall, 1968) and although one animal often looks in the direction that another is looking, there are no reports of one ape deliberately attempting to direct the attention of another to an object. Thus the chimpanzee, unlike the human child, does not seem cognitively preequipped to spontaneously develop the pre-symbolic gestural indication behavioral schema seen during the pre-speech period in the human child.

Following the emergence of indicative behaviors, the human child goes on to learn to specify particular objects through the use of symbols and is thus able, in more complex situations, to specify one of a large number of outcomes. The normal human child comes to do this without any specific intentional training on the part of conspecifics -- but the chimpanzee does not.

When humans intervene and try to obtain behavior in the chimpanzee which is equivalent to that seen in the human child in Figure 17.1 -- what happens? Do humans simply provide a different and more complex model than other apes, thereby enabling the chimpanzee infant spontaneously to develop skills that he would not when surrounded only with conspecifics as models? Is it true that simply by providing a human model and a nonvocal signal system, one creates the conditions sufficient for the spontaneous appearance of indicative language in the chimpanzee?

In each of the ape signing projects the apes have been taught signs by a method termed *molding and guidance* (Fouts, 1973; Gardner & Gardner, 1978; Patterson, 1978; Terrace *et al.*, 1979). The purpose of this method is to induce the ape to produce a symbol on occasions when it would not normally do so spontaneously. If, for example, the ape desires an apple which is in the experimenter's possession, the extended hand gesture (which the ape does without training) is not accepted as a sufficient communication. Instead, the experimenter models APPLE and then molds the ape's hands into the apple configuration repeatedly (using less pressure each time) until the ape signs APPLE by himself, whereupon the ape is given the apple. During the training of any given sign only the object to be associated with that sign is shown to the chimpanzee. Pragmatically, this renders the performance of the sign redundant and ritualistic. The function of the sign APPLE in this context is not to specify APPLE as opposed to some other food, rather it is to satisfy the experimenter's contingencies for giving the apple. The meaning of the sign APPLE, in a pragmatic sense, is simply GIVE that which you have and are withholding from me! When the apple is replaced with a banana, the pragmatics of the situation do not change. The chimpanzee still desires the food which is in the experimenter's possession and the transfer of this food remains the center of the chimpanzee's attention. Now, however, the chimpanzee must learn a different sign in order to effect transference of the food from the experimenter to himself. The pragmatic function of the new sign is not to express a desire for banana as opposed to apple, this is precluded during training since only one fruit or the other is present. The chimpanzee must learn to use one sign to effect transference in the presence of apple and another sign to effect transference in the presence of banana. In neither case, however, does the use of the sign specify any information that is not already presupposed by both participants. The signs are learned as ritualistic request behaviors which, when emitted, fulfill arbitrary contingencies. It is difficult to see, given this type of training, how the chimpanzee would come to do more than form simple associations between signs and objects. A description of a sign training session with Washoe illustrates this problem.

Two kinds of questions were used in an attempt to elicit a response. (a) Question: The experimenter asked Washoe in ASL (American Sign language), "What's this," referring to an exemplar. (b) Implicit question: The experimenter either pointed to an exemplar or held an exemplar in front of Washoe. A correct response was Washoe's correctly forming and completing the sign which the experimenter attempted to teach her in each session. . .

At the beginning of the training session the experimenter gave Washoe a piece of candy to attract her attention, and to attempt to prevent her from avoiding him during a training session. After Washoe ate the candy, the experimenter began training. The experimenter would present an exemplar to Washoe and ask her what it was and then show Washoe the correct sign for the exemplar by using the training procedure for that sign. When the correct response occurred, Washoe was again given candy. The candy was within Washoe's view or shown to her throughout the training session in order to keep her attention on the general task and to increase her cooperation with the experimenter. As the training session progressed, and correct responses were beginning to appear without prompting, Washoe was required to make several correct responses before she was given candy. . .

If Washoe ignored the experimenter or tried to divert him from the task by engaging in mischievous behavior, or by hiding from him during the training session, he would yell at Washoe or threaten her in order to have her return and cooperate in the training session. . .

An experimental sign was judged to be acquired when Washoe began to use it correctly during a session and maintained the use of that sign to the end of the session (Fouts, 1972, pp. 516–517).

As the above description shows, signs were learned as paired-associate responses. Washoe was repeatedly required to move her hands into a particular configuration when shown a particular object. Fouts (1972) asks Washoe what things are by signing WHAT THIS? He also states that Washoe's response *refers* to the object which he shows her (such as an airplane, a telephone, a fork, etc.). It is clear, however, that the training conditions are not conducive to a referential reply. What Fouts is in fact asking Washoe is, WHAT SIGN GOES WITH THIS? If Washoe is correct, she receives candy. Candy or escape from threat function as motivators to learn a series of paired-associate responses. The discriminative stimulus for the emission of a given paired associate becomes the specific object-showing behavior of the experimenter. What Washoe actually learned was to sign TELEPHONE, for example, when her attention was oriented (usually by pointing on the part of the caretaker) to phone, FORK when her attention was oriented to forks, etc. What Washoe was not taught, and consequently did not learn, was that a sign like FORK could be used to specifically identify and to refer to forks when the referent was unclear and she needed to convey to another individual a particular message about forks as opposed to some other item. In fact, Washoe's sign training did not provide her with opportunities which encouraged sign use as a means of conveying nonredundant messages to the experimenter. Therefore it is not surprising that Washoe and Nim (who was trained in a very similar way) used multisign utterances in a way which elaborated or qualified their single-sign utterances (Terrace *et al.*, 1979). When the single utterances themselves are not acquired as referents for objects, but simply as a paired associates to objects, multisign utterances are more likely to reflect confused associates than syntactical relationships. Thus apes appear to be able to form associative labels between symbols and objects or events but there is no clear evidence that these symbols then become representational spontaneously, i.e., that these chimpanzees are able to use them in an indicative or declarative manner.

A point of repeated confusion in the ape-language reports has been the prevailing assumption that the ability of an ape to associate a sign or symbol with a given object or activity is equivalent to *naming* that object or activity in a referential sense (Fouts, 1973; Gardner & Gardner, 1978; Patterson, 1978; Premack, 1976). The numerous difficulties caused by this approach have been pointed out elsewhere (Savage-Rumbaugh, Rumbaugh, & Boysen, 1980). Recently data have been presented which demonstrate that the formation of associative labels between objects and symbols is not equivalent to referential "naming." (Savage–Rumbaugh, Rumbaugh, Smith, & Lawson, 1980). The chimpanzee, Lana, demonstrated clear object-label associations when tested under blind conditions (Rumbaugh, 1977), yet when shown the symbols for the items (as opposed to the items themselves) she could *not* sort the symbols into food and tool groups, suggesting that the symbols did not really represent the items to her. Two other chimpanzees, trained in a different manner, Sherman and Austin, could group the symbols alone into classes of foods and tools.

In our view, the limitations discussed above are the inherent results of training procedures which attempt to endow signs and symbols with meaning by withholding a single desired food, object and activity, until an appropriate symbol is either emitted or selected. When a single item or event (such as candy or the resumption of a chase game) that a chimpanzee clearly desires is withheld until a symbol is produced, that symbol ceases to serve an informative function and is forced to assume a ritualistic request function. Since most apes have been taught their symbols in this way (Fouts, 1973; Gardner & Gardner, 1978; Patterson, 1978; Premack, 1976; Rumbaugh & Gill, 1976), it is perhaps not surprising that such apes continue to use them in that manner. Likewise, since most name training procedures stress the ability to produce the correct sign or symbol when shown an object by the experimenter, it is not surprising that such naming does not function indicatively since the chimpanzee is not attempting to draw attention to the object by naming. It is the experimenter who is soliciting the chimpanzee's attention, and the chimpanzee's symbol production under these circumstances contains no pragmatic function. One might expect that the symbol would be learned in an associative manner since the sole constraints of the task are to give the sign in response to being shown the object.

Children, by contrast, readily exhibit both indication and volition during the one word stage (Bates, 1979; Greenfield & Smith, 1976). Greenfield and Smith's (1976) intensive study of two children from the emergence of their first meaningful words through the establishment of word combinations revealed that not only is the indicative mode present as early as 10 to 12 months, but indicative symbol use actually precedes volitional symbol use. Both subjects in Greenfield and Smith's study began spontaneously pointing at objects and uttering what Greenfield termed "nonsense syllables" prior to the onset of one word speech. It appears that the child makes a spontaneous transition from noncommunicative or "self-involved" pointing to indicative pointing, that is, pointing for the purposes of directing another's attention. In our experience, the chimpanzee typically falls short at this point. Although chimpanzees who are taught to sign have been observed to point at objects, these chimpanzees do not point at objects and look back to adults for confirmation. They do not point at objects in order to show these objects to adults. Noncommunicative pointing also occurs in human children and precedes communicative, indicative (pointing, showing, looking back and forth between adult and indicated object) by three to four months. True communicative indicative pointing appears just prior to the onset of the one word stage. It is important to note that it is not only the onset of the one word stage that is lacking in the chimpanzee but also the onset of indicative communicative pointing. Bates (1976) describes the behavior of a human child, Marta, at

12 months noting that she

> shows, gives, and points to objects and awaits the appropriate adult response. If no response is forthcoming, she will repeat the behavior with greater insistence, or possibly seek another audience. *At this stage she has yet to utter her first word*, and shows no other signs of symbolic behavior. Hence, Marta has full control of the protodeclarative while she is still at sensorimotor stage 5. . . Hence the protodeclarative, for both subjects, (Marta and Carlotta) is a procedural rather than symbolic structure, beginning during the fifth sensorimotor stage (Bates, 1976, p. 62, italics added).

Although chimpanzees do evidence many behaviors typically associated with the fifth sensorimotor stage, as noted earlier, they do not engage in indicative pointing. "Showing," for the chimpanzee, remains a behavior closely tied to an overt demonstration of possession and does not spontaneously evolve into a mechanism of interindividual indicating or a joint attention orienting.

We find the following constellation of behaviors, all appearing rapidly in the human child: (a) giving to others, (b) indicative pointing for others as opposed to pointing for self (Werner & Kaplan, 1963), (c) use of nonsense vocalization to orient the attention of others to an indicated object. None of these behaviors appear spontaneously in the chimpanzee. Nor is there any evidence to suggest that the chimpanzee comprehends the indicative function of such behaviors when they are displayed to him by human experimenters.

It appears, then, that in the human child there exists a cognitive structure which employs and coordinates reference, cooperative giving, and joint orienting of attention in interindividual behavioral interactions which precede language. This cognitive structure is absent in the chimpanzee. In chimpanzees, it is not simply "words" or vocal speech that are absent, rather it is this entire complex structure which serves to orient, organize, and regulate cooperative referential behaviors between individuals. Gardner and Gardner (1978) and Terrace *et al.* (1979) ignored the absence of these behaviors in Washoe and Nim and proceeded straightaway to teach them symbols in the form of ASL signs. With no cognitive structure for interindividual indication, however, these apes were poorly prepared to view signs as referential, nor were they prepared to use them in an indicative sense for purposes of intentional communication with others. Hence, they learned and used signs in ways in which they were cognitively prepared to do. They used signs to initiate chase or tickle games just as they used "showing off" behaviors to get social attention. They learned to associate signs with objects because they had to and they learned that combining these signs also got attention. They remained, however, unable to use indication or reference and unable to understand what others were doing when they used symbols to indicate their wishes. They were never really able to engage in the cooperative, attention-sharing, object-giving domain of one word speech. Yet because they combined signs in three, four, and even five word utterances, these ape language researchers were mistakenly led to conclude that their language skills were representational and possibly even syntactically structured.

What these ape language researchers did not realize is that the ability to indicate intentionally, through use of symbols, implies far more than the acquisition of a simple object–label association. Such associations provide little more than the raw materials for such a skill. Locke (1980) makes a similar distinction with regard to human children between the learning of associations and the emergence of names and points out that while the distinction between these two domains may at times be unclear if one is actually observing a child, it is, conceptually, central to the language issue. Locke (1980) views this as the point of emergence of true language skills, noting that

> . . . the structure of naming is almost propositional. For the first time an action on the part of the child is capable of being judged true or false; the semantic force of his actions being open to determination by arbitrary social agreement, and not by nonarbitrary biological states. Secondly, such mean propositional relationships, which by nature do not implicate the child in their structure, provide the basis for the development of objective knowledge. Prior to this point, the child has learnt the value of an object in terms of the relationship it has to himself through the mediation of his actions upon it. From this point onward there exists the possibility of his giving value to objects which stem from their relationships with other objects; relational systems having an existence independent of his own. . . he is using language proper for the first time (pp. 121–122).

Locke describes the transition from association to naming as passing through stages.

> Firstly, there is a period of teaching in which the mother always initiates the interaction, showing some object to the child and demanding some particular response. . . Secondly, a period seemingly transitional between the first and the third in which the initiator of the interactions passes to the child, such that he "names" an object which he singles out during the "game" and "names" it without being asked to. And thirdly, the child begins the game himself, singling out and naming objects without prompting. The *impression* that arises from looking at these developments in the childs use of his "words" *suggests* that there has occurred a change in his knowledge, such that it now admits an understanding that objects have names. It is difficult to go beyond using notions such as "impression" and "suggestion," for hard evidence that the child attains the principle that objects have names is difficult to find. . . But if it is accepted that on the basis of first experience the child *has* come to possess some general principle, then it has to be admitted that he has not learnt this principle, but has *created* it for himself out of that first experience. While it has been the mother's intention that her child should learn that objects have names, and what these names are in specific cases, she is in fact incapable of teaching him this. All that the mother can do is to teach the child to associate a specific sound with a

specific object and all that the child can *learn* are those specific associations. It has to be left to the child to discover creatively the principle of naming that is inherent in what he has been taught. The mother provides the data in such a way that given his spontaneous abilities it is perhaps inevitable he comes up with the right answer (Locke, 1890, pp. 118-120, italics in the original).

Herein lies the crux of the issue for ape language. Do chimpanzees, when taught by their "mothers" (human teachers) to associate a sound (sign, lexigram, etc.) with an object discover the *inherent principle of naming*? Based on the views which appear in the earlier sections of this paper, it is clear that chimpanzees do not. And one must ask after all, why is it that we would even expect them to, since they do not attach names to things in the wild? However, apes can learn that objects have names, but only if they are taught far more than object-label associations. By attempting to teach such a concept to apes we have gained a better picture of the kind of analytic inferential processes the human child must spontaneously bring to the problem in order to form the concept with very little help.

A very different language training approach has been underway since 1975 at the Language Research Center in Atlanta, Georgia with two male chimpanzees, Sherman and Austin. Although it is impossible to reasonably describe the training which these chimpanzees have received in the remaining space allotted here, it should be stressed that the molding-modeling techniques employed with other apes have not been used with Sherman and Austin.

Instead of being taught to associate an object with a symbol and to be able to produce that symbol when shown that object, they have been taught to use symbols to ask for things they need in situations where a number of alternatives are always available. The symbol is never redundant with the nonverbal information, instead it always serves to add to, in an informative manner, the nonverbal context. Their training has also, in contrast to the training of other apes, emphasized the receptive aspects of symbol acquisition (Savage-Rumbaugh, 1979, 1981).

Sherman and Austin have communicated about specific foods absent from view but hidden in containers, about tools which are out of their sight, about events which have frightened them, and about the classes to which objects belong. They have communicated such things to each other with no teacher present, and they have communicated such things to teachers who are "blind" in the sense that they have no advance knowledge of what the chimpanzee will communicate (Savage-Rumbaugh, Rumbaugh, & Boysen, 1978a, b, c; Savage-Rumbaugh & Rumbaugh, 1980, 1982).

If Sherman and Austin are able to communicate these sorts of things, why are other apes unable to do the same? Why do they imitate their teachers so frequently while Sherman and Austin rarely imitate? (Greenfield & Savage-Rumbaugh, 1982). Sherman and Austin were *not taught to imitate* their teachers. Instead they were taught to comprehend and respond to symbol usage by others as well as how to use symbols to affect the behaviors of others. That is, they were taught language as a functional operative skill, not just as a series of hand movements to be made when the teacher wanted.

In the human child, receptive competence precedes productive competence.   Children generally understand and respond to parental requests and comments for several months before they begin to engage in verbal requests and comments themselves (Benedict, 1979). Responding to the symbolic utterances of others is a *crucial* language skill, yet it has been completely overlooked in the tutoring of other apes. For example, tapes made of Nim by Terrace reveal that Nim had no comprehension of such simple and well learned signs as "apple" and "banana." If both fruits were placed in front of Nim, he could not hand the teacher the one she asked for even though he could easily produce either sign when shown these fruits. By contrast, Sherman and Austin have no difficulty giving the teacher objects that are requested symbolically. They clearly have a well developed receptive competence that was completely absent in Nim, because of the method of instruction employed with Nim. It is receptive competency that makes communication about things possible and this ability has been skillfully mastered by Sherman and Austin. They do not only "talk," more importantly, they demonstrate an understanding of what others say and a willingness to cooperate with others. Consequently, when they do talk, their exchanges are of a quite different sort than those which characterized Nims imitative conversations.

Because Sherman and Austin received symbol training which emphasized the functional indicative aspect of symbols and receptive comprehension of symbols as the symbols themselves were acquired, they developed a cognitive comprehension of symbols which has demonstrably gone beyond that of other apes (Savage–Rumbaugh, Rumbaugh, Smith, & Lawson, 1980).

This assertion is strongly supported by an experimental study designed explicitly to determine whether or not symbols, as used by chimpanzees, represented reality.   In this study we used three chimpanzees, Lana, Sherman, and Austin.   As discussed above, Sherman and Austin's symbol acquisition training was very different from that received by Nim and other signing apes.   In addition, it was also very different from the training which Lana had received in our own lab (Rumbaugh, Savage–Rumbaugh, & Scanlon, 1982). Lana had been taught "vocabulary" much as had Washoe, Nim, and other apes, even though she was using a different communicative medium (geometric symbols which were located on an electronic computer–controlled keyboard). Objects were shown to Lana and she was to light (by touching) the symbol for that object. It was presumed that when Lana could accurately label objects and activities that she "knew their names" and that she could understand the symbols which she produced, just as it was similarly assumed for Washoe, Nim, etc. Lana also imitated the symbol usage of her trainer, and modeling the correct response was an important component of her training. Consequently, although Lana's medium of symbolic communication differed from that of Washoe, Nim, etc., the training strategies used with Lana were quite similar. In addition, Lana produced behaviors which were very similar to the sorts of "combinations" and "conversations" described for Nim and Washoe. For example, she generalized beyond the original training contexts, she produced contextually appropriate novel utterances which varied in length from 2 symbols to 8 symbols, and she used symbols in ways other than originally taught.   Like Nim, she was prone to imitate her teacher's utterances, she said many things which were probably far in excess of the capacity of comprehension, and her longer utterances did not add to the information carried in shorter utterances.   Without the opportunity to directly compare Lana's symbol usage with that of signing apes, it is impossible to be assured of equivalent capacities -- however, it is clear that Lana was a sophisticated symbol user.

The difference between Lana's training, and that received by Sherman and Austin is emphasized because in the study explicitly designed to test the representational capacity of their symbols, Lana performed quite differently than Sherman and Austin.   Furthermore, her performance strongly suggested that her different symbol use skills did not reflect an underlying common dimension of representationality -- as did Sherman and Austin's performance.

The basic idea behind the assay of representational function was to determine whether or not the chimpanzees could answer questions about objects when the questions were presented symbolically and the object of question was not present for sensory reference.   The ability to make such judgements about a symbol (when the referent of that symbol is absent) requires that the ape produce some type of image of the referent in order to answer such questions, since these questions are not about the symbol per se, but about the object for which the symbol stands.

Such a test must insure a number of things, 1) that inadvertent cueing from the experimenter is not possible; 2) that the chimpanzee has no previous training history on these particular "test" questions; 3) that the questions be asked for the first time with the referent absent; 4) that the chimpanzee understand the symbolic questions; 5) that enough different questions be asked to ascertain that the chimpanzee is clearly not guessing or responding in a chance manner.

With these constraints in mind, a test was designed which permitted us to question Sherman, Austin, and Lana about the functional categories to which items in their vocabulary belonged -- first with real objects, and secondarily with symbols.   The chimpanzees were taught the generic symbols "food" and "tool" with 3 exemplars of each class, these exemplars being drawn from their common vocabulary (Figure 17.2).   With only three exemplars for each category, the chimpanzees could learn 6 separate associationistic responses, or they could learn that the food and tool lexigrams referenced classes.   In order to learn the latter, they would need to draw a functional equivalence among 3 different edibles and also one among 3 different tool-like objects.   Additionally, they would need to assign to the lexigrams "food" and "tool," their ideas of equivalence.   If they only learned 6 associations, they would, when shown a specific food or tool not used during training, not know how to classify this new item. However, if they formed functional equivalences and assigned symbols to these functional equivalence concepts, new items would present no problem.

In a blind test situation, with new items, Sherman and Austin readily labeled each of them as food or tool.   Lana became completely confused and could not accurately characterize new items as belonging to one class or the other (Figure 17.2).   In order to determine whether Lana's deficiency was the result of an inability to a) draw the correct functional equivalences or b) an inability to assign her concept of equivalence to the lexigrams --

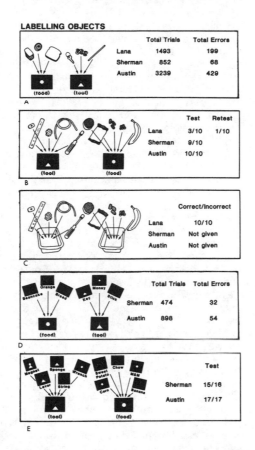

FIG. 17.2. Panel A: The chimpanzees were taught to select "food" when shown the 3 illustrated foods (bread, orange, and beancake) and "tool" when shown the 3 illustrated tools (money, key, and stick). Panel B: The chimpanzees were tested to determine if their concepts of "food" and "tool" generalized beyond the training items. Sherman and Austin displayed functional generalization, but Lana did not. Panel C: Lana was given an additional test to determine whether or not she could categorize training items simply by sorting the items into different bins. This task was performed successfully. Panel D: Sherman and Austin were taught to categorize the lexigrams for beancake, orange, bread, money, key, and stick. Panel E: Sherman and Austin were then tested with the following lexigrams: M&M, banana, wrench, show, corn, sweet potato, magnet, carrot, pudding, pineapple, straw, lemonade, sponge, strawberry drink, string, Austin's room, and lever. In each case they were asked whether the item was a "food" or a "tool." Each question was posed only once, by showing the chimpanzee the lexigram for each of the above items. The experimenter did not know which lexigram was being shown to the chimpanzee.

we administered a simple sorting test to Lana. She was asked to sort the foods and tools (which she had labeled incorrectly) into separate bins, again in a blind setting. Lana had no difficulty sorting these untrained foods and tools in separate bins, thereby revealing that her difficulty lay not at the conceptual level, but at the representational level (Figure 17.2). The lexigrams "food" and "tool" had come to stand for general categories in Sherman and Austin's case, but not in Lana's case. This was true even though their training with the "food" and "tool" lexigrams had been identical. However, their different training histories *prior to* the introduction of "food" and "tool" lexigrams had been quite different and this difference dramatically influenced the way in which concepts were attached to lexigrams.

This study demonstrates that complex symbol use skills, including the formation of novel combinations, naming generalizations to nontrained items, spontaneous utterances, and the solving of complex tasks -- are not necessarily indicative of symbolic encoding. Lana (Rumbaugh, 1977) demonstrated all of the above, but did not symbolically encode the lexigrams "food" and "tool." It is important to note that she used the lexigram for "food" in other than the training situation -- but apparently still did not encode it in a manner similar to that of Sherman and Austin.

We continued this study with Sherman and Austin alone since it seemed fruitless to ask more complex questions about Lana based on her lack of success in the first phase of the study. With Sherman and Austin we replicated the above results using photos of objects instead of real objects and then moved to the critical question -- could they make categorical judgements about lexigrams alone? In order that they should fully understand what we were asking of them, we repeated the original "food" versus "tool" training with three exemplars, only now instead of showing them 6 training foods and tools, we showed them only the lexigrams which stood for those foods and tools (Figure 17.2).

We then tested them with other lexigrams which they had never previously been asked to categorize. Five of these lexigrams represented items about which they had been asked to make a food versus tool judgment when shown the real item during the previous phases of testing. However, the remaining 11 items in Sherman's case, and 12 items in Austin's case, had never been categorized by these chimpanzees before. That is, no one had ever asked them whether such an item was a food or a tool, moreover, no one had ever told them whether such an item was a food or a tool. When they were asked this question for the first time, they were shown only the lexigram, not the real object. Neither they, nor any reader, could upon seeing only the lexigram determine whether the object which that lexigram stood for was a food or a tool unless they had some knowledge as to what the lexigram represented (Figure 17.2). This test was given with the experimenter blind and only trial one data was collected.

Until data are presented on similar symbolic encoding capacities in chimpanzees who have been trained to use signs, there is reason to question whether or not they are encoding the signs at a symbolic level. Perhaps Nim signed as a "way out," not because he was a chimpanzee, but because he never really learned that signs *stood* for things -- as opposed to being just a means of obtaining things.

Data collected under controlled test conditions is seldom reported for signing apes. Gardner and Gardner (1978) have shown that Washoe could label 32 items under controlled conditions, but they have not identified which items these are. Terrace has presented no blind or controlled tests with Nim's skills. Without data that have been collected in a controlled manner, it is not possible to make any comparisons across laboratories or methodologies. The significant differences revealed between chimpanzees reared in the same laboratory, using a similar system, doubly emphasizes the importance of reporting training histories in a manner that includes controlled tests.

This study also emphasizes the caution which must be maintained in presuming that symbol production is equivalent to symbolic encoding. Symbolic encoding does exist in chimpanzees, and can be demonstrated, however, it is not a necessary concommitant of symbol-object pairing. Those who rush to conclude that chimpanzees are incapable of syntax, before they, in fact, ascertain that the chimpanzee has learned to symbolically encode individual symbols, are assuring themselves of a negative answer, for syntactical competence depends upon an ability to comprehend the nature of the relationships (as indicated by order) between a number of symbolically encoded items. As such, it reflects a second-order of symbolic encoding capacity.

In the remainder of this paper we wish to describe the *first* skill which Sherman and Austin have acquired in a manner that can be described as spontaneous, i.e., a language skill which has appeared without instruction on the part of their human teachers because we, like human mothers, had no way of teaching it. This is the skill of indication or naming. By this we do not mean the ability to select a lexigram or sign when shown an object. Lana, Washoe, Koko, Ally, Nim, and other chimpanzees could do this. Such a skill alone, however, implies only that the chimpanzee can successfully produce the correct associations when the human teacher structures the word game. It implies nothing regarding true indication or naming. Sherman and Austin, however, did something quite different. After accomplishing the representational usage of the lexigrams "food" and "tool" as described above, Sherman and Austin began to take the naming games out of the teacher's hands. They began to initiate and structure exchanges. Instead of waiting for the teacher to ask that items be given or labeled, they began to name items and to show them to the teacher (either by pointing or giving). In so doing, they took the "teaching" out of our interaction in the game. They incorporated the teacher's role into their own behavior in a completely appropriate manner. They did not exhibit a stereotyped imitative version of the teachers' role; rather, they incorporated the *function* of the teacher's initiating and singling out role into their own behavior. Thus, when the teacher entered the room with a group of objects, the chimpanzees initiated the interaction by singling out, naming, and showing or giving objects themselves instead of waiting for the teacher to decide the game and initiate the instruction by using symbols, showing objects, etc. For example, they said "straw" at the keyboard and pointed to the straw, then "blanket" and picked up their blanket. They would at times briefly touch an object before naming it *without* looking at the teacher as they touched it, as though they were choosing that item, singling it out from the others to be named. After they named it, they would again point at the object (now deliberately and with a more expressive gesture) which they had just named and they would look at the teacher, or they would simply pick up the object and give it to the teacher.

This behavior first appeared between the chimpanzees themselves in the food sharing context and then initially it was exhibited only by Sherman, the dominant animal (but it soon began to appear in Austin's symbol usage also). If Austin were asking Sherman for foods and could not find the correct lexigram or if he were uncertain and hesitant, Sherman began to make the selection for Austin and to give him the food. Thus, Sherman would, for example, state "orange" and then give both Austin and himself a piece of orange. Austin encouraged this by reaching toward the keyboard, as though preparing to respond, then drawing back and glancing at Sherman. Sherman would then light a symbol. They were not taught to do this, in fact, they were encouraged to do just the opposite. When Austin was in the role of food requestor, and Sherman in the role of food provider, they were encouraged to behave according to their roles at the time.

It is important to note that had we intentionally set out to teach the chimpanzees to do this, we would have encountered considerable difficulty in structuring a task which conveyed to them what it was they were to do. This is because this sort of indication necessitates a comprehension of the necessary correspondence between: (1) the item chosen from among the group as the one to be named at any given time; (2) the name or lexigram symbol selected at the keyboard; and (3) the item which is finally selected and given. From the chimpanzees' point of view, an error could reflect an incorrect initial choice, an incorrect symbol selection, or an incorrect final choice. The teacher, however, has no way to indicate that an error has been made until all three portions of the behavioral act are completed. Therefore, to teach this skill *de nouveau* would be impossible since each incorrect response could mean that the first, second, or third portion of the behavior was incorrect, or that the correspondence between one and two was correct, but not between one and three.

The blind test procedure used to document this skill imposed a far greater burden on representational memory capacity than did the freer teaching setting in which the teacher, the keyboard, and the chimpanzee were much closer together. During the blind test, after the chimpanzee looked at a group of five objects, he had to remember his selection while he walked to the keyboard. Once he was at the keyboard, he could no longer see these objects and he had to select the symbol simply by recalling the object he had seen and chosen to select.

After the chimpanzee had selected the symbol, he came back out of the room and selected the real object or photograph. As he was selecting the object, he could not see the symbol he had lighted. He had to recall the symbol he had chosen.

Both Sherman and Austin demonstrated a readiness to perform the task and did so easily and accurately during a 53 trial blind test. Sherman correctly indicated which object he was going to give on 50 trials. Only once did he say he was going to give an item which was not present in the tray, and on that trial he gave nothing. On two trials he confused the lexigrams "shot" and "key." Austin correctly indicated which item he was going to give on 46 trials. Austin selected a symbol to indicate an item which was not present on 4 trials.

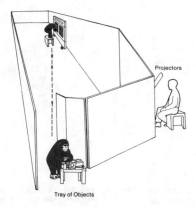

FIG. 17.3. The chimpanzee looks at the tray of objects to decide which one he will take to the experimenter. (The objects on the tray change each trial.) After deciding, the chimpanzee goes to the keyboard and states which object he is going to select.

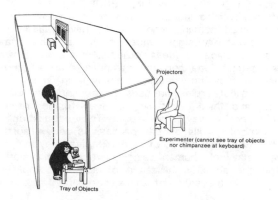

FIG. 17.4. The chimpanzee returns to the tray of objects and picks the one he has stated at the keyboard.

The appearance of spontaneous indication in Sherman and Austin suggests that they have achieved a level of symbolic functioning that is fully representational and equivalent to the onset of true naming as described for human children by Locke (1980). There are no data which would indicate that other language-trained apes have achieved true naming

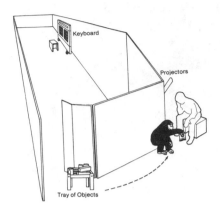

FIG. 17.5. The chimpanzee then takes this object to the experimenter who checks his projectors to see if the chimpanzee brought the item which he had said he was going to bring.

capacities.   Supporting evidence for this view is also to be found in Sherman's and Austin's ability to use symbols when making categorical decisions about other symbols, a skill that also requires true naming capacities and one that has not been reported for other apes (Savage–Rumbaugh, Rumbaugh, Smith, & Lawson, 1980).

The use of symbols to indicate things to one another seems like such a simple thing to adult human beings.   Many have failed to be impressed with the true evolutionary significance of this accomplishment.   Once a symbol can take the place of ideas, objects, and events, man inhabits two worlds -- the tangible one and the symbolic one which is of his own making.   While he may now easily maneuver in and out of both and contemplate the intricate rules he uses to combine symbols and thus to construct second and third order worlds even further removed from the present, he did not always do so.   The acquisition of functional communicative symbolic skills by chimpanzees is an extraordinary thing for it moves these two members of the species beyond their biological heritage in ways that only human beings have ever traveled before.   Only by understanding the actual significance of such accomplishments can we make true progress in the study of animal cognition.

ACKNOWLEDGMENTS

This research was supported by a grant from the National Institute of Child Health and Human Development (HD–06016) and from the Division of Research Resources (RR–00165), National Institutes of Health.

REFERENCES

Bates, E. *Language and context, the acquisition of pragmatics.* New York: Academic Press, 1976.

Bates, E. *The emergence of symbols, cognition and communication in infancy.* New York: Academic Press, 1979.

Benedict, N. Early lexical development: Comprehension and production. *Journal of Child Language,* 1979, *6,* 183–200.

Fouts, R. S. The use of guidance in teaching sign language to a chimpanzee. *Journal of Comparative and Physiological Psychology,* 1972, *80,* 515–522.

Fouts, R. S. Acquisition and testing of gestural signs in four young chimpanzees. *Science,* 1973, *180,* 973–980.

Gardner, B. T., & Gardner, R. A. Evidence for sentence constituents in the early utterances of child and chimpanzee. *Journal of Experimental Psychology: General,* 1975, *104,* 244–267.

Gardner, R. A., & Gardner, B. T. Comparative psychology and language acquisition. In K. Salzinger & F. Denmark (Eds.), *Psychology: The state of the art.* New York: New York Academy of Sciences, 1978.

Gray, H. Learning to take an object from the mother. In A. Locke (Ed.), *Action, gesture and symbol: The emergence of language.* London: Academic Press, 1978.

Greenfield, P., & Savage–Rumbaugh, E. S. Perceived variability and symbol use: A common language–cognition interface in children and chimpanzees. Paper presented at the IXth meeting of the International Primatological Conference, Atlanta, Georgia Society. August, 1982.

Greenfield, P. M., & Smith, J. H. *The structure of communication in early language development.* New York: Academic Press, 1976.

Locke, A. *The guided reinvention of language.* London: Academic Press, 1980.

Miles, L. W. Discussion paper: The communicative competence of child and chimpanzee. In S. R. Harnad, H. D. Steklis, & J. Lancaster (Eds.), *Origins and evolution of language and speech.* (Vol. 280.) New York: The New York Academy of Sciences, 1976.

Patterson, F. The gestures of a gorilla: Language acquisition in another pongid. *Brain and Language,* 1978, *5,* 72–97.

Premack, D. *Intelligence in ape and man.* Hillsdale, N.J.: Erlbaum, 1976.

Rumbaugh, D. M. (Ed.) *Language learning by a chimpanzee: The Lana project.* New York: Academic Press, 1977.

Rumbaugh, D. M., & Gill, T. The mastery of language–type skills by the chimpanzee (*Pan*). In S. R. Harnad, H. D. Steklis, & J. Lancaster (Eds.), *Origins and evolution of language and speech.* (Vol. 280.) New York: The New York Academy of Sciences, 1976.

Rumbaugh, D. M., Savage–Rumbaugh, E. S., & Scanlon, J. L. The relationship between language in apes and human beings. In J. King & J. Fobes (Eds.), *Primate behavior.* New York: Academic Press, 1982.

Sanders, R. J. *The influence of verbal and nonverbal context on the sign language conversations of a chimpanzee.* PhD Thesis, Columbia University, 1980.

Savage, E. S. *Mother-infant behavior in group-living captive chimpanzees.* PhD Thesis, University of Oklahoma, 1975.

Savage-Rumbaugh, E. S. Symbolic communication – its origins and early development in the chimpanzee. *New Directions for Child Development,* 1979, *3,* 1-15.

Savage-Rumbaugh, E. S. Can apes use symbols to represent their world? In T. A. Seboek & R. Rosenthal (Eds.), *The Clever Hans phenomenon: Communication with horses, whales, apes, and people.* (Vol. 364.) New York: The New York Academy of Sciences, 1981.

Savage-Rumbaugh, E. S., & Rumbaugh, D. M. Language analogue project, phase II: Theory and tactics. In K. Nelson (Ed.), *Child language.* (Vol. 2.) New York: Gardner, 1980.

Savage-Rumbaugh, E. S., & Rumbaugh, D. M. A pragmatic approach to chimpanzee language studies. In H. E. Fitzgerald, S. A. Mullings, & P. Gage (Eds.), *Child nurturance.* (Vol. 4.) New York: Plenum, 1982.

Savage-Rumbaugh, E. S., Rumbaugh, D. M., & Boysen, S. Describing chimpanzee communication: A communication problem. *The Behavioral and Brain Sciences,* 1978, *1,* 614-616. (a)

Savage-Rumbaugh, E. S., Rumbaugh, D. M., & Boysen, S. Linguistically mediated tool use and exchange by chimpanzees (*Pan troglodytes*). *The Behavioral and Brain Sciences,* 1978, *1,* 1-28, 539-554. (b)

Savage-Rumbaugh, E. S., Rumbaugh, D. M., & Boysen, S. Symbolic communication between two chimpanzees (*Pan troglodytes*). *Science,* 1978, *201,* 641-644. (c)

Savage-Rumbaugh, E. S., Rumbaugh, D. M., & Boysen, S. Do apes use language? *American Scientist,* 1980, *68,* 49-61.

Savage-Rumbaugh, E. S., Rumbaugh, D. M., Smith, S. T., & Lawson, J. Reference: The linguistic essential. *Science,* 1980, *210,* 922-925.

Terrace, H. S. Introduction: Autoshaping and two-factor learning theory. In C. M. Locurto, H. S. Terrace, & J. Gibbon (Eds.), *Autoshaping and conditioning theory.* New York: Academic Press, 1981.

Terrace, H. S., Petitto, L. A., Sanders, R. J., & Bever, T. G. Can an ape create a sentence? *Science,* 1979, *206,* 891-900.

van Lawick-Goodall, J. The behavior of free-living chimpanzees in the Gombe Stream Reserve. *Animal Behavior Monograph,* 1968, *1,* 161-311.

Werner, H., & Kaplan, B. *Symbol formation.* New York: Wiley, 1963.

18 ABSENCE AS INFORMATION: SOME IMPLICATIONS FOR LEARNING, PERFORMANCE, AND REPRESENTATIONAL PROCESSES

Eliot Hearst
Indiana University

I. INTRODUCTION

When Kurt Koffka (1935) asked his readers to consider why we normally "see things and not the holes between them," individual reaction differed as to whether the question was profound, trivial, misleading, or meaningless. A similar profile of opinions would probably greet a contemporary student of learning and cognition who wondered aloud why we seem to more easily associate things with each other than with the temporal or spatial holes that surround or separate them. If he or she were a persistent person, this psychologist might argue that the selfsame kind of associative bias affects the thinking of researchers, so that certain possibilities for learning have been neglected in the laboratory. Until fairly recently, for example, the study of Pavlovian conditioning in animals has concentrated on phenomena produced by closely contiguous occurrences of conditioned stimuli (CSs) and unconditioned stimuli (USs), and not on phenomena engendered by delivering USs only in the absence of some definite stimulus or by presenting discrete CSs that are never followed shortly by another event.

On the surface, cases in which the absence of something serves as a signal, or is itself signaled by the presentation of a particular stimulus, may appear to have weak ecological validity. The international spy does not indicate the need for a secret meeting by removing a red flower from the pot on his window sill, and the baseball coach does not use a failure to touch his cap as a sign for players to swing the bat or steal a base. Billboards do not announce times when a special bargain is unavailable, and beacons do not start flashing when a danger is over. Like the figure-ground principle that Koffka was exemplifying in his question, nonoccurrences and omissions of some event or behavior seem generally less salient, memorable, or informative than occurrences or additions. Without the establishment of special conditions to overcome or replace attention to and control by positive happenings, efficient learning and performance based on absence seem comparatively hard to achieve. Often theorists find it useful to treat absence in terms of prevailing states or circumstances -- situational cues, contexts, background stimuli -- which, to continue the Gestalt metaphor, represents a way of altering or reversing figure-ground relations.

311

At any rate, control by the absence of something constitutes an apparently significant, but relatively unexplored topic that pertains to many areas of psychology -- perception, conditioning, memory, cognition, and so forth. In a more general sense, "absence" enters the chapters of this book in several guises and is related to some important issues in animal and human cognition. A few examples seem worth mentioning. As one instance, Terrace (Chapter 1, this volume) states that explanations in terms of internal representations constitute the essence of cognitive approaches to learning and performance; in particular, central encodings of past events are assumed to guide behavior when relevant external cues are absent, as in the Olton (radial) maze, delayed matching-to-sample procedures, and other tasks for studying animal memory. Furthermore, the learning (competence) -- performance distinction in psychology arose originally from the realization that special assays are often needed to reveal hidden knowledge gained by organisms from prior experiences. A major lesson from such work corresponds to the well-known epigram: Absence of evidence is not evidence of absence. Finally, recent research on human information processing has examined a variety of pertinent topics, which have not yet received satisfactory theoretical integration despite their surface similarities: search for presence vs. absence of certain features, the difficulty of making judgments on the basis of negation and nonoccurrence, the bias of subjects towards confirmation rather than disconfirmation of their hypotheses, the relative utilization of positive and negative instances in concept learning, and the potential value of attaching positive labels to nonoccurrences as a way of increasing their salience and effectiveness (see Bourne, Ekstrand, Lovallo, Kellogg, Hiew, & Yaroush, 1976; Einhorn & Hogarth, 1981; Fazio, Sherman, & Herr, 1982; Hearst, 1978; Newman, Wolff, & Hearst, 1980; Nisbett & Ross, 1980).

For a long time my colleagues and I have been directly or indirectly concerned with "nothing" -- the behavioral effects and functions of absences or nonoccurrences of events. Such topics as avoidance learning (Hearst, 1969), conditioned inhibition (Hearst, 1972)latent learning (Hearst, 1975, 1978), and omission training (Hearst, 1979) have attracted our attention over the years. In this chapter I first briefly review some of Peter Kaplan's and my recent work with pigeons placed on trace conditioning procedures -- in which an empty period intervenes between a Pavlovian CS and US -- and then I focus on animal and human research concerning the feature-positive effect, which refers to the superior discrimination learning shown by subjects for whom the unique feature distinguishing two kinds of trials is present on positive rather than negative displays. All the results seem to indicate special difficulties in the effective use of stimulus absence and may have implications for our understanding of how experience with absence could be represented in memory. The issue of whether similar underlying mechanisms are responsible for parallel effects found in animals and humans will also be discussed, but it would be premature to offer any strong conclusions along these lines. My intention in this chapter is more to raise questions than to answer them.

II. TRACE CONDITIONING, CONTIGUITY, AND GAP FILLING: MAKING ABSENCE INFORMATIVE

The trace conditioning procedure is one basic arrangement developed by Pavlov for studying CS-US associations. Animals typically display weak or nonexistent learning when the CS and US are separated by an unfilled interval of more than a few seconds; any successful conditioning is usually attributed to neural traces of the CS that persist until US presentation and provide contiguity between the two events. Thus trace conditioning should be and is better with relatively intense CSs, which are presumed to generate longer neural after-effects.

However, Kaplan and Hearst (1982, 1983) demonstrated good conditioning in pigeons to a trace CS that ended 12-60 sec before US, provided that the period between the US and the action of the *next* CS (that is, the intertrial interval, ITI) was itself filled with some auditory or visual stimulus.   Of course, research on serial conditioning had previously shown that filling the gap between the offset of CS and the onset of the next US also yields consistent evidence of conditioned responses (CRs) to the CS. However, serial conditioning arrangements place the gap filler -- an explicit, external stimulus -- in actual temporal contiguity with the US, and the successful outcomes did not seem to us as interesting as for the case in which the CS-US interval remains unfilled but stimulation during other segments of conditioning trials is introduced.

In Kaplan and Hearst's work the birds received trials consisting of 12-sec illuminations of a green disc (CS) followed *x* sec later by inevitable delivery of grain (US).   This procedure is an example of auto-shaping or sign-tracking (see Hearst & Jenkins, 1974; Schwartz & Gamzu, 1977) and, in our different experiments, the interval between CS offset and US onset was 12, 30, or 60 sec. The ITI was of variable duration, averaging out to a value five times the interval between CS and US.   No approach or pecking behavior developed toward the CS when both the CS-US gap and ITI were empty -- that is, standard trace conditioning was, as expected, not effective. In contrast, strong approach to the CS appeared when the two time periods were differentiated by the addition of an external stimulus, either presented *only* during the CS-US gap (which replicates the other work on serial conditioning) or *except* during the gap.   Figure 18.1 displays results for different groups trained with 12-sec intervals between CS offset and US onset. Analogous findings were obtained with 30-sec and 60-sec CS-US intervals in other experiments.

In this set of experiments at least, the failure of standard trace conditioning could not have been due to the mere absence of external stimulation during the CS-US gap. Lack of contiguity between CS and US does not preclude conditioning to CS, provided that the period of stimulus absence is made informative by appropriate differential stimulation at other times (in fact, Kaplan has found that trace conditioning is also successful if the ITI and CS-US gap are both left unfilled, but the former is made 10 or more times longer than the latter in our otherwise standard situation). Although some exteroceptive or temporal basis for discriminating CS-US gaps from intertrial periods (cf. Mowrer & Lamoreaux, 1951) seems necessary for conditioning to occur when CS and US are not closely

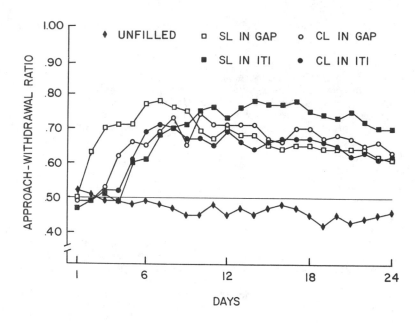

FIG. 18.1. Mean approach-withdrawal ratios to CS (a green key light) across sessions for birds in each of five different groups. One group (UNFILLED) received standard trace conditioning. Two other groups had either a signal light (SL) or clicking sound (CL) presented during the CS—US gap. The final two groups had either the SL or CL presented during the intertrial interval. Approach-withdrawal ratios above 0.50 indicate approach behavior toward CS. (Adapted from Kaplan & Hearst, 1982.)

contiguous, the outcomes of subsequent pigeon experiments performed in our laboratory indicate that such discriminability is not sufficient for the occurrence of strong conditioning; for example, the differential external stimulus conditions in the ITI must remain in force until the offset of CS and cannot terminate at CS onset. This somewhat surprising result appears important for assessing alternative explanations of the overall findings, which are beyond the scope of the present chapter. Interested readers will find relevant discussions of the possible role of second-order conditioning and general or local contextual influences, as well as some extensions to inhibitory conditioning, in prior articles (Kaplan & Hearst, 1982, 1983).

Our general conclusion is that "absence" and contiguity need to be conceptualized in a more relativistic manner than previously, because their effects depend on the detailed serial structure and temporal organization of individual conditioning episodes. We find it rather remarkable that various phenomena produced by experimental analysis of such a simple, fundamental, and classic arrangement as the trace paradigm are not yet well understood; the addition of only one more external stimulus introduces a complicated set of outcomes, contingent on the exact placement of that stimulus. Nevertheless, long periods of stimulus absence can clearly be bridged under appropriate circumstances.

III. DISCRIMINATIONS BASED ON PRESENCE VERSUS ABSENCE OF A FEATURE

A. THE BASIC PHENOMENON

Jenkins and Sainsbury (1969, 1970) reported that when a single distinguishing feature (say, a small dot) is present on positive, food-reinforced visual displays (S+) and is absent on negative, nonreinforced displays (S-), pigeons show much better discrimination learning than under the reverse conditions. This feature-positive (FP) superiority is often enormous in pigeons; throughout Jenkins and Sainsbury's studies, only one out of approximately 50 pigeons ever mastered a feature-negative (FN) discrimination. Although the FP vs. FN difference fails to appear with use of a very salient feature like a line across the key or an extremely bright dot, and the magnitude of the outcome is rarely as huge as in Jenkins and Sainsbury's research (we know, of course, that infrahuman subjects can eventually learn a variety of discriminations corresponding to Pavlov's conditioned-inhibition paradigm: A+, AX-), the basic phenomenon has rather broad generality across animal species, types of reinforcers and stimulus modalities, and discrimination procedures (see below and Hearst, 1975, 1978).

The lefthand sections of Figure 18.2 display one detailed example of a feature-positive effect (FPE) obtained with pigeons in our laboratory. All trials involved the illumination of three horizontally-spaced keys and the feature was a key of one color (say, red) illuminated along with two keys of a different color (say, green) on trials containing the feature. The red light could appear on any one of the three keys. Feature-absent trials comprised illumination of the same three keys, but all green. Although reinforcement would occur for responses to any key on positive trials, within a short time the FP subjects began to restrict their pecking to the red key and mastered the overall discrimination (red keys are not present to peck on negative trials). In contrast, the FN subjects quickly came to direct their pecking at the green keys and, over 12 sessions of training, showed virtually no discrimination between positive and negative trials (green keys are present on both kinds of trials). One way of summarizing the results is. to say that birds in both groups soon channel their behavior toward the best positive predictor of food in the situation -- the 100%-reinforced red key in the FP group and the 50%-reinforced green keys in the FN group.

Worthy of note was the failure of FP subjects to reverse their discrimination when later placed on the FN procedure. They did stop pecking at the distinctive feature within a few sessions, but continued to respond equally to keys of the other color on positive and negative trials for the duration of the FN phase. On the contrary, subjects switched from FN to FP (lower panel, Figure 18.2) eventually mastered the FP

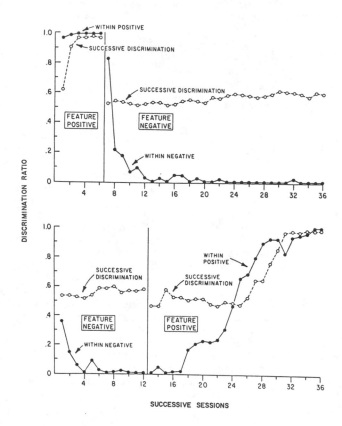

FIG. 18.2. Median discrimination ratios for groups of pigeons trained with a distinguishing feature (key color) either on positive, i.e., reinforced trials (feature positive) or on negative, i.e., nonreinforced trials (feature negative). The index for the development of the successive discrimination (positive vs. negative displays) is the ratio of the number of responses on the positive display to the total number of responses on both displays; this index approaches 1.0 as responses on the negative display approach zero. The index for the development of a discrimination between common and feature key colors within the display containing both key colors is given by the ratio of the number of responses on the feature key to the total number of responses on that display. On positive trials the first peck at any key after 5 sec produced grain. All negative trials lasted 5 sec. After initial exposure to either the feature–positive or feature–negative condition, subjects were switched to the other condition. (From Hearst, 1975.)

discrimination, although an appreciable retardation of such learning occurred as a result of their prior exposure.

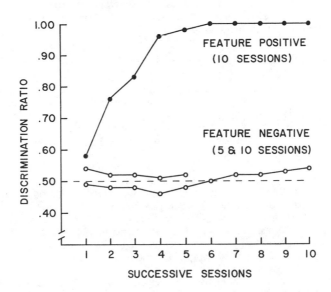

FIG. 18.3. Median discrimination ratios for feature-positive and feature-negative pigeons trained for 5 or 10 sessions on a discriminative auto-shaping procedure. The discrimination ratios were calculated by dividing the number of key pecks on positive trials by the total number of pecks on both positive and negative trials.

The use of an operant task is not necessary to produce a robust FPE in pigeons. In fact, almost all our research on the topic has involved simple auto-shaping procedures in which every positive trial (CS+) is followed by access to grain and every negative trial (CS−) by no US. Figure 18.3 illustrates standard results from an arrangement to which we return shortly when describing recent, unpublished work (see Figure 18.6). All groups in these experiments included 6-9 subjects. The two types of displays involve illumination of a single key for 6 sec, with grain US immediately following half the daily 40 trials, separated by ITIs averaging 60 sec. The CS+ in the FP group is illumination of a green key with a small white square at its center, and the CS− is illumination of a solid green key; for the FN group, these conditions are reversed.

Almost all FP subjects master the discrimination within five sessions, but even after 10 sessions FN subjects usually exhibit little evidence of differential responding to CS+ and CS−. Though we have not actually measured where birds' key pecks are directed, it is highly probable that the FP subjects soon come to peck directly at the white square and the FN subjects at some green section of the key -- the best positive predictors of food, respectively (see Hearst, 1975, 1978; Jenkins &

Sainsbury, 1969, 1970). Incidentally, after approximately 10 sessions of FN training on this specific procedure, many birds do begin to show improved performance -- although a majority of them do not attain discrimination ratios above 0.90, even after a total of 20 days on the FN arrangement. The FP subjects are typically close to 1.00 following 5 days of training and they remain at that level indefinitely.

B. GENERALITY OF THE FPE: HUMANS, TOO

Although the FPE could arise for different reasons in various situations, organisms, and tasks, our interest in the phenomenon has been sustained and even facilitated by several seemingly parallel findings.   The effect is not an isolated experimental curiosity that appears only in pigeons pecking at small objects.   Analogous outcomes have been obtained during CER training in rats with visual and auditory distinctive or common elements (Reberg & LeClerc, 1977); during avoidance conditioning with cats (Diamond, Goldberg, & Neff, 1962); and under circumstances in which the distinguishing feature precedes and does not overlap common trial stimuli in pigeons, rats , and rabbits (Bottjer & Hearst, 1979; Reberg & LeClerc, 1976; Terry & Wagner, 1975; Thiels, 1982; Wolff & Hearst, 1975).   The FPE has been reported in several experiments with young children (see a review in Newman et al., 1980).   Moreover, we recently carried out a series of studies with adult human beings that, at least on the surface, revealed a similar phenomenon.

Several of these experiments (Newman et al., 1980, Experiments 1-3) used standard simultaneous or successive concept-learning procedures in which college students had to classify sets of 3 or 4 letters or symbols into Good or Not Good categories.   They were not told beforehand which rules or attributes might be involved in successful categorization.   The only basis for distinguishing between the two types of items was simply whether a particular letter or symbol was present or absent, and different subexperiments examined (a) immediate vs. delayed feedback about the correctness of the subject's response, and (b) variation vs. uniformity of the common, irrelevant elements from trial to trial.

For the FP subjects combinations of elements including the feature were Good, whereas for the FN subjects they were Not Good.   In the nomenclature for rules governing concept-learning tasks, the former would be considered an example of Affirmation and the latter of Negation; and there is apparently no empirical or theoretical reason to expect a big difference between the two (Bourne, 1970; Neisser & Weene, 1962).

However, in every one of our experiments a clear and significant superiority was exhibited by the FP group, in terms of number of trials before subjects could verbalize the correct solution.   When results from all the different experiments along these lines are combined, approximately 90% of the FP subjects solved the problem and only 25% of the FN subjects.

To reduce the likelihood that our instructions had strongly biased subjects toward attending to Good and disregarding Not Good displays -- which would obviously place the FN subjects at a disadvantage -- we performed another human experiment, with a stimulus-reinforcer display panel that could easily be adapted for pigeon work, too (Experiment 7 of Newman *et al.*'s paper describes the strong FPE obtained in analogous research with pigeons). The 4-key displays, containing different geometric symbols or a blank space, were presented for 5 sec each, separated by 7.5-sec ITIs. Subjects were told that presses anywhere on a correct display would earn them 5 points on a counter located beneath the keys, whereas presses anywhere on an incorrect display would cost them 5 points. The instructions encouraged them to gain as many points as they could, not only by pressing correct displays but also by refraining from pressing incorrect displays. Each subject received a total of 144 trials, half containing the feature (one particular symbol) and half not. The first 72 trials conformed to either the FP or FN arrangement, and then subjects were transferred to the reverse discrimination for 72 trials involving the same feature, without any interruption in the experiment.

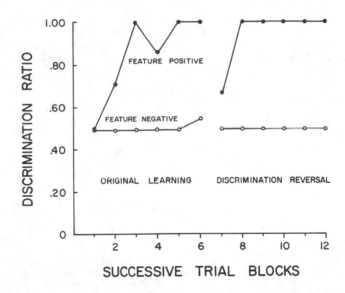

FIG. 18.4. Median discrimination ratios (proportion of responses on positive trials relative to responses on all trials) over 12-trial blocks for the feature-positive (FP) and feature-negative (FN) subjects in human experiments involving a 4-key display. After the sixth block, FP subjects were transferred to the FN procedure (but are still indicated by solid circles), whereas the FN subjects were transferred to the FP procedure (but are still indicated by open circles). (From Newman et al., 1980.)

Figure 18.4 displays the FPE obtained in this experiment. In the first phase, 5 of the 6 subjects in the FP condition eventually attained a discrimination ratio of 1.00, whereas only 1 of the 6 FN subjects ever did better than 0.71 during a single trial block. All five subjects who had learned the FP discrimination in the first phase successfully reversed their strategy and quickly mastered the FN discrimination. In contrast, only 1 or 2 of the subjects switched from FN to FP ever displayed any evidence of learning the latter discrimination. This overall result is typical of other (mostly unpublished) feature-reversal experiments with human subjects exposed to various similar tasks in our laboratory, and the finding differs from the results with pigeons shown in Figure 18.2. Humans mastering FP reverse easily when placed on FN; but once any subject has failed to learn within 50 or 60 trials, he or she is extremely unlikely to achieve a correct solution, even if subsequently placed on a much easier (i.e., FP) discrimination.

FIG. 18.5. Examples of feature (left) and nonfeature (right) displays in the "smoke" experiment with human subjects. All displays included four quadrants, three containing pictures and the fourth blank. On feature displays, smoke (two spiral lines) rose from the chimney or smokestack of one of the pictures, as in the upper right quadrant of the left display. Only one type of display was presented on any trial, i.e., a successive-discrimination procedure was in force throughout the experiment. (From Newman et al., 1980.)

In a final experiment with human beings, which we consider the simplest and perhaps the theoretically most convincing of the series, we employed 4-quadrant displays incorporating line drawings of boats, houses, and locomotives as stimulus materials. Each of these pictures occurred an equal number of times on each of the four quadrants (see Figure 18.5). We still wondered whether the instructions and reinforcing outcomes in prior experiments had failed to make it sufficiently clear to subjects that correct performance on negative trials of a successive discrimination was as critical as correct performance on positive trials. Therefore, in this study -- which resembled a probability-learning experiment (see Estes, 1976) and had characteristics similar to the above pigeon research because two stimulus events occurred independently of the subject's behavior -- a brief red light followed half the trials, regardless of what the subject did. A single poker chip was delivered to the subject for correct predictions about *either* the appearance or the nonappearance of the light.

Subjects were instructed beforehand that only some displays would be followed by the light, and that their task was to predict, during the 5-sec duration of a display, whether the light would or would not appear. For FP subjects the presence of "smoke" in the display meant that light would occur, whereas the absence of smoke signaled the nonoccurrence of the light. Conditions were reversed for FN subjects. Of course, "smoke" was equally paired with the house, boat, or locomotive on feature trials.

Seven of the 8 FP subjects verbalized the correct solution long before the 96-trial session elapsed, whereas only 1 of the 8 FN subjects was successful within the maximum of 96 trials. When the presence of the distinguishing feature was followed by the presence of the light, and the absence of the feature by the absence of the light, learning was easy. However, learning was rare when the absence of the feature was followed by the presence of the light and the presence of the feature by the absence of the light.

Thus, our experiments with human beings demonstrated a robust FPE in a variety of tasks. Subjects acquire a discrimination readily if the distinguishing feature appears on displays associated with a positive event (gain of points; illumination of a light; assignment to the *Good* category), but perform poorly if the feature appears on displays associated with a negative event (loss of points; nonoccurrence of a light; assignment to the *Not Good* category). Subjects who had learned the FP discrimination showed easy reversal and quickly mastered the FN discrimination. On the other hand, the FP discrimination was difficult for subjects who had already failed to learn the FN discrimination.

C. LEARNING VERSUS PERFORMANCE

Looked at superficially, the FPE has considerable reliability and generality. Furthermore, the direction, if not the magnitude of the effect, is basically consistent with several theoretical accounts, including the interpretation of conditioning offered by Rescorla and Wagner (1972; see Hearst, 1978, and Jenkins, 1973, for discussions of various explanations). Despite these points, we have long suspected that, in standard pigeon experiments at least, the failure of FN subjects to exhibit clear differences in behavior between positive and negative trials during training may reflect a deficit in performance rather than learning. After all, FN birds display a rapid reduction in their initial responding to the distinctive feature and soon peck only at some common feature (see Figure 18.2 above; Hearst, 1975; Jenkins & Sainsbury, 1969, 1970), which is a clear indication that they differentiate among the various elements of the entire display, that is, they have "noticed" the feature.

The propensity of pigeons to direct their behavior toward the best positive predictor of an appetitive US -- if the signal is visual and easily localized (see Hearst & Jenkins, 1974) -- could serve to mask any differential behavioral tendencies that actually exist toward the positive and negative displays.   The strength of this sign-tracking is so great that pigeons continue to approach and contact a positive signal even if (a) such behavior is programmed to cancel otherwise inevitable access to grain (the so-called omission effect; see Jenkins, 1973, and Peden, Browne, & Hearst, 1977), and (b) the signal is followed by grain only 10-50% of the time (Gibbon & Balsam, 1981).   On this basis, it may not be surprising that FN birds continue to respond to some common element of the displays on every presentation -- the element is followed by grain on half the trials -- and that they do not alter their behavior when the distinctive feature is also present.

To unmask any hidden learning in the FN case, a variety of procedures for reducing the predictiveness of the trial signal seemed potentially suitable to us.   During the past year we have investigated several such manipulations, the most obvious of which involved removal of all grain deliveries in the situation after 5-10 days on the standard FN procedure. In addition, we examined effects of: adding extra USs in the ITI; giving USs only in the ITI, never within 10 sec of any trial; and shortening the mean ITI to 15 sec.   For purposes of comparison and to check on whether any kind of major change in the procedure would unmask FN learning, other groups of FN subjects received presentations of unexpected red key lights in the ITI, or were exposed to considerably longer ITIs (mean of 120 sec) than had been used during original FN training.

In this series of experiments the target group of birds was first placed for 5 or 10 days on the FN procedure described above in connection with Figure 18.3.   During this period relatively few birds ever show any evidence of learning the discrimination between a green CS+ and a green-plus-white-square CS-. Figure 18.6 demonstrates strikingly that the pigeon's failure to display differential behavior on a FN discrimination in this situation does not represent a lack of appropriate learning, but rather a deficiency in performance.   When placed on extinction (no reinforcement after either CS+ or CS-), virtually every subject in this and several other similar experiments soon began to display excellent discriminative performance.   The data of Figure 18.6 show that replacement of subjects on the regular FN procedure after this period of extinction leads to a deterioration of performance (but usually not back to its prior levels around 0.50).   However, a second extinction phase reinstates excellent performance.

In other words, after 5-10 days of FN training, evidence of discrimination learning is better under conditions of no reinforcement at all than with differential reinforcement for responding to CS+ and CS-. This finding is reminiscent of the gradual steepening of excitatory generalization gradients that occurs in operant studies after complete removal of reinforcement (see Hearst, 1969, and Honig & Urcuioli, 1981), and it recalls Hull's (1934) observation that extinction "resuscitates" appearance of a goal gradient (stronger responding the closer subjects are to their goal). The general outcome also brings to mind the evidence for learning that was revealed by various test procedures in classic studies of behavior during "presolution" stages of discrimination acquisition -- experiments that

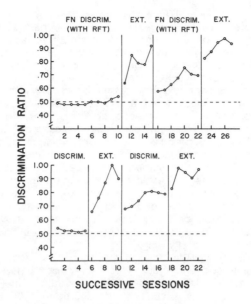

FIG. 18.6. Median discrimination ratios for pigeons originally trained for 5 or 10 sessions on the FN procedure of Figure 18.3. Then each group was successively exposed to five days of complete extinction, seven days back on the FN discrimination, and five more days of extinction.

were performed to differentiate between continuity and noncontinuity explanations of animal discrimination learning (see Amsel, 1967, e.g., p. 15; Sutherland & Mackintosh, 1971). In many such studies of selective attention, appropriate assays indicated that subjects had actually learned something about other aspects or features of the discriminative stimuli while they appeared to be responding only on the basis of one particular characteristic, e.g., spatial position.

The extinction results in Figure 18.6 suggest that FN subjects really learn to discriminate between positive and negative trials as fast as do FP subjects (within 5 sessions; see Figure 18.3), but they fail to exhibit this learning while still exposed to the standard discrimination procedure. Space does not permit description and analysis of the effects produced by the other procedural manipulations listed above as potential candidates for revealing evidence of discrimination learning in FN subjects. However, the results indicate clearly that complete removal of reinforcement from the situation is not essential for unmasking prior learning. After 5 days on the FN procedure, introduction of extra unsignaled US deliveries or of a negative CS–US contingency, as well as reduction of the mean ITI, also reveal much greater responding to CS+ than to CS– in subjects that, as usual, had not previously exhibited differential performance to the two stimuli. On the other hand, changes in the general procedure that do not reduce predictiveness of the trial signal (inserting occasional red key lights in the ITI, doubling the mean ITI) fail to unmask FN learning.

Thus, the extent of a subject's learning a particular presence vs. absence discrimination, or the degree of attention to specific features or dimensions, may not be well revealed in standard measures of acquisition, but can be manifested under certain other conditions or with different response indices. Conceivably, many differences of this kind between learning and performance could be encompassed within the framework of signal detection theory (see Honig & Urcuioli, 1981, pp. 413-414), as a reflection of changes in the subject's criterion (bias) to respond. The strong sign-tracking exhibited in our pigeon research presumably goes hand in hand with a very permissive criterion, which should become much stricter under circumstances in which the predictiveness of the common elements appearing on trials is somehow reduced. To evaluate this kind of interpretation, as well as to establish the generality of, say, extinction procedures for revealing FN discrimination learning, a worthwhile plan would be to undertake parallel experiments with (a) aversive USs (e.g., in the rat CER arrangement), or (b) diffuse auditory and visual stimuli that do not elicit sign-tracking, or (c) introduction of penalties for responding to CS– from the very start of training.

I seriously doubt whether our human subjects who showed no FN learning would have suddenly begun to discriminate between the positive and negative displays if all feedback were removed for their responses. In the tasks we used with human beings, the solution of FP or FN problems seems mainly to entail discovery of the distinctive feature; once the (few) subjects who solved the FN problem noticed the differential presence of the feature on negative vs. positive trials, their major obstacle was apparently surmounted. The question is why identification of the feature is so rare in human subjects when it appears on negative rather than positive displays (see a relevant discussion in Newman et al., 1980). On the other hand, in the arrangements we used with pigeons, both FN and FP subjects apparently notice the distinctive feature from the start of training, but FN birds cannot easily stop responding to the highly predictive common elements.

Currently, we do not know whether all reported instances of failures to learn FN discriminations in pigeons represent a performance, rather than learning or detection deficit. For a more appropriate comparison with human results, the extinction behavior of FN pigeons after unsuccessful training on the complex 4-key symbol discrimination mentioned in connection with Figure 18.4 should be investigated. Pertinent is the facility of human beings to reverse their discriminative performance when switched from FP to FN –– which was not the case for pigeons (see, for example, Figure 18.2). This outcome implies that once our human subjects discovered the predictive feature they were guided by a verbally-mediated, relatively flexible rule (e.g., "it's whether the smoke is present or not") that can readily be transferred to the opposite discrimination. In contrast, the performance of our pigeons seemed rather rigidly controlled by characteristics of and relations between specific external stimuli and not by analogous general rules –– at least in situations that foster sign-tracking tendencies and do not entail prior exposure to many discrimination reversals or similar problems. This conclusion parallels Kendler and Kendler's (1975) interpretation of species and developmental differences in discriminative transfer of other kinds.

IV. ABSENCE: DETECTION, ASSOCIATION, REPRESENTATION, AND BEHAVIORAL CONTROL

It is not so simple to treat "absence" in a psychologically meaningful way. Innumerable things are always simultaneously absent and yet we obviously differentiate between the absence of a US and of a CS, between the absence of a friend and the absence of thunder. Subjects in our studies have to identify which of the very large set of absent features or events is relevant. The nonoccurrence of countless stimuli is normally the prevailing state in natural and experimental settings -- a condition interrupted by relatively brief and rare events whose occurrence is therefore potentially more informative than their nonoccurrence.

Newman et al. (1980) considered the possibility that organisms capable of learning may have been equipped via evolution with a strong bias toward monitoring and responding to the presence of events rather than their absence. A biological system of that kind would require a comparatively simple and economical set of physiological structures and processes. (The presence of powerful internal cues corresponding to, say, hunger and thirst drives would instigate appetitive search patterns, not the absence of food or water in the immediate environment.) Moreover, organisms capable of learning may more easily form and/or use associations between the occurrences of two events than between the nonoccurrence of one event and the occurrence of another. On this view, the tendency for animals and humans to utilize positive rather than negative information may not represent a well-learned general strategy derived from extensive past experiences of individual organisms, as some writers (e.g., Bourne, 1974; Bruner, Goodnow, & Austin, 1956) have suggested on the basis of research concerning human information processing. Instead, the tendency might represent a strong biological predisposition with considerable survival value.

Related to this general idea is Wolff's (1983; see also Sperling & Perkins, 1979) finding that even when the predictiveness of common features on the response key is arranged to be no stronger than general contextual cues -- by presenting a green light continuously on the key -- mere presentation and removal of a small white square on the key still yields a definite FPE in pigeons. This outcome argues against a primary role for relative cue predictiveness in explaining many of the effects we reported in pigeons, and suggests that my above emphasis on the control exerted by common elements presented on trials may be misplaced or unwarranted. Perhaps, in some cases, the removal of one element from an otherwise fixed set of stimuli may simply be less discriminable than the addition of that element. Human beings seem to notice quickly the new mustache that an acquaintance has grown, but have a harder time identifying the facial change brought about by the removal of a mustache ("something's different, but what?").

However, the notion that the associative learning, choice performance, and decision making of organisms are guided by certain widespread natural or learned biases with respect to processing events vs. non-events is compatible with findings of Allan and Jenkins (1980). They reported that if two relatively rare events have occurred together, human beings assume the second would not occur in the absence of the first. The bearing of such a bias on the development and strength of superstitious behavior is direct. "In everyday circumstances we seek the causes of events rather than of prevailing states or non-events. Further, in looking for the cause of an event, we naturally seek out another event rather than a prevailing state" (Allan & Jenkins, 1980, p. 10).

In the light of our recent findings with pigeons, analysis of learning vs. performance effects appears needed in situations that presumably exhibit a FPE. I should also mention the need for greater precision in formulating criteria for deciding what a "feature" is (see also Hearst, 1978). Theoretical integration of FPE research in various organisms and arrangements awaits work along these lines, as well as evaluation of the fact that human subjects were informed or could infer that there was a *perfect* solution to their discrimination tasks. The arrangements with young children and animals have not been designed to encourage such an inference.

It seems intuitively clear that the absence of something normally exerts strong behavioral control only when there is good reason for the subject to expect the occurrence of that event. A student who never comes to class would hardly be conspicuous by his absence. Once a subject has become habituated to regular repetitions of a particular stimulus, omission of that stimulus will evoke a strong orienting response (Sokolov's, 1963, missing-stimulus effect). Analogously, in discussing the use of negation and the possible referents of *no, not, never,* and *nothing* in verbal behavior, Skinner (1957, pp. 322–326) basically agreed with Bertrand Russell that whenever we say *it is NOT raining* there must be good justification for our saying *It IS raining;* "not" is a qualifier or, in Skinner's terms, an autoclitic. Several influential past and current theories of aversive learning (see Dinsmoor, 1977; Overmier, 1979) have proposed fear reduction or termination of conditioned aversive stimuli as the source of reinforcement for avoidance behavior, at least partly because the mere absence of something (the US) presumably could not itself serve that function -- US absence would only lead to extinction of the response consistently preceding it. However, if the subject has come to expect shock at certain times, then the absence of shock could represent improved conditions, and associated stimuli should act as safety signals (cf. Gibbon, 1972). Finally, according to the Rescorla–Wagner (1972) model, conditioned inhibition develops to a stimulus or element paired with no–US, but only after excitatory control by concurrent discrete or contextual stimulation has been established (the subject expects USs at that time or in that situation).

A plausible conjecture is that the influence of "nonoccurrence" ought to be enhanced by somehow increasing the salience or informativeness of the non-event. In Kaplan and Hearst's work demonstrating facilitative effects of ITI fillers in trace conditioning (see Figure 18.1 above), the absence of explicit external stimulation was made a relatively rare, distinctive circumstance that preceded and predicted US occurrence. Specific instructions can promote the detection and utilization of nonoccurrence by human subjects. For instance, FN learning greatly improves if subjects are told beforehand that the presence *or* absence of some feature could be important, or that information on the *Not Good* items is critical (see Newman *et al.*, 1980).

An alternative to accentuating stimulus absence would be, wherever possible, to provide positive markers as substitutes, along the lines of Nisbett and Ross's (1980, pp. 48–49; see also Fazio *et al.*, 1982) suggestion that the concreteness, memorability, and informativeness of the nonoccurrence of certain human behaviors can often be augmented by encoding them in terms of the presence of something, e.g., the absence of sexual responsiveness is interpreted as frigidity.   The absence of a US following negative trials in a standard animal discrimination task frequently involves mere reinstatement of the ITI condition; learning might be better if a definite external cue, of no intrinsic value in itself, replaced simple nonreinforcement.   See Hearst, 1978, pp. 81–82, and Lubow, Weiner, and Schnur, 1981 , pp. 33–34, for further assessment of this possibility; Peterson's discussion (Chapter 8, this volume) of the differential-outcomes paradigm may be pertinent in this connection, too.   Furthermore, there is evidence (Newman *et al.*, 1980) indicating that use of positive consequences like blue vs. green lights, rather than polar extremes like a red light vs. no red light, would enhance human performance relative to that on the FN tasks we used.

Nevertheless, the fact remains that a variety of discriminations in animals seem to be mastered on the basis of mere presence vs. absence of some event or consequence.   Salient CSs, not differentiated solely by a single distinguishing feature, and biologically significant consequences are usually involved in these cases.   Thus, either the occurrence or nonoccurrence of US can serve as an effective cue for correct responding on a subsequent trial, presumably mediated by the different internal states they elicit (e.g., Capaldi, 1971).   Stay vs. shift strategies (see Kamil, Chapter 29, this volume) also entail control by preceding presence vs. absence.   Presence may be represented as a collection of specific event attributes, but how may absence be encoded in instances of these kinds?   Only a few writers in the field of animal learning have explicitly considered such issues.   Konorski (1948) claimed that during discrimination learning both excitatory and inhibitory connections are established with a single neural center, the one where the US is represented.   However, in 1967, Konorski revised his views and concluded that all acquired connections are basically excitatory, but certain types involve linkage with US centers and others with no-US centers.   Besides mentioning the relevance of "on-units" and "off-units" as important, correlated neural mechanisms of perception, Konorski speculated that links with US centers are stronger than links with no-US centers.   However, he did not offer any definite justification for this speculation (see Dickinson & Boakes, 1979; Konorski, 1972; Rescorla, 1979).

Without any further amplification, the basic nature of a no-US center remains clear.   Are there different no-US centers for every conceivable US or are they more broadly grouped, say, appetitive vs. aversive?   And how would absence of a CS be represented?   Like some possibilities alluded to earlier in this chapter, Dickinson (1980, p. 100) has suggested that US absence is encoded rather vaguely, not as omission of an event with particular properties, but in terms of relatively undifferentiated emotional or motivational states. Consequently, he proposed that omission of expected food and the presentation of shock may excite some common representation, different from that activated by the omission of expected shock and the presentation of food. Behavioral findings (see, for example, Dickinson & Dearing, 1979) indicate that appropriate reciprocal relations do seem to exist between these opposed motivational systems. Nevertheless, we lack any deep empirical and theoretical analyses of how event nonoccurrence might be effectively encoded by animals and humans —— a gap in our knowledge that demands some filling in the future.

V. CONCLUDING COMMENTS

This chapter examined several phenomena collectively linked by their apparent reliance on stimulus absence as a basis for effective learning or performance. Pavlovian trace conditioning with a long unfilled gap between CS and US was successful only when this empty interval was made informative by clearly differentiating it from other segments of conditioning trials, especially the intertrial interval. Studies of the feature–positive effect revealed that, under a variety of discrimination arrangements, animal and human subjects perform considerably better when the presence of some specific feature of a compound stimulus is the signal for a positive event than when absence of the feature signals the positive event. This asymmetry in the way positive and negative information are processed in our specific experiments may arise via different mechanisms for animals vs. humans, because post–training assays revealed that our FN pigeons had apparently learned about the relation between feature presence and nonreinforcement even though they did not perform appropriately during actual discrimination learning sessions. On the other hand, humans were unlikely to identify the distinguishing feature when it appeared only on negative trials. In our tasks, the pigeon's deficit was presumably one of performance, not learning or detection, whereas the human's difficulty lay in discovery of the relevant cue.

Of course, one should not infer that pigeons "learn" FN discriminations and identify visual features better than human beings. The exact nature of any possible species difference requires considerable research designed to arrange more comparable tasks (e.g., types of features, the consequences on negative trials) and a variety of performance assays, including those associated with extinction and discrimination reversal procedures. Successful human subjects seem to formulate a rather flexible, verbal rule that enables them to quickly learn a reversal of the original discrimination. In contrast, pigeons do not show analogous transfer; they appear unduly controlled by characteristics of and relations between particular external stimuli.

I briefly discussed some issues pertinent to the conceptualization and potential encoding of "absence," and mentioned the possibility that organisms capable of learning may possess a predisposition to more easily form and/or utilize associations between the occurrences of two events than between the nonoccurrence of one event and the occurrence of another. Whether or not this broad speculation is warranted, trace conditioning procedures and discrimination arrangements based on presence vs. absence of a single feature represent fairly simple paradigms -- whose further developmental and comparative study might tell us a great deal about the role of stimulus absence in guiding learning and performance in more complex situations.

If one of the major themes of this chapter is valid, researchers must actively resist the tendency to discount information derived from the nonoccurrence of some experimental outcome or prediction. The nature of both conditioning and cognition may be illuminated as much by consistent failures to learn, perform, remember, or discriminate as by successes.

ACKNOWLEDGMENTS

Most of our research described in this chapter was supported by National Institute of Mental Health Grant MH 19300. A James McKeen Cattell Fellowship, held during my 1981–1982 sabbatical, allowed time to pursue some parallels between aspects of human information processing and phenomena of animal learning and behavior. I acknowledge Herbert M. Jenkins's substantial influence on my thinking, and I thank Sarah W. Bottjer, Russell Fazio, Dexter Gormley, Peter Kaplan, Joseph Newman, Jim Sherman, Edda Thiels, and William T. Wolff for their valuable criticisms and help.

REFERENCES

Allan, L. G., & Jenkins, H. M. The judgment of contingency and the nature of the response alternatives. *Canadian Journal of Psychology*, 1980, *34*, 1–11.

Amsel, A. Partial reinforcement effects on vigor and persistence: Advances in frustration theory derived from a variety of within-subjects experiments. In K. W. Spence & J. T. Spence (Eds.), *The psychology of learning and motivation*. (Vol. 1.) New York: Academic Press, 1967.

Bottjer, S. W., & Hearst, E. Food delivery as a conditional stimulus: Feature-learning and memory in pigeons. *Journal of the Experimental Analysis of Behavior*, 1979, *31*, 189–207.

Bourne, L. E. Knowing and using concepts. *Psychological Review*, 1970, *77*, 546–556.

Bourne, L. E. An inference model of conceptual rule learning. In R. Solso (Ed.), *Theories in cognitive psychology*. Potomac, Maryland: Erlbaum, 1974.

Bourne, L. E., Ekstrand, B. R., Lovallo, W. R., Kellogg, R. T., Hiew, C. C., & Yaroush, R. A. Frequency analysis of attribute identification. *Journal of Experimental Psychology: General*, 1976, *105*, 294–312.

Bruner, J. S., Goodnow, J. J., & Austin, G. A. *A study of thinking*. New York: Wiley, 1956.

Capaldi, E. J. Memory and learning: A sequential viewpoint. In W. K. Honig & P. H. R. James (Eds.), *Animal memory*. New York: Academic Press, 1971.

Diamond, I. T., Goldberg, J. M., .& Neff, W. D. Tonal discrimination after ablation of auditory cortex. *Journal of Neurophysiology*, 1962, *25*, 223–235.

Dickinson, A. *Contemporary animal learning theory*. Cambridge, England: Cambridge University Press, 1980.

Dickinson, A., & Boakes, R. A. (Eds.) *Mechanisms of learning and motivation: A memorial volume to Jerzy Konorski*. Hillsdale, N.J.: Erlbaum, 1979.

Dickinson, A., & Dearing, M. F. Appetitive–aversive interactions and inhibitory processes. In A. Dickinson & R. A. Boakes (Eds.), *Mechanisms of learning and motivation: A memorial volume to Jerzy Konorski*. Hillsdale, N.J.: Erlbaum, 1979.

Dinsmoor, J. A. Escape, avoidance, punishment: Where do we stand? *Journal of the Experimental Analysis of Behavior*, 1977, *28*, 83–95.

Einhorn, H. J., & Hogarth, R. M. Behavioral decision theory: Processes of judgment and choice. *Annual Review of Psychology*, 1981, *32*, 53–88.

Estes, W. K. The cognitive side of probability learning. *Psychological Review*, 1976, *83*, 37–64.

Fazio, R. H., Sherman, S. J., & Herr, P. M. The feature–positive effect in the self-perception process: Does not doing matter as much as doing? *Journal of Personality and Social Psychology*, 1982, *42*, 404–411.

Gibbon, J. Timing and discrimination of shock density in avoidance. *Psychological Review*, 1972, *79*, 68–92.

Gibbon, J., & Balsam, P. Spreading association in time. In C. M. Locurto, H. S. Terrace, & J. Gibbon (Eds.), *Autoshaping and conditioning theory*. New York: Academic Press, 1981.

Hearst, E. Aversive conditioning and external stimulus control. In B. A. Campbell & R. M. Church (Eds.), *Punishment and aversive behavior*. New York: Appleton–Century–Crofts, 1969.

Hearst, E. Some persistent problems in the analysis of conditioned inhibition. In R. A. Boakes & M. S. Halliday (Eds.), *Inhibition and learning*. New York: Academic Press, 1972.

Hearst, E. Pavlovian conditioning and directed movements. In G. Bower (Ed.), *The psychology of learning and motivation*. (Vol. 9.) New York: Academic Press, 1975.

Hearst, E. Stimulus relationships and feature selection in learning and behavior. In S. Hulse, H. Fowler, & W. K. Honig (Eds.), *Cognitive processes in animal behavior*. Hillsdale, N.J.: Erlbaum, 1978.

Hearst, E. Classical conditioning as the formation of interstimulus associations: Stimulus substitution, parasitic reinforcement, and auto-shaping. In A. Dickinson & R. A. Boakes (Eds.), *Mechanisms of learning and motivation: A memorial volume to Jerzy Konorski*. Hillsdale, N.J.: Erlbaum, 1979.

Hearst, E., & Jenkins, H. M. *Sign-tracking: The stimulus-reinforcer relation and directed action*. Austin, Texas: The Psychonomic Society, 1974.

Honig, W. K., & Urcuioli, P. J. The legacy of Guttman and Kalish (1956): 25 years of research on stimulus generalization. *Journal of the Experimental Analysis of Behavior,* 1981, *36,* 405–445.

Hull, C. L. The rat's speed-of-locomotion gradient in the approach to food. *Journal of Comparative Physiology,* 1934, *17,* 393–422.

Jenkins, H. M. Noticing and responding in a discrimination based on a distinguishing element. *Learning and Motivation,* 1973, *4,* 115–137.

Jenkins, H. M., & Sainsbury, R. S. The development of stimulus control through differential reinforcement. In N. J. Mackintosh & W. K. Honig (Eds.), *Fundamental issues in associative learning.* Halifax: Dalhousie University Press, 1969.

Jenkins, H. M., & Sainsbury, R. S. Discrimination learning with the distinctive feature on positive or negative trials. In D. Mostofsky (Ed.), *Attention: Contemporary theory and analysis.* New York: Appleton-Century-Crafts, 1970.

Kaplan, P. S., & Hearst, E. Bridging temporal gaps between CS and US in autoshaping: Insertion of other stimuli before, during, and after CS. *Journal of Experimental Psychology: Animal Behavior Processes,* 1982, *8,* 187–203.

Kaplan, P. S., & Hearst, E. Trace conditioning, contiguity, and context. In M. L. Commons, R. J. Herrnstein, & A. R. Wagner (Eds.), *Quantitative analyses of behavior: Vol. III, Acquisition.* (Vol. 3.) Cambridge, Mass.: Ballinger, 1984, 187–203. In press.

Kendler, H. H., & Kendler, T. S. From discrimination learning to cognitive development: A neobehavioristic odyssey. In W. K. Estes (Ed.), *Handbook of learning and cognitive processes.* (Vol. 1.) Hillsdale, N.J.: Erlbaum, 1975.

Koffka, K. *Principles of Gestalt psychology.* New York: Harcourt, Brace, & World, 1935.

Konorski, J. *Conditioned reflexes and neuron organization.* Cambridge, England: Cambridge University Press, 1948.

Konorski, J. *Integrative activity of the brain.* Chicago: University of Chicago Press, 1967.

Konorski, J. Some ideas concerning physiological mechanisms of so-called internal inhibition. In R. A. Boakes & M. S. Halliday (Eds.), *Inhibition and learning.* New York: Academic Press, 1972.

Lubow, R. E., Weiner, I., & Schnur, P. Conditioned attention theory. In G. Bower (Ed.), *The psychology of learning and motivation.* (Vol. 15.) New York: Academic Press, 1981.

Mowrer, O. H., & Lamoreaux, R. R. Conditioning and conditionality (discrimination). *Psychological Review,* 1951, *58,* 196–212.

Neisser, U., & Weene, P. Hierarchies in concept attainment. *Journal of Experimental Psychology,* 1962, *64,* 640–645.

Newman, J., Wolff, W. T., & Hearst, E. The feature-positive effect in adult human subjects. *Journal of Experimental Psychology: Human Learning and Memory,* 1980, *6,* 630–650.

Nisbett, R. W., & Ross, L. *Human inference: Strategies and shortcomings of social judgment.* Englewood Cliffs, N.J.: Prentice-Hall, 1980.

Overmier, J. B. Theories of instrumental learning. In M. E. Bitterman, V. M. LoLordo, J. B. Overmier, & M. E. Rashotte (Eds.), *Animal learning: Survey and analysis*. New York: Plenum, 1979.

Peden, B., Browne, M. P., & Hearst, E. Persistent approaches to a signal for food despite food omission for approaching. *Journal of Experimental Psychology: Animal Behavior Processes*, 1977, *3*, 377–399.

Reberg, D., & LeClerc, R. *Pavlovian discriminations based on distinguishing stimuli*. Research Bulletin No. 367: ISSN 0316–4675. London, Canada: University of Western Ontario, Dept. of Psychology, 1976.

Reberg, D., & LeClerc, R. A feature positive effect in conditioned suppression. *Animal Learning and Behavior*, 1977, *5*, 143–147.

Rescorla, R. A. Conditioned inhibition and extinction. In A. Dickinson & R. A. Boakes (Eds.), *Mechanisms of learning and motivation: A memorial volume to Jerzy Konorski*. Hillsdale, N.J.: Erlbaum, 1979.

Rescorla, R. A., & Wagner, A. R. A theory of Pavlovian conditioning: Variations in the effectiveness of reinforcement and nonreinforcement. In A. H. Black & W. F. Prokasy (Eds.), *Classical conditioning II: Current research and theory*. New York: Appleton–Century–Crofts, 1972.

Schwartz, B., & Gamzu, E. Pavlovian control of operant behavior. In W. K. Honig & J. E. R. Staddon (Eds.), *Handbook of operant behavior*. Englewood Cliffs, N.J.: Prentice–Hall, 1977.

Skinner, B. F. *Verbal behavior*. New York: Appleton–Century–Crofts, 1957.

Sokolov, Y. N. *Perception and the conditioned reflex*. New York: Pergamon Press, 1963.

Sperling, S. E., & Perkins, M. E. Autoshaping with common and distinctive stimulus elements, compact and dispersed arrays. *Journal of the Experimental Analysis of Behavior*, 1979, *31*, 383–394.

Sutherland, N. S., & Mackintosh, N. J. *Mechanisms of animal discrimination learning*. New York: Academic Press, 1971.

Terry, W. S., & Wagner, A. Short-term memory for "surprising" vs. "expected" unconditioned stimuli in Pavlovian conditioning. *Journal of Experimental Psychology: Animal Behavior Processes*, 1975, *104*, 122–133.

Thiels, E. A sequential feature–positive effect: Approach–withdrawal behavior to individual elements. Unpublished manuscript, Indiana University, 1982.

Wolff, W. T. *Discrimination learning based on presence and absence of predictive stimuli: An analysis of the feature positive effect*. PhD Thesis, Indiana University, 1983.

Wolff, W. T., & Hearst, E. *A feature-positive superiority in discriminations involving a temporally prior feature*. Paper presented at the meeting of the Psychonomic Society, November, 1975.

19   DO PIGEONS DECOMPOSE STIMULUS COMPOUNDS?

Donald A. Riley
University of California, Berkeley

I.   INTRODUCTION

This chapter has two purposes. The first is to address an important issue in the analysis of delayed matching–to–sample behavior in pigeons. If a pigeon is shown a display on a key containing both color and line orientation information, and has learned that a separate retention test may be given for either attribute, can it detect and store the information about both as efficiently, as either alone, as though it were processing both kinds of information in a parallel fashion, or is the compound processed less efficiently, as though sequentially? Are there conditions in which the relevant information contained in a compound stimulus is obtained as quickly and accurately as the information contained in either element when they are presented alone and others where such information is obtained less quickly and accurately? As we shall see, the answer to this question is revealed by manipulations of the stimulus arrangement and the instructions given the pigeon by the reinforcement contingencies.

The second purpose of this chapter is to comment on the broader problem of the generality of the findings discovered in the Element/Compound matching–to–sample paradigm. It seems likely that when animals, such as pigeons, view the world about them, they perceive objects and discrete events in that world as such, and respond appropriately to them. This position, which will be recognized as a central tenet of Gestalt psychology, asserts that, for pigeons, such objects as pieces of grain, other pigeons, places to roost, etc., are perceived, organized and remembered as units, distinct from the rest of the stimulus surround.

Such assumptions not only agree with common sense, they are also implied by the research of Herrnstein (Chapter 14, this volume) and his associates. Their research has found that pigeons can learn to categorize pictures of object classes, such as people, pigeons, trees, and fish. Because of the positive transfer to new instances of objects of the same class, they have argued that the pigeon must group attributes of objects into wholes or parts of wholes. Although the evidence that pigeons can perform complex categorization is persuasive, it is not yet clear that it is the object properties of these collections of stimuli that control the pigeon's behavior, rather than some other set of simpler stimulus attributes (e.g., intensity). Nor is it yet clear that the principles involved in natural categorizing are the same as those involved in Compound delayed matching–to–sample.   Concept acquisition in Herrnstein's task is rapid relative to the prolonged training necessary to perform well in the delayed matching–to–sample task.   To some extent these differences in acquisition speed probably reflect differences in the difficulty of learning a simple yes–no discrimination versus the more complex conditional discrimination of the delayed matching–to–sample.   Also, as the work of Wright, Santiago, Sands, and Urcuioli (Chapter 20, this volume) has suggested, the use of many different instances in the category task, as opposed the repeated use of the same instances in different combinations in the Compound delayed matching–to–sample task may be responsible for some of this difference.   Finally, these differences in ease of acquisition may be attributable to the differences between naturally occurring categories and acquired concepts.   Regardless of these uncertainties, alternative and complementary approaches to the problem of how the pigeon organizes the material on the key should aid in the understanding of visual memory. The Compound delayed matching–to–sample task has proved useful in understanding how the pigeon processes information about two attributes of a compound stimulus and thus sheds light on how stimulus organization affects the contents of working memory.

II.   BACKGROUND

The first work using Compound delayed matching–to–sample showed that pigeons can match both elements of a compound sample stimulus, but not as well as either element presented alone (Maki & Leuin, 1972; Maki, Riley, & Leith, 1976).   The sample elements in these experiments, which appeared on the center key of a three key display, were red or blue colored fields and white lines that varied in angular orientation, either vertical or horizontal.   Compound samples were made by simultaneously illuminating one value from each dimension, so that the line would appear on the colored field.   The test stimuli, which appeared on the two side keys, were, for both element and compound samples, the two elements from one of the two dimensions, with the tested dimension varying between trials.   These test stimuli were simultaneously presented on the two side keys.   Pecking the Correct key led to food.   Further research found that the magnitude of the element–compound difference is influenced by sample duration, the effect being non–existent at short sample durations where performance on either task is at chance, and also very small or absent at long sample durations where performance on both tasks seems to converge on a common asymptotic performance value (Maki & Leith, 1973).

These facts are of interest because they are consistent with a number of assumptions about the nature of Compound delayed matching–to–sample performance in pigeons.   As stated by these authors, the slow information uptake over several seconds, the apparent convergence of element and compound matching functions at long sample durations, and the superiority of element matching performance, all indicate that the animal is a limited capacity system that can process (i.e., detect and store information about) only a small amount of information at one time.   This account goes on to state that when an animal is forced to process two elements at the same time, it does so either by rapid switching of attention between the two elements or by allocating a portion of the total processing capacity to each element, resulting in poorer performance on compound samples.

This interpretation of the facts implies that the animal can and does decompose the stimulus compound into those attributes that are made relevant by the reinforcement contingencies. That is, each compound of white lines on a colored surround is assumed to be processed as two components: lines, either vertical or horizontal, and colors, blue or red, for example. It is this assumption of the separate processing of components that is the central focus of this paper.

III. DECOMPOSITION

What does it mean to decompose a stimulus? In the most extreme sense it means that the subject performs operations on some property of the stimulus without reference to other properties. Garner (1974) in his research on how humans organize stimulus compounds has designated such compounds as Separable. Compounds in which the perception of each element is influenced by each other element and in which the elements resist individual analysis by the Subject are referred to as Integral. Yet other compounds have intermediate properties. While some principles have emerged in the research with humans, the question of how humans will organize new untried stimulus compounds remains largely empirical. Modest as this assessment is for the human research, far less is known about the conditions of stimulus decomposition in animals. Nevertheless, there are some reasonable starting points. One would be to examine with animals, the same sorts of compounds that Garner and others have used in research with humans. Robert Cook, working in our laboratory at Berkeley is at the present time doing just this. Another approach which reflects more accurately what we have done has relied on a mixture of common sense and Gestalt psychology.

Using this approach, one might reasonably argue that whether or not the animal decomposes the stimulus compound is going to depend on, among other things, the nature of the compound. For example, decomposition may be more likely to occur if the various parts of a compound stimulus can be readily analyzed into different figural components. Thus, if the pigeon is required to detect and remember both a color and a line orientation, these might be decomposed if the orientation is part of one figure, such as a white vertical line, and the color is part of another figure, such as an adjacent red semi-circle. In such cases decomposition might be facilitated because each of the two relevant attributes is a component of a different figure. On the other hand, decomposition might be less likely to occur if the two relevant attributes are components of the same figure, such as a colored bar in a specified orientation. In the literature on human perception and memory, various hypotheses concerning the physical arrangements that might lead to storing different attributes of a compound in a unified (i.e., non-decomposed) fashion have been made which have emphasized organizational principals (Spyropoulos & Ceraso, 1977), integral and configured stimulus dimensions (Garner, 1974), and the action of focal attention on elements presented in the same place and time (Treisman, 1977). If, then, a satisfactory preparation is identified that distinguishes between conditions that lead to decomposition and those that do not, determining the adequacy of these apparently alternative accounts of unification should prove an interesting challenge.

This chapter is a first step in this direction. It explores whether the known facts point to conditions under which decomposition does and does not occur. Before considering this evidence, however, we must first consider the criticisms and alternatives to the limited capacity hypothesis as an account of element–compound differences. Such criticisms may have implications for our analyses of stimulus decomposition.

IV. CRITICISMS OF THE LIMITED CAPACITY HYPOTHESIS

The superiority of element over compound matching is consistent with both the assumption of decomposition and the limited capacity hypothesis, but by itself; it is not compelling evidence for these hypotheses. Three sorts of arguments against the relevance of this fact for the limited capacity hypothesis have been offered. The first two are also evidence against the decomposition hypothesis. They are: (a) that the effect is not repeatable, (b) that it can be attributed to differences between the compound sample and the typically used element tests, differences which are not present with element sample/test arrangements, and (c) that the effect reflects differences in processing requirements not associated with limitations on information uptake.

The first criticism, that the effect is not repeatable, stems from a careful attempt at replication by Farthing, Wagner, Gilmour, and Waxman (1977). Because we have been able to repeat the effect in a number of different studies, we think their failure may reflect differences between their stimulus displays and ours. Like the early work from our laboratory, Farthing et al. produced compound samples by simultaneously illuminating a line cell and a color cell in an in-line projector. Their line cell differed from ours in having only one line rather than two or three. Why this should make a difference is a matter of conjecture (cf. Riley, Cook, & Lamb, 1981), but in any event we and others have obtained the element–compound difference with several other kinds of displays (Lamb & Riley, 1981; Roberts & Grant, 1978; Cook, Riley, & Brown, 1982).

The second criticism is that the test stimuli, which consist of elements, are more similar to element samples than to compound samples, and that this difference can account for the superiority of element over compound matching without recourse to the limited capacity hypothesis. This alternative, called the generalization decrement hypothesis, was first addressed by Maki, Riley, and Leith (1976) who, in a test of this hypothesis found no improvement in compound delayed matching-to-sample when compound samples were followed by comparison tests which provided redundant test information (e.g., RV sample, RV + vs. GH−tests).[1] The superiority of performance with element as opposed to compound samples regardless of whether the tests following compound samples were element or compound was regarded as evidence against the generalization decrement hypothesis, for even when one of the test stimuli exactly matched the compound sample, performance did not improve. Roberts and Grant (1978) subsequently showed, however, that if the negative comparison test had only one element, compound sample matching was as good as or better than element matching (e.g., RV sample; RV vs. G−). Their analysis of this problem assumes that since all test stimuli have been associated with reinforcement, all tend to elicit pecking. Therefore, adding test elements to both the positive and negative comparison stimuli tends to increase the probability of a response to either. Thus, if compound samples are followed by element tests, performance suffers because of dissimilarity between sample and test (generalization decrement). But, if redundant compound tests follow compound samples, as in Maki *et al.*, then performance suffers because of the increased elicitation of pecks to both positive and negative test stimuli. This analysis requires no assumption of decomposition during the processing of sample information. It does, however, require that test elements are additive, an assumption which, in turn, implies that the compound test stimuli are elemental, that is, decomposed and then summed. While it is possible that test elements might be decomposed even though sample elements are not, this conjunction of events seems unlikely. One would have to assume that there are processes that allow for decomposition of test stimuli that are not operating during sample presentation. Such processes have not yet been suggested. Nevertheless, their experiments, while not ruling out divided attention and the decomposition of compound samples, provide an alternative explanation.

D'Amato and Salmon (Chapter 9, this volume) present similar evidence from experiments done by Cox and D'Amato (1982) using Cebus monkeys. Like Roberts and Grant, they found superior matching to sample when a compound sample was followed by a compound positive comparison stimulus and an element negative comparison stimulus. These data are inconclusive, however, as a test between the decomposition hypothesis and the generalization decrement hypothesis for the same reasons that Roberts and Grant's results are inconclusive. That is, Roberts and Grant's explanation of the consequences of different test procedures assumes decomposition of the test compound. While decomposition of test stimuli but not sample stimuli is possible, it does not seem likely.

Three other kinds of evidence are reported by Cox and D'Amato, all of which are said by them to favor the generalization decrement interpretation rather than decomposition, or, by inference, the limited capacity hypothesis. First, in a simultaneous matching−to−sample task (where both the sample and test stimuli are shown at the same time) using element tests, element sample matching is superior to compound sample matching despite long sample durations and no obvious memory load. This

---

[1]R = red, V = vertical, G = green, H = horizontal. RV is a compound of red and vertical, etc. + indicates the reinforced choice; − indicates the nonreinforced alternative.

observation appears to offer strong support to the generalization decrement interpretation. Whether the same results will be found with pigeons, where all of the work supporting the decomposition hypothesis has been found remains to be seen. That research from D'Amato's laboratory may be tapping different processes is also suggested by an earlier finding (D'Amato & Worsham, 1972) that monkeys take in information at rates much higher than those typically found in research with pigeons, and show no improvement with sample duration. Other investigators have failed to repeat these observations, but all of the experiments differ in important ways (Devine, Jones, Neville, & Sakai, 1977; Herzog, Grant, & Roberts, 1977). Certainly, it is possible that conditions will emerge in which decomposition is not the rule. Cox and D'Amato may have found such a condition.

Cox and D'Amato also found that memory loss during a delay interval appears to occur at the same rate following element samples and compound samples. They argue that since retaining two items in memory should be more difficult than retaining one, divergent retention functions should be observed if the animal were decomposing the sample. The parallel retention functions they obtained for different delay intervals suggest that only one unit is retained in each case. This, in turn, indicates that the compound sample is processed as a unit, and that the element–compound difference is attributable to generalization decrement. Plausible as this argument is, it rests on two questionable assumptions. One is that the two units of a compound would be forgotten more rapidly if processed separately. But because interference between items in memory will almost surely depend on factors such as the similarity between these items, this assumption must be tested empirically by sequential presentation of the items. The other questionable assumption is that memory losses, as revealed by percent correct, are linear over the percent scale. That is, that a drop from (e.g.) 99 to 90% is equal to a drop from 89 to 80%. While this assumption is almost surely false, we don't know whether the departures from linearity are great enough to affect the interpretation of the data. In any event, the meaning of these parallel retention functions is not clear.

Finally, Cox and D'Amato present interesting evidence that a between-dimensions comparison test produces performance superior to a within-dimension test following an element sample, but not following a compound sample (examples of between dimension tests would be: Element sample: Red, Test: Red+ vs. Vertical–; Compound sample: Red Triangle, Test: Red+ vs. Vertical. At this point it is not clear how this effect is to be explained, but it almost surely cannot be regarded as critical evidence for or against either hypothesis. A decomposition analysis might assume that the failure to find superiority of interdimensional tests following compound samples is because, in the interdimensional test, the negative value reinstates the memory of having just seen a stimulus value on that dimension and so confuses the animal. Cox and D'Amato attribute the result to generalization between the compound sample and the negative test. Since no independent test of either hypothesis has been carried out, the meaning of these facts is unclear. Although Cox and D'Amato have examined a variety of kinds of evidence, the strongest evidence supporting the generalization decrement position is their finding of element–compound differences during simultaneous matching to sample. Unless and until further analyses of this finding allow a different interpretation, this fact stands as evidence that at least under some circumstances, monkeys show element–compound differences that are most easily attributable to generalization decrement.

The third alternative view of element–compound differences has been proposed by Grant (1981). He argues that a variety of experiments from his, Maki's, and Roitblat's laboratories indicate that the pigeon in delayed matching–to–sample maintains a memory of what it must do at the time of the test. Grant compared single, double, and triple sample presentations when the same sample was used in the multiple presentations and when different samples were used. In both "same" and "different" presentations, a common test stimulus was correct. Samples were colors (red vs. green), pecking responses to the sample key (1 vs. 20), and presence vs. absence of a food reinforcer prior to the test. The comparison tests were always red vs. green. Grant's data show that when more than one type of sample predicts the same test stimulus, multiple "different" samples add together (i.e., improve test performance) in what appears to be the same way as multiple "same" samples. This suggests, he argues, that the different samples are all recoded as an instruction to remember the relevant properties of the test key and peck at it. The notion that the subject remembers tests, rather than samples, is not consistent with the hypothesis that the pigeon's memory is mediated by a trace of the sample stimulus, as is suggested by the information overload hypothesis (Riley, Cook, & Lamb, 1981). Grant also finds that elements presented sequentially are retained in short-term memory no better than when the same elements are presented simultaneously as a compound. This fact, he says, is not consistent with the limited capacity hypothesis, which would seem to require that the greater time given in the sequential task would result in superior performance.

I shall comment on two separate issues: (a) the relevance of Grant's data and that of other investigators to the issue of whether, on seeing a sample, the pigeon encodes sample or test information and (b) the relevance of Grant's data to the limited capacity account of element–compound differences. First, Grant's finding of no difference in the way pigeons process multiple "different" and multiple "same" stimuli is consistent with the hypothesis of a common code. Whether the common encoding is or is not prospective (i.e., of test information) or retrospective (i.e., of sample information) is not clear. Elsewhere, my associates and I (Riley, Cook, & Lamb, 1981) have also argued that a number of facts, such as those found by Roitblat (1980), support the test stimulus encoding hypothesis, but that the work of Urcuioli and Nevin (1975) and Zentall and Hogan (1978) suggests that pigeons can encode sample information. Thus, it is a distinct possibility that pigeons encode either sample or test stimuli depending on the circumstances. One critical factor would seem to be the nature of the sample–test stimulus mapping. It is likely that many samples mapping onto one test, as in Grant's work, would support test stimulus encoding, whereas the opposite would support sample stimulus encoding. At present, because of the lack of relevant evidence, this issue must remain open.

Consider next Grant's evidence against the hypothesis that the superiority of element over compound in delayed matching–to–sample reflects a limited capacity at the time of encoding. Grant compares two kinds of many to one mapping in which three samples (red key, 20 pecks and food) predict food if a red test key is pecked, and three others (green key, one peck and no food) predict food if a green test key is pecked. The comparison is between the successive presentation and the simultaneous presentation of samples. Grant states that the limited capacity hypothesis requires that performance in the simultaneous task should be poorer than in the successive because the information load should be greater. Because performance is the same in the two conditions, he concludes that some other explanation is needed to account for element superiority. His proposal is that the typical Compound delayed matching–to–sample task requires that the animal encode and remember two instructions (e.g., "peck red" and "peck horizontal") i.e., that the overload is on memory rather than on the encoding of information. Grant argues that the sample durations in the simultaneous condition were sufficiently short (approximately 2.5 to 4.5 sec) to permit information overload. But it is not clear that Grant's data speak to the question. To know whether the comparison he makes is relevant to the question of information overload, we would first have to know that his combination of elements would result in the superiority of element over compound performance, when tested in a Element–Compound delayed matching–to–sample task. We have no evidence on this issue, since elements were never presented alone as samples. Also, Grant's two groups performed near 100% with a 0 sec delay, making a meaningful interpretation of the null difference virtually impossible.

Despite these criticisms of this particular experiment, Grant's hypothesis that the element–compound difference reflects some aspect of a memorial difference rather than an encoding difference may be correct. Grant points out that Roberts and Grant (1978) found no evidence of convergence of performance on element and compound conditions with long (6 sec) sample durations, as the limited capacity hypothesis predicts. This conclusion has been supported by the data of Lamb and Riley (1981), Cook, Riley, and Brown (1982), and Santi, Grossi, and Gibson (1982). To summarize, the limited capacity interpretation of the superiority of element over compound delayed matching–to–sample assumes that the pigeon decomposes the sample stimulus into its relevant dimensions and processes the elements separately. It also assumes that the rate of processing incoming information is limited, an assumption consistent with a number of facts in delayed matching–to–sample experiments with pigeons. A second hypothesis, generalization decrement, accounts for the superiority of element over compound matching by assuming that the difference between the compound sample and the most similar test stimulus is greater than the difference between the element sample and the most similar test stimulus. This hypothesis does not assume limited capacity or decomposition of the compound into the relevant elements. The third account assumes that the differences are to be explained by the number of items the pigeon must retain following compound as opposed to element samples. This hypothesis, like the first, appears to require decomposition, but not necessarily during sample encoding. It does not require limited capacity at encoding, but does assume limited memory capacity.

As we have seen, some facts are consistent with all three hypotheses and some, although interesting, are at present not well enough understood to have clear implications. Certainly, there is some evidence for generalization decrement in the work of Cox and D'Amato. If, however, this hypothesis does not provide a complete account of element–compound differences, then decomposition and capacity limitations must be considered. Such facts as the slow rate of information uptake in pigeons and the failure to find an improvement in matching when redundant test compounds are used, are not predicted by the generalization decrement hypothesis. They are, though, predicted by the limited capacity, decomposition hypothesis. But other facts such as the failure to find element–compound equality in the simultaneous matching-to-sample with monkeys and the failure to find element–compound convergence with long sample durations are not consistent with the limited capacity hypothesis, and so also appear to be inconsistent with the decomposition hypothesis. While clarification of these inconsistencies may be found in a more complete understanding of the facts just alluded to, further evaluation of the alternative interpretations can also be sought in other experimental preparations.

What additional kinds of evidence would bear critically on the question of decomposition of compound sample stimuli during encoding? We have recently used two different classes of experimental manipulations that are relevant to the issues of if, and under what conditions, pigeons decompose stimuli. The first involves variations in the conditions of sample stimulus presentation. Those that have proven most revealing are (a) those related to the number of relevant elements, (b) spatial relations among the elements, and (c) the duration of sample stimuli. The second involves variations in test procedures which appear to influence the way in which the pigeon processes sample information. Reinforcement outcomes correlated with one element of a compound, but not with another, cause the subject to process the correlated element at the expense of the other. If some kinds of displays are sensitive to such testing differences while others are not, then these procedures will also shed light on the conditions which facilitate decomposition.

V. EVIDENCE FROM VARIATIONS IN STIMULUS ARRANGEMENTS

Recently Lamb and Riley (1981) compared the effect of differences in spatial separation of two relevant elements on compound delayed matching-to-sample performance. An element sample consisted of a small (5 mm square) colored patch either red or orange, or three white lines either vertical or horizontal, filling a 5 mm square. These were presented on a 3 cm square sample key. Comparison tests on side keys presented a choice between the two alternatives on the same dimension, and a correct choice was followed by food. A Separated compound stimulus consisted of an element from each dimension presented together. The degree of separation between the two elements varied from one trial block to another within a day. Unified compounds were lines which varied in color as well as in angular orientation. A block of these as well as a block of element samples was also presented each day.

Three of our results are of interest in this discussion: (1) Performance in the Unified compound condition, in which color and line orientation information were carried by the same colored lines, was for two out of three subjects equal to performance in the element task. (The third bird was slightly poorer on the Unified than on the element condition.) These data are consistent with those of Farthing, Wagner, Gilmour, and Waxman and suggest that under conditions in which both elements are embodied in the same physical event, information about two attributes can be processed as easily as one. (2) Two other compounds in which the color and line elements were separated, but by different amounts, resulted in poorer performance than the Unified condition, with increased separation resulting in poorer performance. (3) Long sample durations did not eliminate the differences between the various conditions. How do the different accounts of the element–compound difference fare with these data? First, the limited capacity hypothesis suffers because long sample durations do not reduce the differences between the Separated compound conditions and the Element condition. The multiple–rule hypothesis (Carter & Werner, 1978) that matching difficulty is a function of the number of procedural rules required to describe the task, is discredited because different compounds are matched differently despite the same rule relating samples and tests. Finally, the generalization decrement hypothesis appears inadequate because the sample–test similarity relations do not predict the order of matching accuracy. The Unified compound, unlike the Separated compounds, had no white lines or colored squares, which are the attributes of the test stimuli. Unified compound samples consist only of colored lines. Despite this apparently greater similarity of the Separated samples to the test stimuli, Unified samples are matched better, and are matched about as well as Element samples, which are like the test stimuli. This fact appears to argue that the inferiority of the Separated compounds to the Unified is not because they are less similar to the test stimuli, but because of some other factor. The suggestion here is that this other factor is related to the decomposability of the Separated compounds into their elements. Such an argument must be made with great caution, for the issue of which compound really is more similar to the test stimuli can only be resolved by appropriate scaling experiments, which have not been conducted. Also, if the pigeons do decompose these compounds as I am suggesting, such decomposition is not revealed by changes in the relative performance on elements and compounds with sample duration, as the limited capacity hypothesis predicts. Either such decomposition takes place after the information uptake, or some other factor prevents convergence of performance in the Separated compound conditions with performance in the Unified and Element conditions at long sample durations.

There are other complications that prevent a conclusion that the difference between Unified and Separated is evidence for decomposition of the Separated compounds into their elements. There were two confounds in the difference between the Unified and Separated conditions in Lamb and Riley's experiments either of which alone could cause the observed differences between the conditions. Whereas the Element and Unified conditions appeared in the center of the key, the Separated stimuli were outside the center, particularly one of the stimulus types, the elements of which appeared in two opposite corners of the 2.5 cm square pecking key. Second, the location of the two dimensions in the Separated conditions randomly alternated between the two possible locations. These differences in location and uncertainty of dimensional location could separately or together have produced the differences between the Unified, and Separated conditions and could also account for the asymptotic differences with long sample durations. To determine whether the same kinds of systematic differences could be found in the absence of these confounds. Cook, Riley, and Brown (1982) conducted an experiment which improved and expanded Lamb and Riley's work. Figure 19.1 shows how some of the stimuli appeared on the sample key.

COLOR CONSTANT STIMULI

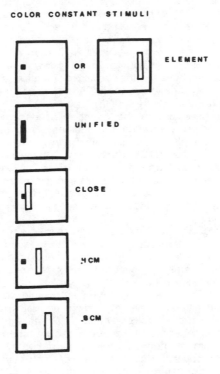

FIG. 19.1. Representative samples from Cook, Riley, and Brown (1982). Shown are samples where the color element's location remains constant while the line element varies. Distances between the elements represent edge-to-edge separation. Lines were white, colors were either red or green, and the surround was dark. Not shown are samples containing horizontal elements nor the corresponding stimulus set in which the line's location remained constant.

First, the stimuli differ from Lamb and Riley's both in that they are simpler, and that the color stimuli are smaller than the line stimuli, to make the two dimensions more equal in difficulty. Second, differences in spatial separation are achieved by keeping one element (e.g., color) in the same place (e.g., the left side of the key) and, in different blocks of trials, varying the distance between this constant element and the second element. For other animals, the location of the line element remains constant and spatial separation is achieved by varying the color element's location on the key. For the unified compounds, location of both elements is constant. These manipulations allow us to examine performance on different kinds of compounds and their dimensions (i.e., color vs. line or constant vs. variable element location).

Five animals were tested with all the compound types, with color changing location for three of the birds and line changing for the other two. All compounds were presented each day in blocks of sixteen trials per compound type. In different experiments, sample exposure was varied by different fixed ratios or by different fixed intervals. The first data shown are from an experiment in which each sample appeared for 20 pecks. The second set is from an experiment that varied sample duration. The performance on color and line tests were not systematically different, but matching of the element in the constant location as opposed to the element in the changing location was different, consequently our emphasis here will be on the latter.

Two analyses are relevant to the question of decomposition. The first compares matching performance on the Element samples, Unified samples, and the Separated sample type in which the elements were spatially contiguous ("Close"). Figure 19.2, which shows these results, indicates that both Unified and Close compounds are equal and slightly inferior to the Elements presented alone. Had the two been different from each other, as in Lamb and Riley it would have argued for different types of processing of the two compounds. Because they are the same, no clear conclusion is possible.

The second analysis compares different spatial separations of compound stimuli. Figure 19.2 allows comparison of performance differences of those widely separated compound stimuli as oppsoed to those that are immediately contiguous, i.e., "Close", with those with increasing spatial separations up to 8 cm. At all spatial separations, one of the elements is matched less well with increasing spatial separation. It is of interest that it is the element which occupies different positions across the different trial blocks which is matched more effectively. Apparently, the pigeon attends to the dimension with the less predictable location, and this element consequently receives preferential processing when the stimuli are separated. It should be emphasized, however, that even at the greatest separation, both elements are matched above chance. Thus, the birds do not simply detect one element and not the other. These results have now been found in five birds in three experiments.

It is our conclusion that the pattern of differences among the compounds is clear evidence against the generalization decrement hypothesis as a complete account of element–compound differences. This hypothesis requires that any differences in performance between the elements and compounds and, in addition, between the compounds themselves be attributable to the similarity relation between the sample and

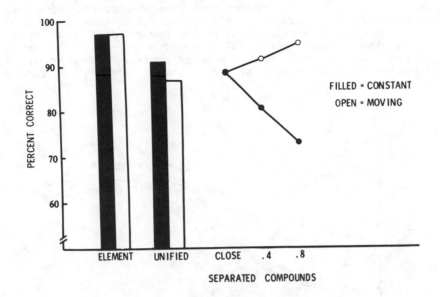

FIG. 19.2. Matching-to-sample performance for 5 birds from FR20 phase of Cook, Riley, and Brown (1982). Shown is performance for Element and compounds. Filled bars and circles represent performance on elements, the location of which remained constant, open bars and circles represent performance on the element that varied in location across blocks.

test. In this experiment, the most informative case is the compound with the largest separation. this compound was exactly the same for all birds, regardless of whether the line or color was the uncertain component of the sample in the other compounds. However, the component that was attended to was determined by the sample properties of spatial separation and locational certainty, and not the sample/test similarity, which was the same for all birds. Thus, some factor other than dissimilarity of sample and test must be at work. That other factor appears to be the separate and unequal processing of two elements.

The other issue that was addressed in these spatial separation experiments was whether the elimination of the confounds present in Lamb and Riley's work resulted in performance on the Separated stimulus conditions being more like that in the Unified and Element conditions. We have already seen that the same difference between this experiment and that of Lamb and Riley (1981), perhaps the elimination of trial-to-trial changes in color and line location and/or placement of all constant location stimuli on the edge of the key, resulted in performance on Close and Unified stimulus compounds being the same. The other question is what happens with long sample durations? Is there evidence of convergence? The answer is shown in Figure 19.3. Even at 12 sec, there is no evidence of convergence between performance on the Elements and the various compounds. We can only conclude that these data do not support the information overload hypothesis. If increasing spatial separation reduces the effectiveness of processing one of the elements, it is apparently not because only a limited amount of encoding can occur when sample durations are short. Rather, the limitation must be elsewhere in the information processing system, such as in storage or other post-stimulus components.

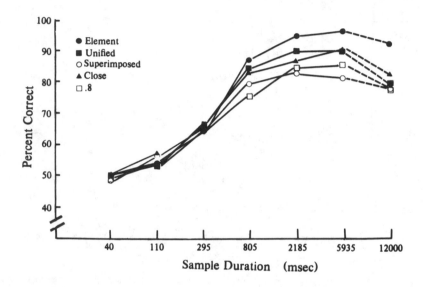

FIG. 19.3. Matching-to-sample performance for 5 birds from sample duration phases of Cook, Riley, and Brown (1982). Solid lines connect data collected in the same phase. Data for the 12 sec sample duration, connected with dotted lines, were collected in a phase that directly followed. (Note, a compound type called Superimposed was also run. It was not discussed in the text nor shown in Figures 19.1 or 19.2.)

VI. THE EFFECT OF PRE-CUEING ON COMPOUND MATCHING TO SAMPLE

The work just described demonstrated selective processing effects by the procedure of allowing one component of a compound to vary its location from one trial block to another while the other component remained in a fixed location. An alternative procedure is to ask about the effect of cueing procedures on selective attention to either element given different kinds of stimulus compounds. Such an experiment has recently been carried out by Lamb (1982), which compared the effects of cueing on Unified and Separated compounds. His cueing procedure employed two successive samples, the first an element (e.g., a white vertical line) which serves as a cue for the succeeding compound (e.g., a white vertical line and a red square). Given the initial vertical line, the test stimuli following the compound were always vertical vs. horizontal with a choice of vertical followed by food. (If, however, the cue is a red square, then following the same red vertical compound, the test would be red square vs. green square.) Performance in this cued condition was compared with performance in a reverse procedure in which the compound was presented first and the element, which again informed the subject about the relevant dimension, was presented second. In addition to these relevant dimension tests, other probe tests on the uncued dimension, as well as between dimension tests, were conducted.

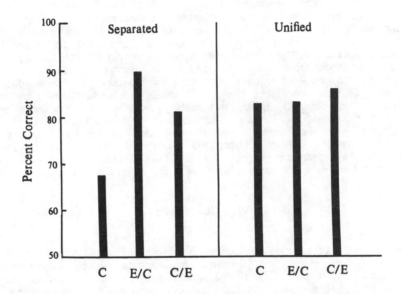

FIG. 19.4. Data from Lamb (1982) showing performance on Unified and Separated compounds when the compounds were presented alone (C), preceeded by an element pre-cue (E/C), or followed by an element post-cue (C/E).

All birds were run repeatedly in all conditions. An analysis of the effects of such cueing procedures in both a Unified compound stimulus condition and a Separated compound stimulus condition are shown for the first two birds to finish the experiment in Figure 19.4. For these two birds, attentional effects appear in the Separated condition, but not in the Unified condition. The columns on the left show the results from the Separated conditions. The first column shows performance with uncued compounds. The second column shows the pre-cueing effect, and the third, post-cueing. Thus, when the element cues come before the compound (column 2), performance is superior relative to when the cue comes after the compound (column 3). The comparable data in the Unified columns on the right do not contain significant differences. These data are consistent with the assumption that the birds are processing the two components of the Separated condition separately. That is, the effectiveness of cueing depends on the nature of the stimulus compound. In our displays, if the two relevant attributes are combined in a single "object," as in the Unified condition, selective processing of one attribute is difficult. If the two attributes are properties of different "objects," as in the Separated condition, then the cueing procedure is effective.

The relevance of Lamb's research to the issue of whether or not pigeons decompose compound stimuli into their components should be clear. The answer is: not necessarily, but if the stimulus arrangements are favorable then decomposition will occur. The cueing task allows these effects to be seen in the Separated arrangements. Under other circumstances, such as in the Unified condition, stimuli are treated more holistically. Finally, the difference between the pre-cued and post-cued results suggests that the decomposition occurs during information uptake, a conclusion inconsistent with the previously described failure to find element-compound convergence with long sample durations (Cook, Riley, & Brown, 1982; Lamb & Riley, 1981). Obviously, our understanding of these effects is not perfect. (Since this material was written, the three other birds in the experiment have finished. Data from all five birds are consistent, with one interesting exception. One of the birds shows the selective attention effect in both the Separated and Unified conditions. Apparently, a pigeon can decompose the Unified stimulus, suggesting that decomposition need not depend on where the pigeon is looking.)

VII. SUMMARY

There appears to be clear evidence of compound stimulus decomposition in both Cook, Riley, and Brown's experiments, which varied spatial separation of elements, and in Lamb's work, which used a cueing procedure. There is also, in Lamb's work, new evidence supporting the limited capacity hypothesis, but the status of this hypothesis is less certain because of the apparent negative evidence. Also, on the issue of decomposition, it seems apparent that the question is not whether it occurs, but what are the conditions that facilitate or inhibit its occurrence? In the search for these factors, useful insights may be gained from the study of research on perceptual organization and memory in humans (cf. Riley, Cook, & Lamb, 1981). Such work suggests the possibility that decomposition may not be a high level cognitive achievement, as suggested by Cox and D'Amato, but a fundamental fact of perception. In that event, its manifestation in pigeons should not be surprising. But one should then also expect a similar outcome with monkeys. If Cox and D'Amato's monkeys did not decompose the stimulus compounds in their displays, why not? One possible answer might be that viewing distance was much greater in their experiments and, because their stimuli were the same size as ours, visual angle would have been much smaller. Spatial proximity seems to discourage decomposition.

Even though there is evidence for decomposition of stimulus compounds, there is no clear-cut corresponding support for the information overload hypothesis. Despite the early suggestions of convergence with long sample durations, the more recent evidence failing to find convergence must be explained for this hypothesis to continue to receive serious consideration. On the other hand, Lamb's data showing a pre-compound cueing effect is consistent with decomposition during stimulus encoding. These discrepancies will have to await further research for clarification.

Finally, Lamb's evidence that selective attention is more likely with Unified than Separated samples points to the role of figural processes in delayed matching-to-sample. It will be recalled that an initial question in this paper was what kind of bearing, if any, the phenomena studied in element and compound delayed matching-to-sample have on object categorizing. I would like to suggest that Lamb's research supports the idea that the analytic study of element-compound differences and compound-compound differences provide vehicles for investigating object perception in animals.

ACKNOWLEDGMENTS

This research was supported by a National Science Foundation Grant BNS 79-08839. I thank M. Brown, R. Cook, and M. Lamb for their effective participation in this research and their helpful comments on this paper.

REFERENCES

Carter, D. E., & Werner, T. J. Complex learning and information processing by pigeons: A critical analysis. *Journal of the Experimental Analysis of Behavior*, 1978, *29*, 565–601.

Cook, R. G., Riley, D. A., & Brown, M. F. The role of element configuration in pigeon matching-to-compound sample performance. Paper presented at a meeting of the Western Psychological Association, Sacramento, Calif. Society. April, 1982.

Cox, J. K., & D'Amato, M. R. Matching-to-compound samples by monkeys *(Cebus apella)*: Shared attention or generalization decrement? *Journal of Experimental Psychology: Animal Behavior Processes*, 1982, *8*, 209–225.

D'Amato, M. R., & Worsham, R. W. Delayed matching in the capuchin monkey with brief sample durations. *Learning and Motivation*, 1972, *3*, 304–312.

Devine, J. V., Jones, L. C., Neville, J. W., & Sakai, D. J. Sample duration and type of stimuli in delayed matching-to-sample in rhesus monkey. *Animal Learning and Behavior*, 1977, *5*, 57–62.

Farthing, G. W., Wagner, J. M., Gilmour, S., & Waxman, H. M. Short-term memory and information processing in pigeons. *Learning and Motivation*, 1977, *8*, 520–532.

Garner, W. R. *The processing of information and structure*. Hillsdale, N.J.: Erlbaum, 1974.

Grant, D. S. Short-term memory in the pigeon. In N. E. Spear & R. R. Miller (Eds.), *Information processing in animals: Memory mechanisms*. Hillsdale, N.J.: Erlbaum, 1981.

Herzog, H. L., Grant, D. S., & Roberts, W. A. Effects of sample duration and spaced repetition upon delayed matching-to-sample in monkeys *Macaca arctoides* and *Saimiri sciureus*. *Animal Learning and Behavior*, 1977, *5*, 347–354.

Lamb, M. R. *Selective attention in pigeons*. PhD Thesis, University of California, Berkeley, 1982.

Lamb, M. R., & Riley, D. A. Effects of element arrangement on the processing of compound stimuli in pigeons *(Columba livia)*. *Journal of Experimental Psychology: Animal Behavior Processes*, 1981, *7*, 45–58.

Maki, W. S., & Leith, C. R. Shared attention in pigeons. *Journal of the Experimental Analysis of Behavior*, 1973, *19*, 345–349.

Maki, W. S., & Leuin, T. C. Information processing by pigeons. *Science*, 1972, *176*, 535–536.

Maki, W. S., Riley, D. A., & Leith, C. R. The role of test stimuli in matching to compound samples by pigeons. *Animal Learning and Behavior*, 1976, *4*, 13–21.

Riley, D. A., Cook, R. G., & Lamb, M. R. A classification and analysis of short-term retention codes in pigeons. In G. H. Bower (Ed.), *The psychology of learning and motivation*. (Vol. 15.) New York: Academic Press, 1981.

Roberts, W. A., & Grant, D. S. Interaction of sample and comparison stimuli in delayed matching-to-sample with the pigeon. *Journal of Experimental Psychology: Animal Behavior Processes*, 1978, *4*, 68–82.

Roitblat, H. L. Codes and coding processes in pigeon short-term memory. *Animal Learning and Behavior*, 1980, *8*, 341–351.

Santi, A., Grossi, V., & Gibson, M. Differences in matching-to-sample performance with element and compound sample stimuli in pigeons. *Learning and Motivation*, 1982, *13*, 240–256.

Spyropoulos, T., & Ceraso, J. Categorized and uncategorized attributes as recall cues: The phenomenon of limited access. *Cognitive Psychology*, 1977, *9*, 384–402.

Treisman, A. Focused attention in perception and retrieval of multidimensional stimuli. *Perception & Psychophysics*, 1977, *22*, 1–11.

Urcuioli, P. J., & Nevin, J. A. Transfer of hue matching in pigeons. *Journal of the Experimental Analysis of Behavior*, 1975, *24*, 149–155.

Zentall, T. R., & Hogan, D. E. Same/different concept learning in the pigeon: The effect of negative instances and prior adaptation to the transfer stimuli. *Journal of the Experimental Analysis of Behavior*, 1978, *30*, 177–186.

V. JUDGMENTS OF SIMILARITY AND DIFFERENCE

20 PIGEON AND MONKEY SERIAL PROBE RECOGNITION: ACQUISITION, STRATEGIES, AND SERIAL POSITION EFFECTS

Anthony A. Wright
Hector C. Santiago
University of Texas Health Science Center
at Houston Graduate School of Biomedical Sciences
Stephen F. Sands
University of Texas at El Paso
Peter J. Urcuioli
Purdue University

I. INTRODUCTION

Forgetting and memory received study in ancient times in terms of learning mnemonic systems required during the course of scholastic training (Highbee, 1977). The actual scientific study of forgetting has come only recently with the pioneering work of Ebbinghaus (1885) and by the study of the delayed response task in lower animals by Hunter (1913). Hunter's work was the first attempt to compare memory performance across species. Unfortunately, the task suffered from the emphasis it placed on the animal's particular body orientation to bridge the time delay. (Ruggiero & Flagg, 1976) If the animal oriented toward the correct choice during the delay, then conceivably perfect performance could be maintained across any memory interval. The delayed matching-to-sample task improved upon this procedure by using an instructional stimulus (sample). Choices were not presented until after the delay interval, and varied randomly in their location. Thus, simple orientation to one side or another could not be used to bridge the delay interval. One variation of the delayed matching-to-sample task (Konorski, 1959; Nelson & Wasserman, 1978) is to present only one choice stimulus and require that the subject (e.g., a pigeon) identify it as either being the same as the sample stimulus (by pecking it) or different from it (by not pecking it). Alternatively, the pigeon could be presented with two report keys. A peck on one could be defined as correct when the stimuli were the same and a peck on the other when they were different (cf. Wright, 1972). This is a Same/Different task. The Same/Different task is also a serial probe recognition task with one list item.

A serial probe recognition task consists of a series of serially presented list items followed by a single test item, the probe. The subject has to decide whether it matches one of the list items and respond "Same," or none of them and respond "Different." There are advantages to conducting animal memory experiments with the serial probe recognition task.   Animal and human memory processes can be directly compared using the same tasks and the same items; humans usually can not be effectively tested in a traditional animal memory task because the tasks are too easy.   The serial probe recognition task has another advantage over traditional animal memory tasks in that it can be used to study the fine-grain nature of memory processing such as serial position effects (Sands & Wright, 1980a, b).

Unfortunately, animals do not always perform well in the serial probe recognition task (Gaffan, 1977; Eddy, 1973; MacPhail, 1980; Roberts & Kraemer, 1981).   We showed that this comparatively poor performance was due to repeating items (Sands & Wright, 1980a, b).   When we eliminated item repeats performance was 86% correct with 10-item list lengths and 81% correct with 20-item list lengths for a rhesus monkey.

Item repeats tend to confuse the subject.   Consider the situation where the subject is performing on a "Different" trial and the items have already been repeated, maybe several times in previous lists.   The correct response is "Different," but having seen the probe item previously confuses the subject and it may tend to think that the item was in the list currently been tested, and respond "Same."   Indeed, with frequent repeats, the subject might tend to respond "Same" on every trial, but bias would eventually shift although accuracy would suffer.   The type of interference we have been refering to is called proactive interference (PI) because the memory of some previously seen item interferes with judgment about memory for a list of items seen later.   Animals are very sensitive to the effects of proactive interference.

Improved performance with an increase of pool size (i.e., the possible items from which list and probe items are selected) can be seen in the data reported by Mason & Wilson (1974), Worsham (1975), Mishkin & Delacour (1975), and Overman & Doty (1980) with monkeys and Herman & Gordon (1974) with dolphins.   Hayes and Thompson (1953) reported similar findings in a delayed response task with chimpanzees.   Using a serial probe recognition task, Gaffan (1974) found high levels of accuracy with lists of 3, 5, or 10 items when monkeys were trained with 300 stimulus objects. In a subsequent experiment, recognition accuracy for 3-item list was reduced by nearly 25% when the pool size was restricted to 6 items (Gaffan, 1977).

The powerful effects of PI are further indicated by two other findings. First, acquisition of delayed matching-to-sample is 3.5 times slower when two repetitive samples are used compared to trial-unique samples (Mishkin & Delacour, 1975).   Second, shifting from trial-unique samples to repetitive samples produces a dramatic loss of delayed matching-to-sample accuracy even though pre-test performance was both highly accurate and stable. (Mishkin & Delacour, 1975; Overman & Doty, 1980)

The powerful effects of proactive interference on monkey serial probe recognition performance are shown in Figure 20.1 for a single monkey, Oscar, tested shortly after it had learned the serial probe recognition task and tested in an identical manner three years later after it had extensive experience in the serial probe recognition task.   Briefly, the procedure was a 3-item (fixed list length) serial probe recognition task with 1-sec item presentation, 1-sec inter-item interval, and 1-sec probe delay (see Sands & Wright, 1980b for more detail).   The condition of low

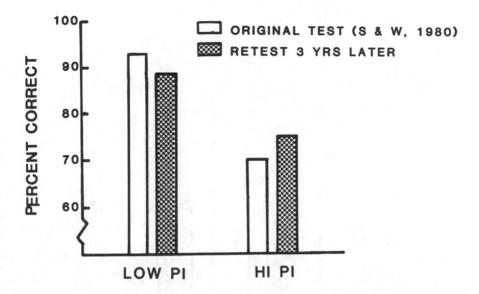

FIG. 20.1. Performance by monkey Oscar on identical proactive interference (PI) tests separated by three years. In the LOW PI condition, the three list items and probe item of each trial were drawn from a 210 item pool and items were not repeated during a session. In the HI PI condition, items were drawn from a six item pool and were repeated many times.

proactive interference was produced by constructing the three item list from a 210 item pool so that the items in each of the 60 trials (30 Same and 30 Different) were different from the items of all of the other trials. The condition of high proactive interference was produced by constructing the three item list from a pool of only six items (the six best discriminated items from a former experiment, Sands & Wright, 1980b, Experiment 1). The second test shown in Figure 20.1 was conducted with the same procedures and the same items as the former test. In both tests, performance is much better under the low PI condition than in the high PI condition. More recently, we have conducted these tests with four other monkeys (Felix, Joe, Linus, Max); their average performance is shown in Figure 20.2. These monkeys are at about the same stage of training as Oscar was in its original test. In the low PI condition, 231 items were used to compose 33 Same and 33 Different trials (four items are used in each Different trial and three in each Same trial). In the high PI condition, the same six items were used as were used to test Oscar. Each condition was tested four times, but not in consecutive sessions. As with Oscar, performance in the low PI is much better than in the high PI condition $[t(24) = 7.1, p < 0.01]$.

These experiments demonstrate that proactive interference has a strong disruptive effect on serial probe recognition performance after it

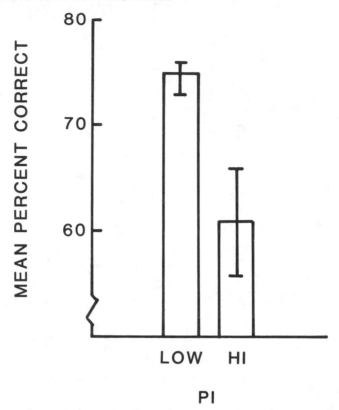

FIG. 20.2. Average performance by four monkeys (Felix, Joe, Linus, Max) on three-item serial probe recognition tests with trial unique stimuli (LOW PI) and repeating stimuli from a six item pool (HI PI).

has been acquired. It can be expected to have an even greater detrimental effect on serial probe recognition acquisition. Therefore, when we began training our monkeys and pigeons we minimized PI by making the items in each trial unique from those in every other trial, and by avoiding items which appeared similar to one another. Proactive interference can occur through generalization of effects among similar items as well as among identical items.

II. PIGEON AND MONKEY SERIAL PROBE RECOGNITION ACQUISITIONS WITHOUT PROACTIVE INTERFERENCE

One of our goals was to test pigeons and monkeys with variable list lengths of one to six items. Therefore, after the pigeons and monkeys had adequately acquired the Same/Different task (which entailed some 18 months of training), we placed them directly in the serial probe recognition task with variable list lengths of one to six items.

A. PIGEON AND MONKEY SERIAL PROBE RECOGNITION EXPERIMENTS WITH 1-6 ITEM VARIABLE LIST LENGTHS

1. Procedure.   The number of list items in each trial could be 1, 2, 3, 4, or 6 items.   The experiment consisted of 216 unique trials (108 Same trials and 108 Different trials) which were constructed from 864 unique items.   Because the Kodak Carousel projectors held only 140 items these trials were distributed into six different carousel projector trays.   They were actually pairs of trays but are referred to here as tray 1, 2, 3, etc. The list and probe items were in separate trays and projected onto separate screens.   List items were projected onto an upper screen and probe items onto a lower screen.   Image size was 18.4 cm horizontal by 12.4 cm vertical.   The screens were spaced 16 cm center to center and were at a distance of approximately 50 cm from the monkeys and 62 cm from the pigeons.   The pigeons viewed the stimuli through a large (8.75 cm square) and thin Plexiglass center pecking key.   Two choice keys were located on either side of the center pecking "window."   A peck on the red right one was a Same response and a peck on the green left one was a Different response.   Monkeys moved a lever to the right (Same) or left (Different) to indicate their choice responses.   Pigeons initiated trials by a center key peck and monkeys by pressing down on a 3-position lever. Both were cued by a ready "clicker" signal. The six different trays of items were run in blocks with different orders within each block.   The items were presented for 1-sec for the monkeys and 2-sec for the pigeons with a 1-sec delay between list items.   Following the last list item a probe item was presented following a delay of 1-sec for the monkeys and a 0.5-sec for the pigeons.   A 3-sec intertrial interval separated successive trials.   All correct responses were rewarded (grain for pigeons, orange juice or banana pellets for monkeys).   All incorrect responses were punished with a time out (with house light turned on) of 10-sec for monkeys and 6-sec for pigeons.   The entire set of 216 trials were usually conducted daily with each of the monkeys and pigeons.   Occasionally, only a portion of the set was conducted, and training resumed the next day at the position in the set where it had been terminated the day before.

2. Results.   Figure 20.3 shows the average performance for the three monkeys (Joe, Max, and Linus) and the three pigeons which were run in this experiment.   The monkeys began at better than chance performance and showed a gradual and steady increase in performance eventually reaching better than 90% correct after 7,200 trials.   Their initial performance of 63% correct is not terribly surprising, considering that just previous to this experiment they were performing very accurately in a Same/Different task and had revealed nearly perfect transfer performance in a test with novel items.   The pigeons, on the other hand, showed no signs of acquiring 1-6 item variable list length serial probe recognition task even after more than 2,300 trials.   Their performance stayed near chance performance (52% correct).   The pigeons too had performed accurately in the Same/Different task and had been given training identical to the monkeys.   The pigeons, however, did not transfer well to novel

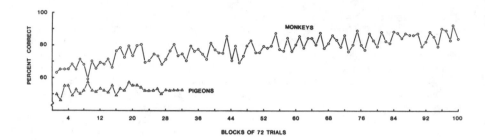

FIG. 20.3. Mean performance by three monkeys and three pigeons on a variable list length (1–6 items) serial probe recognition task.

items as did the monkeys.

B. PIGEON SERIAL PROBE RECOGNITION EXPERIMENT WITH 1-3 ITEM VARIABLE LIST LENGTHS

In order to make the task easier for the pigeons, the number of list items was reduced. We extracted from the 1–6 variable list length serial probe recognition task those trials containing 1, 2, or 3 list items so that the items would be familiar to the pigeons, and trained them only with these trials. Figure 20.4 shows that with the shorter list lengths the pigeons showed some signs of acquiring the task over the 6000 trials, but their acquisition was slow and seemed to level off at about 65% correct. Therefore, once again we changed the procedure and this time we changed it to a fixed list length of three items.

C. PIGEON SERIAL PROBE RECOGNITION ACQUISITIONS WITH FIXED LIST LENGTHS

In a pilot experiment, we had separated the variable list lengths into groups of fixed list lengths of 1, 2, and 3 items, and trained the pigeons on blocks of trials with list length fixed within a block. Somewhat to our surprise there was considerable interference among the different list lengths. Training on a list length of three items (LL3) seemed to interfere

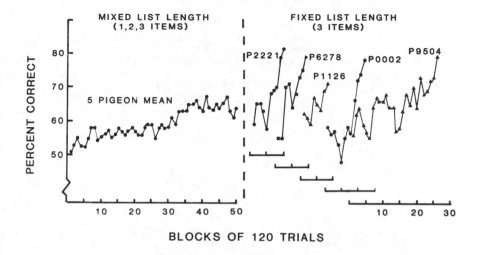

FIG. 20.4. Mean performance by five pigeons on a variable list length serial probe recognition task with 1–3 list items (left–hand half), and individual performance and acquisition by the same five pigeons on a fixed list length serial probe recognition task with three list items (right–hand half).

with performance on the shorter list lengths and vice versa.   Therefore, we conducted the serial probe recognition acquisition with only list length 3, the longest and probably the most difficult list length.   The individual acquisition functions for five pigeons are shown in the right–hand portion of Figure 20.4.   Figure 20.4 shows that they rapidly acquired the the task with fixed list length of three items, much more rapidly than they were formerly acquiring the 1–3 variable list–length task.   They were trained on LL3 until they performed 88% correct or better on two, 20–trial sessions (where each point in Figure 20.4 is an average of six 20–trial sessions). They were then trained on list–length 4 (see Figure 20.5) and acquired that rapidly too, requiring approximately 1,500 trials on the average to reach criterion.   They were then tested to determine the degree to which training on different list lengths can disrupt performance.

D. PIGEON SERIAL PROBE RECOGNITION PERFORMANCE DISRUPTED BY VARIABLE LIST LENGTHS

Disruption by different list lengths was measured by interpolating three 20–trial sessions of a different list length between two sessions on LL4. A performance decrement on the second LL4 session relative to the first indicated disruption by training on a different list length.   On each of four

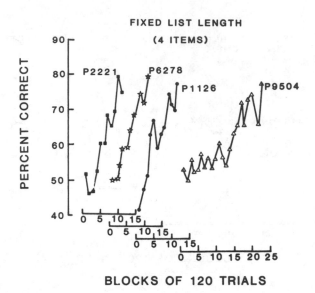

FIG. 20.5. Individual performance and acquisition by four pigeons on a fixed list length serial probe recognition task with four list items.

days they were given three consecutive 20-trial sessions on the original LL4, followed by three consecutive 20-trial sessions on the intervening LL, followed again by three consecutive 20-trial sessions on the original LL4. The control condition, the first test shown in Figure 20.5 but the one tested last, controlled for the possibility that a change in the items, rather than a change in the list length, during the three interpolated sessions would disrupt performance on the original LL4. In this condition, we used the items from the LL3, LL2, and LL1 trials so that they would not be totally unfamiliar items. Figure 20.6 shows the average performance from the three pigeons which were run in this experiment. The unfilled histogram in Figure 20.6 is the average performance on the LL4 session just prior to the intervening list length training. The stippled histograms in Figure 20.6 are for the first session performance on LL4 just following training on the intervening list length. Changing the items, but keeping the list length the same produces a small but significant performance decrement [$t(2) = 4.5$, $p < 0.05$]. The disruptive effect is more pronounced when the list length is changed, and the magnitude of the effect appears to be related to the degree of the list length change. Intervening training with list lengths of one and two items produced more than a 20% decrement in performance [$t(2) = 7.7$, $p < 0.02$; $t(2) = 12.3$, $p < 0.01$ for LL1 and LL2 respectively]. This disruptive effect of intervening training with a different list length appears to explain why the pigeons did not acquire the variable list length task of 1-6 items and showed very slow acquisition of the variable list length task with 1-3

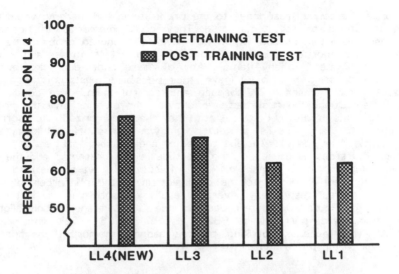

FIG. 20.6. Mean percent correct on list length (LL) of four items before (PRE) and following (POST) training on a different list of either the same length (new LL4) or different length (LL3, LL2, LL1).

items. Indeed, the disruptive effects of variable list length are likely to be even greater when the trials are intermixed than when they are run in blocks.

It is not clear why such disruption occurs with pigeons. It may have to do with their tendency to depend upon a stereotype sequence of events or rhythm of the trial as it progresses. A disruption which might have a similar basis is the disruption caused in delayed matching-to-sample acquisition with variable delays (Berryman, Cumming, & Nevin, 1963; Perkins, Lyderson, & Beaman, 1973).

III. TESTING FOR MEMORY STRATEGIES:
PROBE MEMORIZATION VERSUS RELATING PROBE TO LIST MEMORY

A. PIGEONS

Acquisition of LL4 and tests of the list-length effect entailed presenting the same 20-trial session (LL4) many times to the pigeons. Before continuing our tests with the pigeons, we tested the pigeons (and monkeys too) to determine whether or not they had adopted any short-cuts to comparing the probe item to their memory of the list items, memory scanning. There are at least three possible short-cuts: (1) They could memorize the correct responses to the probe items and thereby would not have to attend to list items.   (2) They could memorize the correct responses to list items and would not have to attend to probe items. (3) They could memorize the sequence of right and left responses and would not have to attend to any items.   Any of these short-cuts circumvents memory processing, and so whenever their possibility exists it is important to assess their presence or absence and control them.   We conducted two tests.   During the first test, we showed the pigeons only the probe items to determine whether or not they had memorized the correct responses to them.   The list projector light was turned off, but everything else remained as normal.   The list projector and adjoining shutters stepped through the slides as usual, only no list slides were presented and the list screen (upper one) was dark.   Responses were defined as correct/incorrect according to the original training, and reinforcement was given for correct responses.   Average results are shown in the left-hand histogram of Figure 20.7.   Their performance (56% average) is only slightly better than chance performance: P6278 – 59%, P1126 – 52%, and P2221 – 57%.   Thus, generally speaking, pigeons had not memorized the correct responses to the probe items.

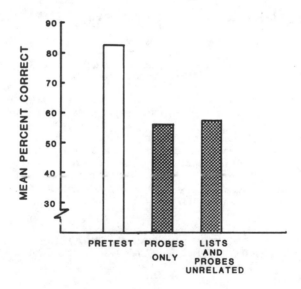

FIG.   20.7.   Mean performance for three pigeons in two separate tests to determine control by probes and lists in a serial probe recognition task.

The second test controlled for the possibility that the dark screen was itself disrupting to performance by changing the task milieu.   Substitutions were made for the list items in each trial.   They were items familiar to the pigeons so that their novelty would not confound interpretation of the test.

None of the list items matched the probe items, but reinforcement was maintained according to the original LL4 contingencies. Performance was 57% on the average: P6278 – 57%, P1126 – 55%, P2221 – 59%. Thus, this second test also demonstrates that the pigeons generally did not perform the task by memorizing the correct responses to probe items.

B. MONKEYS

Monkeys, unlike pigeons, did adopt the strategy of memorizing the correct responses to the probe items. This is a very efficient performance strategy which we have found that humans tend to employ in this task too. The monkeys had been trained with variable list lengths of 1–6 items and had seen the 216 trials of this task at least 35 times (Joe 40 times, Linus 36 times, Max 35 times) prior to testing. Their performance over the last 10 exposures was: Joe 91%, Linus 78%, Max 83%. The probe items from the 216 trials were presented without the list items. The mechanical noises, time intervals and reinforcement contingencies were the same as during training only the carousel projector light in the list projector was not turned on. The results show that. the monkeys' performance can largely be accounted for by a strategy of their memorizing the correct responses to the probe items; average performance on this first test shown in the right hand histogram of Figure 20.8 was 74% correct (Joe 85%, Linus 70%, Max 68%).

A second test showed that the sequence of left/right responses was unimportant. Probe items were rearranged so that their order was different from the order used in original training and presented in an otherwise identical manner to those of the first test. The correct responses, as defined by original training to the probe items, were reinforced. If order was unimportant then performance in this second test should be unchanged relative to the first test, and it was [t = 0.09, p > 0.90]. Their average performance was 74% correct (Joe – 89%, Linus – 72%, Max – 61%).

A third test showed that unrelated list items would not disrupt the monkeys' "probe" strategy. This test was identical to the second test conducted with the pigeons. Previously familiar items were used to compose the lists, and the list length and reinforcement contingency associated with a particular probe item was maintained as during original training. The results showed that they ignored the list items, disregarded the list-probe relationship and responded 79% correctly to the probe items (Joe 91%, Max 79%, Linus 67%).

A fourth test showed that the monkeys' would not use a "list" strategy, even when they had no opportunity to use their "probe" strategy. Only the list items were presented. The light in the probe item projector was turned off. Reinforcement according to original training with list items was maintained. Average performance was 55% correct (Joe 25%, with 47% of the trials aborted because no choice was made during the 2 second choice period, Max 44%, with 5% aborts, and Linus made 85% aborts so his data are not included). Clearly, the monkeys could not use the list items to cue correct responses.

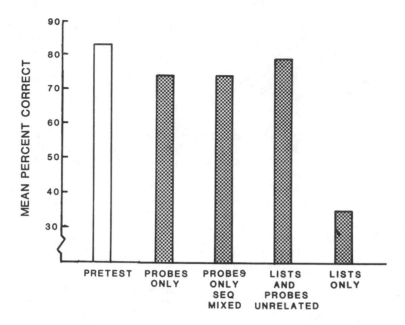

FIG. 20.8. Mean performance for three monkeys (Joe, Max, Linus) in four separate tests to determine the degree to which individual items controlled choices in a serial probe recognition task.

1. Conclusions. These results show that monkeys, but not pigeons, will memorize the correct responses to the probe items, and will short cut the memory process. This is a potential problem for memory work with monkeys. We have not encountered this problem in our past work (Sands & Wright, 1980a, b, 1982) and we have guarded against it since these findings. In our earlier work, lists and probes were changed on a regular basis to force the monkey into a relational strategy. All 210 items were tested in every serial position and the 10-item list experiment was composed of 16,880 trials. The list items were cycled through four times in a session, but the items (e.g., 10 for a 10 item list) displayed on a trial varied from cycle to cycle. For example, on the second cycle through the list items, the last four items of a previously presented trial might be combined with the first six items of what previously was the next trial to make a new list of 10 items. The probe items were completely different from cycle to cycle with Same/Different designation and serial position tested varying randomly. We changed all of the items on a regular basis too. For example, in our 10-item list study, thirty separate sets of trays were used. No sets of trays were tested consecutively in order to avoid the possibility that the monkey might gain familiarity with a particular set.

The robust primacy and recency effects of the serial position functions demonstrated that this monkey had not memorized the correct responses to the probe items. Likewise, the probe-delay effects on the serial position function shown in the next section of this chapter demonstrate that in these experiments the monkeys had not memorized the correct responses to the probe items either.   Accuracy would not vary with serial position if they responded only on the basis of their memory of the correct responses to the probe items and disregarded list items. Likewise, such a strategy would not have produced the systematic changes in the serial position functions that we observed with probe delay.

2. *Strategies Continued.*   Short-cuts to memory processing that monkeys adopt are interesting in their own right.   When pigeons and monkeys are compared in this regard the monkeys readily adopt new and more efficient strategies than do pigeons and in some cases, short cuts. The monkeys are more flexible in their approach to the task than are pigeons.   Another example of this flexibility and comparatively rapid strategy change by monkeys is the strategy of a monkey (Sands & Wright, 1982) to make "quick" Different responses on Different trials based upon item familiarity.   Reaction times were tested in a variable list length task to analyze the monkey's memory scanning strategies (cf.   Sternberg, 1966). We were successful in modifying this strategy by presenting the probe items in a previous list.   The monkey modified its strategy and appeared to scan its memory on Different trials just as it has done (and continued to do) on Same trials.   Another example where monkeys employed more flexible strategies than pigeons was in the transfer of Same/Different performance (Wright, Santiago, Urcuioli, & Sands, 1984). Both monkeys and pigeons initially learned the Same/Different task by learning the correct responses to individual items of the item pair, rather than learning the Same/Different concept.   Daily item changes produced a change in the monkeys' strategy and they revealed near perfect transfer to novel pairs.   This result means they had learned Same/Different concept. Pigeons showed much poorer transfer which meant that they were much slower than monkeys in learning the Same/Different concept, or possibly incapable of fully learning it (cf. Chapter 22, this volume, by Zentall, Hogan, & Edwards).

IV. SERIAL POSITION EFFECTS AS A FUNCTION OF PROBE DELAY

Our discovery that monkeys would adopt a short-cut to memory processing meant that we would have to thwart these short-cuts in order to test the monkey's list memory.   There are several approaches to preventing them from memorizing the correct probe responses: (1) repeating the session unchanged, but testing at regular intervals for probe memorization and retraining on a new tray whenever there is significant probe memorization, (2) presenting each probe item twice in the session, once on a Same trial and once on a Different trial, (3) presenting novel items each day, and (4) mixing the items each session or each day.   Any one of the four would probably work.   Recently we have successfully employed the second method, although we do not present results using it here. We used the fourth method previously (Sands & Wright, 1980a, b). We have used the first method to collect the data reported in this section. We tested for serial position effects with a list length of four items.   We varied probe delay (time between last list item and probe item) and found systematic changes in the serial position function for pigeons, monkeys and humans.

The "U" shaped serial position function is one of the hallmarks of human memory processing. The initial and terminal items of the serial lists are remembered better than the middle list items. Good memory of the initial list items is referred to as the primacy effect, and good memory of the terminal list items are referred to as the recency effect The "U" shaped serial position function figures prominently in theories of memory and has been interpreted as an interaction of proactive (PI) and retroactive (RI) interference (Postman & Phillips, 1965) and as a dual process of a long–term and short–term storage with rehearsal transfering information from short to long–term storage (Waugh & Norman, 1965; Atkinson & Shiffrin, 1968). Previously, we showed a primacy effect for the rhesus monkey in a recognition procedure. To our knowledge, this was the first clear cut demonstration of a primacy effect for any animal. The form of the serial position function with its primacy and recency effects probably reveals something basic about the strategies and capabilities of the organism's memory processing. We therefore wanted to compare human, monkey, and pigeon serial position functions under identical conditions to see what similarities or differences would emerge. Throughout our research we continuously fiddle with parameters in the hopes of discovering those variables which are really important. In just this manner we discovered the importance of probe delay on the form of the serial position function.

A. PROCEDURES

Two monkeys were tested on two 20–trial sequences with a fixed list length of four items each session. Ninety unique items (from a 3000 item pool) were used to construct each 20–trial sequence of 10 Same trials (40 items) and 10 Different trials (50 items). The order of probe delay testing was: 2, 10, 0, 20, 30, 2, 1, 10, 30, 20, 2, 30, 10, 2, 0, 1, 20, 0, 1, 30, 10, 1, 20, 0 sec. Other experimental parameters were unchanged: 1–sec viewing time, 1–sec inter–stimulus interval, 3–sec intertrial interval, 10–sec time out for incorrect responses and the probability of reinforcement for correct responses was 1.0 (Tang orange juice was given on 60% of the trials and banana pellets on 40% of them). After the Probe–delay tests, Probe–only tests were conducted with each set of items. Monkey Joe had largely memorized the probe items (85% and 84% of the two Probe–only tests). We have since determined that this monkey will memorize the correct responses to probe items in an average of 12 sessions. Thus, the majority of serial position function changes for this monkey probably come from earlier sessions before he had memorized the correct responses to probe items.

Four pigeons were tested on the same list length (LL4) in a manner very similar to the monkeys except that only one sequence of 20 trials were used. The probe delays were 0, 0.5, 1, 2, 6, and 10 sec (conducted in four randomized blocks), 2–sec viewing time, 1–sec interstimulus interval, 3–sec intertrial interval, 6–sec long time out for incorrect responses, and all correct responses reinforced with access to mixed grain.

A 13 year old male human subject was tested in the same experiment with the identical parameters as used with the monkeys except that the items used were drawn from only a 6-item pool (the same items used in the previously described experiment on high proactive interference with the monkeys). It was necessary to repeat items and create high proactive interference because this subject would memorize the correct probe responses after only one exposure and revealed a ceiling effect even on the first session. Each of the following probe delays was tested twice with the 20-trial, Hi-PI session: 0, 1, 5, 10, 20, 30, 40 sec. The order of testing was quasi-randomly mixed and different for the two blocks of probe delays.

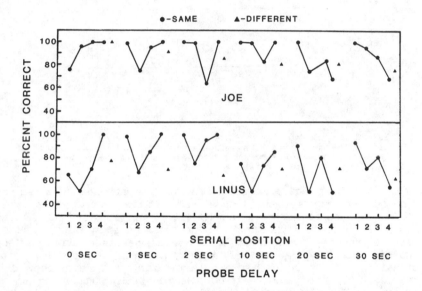

FIG. 20.9. Serial position functions (circles) from Same trial performance and Different trial performance (triangles) for two monkeys. The six serial position functions are for different probe delays, the time between the last list item (#4) and the probe item.

B. RESULTS

The serial position curves for the different probe delays are shown in Figures 20.9, 20.10, and 20.11 for the monkeys, pigeons, and the human respectively. The important results for this series of experiments is that monkeys, pigeons, and human showed the same systematic changes in the form of the serial position function with changes in the probe delay. At the shortest probe delay (0-sec) the serial position functions show an

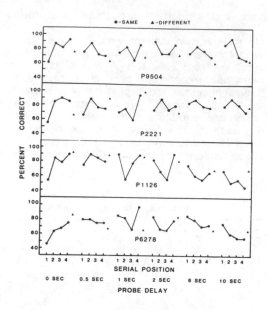

FIG. 20.10. Serial position functions (circles) from Same trial performance and Different trial performance (triangles) for four pigeons. The six serial position functions are for different probe delays, the time between the last list item (#4) and the probe item.

increase in performance towards the end of the list revealing a recency effect (remembering the last list items well). At intermediate probe delays (1, 2, and 10 sec for monkeys, 1 and 2 sec for pigeons and 10, 20, and 30 sec for the human) the serial position functions show both a primacy effect (remembering the first list items well) as well as a recency effect. At long probe delays (20 and 30 sec for monkeys, 6 and 10 sec for pigeons, and 40 sec for the human) the serial position functions show only a primacy effect, no recency effect. (Neuman–Keuls multiple comparison tests show these trends to be highly significant). In addition to the systematic changes in the serial position curves as a function of probe delay, another important finding is that Figure 20.10 displays the first primacy effect shown for any avian species.

C. DISCUSSION

Interpretation of these systematic changes in the serial position functions with probe delay bears upon the memory mechanism involved in the storage and retrieval of visual information. Good performance on the last list items demonstrated at short probe delays deteriorates to near chance performance at long probe delays. These downward sloping serial position functions at long probe delays are the result of an increase in performance on the first list items as well as a decrease in performance

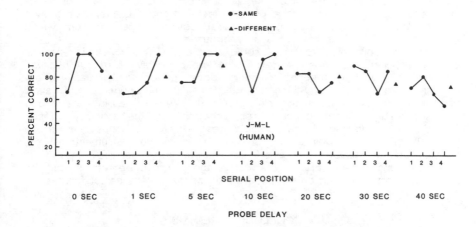

FIG. 20.11. Serial position functions (circles) from Same trial performance and Different trial performance (triangles) for an adolescent human subject. The six serial position functions are for different probe delays, the time between the last list item (#4) and the probe item.

on the last list items. Notice that performance on the first list item is in all cases much better at long delays than at zero delays.

We tend to favor the following theoretical account of these changes in the serial position function. The recency portion tends to dissipate with time (probe delay). This may be due to a passive decay with time (Brown, 1958), similarity of list items (acid bath theory of Posner & Konick, 1966), or interference created by the subject attending to the experimental environment during the delay interval (cf. the Brown–Peterson distractor task). Theoretical interpretation of the rebound of the primacy effect is a little more complicated. The most likely explanation is that it represents a release from the effects of retroactive interference. Retroactive interference (RI) is interference of memory of a previously presented item by items which follow. This form of interference is thought to dissipate quite rapidly allowing recovery of memory of the previously presented item (Melton & Irwin, 1940; Postman, Stark, & Fraser, 1968; Postman & Underwood, 1973). These effects of recovery of the primacy effect fit well with interference theory and the prediction of recovery from RI. Indeed, it was the dearth of such findings that lead to the demise of interference theory. (Keppel, 1968)

Most of the studies have found a greater release from RI effect with recall than recognition procedures (Garskof & Sandak, 1964; Postman, 1965; Postman, Stark, & Fraser, 1968; Sandak & Garskof, 1967), but we know of none using a serial probe recognition procedure. One recall experiment (Postman & Riley, 1959) is distinctive in that it used a serial learning procedure from which serial position functions were obtained. They showed a rebound of the primacy effect much as we have shown here. Most of the studies used a paired-associate design from which serial position functions are not usually obtained.

We feel compelled to make some mention of the role of rehearsal before leaving this discussion of the probe-delay effects, because rehearsal is currently thought to be responsible for the primacy effect (Waugh & Norman, 1965; Atkinson & Shiffrin, 1968). Longer probe delays provide more opportunity for rehearsal, but rehearsal during the probe delay probably cannot account for rebound of the primacy effect. Consider the following: If the subjects can retrieve the items in order to rehearse them, then why can't this same retrieval be instrumental in the correct choice response at short as well as long delays? Indeed, if rehearsal functions to bridge the delay, then primacy effect should be better at short than long delays because of occasional rehearsal failures.

V. CONCLUDING REMARKS

Pigeons acquired serial probe recognition performance with list lengths as long as four items and exhibited primacy as well as recency effects in their serial position functions. The primacy effect figures prominently in many memory theories, and from a comparative memory stand-point it is important that pigeons, like humans and monkeys, exhibited this effect. Pigeons have difficulty performing with variable list lengths and so unless this difficulty can be overcome, their memory-scanning processes cannot be tested and compared to monkeys and humans. Unlike monkeys, however, they do not readily develop short-cuts to the memory task and, for example, memorize the correct responses to the probe items.

Monkeys (and humans too) when given the opportunity, memorize the correct responses to probe items and thereby short-circuit the memory process of comparing the probe item to their memory of the list items. In order to thwart their probe memorization strategy, items were changed on a regular basis before they had a chance to memorize them. The "U" shaped serial position functions and the systematic changes in the serial position function with probe delay demonstrated that all three species were making a memory comparison.

The probe-delay effects reflect what may be an important mechanism of memory processing generally: an absence of a primacy effect at short probe delays, its emergence at intermediate delays, its dominance of the serial position function at long probe delays. These probe-delay effects are demonstrated and reported here for three different organisms: humans, monkeys and pigeons. These results may be important for the human memory literature from the standpoint of development of the primacy effect and may support interference theory and the release from RI. Such results, dealing with the subtle and fine-grained inner workings of the memory system, have heretofore been confined to human memory research. It is of paramount importance that these results have been demonstrated in two different animals, pigeons and monkeys as well as man. The probe-delay findings demonstrate basically similar memory processes in the situations tested here and such similar memory processing may be wideserial probe recognitioned throughout much of the animal kingdom. The particular tasks and tests to reveal these memory processes may have to be devised to take into account the special limitations of the organism under study, much as we did to test our monkeys and pigeons. The single most important factor leading to our successful testing of monkeys and pigeons was the elimination of PI with a large item pool so that items were not repeated. In closing, these and our other experiments demonstrate that the rhesus monkey is a good model of human memory processing and provide the foundation for experiments which are unfeasible and unethical with human subjects. Drug studies and neurophysiological studies can now be performed on a preparation which reveals some of the sensitive inner workings of memory and one which is a good model of the human memory system.

ACKNOWLEDGMENTS

The authors thank Michael J. Watkins for his thoughtful comments on a draft of this manuscript, David Floyd for his conscientious conduct of the monkey experiments and Jacquelyne Rivera for her help with the figures. Portions of this research were part of a doctoral dissertation by Hector C. Santiago to the University of Texas Graduate School of Biomedical Sciences. This research was supported by grant MH-35202 to the authors.

REFERENCES

Atkinson, R. C., & Shiffrin, R. M. Human memory: A proposed system and its control processes. In K. W. Spence & J. T. Spence (Eds.), *The psychology of learning and motivation.* (Vol. 2.) New York: Academic Press, 1968.

Berryman, R., Cumming, W. W., & Nevin, J. A. Acquisition of delayed matching in the pigeon. *Journal of the Experimental Analysis of Behavior,* 1963, *6,* 101–107.

Brown, J. Some tests of the decay theory of immediate memory. *Quarterly Journal of Experimental Psychology,* 1958, *10,* 12–21.

Ebbinghaus, H. B. *Memory: A contribution to experimental psychology.* New York: Dover, 1964. Originally published in 1885 –– translated in 1913.

Eddy, D. R. *Memory processing in Macaca speciosa: Mental processes revealed by reaction time experiments.* PhD Thesis, Carnegie–Mellon University, 1973.

Gaffan, D. Recognition impaired and association intact in the memory of the monkeys after transection of the fornix. *Journal of Comparative and Physiological Psychology,* 1974, *86,* 1100–1109.

Gaffan, D. Recognition memory after short retention intervals in fornix–transected monkeys. *Quarterly Journal of Experimental Psychology,* 1977, *29,* 577–588.

Garskof, B. E., & Sandak, J. M. Unlearning in recognition memory. *Psychonomic Science,* 1964, *1,* 197–198.

Hayes, K., & Thompson, R. Nonspatial delayed response to trial–unique stimuli in sophisticated chimpanzees. *Journal of Comparative and Physiological Psychology,* 1953, *46,* 498–500.

Herman, L. M., & Gordon, J. A. Auditory delayed matching in the bottlenose dolphin. *Journal of the Experimental Analysis of Behavior,* 1974, *21,* 19–26.

Highbee, K. L. *Your memory: How it works and how to improve it.* Englewood Cliffs, N.J.: Prentice–Hall, 1977.

Hunter, W. S. The delayed reaction in animals. *Behavior Monograph,* 1913, *2,* 6.

Keppel, G. Retroactive and proactive inhibition. In T. R. Dixon & D. L. Horton (Eds.), *Verbal behavior and general behavior theory.* Englewood Cliffs, N.J.: Prentice–Hall, 1968.

Konorski, J. A. A new method of physiological investigation of recent memory in animals. *Bulletin de l'Academie Polonaise des Sciences: Serie des Sciences Biologiques,* 1959, *7,* 115–119.

MacPhail, E. M. Short–term visual recognition memory in pigeons. *Quarterly Journal of Experimental Psychology,* 1980, *32,* 531–538.

Mason, M., & Wilson, M. Temporal differentiation and recognition memory for visual stimuli in rhesus monkeys. *Journal of Experimental Psychology,* 1974, *103,* 383–390.

Melton, A. W., & Irwin, J. M. The influence of degree of interpolated learning on retroactive inhibition and the overt transfer of specific responses. *American Journal of Psychology,* 1940, *53,* 173–203.

Mishkin, M., & Delacour, J. An analysis of short–term visual memory in the monkey. *Journal of Experimental Psychology: Animal Behavior Processes,* 1975, *1,* 326–334.

Nelson, K. R., & Wasserman, E. A. Temporal factors influencing the pigeon's successive matching–to–sample performance: Sample duration, intertrial interval, and retention interval. *Journal of the Experimental Analysis of Behavior,* 1978, *30,* 153–162.

Overman, W. H., & Doty, R. W. Prolonged visual memory in macaques and man. *Neuroscience,* 1980, *5,* 1825–1831.

Perkins, D., Lyderson, T., & Beaman, D. Acquisition under mixed–delay and multiple–delay matching–to–sample. *Psychological Reports,* 1973, *32,* 635–640.

Posner, M. I., & Konick, A. W. On the role of interference in short–term retention. *Journal of Experimental Psychology,* 1966, *70,* 237–245.

Postman, L. Unlearning under conditions of successive interpolation. *Journal of Experimental Psychology,* 1965, *70,* 237–245.

Postman, L., & Phillips, L. W. Short–term temporal changes in free recall. *Quarterly Journal of Experimental Psychology,* 1965, *17,* 132–138.

Postman, L., & Riley, D. A. Degree of learning and interserial interference in retention. *University of California Publications in Psychology,* 1959, *8,* 271–396.

Postman, L., & Underwood, B. J. Critical issues in interference theory. *Memory & Cognition*, 1973, *1*, 19–40.

Postman, L., Stark, K., & Fraser, J. Temporal changes in interference. *Journal of Verbal Learning and Verbal Behavior*, 1968, *7*, 672–694

Roberts, W. A., & Kraemer, P. J. Recognition memory for lists of visual stimuli in monkeys and humans. *Animal Learning and Behavior*, 1981, *9*, 587–594.

Ruggiero, F. T., & Flagg, S. F. Do animals have memory? In D. L. Medin, W. A. Roberts, & R. T. Davis (Eds.), *Processes of animal memory*. Hillsdale, N.J.: Erlbaum, 1976.

Sandak, J. M., & Garskof, B. E. Associative unlearning as a function of degree of interpolated learning. *Psychonomic Science*, 1967, *7*, 215–216.

Sands, S. F., & Wright, A. A. Primate memory: Retention of serial list items by a rhesus monkey. *Science*, 1980, *209*, 893–940. (a)

Sands, S. F., & Wright, A. A. Serial probe recognition performance by a rhesus monkey and a human with 10– and 20–item lists. *Journal of Experimental Psychology: Animal Behavior Processes*, 1980, *6*, 386–396. (b)

Sands, S. F., & Wright, A. A. Human and monkey pictorial memory scanning. *Science*, 1982, *216*, 1333–1334.

Sternberg, S. High-speed scanning in human memory. *Science*, 1966, *153*, 652–654.

Waugh, N. C., & Norman, D. A. Primacy memory. *Psychological Review*, 1965, *72*, 89–104.

Worsham, R. W. Temporal discrimination factors in the delayed matching-to-sample task in monkeys. *Animal Learning and Behavior*, 1975, *3*, 93–97.

Wright, A. A. Psychometric and psychophysical hue discrimination functions for the pigeon. *Vision Research*, 1972, *12*, 1447–1464.

Wright, A. A., Santiago, H. C., Urcuioli, P. J., & Sands, S. F. Monkey and pigeon acquisition of same/different concept using pictorial stimuli. In M. L. Commons, R. J. Herrnstein, & A. R. Wagner (Eds.), *Quantitative analyses of behavior: Vol. III, Acquisition*. Cambridge, Mass.: Ballinger, 1984. In press.

21 SERIAL POSITION EFFECTS AND REHEARSAL IN PRIMATE VISUAL MEMORY

Stephen F. Sands
The University of Texas at El Paso
Peter J. Urcuioli
Purdue University
Anthony A. Wright
Hector C. Santiago
The University of Texas

I. INTRODUCTION

One of the most studied and ubiquitous findings in human memory research has been the serial position effect. Serial position effects refer to the finding that when subjects are given a list of items to remember, they recall the initial (primacy effect) and terminal (recency effect) items of the list with higher accuracy than the medial items.

Several theoretical accounts of serial position effects have been suggested. For example, serial position effects have been attributed to intralist interference (Foucault, 1928; Waugh, 1960), differential accessibility of the initial and terminal items (Murdock, 1960; Tulving, 1968), and perceptual–organizational advantages for the initial and terminal items (Asch, Hay, & Diamond, 1960). The most common theoretical account, however, derives from information processing theories of memory (e.g., Atkinson & Shiffrin, 1968).

Information processing theories posit that serial position effects reflect the operation of two distinct memory processes; a temporary short–term memory of limited capacity and a more permanent long–term memory of large capacity. Information held in short–term memory is maintained there by rehearsal processes, and acts to build a more permanent long–term memory representation. The primacy effect occurs because few if any items are maintained in short–term memory at the beginning of the list, and consequently more information about these items can be transferred to long–term memory. The recency effect occurs because items at the end of the list are still present in short–term memory at the time of the retention test. Medial items show the poorest retention because they do not benefit either from differential rehearsal or from storage in short–term memory.

Recently, we obtained serial position functions from a rhesus monkey trained in a serial probe recognition task (Sands & Wright, 1980a, b, 1982). In the serial probe recognition task, the monkey was shown a series of pictures followed by a single "probe" picture. Upon presentation of the probe, the monkey was required to make a Same (in list) or

Different (not in list) judgment. The techniques and conditions which lead to good performance of animals in this task are the topic of another chapter (see Chapter 20, by Wright, Santiago, Sands, & Urcuioli) in this volume.

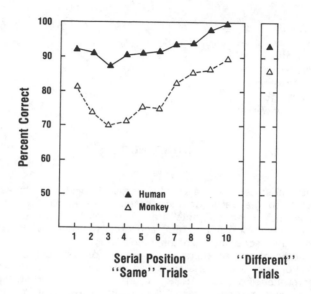

FIG. 21.1. Percent correct as a function of serial position for Same probe trials and percent correct on Different probe trials for human and monkey subjects with 10-item lists. (From Sands & Wright, 1980b. Copyright 1980 by the American Psychological Association. Reprinted by permission.)

In one of our first experiments, we tested a rhesus monkey (Oscar) and a human subject with 10- and 20-item lists. Figure 21.1 shows the 10-item serial position curve and Figure 21.2 shows the 20-item serial position curve. Serial position curves from both the monkey and the human show primacy and recency effects. These data suggest that the determinants of the serial position curves for the monkey and human may be identical, thus opening the way for an intensive examination of how this function is generated.

Our results have been recently replicated in other species using similar tasks. Roberts and Kraemer (1981), for example, tested squirrel monkeys, and humans in a multiple-item delayed matching-to-sample task in which subjects viewed a list of patterns and were then required to chose between a list and a nonlist item. Although overall accuracy was higher for humans and monkeys, both species showed distinct primacy and recency effects. In another list memory study, Lana, a chimpanzee trained to communicate with the Yerkish language, was tested in a recognition task similar to human verbal free recall (Buchanan, Gill, & Braggio, 1981). Lana

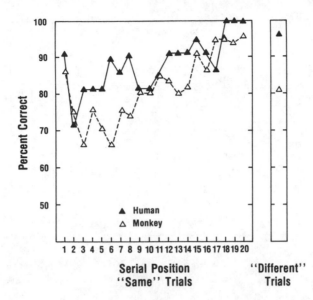

FIG. 21.2. Percent correct as a function of serial position for Same probe trials and percent correct on Different probe trials for the human and monkey subjects with 20-item lists. (From Sands & Wright, 1980b. Copyright 1980 by the American Psychological Association. Reprinted by permission.)

was shown a sequentially presented series of words varying in length from one to eight items and trained to "recall" the list by duplicating the list on a response panel. The serial position curves from this experiment showed strong primacy and recency effects that were nearly identical to those obtained from humans in free recall studies. Taken together, the above results appear to extend the range of species to which theoretical accounts of serial position effects must be applied.

We are now in a position to begin to explore in greater depth the factors responsible for these serial position effects. Of particular interest has been the demonstration of the primacy effect. There have been several list memory studies demonstrating the recency effect but not the primacy effect with pigeons (MacPhail, 1980; Shimp, 1976; Shimp & Moffitt, 1974), rats (Roberts & Smythe, 1979), monkeys (Davis & Fitts, 1976; Eddy, 1973; Gaffan, 1977), and a dolphin (Thompson & Herman, 1977).

The demonstration of a primacy effect in our data and the data of others (Buchanan, Gill, & Braggio, 1981; Roberts & Kraemer, 1981) leads naturally to the question of whether or not it is the result of an active rehearsal process favoring the initial items of the list (Rundus, 1971).

Recently, however, we have obtained data which has led us to question this interpretation. It has been three years since we obtained the serial position curves generated by the monkey (Oscar) shown in Figures

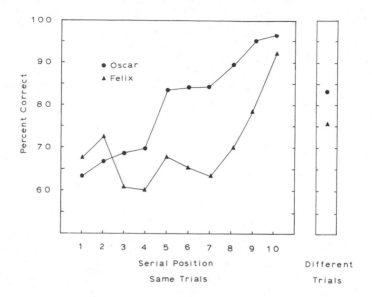

FIG. 21.3. Percent correct as a function of serial position for a monkey (Oscar) after 3 years of training in the serial probe recognition task and a monkey (Felix) recently trained in the serial probe recognition task. Accuracy values have been corrected for guessing and according to this transformation chance performance is 0 percent correct.

21.1 and 21.2. We have continued to test him and have trained an additional monkey (Felix) in the serial probe recognition task.

During this period, Oscar has shown a steady improvement in performance obtaining levels of accuracy equaling or exceeding those of humans familiar with the task, although perhaps less experienced. In spite of his improving accuracy, the primacy effect, which was so pronounced in Oscar's earlier serial position curves, has disappeared. Figure 21.3 shows a recent example of this with serial position curves from a 10-item list recognition task.[1] The recency effect is clearly evident in Oscar's serial position curve, but the primacy effect is absent. For some reason, Oscar appears to have learned to maintain items in memory via mechanisms which do not result in a primacy effect. Serial position curves from the recently trained Felix, however, show both primacy and recency. If Oscar's previously displayed primacy effect was the result of rehearsal, do his more recent data indicate that he has stopped rehearsing?

---

[1] The overall level of choice accuracy shown in this Figure is somewhat lower than we typically obtain from Oscar and Felix because the data were collected in an experiment that explored effects of proactive interference. In this experiment, half of items appearing as Different probes were drawn from the previous list adding considerable difficulty to the same/different discrimination. The level of accuracy obtained from Oscar, however, equals or exceeds data obtained from humans under the same conditions.

Alternatively, does he still rehearse and nonetheless show no primacy? In addition, why would a less well trained monkey (Felix) rehearse items in visual memory, yet the well-trained monkey (Oscar) apparently does not?

The experiment reported here is an attempt to encourage rehearsal by manipulating a variable thought to be important in developing and maintaining rehearsal: the duration of stimulus "off-time" between successively presented items in a list. To the extent that an item maintained in memory can be effectively rehearsed, recognition accuracy should be proportional with stimulus "off-time" between successive items, the interstimulus interval. Furthermore, if the primacy effect is indeed the result of a cumulative rehearsal strategy, then it should be accentuated with longer interstimulus intervals.

II. METHODS AND PROCEDURES

Two male rhesus monkeys (*Macaca mulatta*), Oscar and Felix, served as subjects. Oscar was the subject of several recognition experiments (Sands & Wright, 1980a, b, 1982; Sands, Lincoln, & Wright, 1982) and at the time of testing had experienced well over 200,000 trials in the serial probe recognition task. Felix, the subject of one other experiment (Sands, Lincoln, & Wright, 1982), was less experienced in the serial probe recognition task (40,000 trials) but performed with good accuracy at the time of testing. The two monkeys who served as subjects in this study differed significantly in their stage of development. Oscar, a seven year old male, was a young adult. Felix, a four year old male, was an adolescent. We also tested human subjects whose developmental age approximated that of Oscar and Felix. Two male adult graduate students at the University of Texas at El Paso 24 and 30 years of age and one male and three female adolescent humans 10-12 years of age were tested. It is well known that children gradually develop the ability to utilize rehearsal strategies in memory tasks with verbal materials (Appel, Copper, McCarrell, Sims-Knight, Yussen, & Flavell, 1972) and similar relationship may also hold for pictorial rehearsal.

The techniques used to test the monkeys and humans were identical to those described in Sands and Wright (1980b). Briefly, the monkeys viewed stimuli on two vertically arranged rear projection screens from a primate restraining chair. Behind each screen was a Beseler Cue/See Super 8mm projector for image presentation. The stimuli were a set of 140 color 35mm slides of a wide variety of objects such as fruits, flowers, animals, and people. Stimulus sequences were constructed by re-photographing the 35 mm slides with a Super 8mm movie camera onto Kodachrome (ASA 40) movie film with dark frames interposed between successive frames to control image onset/offset.

A three-position lever (left, right, and down) was mounted on the waistplate of the primate chair and was manipulated by the monkey's right hand. Correct responses were rewarded with either a small squirt of orange juice or a banana flavored pellet. Experimental events and data acquisition were controlled by a Cromemco Z-2D microcomputer system.

All subjects were tested on a 1,200 trial stimulus sequence divided into 20-trial blocks. Selection of list length, list items, same/different trials and serial position were determined in a pseudorandom fashion. Within a 20-trial block, the number of trials at each list length and the number of same/different trials occurred with equal frequency. List items were drawn without replacement from the 140 item pool of pictures every 20 trials. On Same trials, the serial position of the target (match to probe) was sampled with equal frequency in a pseudorandom fashion for each list length. On Different trials, unused slides within the 20-trial block were designated as probe items. The 1,200-trial sequence was divided into five smaller sequences of 240 trials for daily experimental sessions. Both monkeys and humans were given one daily session with one of the five 240-trial sequences.

It is important to note that the stimulus sequence was sufficiently large to ensure that each session presented a unique stimulus sequence. This made it impossible for subjects to learn the order of stimulus presentation or to form stimulus-response associations. Previous studies (see Chapter 20, by Wright, Santiago, Sands, & Urcuioli, this volume) have shown that if such precautions are not taken, monkeys are quite capable of memorizing correct responses to probe items.

A trial began with a 100 Hz "ready" tone that signaled the availability of a trial to the monkey. A downward press of the lever terminated the "ready" tone, and after a 0.5-sec delay, a list containing either 2, 4, 6, 8, or 10 items was presented on the top screen. Each item in the list was presented briefly for 80 msec. The major variable of interest in this experiment was the interstimulus interval between successive items. Interstimulus intervals of 80 and 1000 msec were selected on the basis of previous data on pictorial recognition (Intraub, 1980). One sec following the presentation of the last item of the list, a probe item was displayed on the bottom screen and remained in view until the monkey responded or until a 2-sec response cutoff had elapsed. Upon presentation of the probe item, the subjects then responded either Same or Different by moving the response lever to the right or to the left, respectively. Correct responses were followed by a short tone (3000 Hz, 1-sec duration), a reinforcer and a 4-sec intertrial interval. Incorrect responses or failures to respond within 2 sec were followed by illumination of the houselight, omission of the reinforcer, and a 5-sec intertrial interval. Failures to respond within the cutoff interval were treated as incorrect responses. All human subjects received five sessions using a similar procedure at each rate of presentation. Oscar and Felix received 19 and 17 experimental sessions at each presentation rate, respectively.

The humans were tested under conditions similar to those of the monkeys except that they sat in a chair 100 cm from the stimulus panel and held the response lever in their right hand.

III. RESULTS

Three major findings from this experiment were found; 1) increasing interstimulus interval significantly improved recognition accuracy for the adult humans and monkey (Oscar) but had no effect on the adolescent humans and monkey (Felix), 2) increasing interstimulus interval failed to enhance the primacy effect, and 3) the adolescent humans and Felix showed significant primacy and recency effects.

Overall, the adult humans performed with significantly higher accuracy with the 1000 msec interstimulus interval than with the 80 msec Interstimulus Interval (92% vs. 85%, F[1,90] = 14.1; p < 0.001). Oscar, the highly trained adult monkey, also performed with significantly higher accuracy with the 1000 msec interstimulus interval than with the 80 msec interstimulus interval (81% vs. 77%, F[1,180] = 9.9, p < 0.01). The adolescent humans and Felix, the adolescent monkey, failed to show any significant effects of interstimulus interval (adolescent humans, 70% vs. 69%, F[1,190] = 0.32; Felix, 63% vs. 62%, F[1,160] = 1.1). Analysis of the data from the two adult and four adolescent humans revealed that the results from these two groups were internally consistent under every condition, and consequently the data from the adults were combined and data from the adolescents were combined for more thorough analyses.

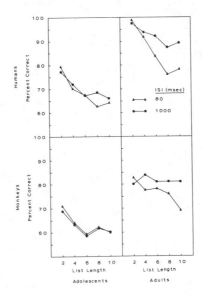

FIG. 21.4. Combined percent correct on Same and Different trials as a function of list length and interstimulus interval for the humans and monkeys.

Figure 21.4 shows the overall accuracy of the humans and monkeys as a function of list length and interstimulus interval. The adult humans showed a significant list length effect (F[4,90] = 15.0, p < 0.001), and post-hoc comparisons revealed performance with list lengths 8 and 10 was of significantly lower than with list lengths 2-6 for the 80 msec interstimulus interval presentation rate. No significant performance effects due to list length were found in the 1000 msec interstimulus interval condition. Oscar also showed a significant list length effect (F[4,180] = 2.9, p < 0.05), and post-hoc comparisons revealed an effect similar to the adult humans; significantly lower accuracy at list length 10 than at list lengths 2-8 with the 80 msec interstimulus interval condition. In addition, no significant list length effects were found with the 1000 msec interstimulus interval condition. The adolescent humans and Felix also showed a significant decrease in accuracy with increasing list length (adolescent humans, F[4,190] = 19.5, p < 0.001; Felix F[4,160] = 12.8, p < 0.001), but unlike Oscar and the adult humans, they did not show any facilitation in accuracy with the 1000 msec interstimulus interval.

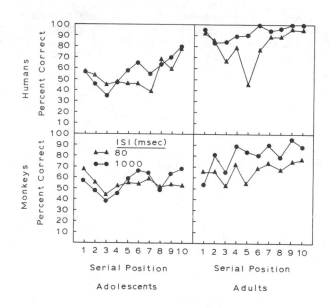

FIG. 21.5. Representative serial position functions on Same trials for the humans and monkeys at 80 and 1000 msec interstimulus intervals.

Figure 21.5 shows representative serial position curves for 10-item lists for the adult and adolescent humans, Oscar, and Felix. For the adult humans, examination of the serial position curves revealed poorer retention of the medial items of the lists with the 80 msec interstimulus interval than with the 1000 msec interstimulus interval (F[1,9] = 3.2, p < 0.01). Post-hoc comparisons showed significantly lower accuracy with the 80 msec interstimulus interval than with the 1000 msec interstimulus interval at serial positions 3–7 for list length 10.

Examination of Oscar's serial position curves showed that increased presentation rate resulted in a loss of retention at the medial and terminal serial positions (F[1,9] = 5.0, p < 0.001). Post-hoc comparisons of the 10–item lists revealed significantly lower accuracy with the 80-msec interstimulus interval than with the 1000 msec interstimulus interval at serial positions 5–10.

Finally, the adolescent humans and Felix, who were unaffected by interstimulus interval, showed results opposite to those of the adult humans and Oscar. Both showed significant primacy and recency effects (adolescent humans, F[1,9] = 5.2, p < 0.001; Felix, F[1,9] = 2.9, p < 0.01).

IV. DISCUSSION

A. SERIAL POSITION EFFECTS

Studies of human free recall and recognition have shown that slow rates of presentation generally facilitate retention of the initial and medial items of the serial position curve while they have little or no effect on the terminal items (Ellis & Hope, 1968; Glanzer & Cunitz, 1966; Murdock, 1962; Raymond, 1969). Recent work with non-verbal visual materials has extended this finding to visual memory (Phillips & Christie, 1977, Experiment 6). Differential effects of presentation rate on the serial position curve are generally cited as evidence for dual-process theories of memory. According to these theories, slow presentation rates increase the time available for rehearsal, and as a result, lead to better performance on the pre-recency portion of the serial position curve. The recency portion, which reflects output from rehearsal independent short-term memory, remains unaffected.

The adult human serial position curves conform to the typical finding in the human literature: increased stimulus off-time facilitated the pre-recency portion of the serial position curves. Oscar's serial position curves, however, were quite different from the humans. Although Oscar benefited from additional stimulus off-time, facilitation was limited to the later items of the sequence. This result is one of the first differences in serial probe recognition performance that we have found between Oscar and the human subjects and suggests that mechanisms other than output from short-term memory may be responsible for Oscar's recency effect. According to the short-term memory/long-term memory analysis, the recency effect should be facilitated rather than inhibited by a short interstimulus interval. Each successive item of the list should result in a backward masking of the previous item (e.g., Rosenblood & Poulton, 1975). This would be true for all serial positions except, of course, the last item of the list.

Furthermore, serial position curves from Oscar and the adult humans lack a primacy effect. Oscar and the adult humans clearly show an advantage with additional stimulus off-time, but this advantage does not appear to be related to the primacy effect. If the primacy effect is the result of a cumulative rehearsal strategy, then the initial items of the list should show an advantage with the longer interstimulus interval. Instead, the one serial position curve from the humans showing a primacy effect was not with the 1000 msec interstimulus interval as predicted but with the 80 msec interstimulus interval at list length 10.

Additional support indicating that the primacy effect is not related to rehearsal can be found in the serial position curves from the adolescent humans and Felix. Both failed to show any significant effects due to interstimulus interval, yet their serial position curves show primacy effects. If anything, the 80 msec interstimulus interval tended to facilitate rather than inhibit the primacy effect as would be predicted by the short-term/long-term memory analysis.

These results point to alternative theoretical accounts of serial position effects in nonhuman and human primate visual memory. One possibility is that initial and terminal items of a list are somehow more distinctive than medial items of the list (Tulving, 1968; Wagner & Pfautz, 1978).

Experiments which systematically manipulate the saliency of the ends of the lists may be informative in this regard. Another possibility is that primacy and recency effects result from a combination of proactive and retroactive interference gradients within the list (Foucault, 1928).

According to this notion, the primacy effect occurs because the initial items are subjected to less proactive interference than later items, and the recency effect occurs because terminal items are subjected to less retroactive interference from later list items. Medial items show the poorest retention because they are subjected to both proactive and retroactive interference. If this is indeed the case, then it might be possible to modulate the form of the serial position curve by manipulating retroactive and proactive interference within a list.

B. POSTSTIMULUS PROCESSING

The major finding of this experiment is that recognition of briefly presented pictures is facilitated with additional stimulus off-time for adult human subjects and the highly trained monkey, Oscar. This result indicates that processing occurs after a brief visual image and this processing aids in maintaining visual memory of these items. This result is in agreement with a large number of human studies showing facilitation of visual memory with increased stimulus off-time (Intraub, 1979, 1980; Lutz & Scheirer, 1974; Phillips & Christie, 1977; Read, 1979; Tversky & Sherman, 1975; Weaver, 1974; Weaver & Stanny, 1978) and extends the finding to at least one adult highly trained rhesus monkey.

An alternative explanation for the post-stimulus processing effect is the differential effects of backward masking. When the interval between items is sufficiently short, presentation of a new picture "masks" the memory of the previous picture. Backward masking predicts that the serial position curves for short interstimulus intervals (80 msec) should result in lower performance at all serial positions excluding the last item of the list, since the last item of the list is not followed immediately by a list item. Serial position curves from the adult humans and Oscar do not show this generalized effect. To the contrary, the short–interstimulus interval curves from the humans show lower performance only at medial items, while Oscar's curves show a lower performance only at terminal items.

An unexpected result of this study was the failure of the adolescent humans and monkey to show any facilitation with additional stimulus off-time. This finding suggests the interesting possibility that rehearsal in visual memory is an ability that may be acquired at later stages of development not only for humans but also for rhesus monkeys. The human developmental literature has shown that active rehearsal strategies do not appear until the later elementary school years (Appel et al., 1972; Hagen, Jongeward, & Kail, 1975). A similar conclusion would be consistent with the results of this experiment. Further support for the developmental hypothesis can be found in the striking similarities between the adolescent humans and Felix in absolute levels of accuracy, the effects of list length and interstimulus interval. Both showed lower levels of accuracy than their adult counterparts, both showed a progressive loss in accuracy with increasing list length, and both were unaffected by interstimulus interval.

If we accept the notion that Oscar is maintaining list items in a manner analogous to rehearsal, then this result has important implications for the nature of rehearsal in visual memory. Some investigators have shown that the ability to rehearse and maintain pictures is enhanced if the pictures can be described by a convenient verbal label (Lutz & Scheirer, 1974; Tabachnik & Brotsky, 1976). Several other investigators, however, have shown that verbal descriptions are not necessarily an essential requirement to the rehearsal of pictures (Phillips & Christie, 1977; Graefe & Watkins, 1980; Intraub, 1979; Tversky & Sherman, 1975). The results from Oscar provide evidence that rehearsal of pictorial materials does not depend exclusively on verbal encoding. Human subjects may employ verbal codes to mediate pictorial recognition. Indeed, one of our human subjects employed a verbal labeling strategy for the pictures in the later stages of testing and overtly rehearsed (verbalized) during the interstimulus interval and probe delay. The other adult human subject also reported the adoption of verbal labels as testing progressed. The advantage to a verbal encoding strategy, however, is not clear since performance of the human subjects remained relatively constant across all experimental sessions even though both reported a shift from visualization to verbalization strategies.

V. SUMMARY

Serial–probe recognition performance of briefly (80 msec) presented pictures was facilitated with a 1000 msec interstimulus interval relative to a 80 msec interstimulus interval in two human subjects and in a single highly–trained rhesus monkey (Oscar). These results demonstrated the benefits of additional time to process or rehearse successively presented pictures. Serial position curves for the adult humans showed an increase in the pre–recency portion of the curve with the 1000 msec interstimulus interval. Serial position curves for Oscar showed an increase in the recency portion of the curve with additional stimulus off time. In addition, serial position curves for Oscar, who in previous studies had shown primacy and recency effects (Sands & Wright, 1980a, b), failed to show a primacy effect. A recently–trained adolescent monkey (Felix) and four adolescent humans, however, did not show any facilitation with the 1000 msec interstimulus interval. Felix and the adolescent humans showed both primacy and recency effects. It is suggested that the primacy effect may reflect some process other than rehearsal or poststimulus processing in primate visual memory.

ACKNOWLEDGMENTS

This research was partially supported by Grant MH35202–01 from the National Institutes of Mental Health.

REFERENCES

Appel, L. F., Copper, R. G., McCarrell, N., Sims–Knight, J., Yussen, S. R., & Flavell, J. H. The development of the distinction between perceiving and memorizing. *Child Development*, 1972, *43*, 1365–1381.

Asch, S. E., Hay, J., & Diamond, R. M. Perceptual organization in serial rote-learning. *American Journal of Psychology*, 1960, *73*, 177–198.

Atkinson, R. C., & Shiffrin, R. M. Human memory: A proposed system and its control processes. In K. W. Spence & J. T. Spence (Eds.), *The psychology of learning and motivation*. (Vol. 2.) New York: Academic Press, 1968.

Buchanan, J. P., Gill, T. V., & Braggio, J. T. Serial position and clustering effects in chimpanzee's "free recall". *Memory & Cognition*, 1981, *9*, 651–660.

Davis, R. T., & Fitts, S. S. Memory and coding processes in discrimination learning. In D. L. Medin, W. A. Roberts, & R. T. Davis (Eds.), *Processes in animal memory*. Hillsdale, N.J.: Erlbaum, 1976.

Eddy, D. R. *Memory processing in Macaca speciosa: Mental processes revealed by reaction time experiments*. PhD Thesis, Carnegie–Mellon University, 1973.

Ellis, N. R., & Hope, R. Memory processes and the serial position curve. *Journal of Experimental Psychology*, 1968, *77*, 613–619.

Foucault, M. Les inhibitions internes de fixation. *Annee Psychologique*, 1928, *29*, 92–112.

Gaffan, D. Recognition memory after short retention intervals in fornix-transected monkeys. *Quarterly Journal of Experimental Psychology*, 1977, *29*, 577–588.

Glanzer, M., & Cunitz, W. H. Two storage mechanisms in free recall. *Journal of Verbal Learning and Verbal Behavior*, 1966, *5*, 351–360.

Graefe, T. M., & Watkins, M. J. Picture rehearsal: An effect of selectively attending to pictures no longer in view. *Journal of Experimental Psychology: Human Learning and Memory*, 1980, *6*, 156–162.

Hagen, J. W., Jr., Jongeward, R. H., & Kail, R. V., Jr. Cognitive perspective on the development of memory. In H. W. Reese (Ed.), *Advances in child development and behavior*. (Vol. 10.) New York: Academic Press, 1975.

Intraub, H. The role of implicit naming in pictorial encoding. *Journal of Experimental Psychology: Human Learning and Memory*, 1979, *5*, 78–87.

Intraub, H. Presentation rate and the representation of briefly glimpsed pictures in memory. *Journal of Experimental Psychology: Human Learning and Memory*, 1980, *6*, 1–12.

Lutz, W. J., & Scheirer, C. J. Coding processes for pictures and words. *Journal of Verbal Learning and Verbal Behavior*, 1974, *13*, 316–320.

MacPhail, E. M. Short-term visual recognition memory in pigeons. *Quarterly Journal of Experimental Psychology*, 1980, *32*, 531–538.

Murdock, B. B., Jr. The distinctiveness of stimuli. *Psychological Review*, 1960, *67*, 16–31.

Murdock, B. B., Jr. The serial position effect of free recall. *Journal of Experimental Psychology*, 1962, *64*, 482–488.

Phillips, W. A., & Christie, D. F. M. Components of visual memory. *Quarterly Journal of Experimental Psychology*, 1977, *29*, 117–133.

Raymond, B. Short-term storage and long-term storage in free recall. *Journal of Verbal Learning and Verbal Behavior*, 1969, *8*, 567–574.

Read, J. D. Rehearsal and recognition of human faces. *American Journal of Psychology*, 1979, *92*, 71–85.

Roberts, W. A., & Kraemer, P. J. Recognition memory for lists of visual stimuli in monkeys and humans. *Animal Learning and Behavior*, 1981, *9*, 587–594.

Roberts, W. A., & Smythe, W. E. Memory for lists of spatial events in the rat. *Learning and Motivation*, 1979, *10*, 313–336.

Rosenblood, L. K., & Poulton, T. W. Recognition after tachistoscopic presentations of complex pictorial stimuli. *Canadian Journal of Psychology*, 1975, *29*, 195–200.

Rundus, D. Analysis of rehearsal processes in free recall. *Journal of Experimental Psychology*, 1971, *89*, 63–77.

Sands, S. F., & Wright, A. A. Primate memory: Retention of serial list items by a rhesus monkey. *Science*, 1980, *209*, 893–940. (a)

Sands, S. F., & Wright, A. A. Serial probe recognition performance by a rhesus monkey and a human with 10- and 20-item lists. *Journal of Experimental Psychology: Animal Behavior Processes*, 1980, *6*, 386–396. (b)

Sands, S. F., & Wright, A. A. Human and monkey pictorial memory scanning. *Science*, 1982, *216*, 1333–1334.

Sands, S. F., Lincoln, C. E., & Wright, A. A. Pictorial similarity judgments and the organization of visual memory in the rhesus monkey. *Journal of Experimental Psychology: General*, 1982, *111*, 369–389.

Shimp, C. P. Short-term memory in the pigeon: Relative recency. *Journal of the Experimental Analysis of Behavior*, 1976, *25*, 55–61.

Shimp, C. P., & Moffitt, M. Short-term memory in the pigeon: Stimulus-response associations. *Journal of the Experimental Analysis of Behavior*, 1974, *22*, 507–512.

Tabachnik, B., & Brotsky, S. J. Free recall and complexity of pictorial stimuli. *mc*, 1976, *4*, 466–470.

Thompson, R. K. R., & Herman, L. M. Memory for lists of sounds by the bottle–nosed dolphin: Convergence of memory processes with humans? *Science*, 1977, *153*, 501–503.

Tulving, E. Theoretical issues in free recall. In T. R. Dixon & D. L. Horton (Eds.), *Verbal behavior and general behavior theory*. Englewood Cliffs, N.J.: Prentice–Hall, 1968.

Tversky, B., & Sherman, T. Picture memory improves with longer on time and off time. *Journal of Experimental Psychology: Human Learning and Memory*, 1975, *1*, 114–118.

Wagner, A. R., & Pfautz, P. L. A bowed serial-position function in habituation of sequential stimuli. *Animal Learning and Behavior*, 1978, *6*, 395–400.

Waugh, N. C. Serial position and memory span. *American Journal of Psychology*, 1960, *73*, 68–79.

Weaver, G. E. Effects of poststimulus study time on recognition of pictures. *Journal of Experimental Psychology*, 1974, *103*, 799–801.

Weaver, G. E., & Stanny, C. J. Short–term retention of pictorial stimuli as assessed by a probe recognition technique. *Journal of Experimental Psychology: Human Learning and Memory*, 1978, *4*, 55–65.

22 COGNITIVE FACTORS IN CONDITIONAL LEARNING BY PIGEONS

Thomas R. Zentall
University of Kentucky
David E. Hogan
Northern Kentucky University
Charles A. Edwards
University of Kentucky

I. INTRODUCTION

Behaviorism developed as a reaction to the subjective introspection of the structuralists (e.g., Titchener). With behaviorism came not only objective data-based research which emphasized control of extraneous variables, but also an attempt to describe all behavior in terms of external (i.e., observable) stimulus and response events. Even when they were forced to retreat into the organism, behaviorists such as Hull (1943) postulated mechanisms that were potentially observable (e.g., the fractional anticipatory goal response). Occasionally the behaviorist mantle was adopted by someone like Tolman (1932) who felt comfortable with purely internal, stimulus–stimulus associations, but this was the exception rather than the rule.

For all of its methodological contribution, behaviorism has tended to result in an arbitrarily restricted view of behavior. Although the language of behaviorism freed us from the vicissitudes of introspection, the price of that freedom was a Whorfian trap. Many of us became unduly constrained by our newly-acquired stimulus–response language; a language that deterred us from thinking outside of its narrow limits. The language of behaviorism did not *prevent* us from asking if a more cognitive terminology were needed; language plays a much more subtle role in our behavior. Instead, it often led us to design experiments that were focused on an overly constrained set of issues.

Recently we have been moving beyond that limiting terminology. We continue to recognize the need for objective data gathering and careful control, but we acknowledge that by allowing cognitive terminology into our language we are better able to explain phenomena that fall outside of the realm of traditional stimulus–response theory. The new cognitive approach combines the rigorous experimental design of the behaviorists with the empirically testable speculation of the structuralists. When this new approach is applied to animal research, it encourages us to use terms that were originally applied to cognitive processing by humans (e.g., working memory, rehearsal, coding, and concept learning). In so doing, it encourages us to look for analogous behavior in other organisms and more important it allows us to design experiments that will test the implications of these cognitive terms.   A cognitive approach to animal learning assumes that animals are capable of actively processing stimulus input and that specification of the physical properties of a stimulus may be insufficient to predict the effect that stimulus will have on behavior.   The relation which that stimulus has to other events (e.g., other stimuli, responses, and outcomes) may determine what stimulus processing "strategy" the animal adopts.   In the final analysis, however, whether this cognitive approach proves useful in characterizing the behavior of animals will be determined not by rational argument, but by empirical test.

In the present chapter we examine the components of the conditional discrimination learning paradigm to determine and characterize the cognitive factors that are involved.

II.   THE CONDITIONAL DISCRIMINATION LEARNING PARADIGM

A conditional discrimination can be described as follows: In the presence of one or two (or more) conditional (or sample) stimuli an organism is given a choice between two (or more) choice or comparison stimuli.   In the presence of one sample, one of the comparison stimuli is correct (i.e., a response to it is reinforced); in the presence of the other sample, the other comparison stimulus is correct.

Acquisition of a conditional discrimination involves three necessary components: (a) discrimination among the sample stimuli, (b) discrimination among the comparison stimuli, and (c) development of mapping rules or associations between the conditional stimuli and the choice stimuli.

The role of sample and comparison discriminability in conditional discrimination learning has been documented by Carter and Eckerman (1975).   The easier is the successive discrimination among sample stimuli, the more rapid learning will be.   Similarly, the easier is the simultaneous discrimination among comparison stimuli, the more rapid learning will be. Carter and Eckerman's data also suggest that sample discriminability affects the difficulty of conditional discrimination acquisition more than comparison discriminability.   The larger role of sample discriminability is probably due to the fact that simultaneous discriminations are generally easier to learn than successive discriminations (discrimination among sample stimuli requires successive discriminations, i.e., only one sample is presented on each trial, whereas discrimination among comparison stimuli involves simultaneous discriminations).

The purpose of the present chapter is to examine factors that influence the third component of conditional discrimination learning, i.e., the rule that relates the sample and comparison stimuli.   We will examine (a) stimulus factors (i.e., does the physical similarity between sample and comparison affect what is learned), (b) response factors (i.e., does the relation between responses to one sample and responses to another sample affect what is learned), and (c) outcome factors (i.e., does the relation between the outcome following a correct response to one comparison and the outcome following a correct response to another comparison affect what is learned).

III. STIMULUS FACTORS

A. ASSOCIABILITY

It has been argued that the rate of acquisition of a conditional discrimination is independent of the relation between sample and comparison stimuli and, as mentioned earlier, dependent only on the discriminability of the successively presented sample stimuli, and the simultaneously presented comparison stimuli (Carter & Eckerman, 1975).

This position is a more explicit statement of a hypothesis earlier proposed by Skinner (1950). In its more general form this hypothesis has been called the equivalence-of-associability assumption (Seligman, 1970).

The assertion is that the associability of a CS with a US, for example, is independent of the relation of stimulus attributes between them (i.e., whether the CS and US "naturally" go together).

Evidence against the equivalence-of-associability assumption has come from a variety of sources (see Hinde & Stevenson-Hinde, 1973; Seligman & Hager, 1972). In many cases the negative evidence has come from learning that has involved associations between stimuli (or responses) and biologically meaningful outcomes (e.g., the association between visual, auditory, or gustatory stimuli and X-irradiation or poisoning, Garcia & Koelling, 1966, or the association between running, rearing, or turning-around and electric shock, Bolles, 1969). There is more general evidence, however, against the equivalence-of-associability hypothesis, evidence that involves associations between events that have minimal distinctive biological meaning. Dobrzecka, Szwejkowska, and Konorski (1966), for example, showed that dogs which learned a conditional left-right response involving auditory cues that differed both in location (front-back) and quality (metronome-buzzer), did so exclusively on the basis of location cues, whereas dogs which learned a go/no-go response (single location) that involved the same auditory cues, did so primarily on the basis of the quality of the cues. Thus, in a task that emphasized differential response location, learning apparently involved spatial attributes of the stimuli. And, in a task that emphasized nondifferential response location, learning involved predominantly qualitative attributes of the stimuli. Simple stimulus generalization can be ruled out as an explanation for these results because the stimulus dimension (left-right) and the response dimension (front-back) were literally orthogonal. Instead, a higher-order associative process appears to be involved; one that implicates the abstract or relational categorization of events.

B. IDENTITY

If animals can develop a concept of spatially differentiated events, can they also develop a concept of physical identity? More specifically, in a conditional discrimination, can similarities in the non-spatial attributes of conditional and test stimuli affect the readiness with which the task is learned? According to Carter and Eckerman (1975), the physical similarity between conditional stimuli (samples) and test stimuli (comparisons) plays no role in conditional discrimination learning by pigeons. Carter and Eckerman trained pigeons on either a matching–to–sample task (a task in which an identity relation exists between the sample stimulus and the correct comparison stimulus) or a symbolic matching–to–sample task (a task in which no identity relation is present). The pigeons exposed to the matching task were trained with either hues or shapes. The pigeons exposed to the symbolic matching task were trained with either hue samples and shape comparisons or shape samples and hue comparisons. Since the hue discriminations were substantially easier to learn than the shape discriminations, Carter and Eckerman independently assessed the relative difficulty of simultaneous and successive hue and shape discriminations and "corrected" the matching and symbolic matching acquisition functions accordingly. The corrected data indicated there was little difference in rate of acquisition between the matching and symbolic matching tasks.

In response to Carter and Eckerman (1975), we proposed that (a) a within subject design might be more sensitive to task–acquisition differences, (b) task–acquisition differences might be greater if the number of examples of matching and symbolic–matching were greater (i.e., if more exemplars of the identity and nonidentity relation were provided), and (c) the need to correct the acquisition functions for differential discriminability of stimuli should be avoided (Hogan, Edwards, & Zentall, 1981). When we altered our design accordingly by (a) training individual pigeons on both matching and symbolic matching trials, (b) increasing the number of stimuli to be matched (and to be symbolically matched) from two to four, and (c) using only highly discriminable hues as stimuli, we found that the matching task was learned significantly faster than the symbolic matching task.

Further evidence for the role of sample comparison–identity relations in conditional discrimination learning was found in a series of transfer studies (Zentall & Hogan, 1974; 1976; 1978). In these experiments pigeons were trained on a matching–to–sample task or an oddity–from–sample task (responses to the stimulus that does not match the sample are defined as correct) and were then presented with a second task (matching or oddity) with new stimuli, in a two by two design (i.e., half of each training group was transferred to a matching task, the other half to an oddity task). In all three studies, pigeons transferred to the same task as during training (matching to matching and oddity to oddity) learned the second task faster than birds transferred to a different task (matching to oddity and oddity to matching). This transfer effect was found (a) when training involved stimulus hues (red and green) and transfer involved novel hues (yellow and blue) (Zentall & Hogan, 1974; 1978), (b) when training involved brightness differences and transfer involved stimulus hues (red and green) (Zentall & Hogan, 1974b), and (c) when training involved stimulus shapes (circle and plus) and transfer involved stimulus hues (red and green) (Zentall & Hogan, 1976; 1978). An example of the transfer effects can be seen in Figure 22.1. The transfer differences can be attributed to the acquisition of generalized concepts "same" and "different," because the transfer stimuli were all novel and the relation between the acquisition task and transfer task was the only factor that distinguished the same–task group (nonshifted) from the different–task group (shifted).

The data cited above suggest that matching training results in at least

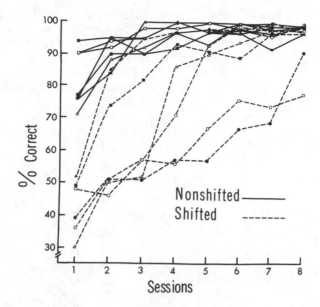

FIG. 22.1. Acquisition of the transfer task for nonshifted pigeons (trained on red–green matching, and transferred to yellow–blue matching, or trained on red–green oddity, and transferred to yellow–blue oddity) and shifted groups (trained on red–green oddity, and transferred to yellow–blue matching, or trained on red–green matching, and transferred to yellow–blue oddity) (from Zentall & Hogan, 1978).

partial acquisition of the concept "same" and oddity training results in at least partial acquisition of the concept "different," because similar positive effects were found for both of the like–task transfer groups. But because of the symmetry of the tasks (and of the concepts) it is possible that the transfer effects could have been produced by the development of a single concept (either same or different).

To assess the source of concept transfer Zentall, Edwards, Moore, and Hogan (1981) trained birds on either a matching or an oddity task involving four stimuli. The training trials were selected such that each sample stimulus appeared with only one (of two) of the three remaining stimuli as nonmatching alternatives (e.g., with A as a sample, either B or C could serve as the nonmatching alternatives but not D). Through counterbalancing the stimulus that was not presented as a nonmatching alternative with a particular hue as sample, did appear with that hue as comparison on other trials. Thus, task acquisition required that each bird learn to discriminate each stimulus from each of the others (e.g., in the case above, with D as a sample, A was one of the two stimuli that could serve as a nonmatching alternative). On test trials the familiar but never-presented nonmatching stimulus (new stimulus) replaced either the matching alternative (e.g., with A as a sample D and B were presented as comparison stimuli) or the nonmatching alternative (e.g., with A as a sample A and D were presented as comparison stimuli). Since no previous association had been established between the sample and the new stimulus, performance on these test trials should indicate the degree of control developed during training by the matching and the nonmatching comparison stimuli.

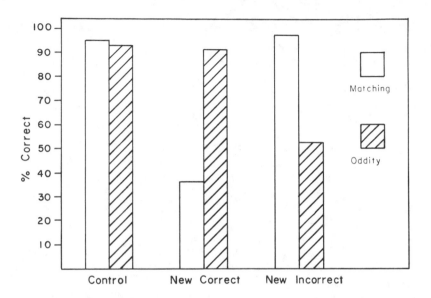

FIG. 22.2.   Test   performance   on   Control   trials,   New   Incorrect comparison trials, and New Correct comparison trials by pigeons trained on matching and oddity tasks.   New trials involved new configurations of familiar stimuli (from Zentall, Edwards, Moore, & Hogan, 1981).

According to the traditional view of conditional discrimination learning (e.g., Carter & Eckerman, 1975), pigeons' choice behavior should be controlled exclusively by the correct comparison stimulus. Pigeons trained with a matching task should be controlled by the matching comparison stimulus, and pigeons trained with an oddity task should be controlled by the nonmatching comparison stimulus. The results indicated, however, that for both the matching and oddity tasks pigeons learned the relation between the sample and the matching comparison stimulus. For matching birds, performance was disrupted only when the correct (i.e., matching) comparison stimulus was replaced. But for oddity birds, performance was disrupted only when the incorrect comparison (i.e., matching) stimulus was replaced (see Figure 22.2). Apparently, matching birds learned to approach (peck) the matching comparison stimulus, whereas oddity birds learned to avoid (not peck) the matching comparison stimulus. Thus, for both matching and oddity acquisition "same" appears to be the determining concept.

The magnitude of the concept learning effect found by Zentall et al.

(1981) was considerably greater than that found in most of our concept transfer experiments with novel transfer stimuli. This finding suggests that the presentation of novel transfer stimuli may have partially obscured concept transfer effects in earlier research. Suggestive support for reduced concept transfer due to a "neophobic" response to the novel transfer stimuli was found by Zentall & Hogan (1978). When pigeons were given single–stimulus presentations of the transfer stimuli during training, they showed larger transfer effects.

The research reported in this section provides strong evidence that pigeons are capable of abstracting an identity relation. These data add to the growing literature that has demonstrated concept learning by pigeons (see Chapter 14, this volume, by Herrnstein).

IV. RESPONSE FACTORS

It is well known that acquisition of matching–to–sample is facilitated by requiring pigeons to peck at the sample prior to presentation of the comparison stimuli (Zentall, Hogan, Howard, & Moore, 1978). Furthermore, within the limits studied, the rate of acquisition increases monotonically with the number of sample pecks required (Sacks, Kamil, & Mack, 1972).

A sample peck requirement presumably ensures that the pigeon observes the sample. The more pecks required, the longer is observation of the sample. (In principle one can make a distinction between observing time and number of responses, but practically they may not be distinguishable because they are confounded.)

In addition to the finding that rate of acquisition increases as number of required sample pecks increases, there is also evidence that differential response requirements to the sample can facilitate acquisition of matching–to–sample (Cohen, Looney, Brady, & Aucella, 1976; Cohen, Brady, & Lowry, 1981; Urcuioli & Honig, 1980; Zentall, Hogan, Howard, & Moore, 1978).

For example, Cohen et al. (1976) found that pigeons trained to peck one sample 16 times to illuminate the comparison stimuli, and to peck the other sample twice with at least a three sec interresponse time, learned the task faster than pigeons trained to respond to the two samples nondifferentially.

The simplest explanation for this phenomenon is that the differential response requirement provided an additional independent cue in the form of response–produced feedback. The response–produced feedback could then serve as a cue to be used at the time of comparison choice. The fact that pigeons can learn a conditional discrimination solely on the basis of the number of pecks needed to illuminate the comparison stimuli (i.e., no differential visual sample cues are provided) suggests that differential response–produced feedback can serve as an effective independent conditional cue (Lyderson & Perkins, 1974).

Alternatively, it is possible that matching performance is facilitated by differential sample behavior because the visual sample and the sample-specific behavior combine to form a configural cue more salient than the visual cue itself. According to this hypothesis, neither cue by itself should control correct performance.

Finally, the differential response requirement may facilitate acquisition by enhancing the discriminability of the sample stimuli.   In an ingenious experiment, Urcuioli and Honig (1980, Experiment 4) found that when sample–specific behavior cannot serve as a discriminative cue for correct choice, differential sample behavior can still facilitate the discrimination of hard to discriminate samples.   In this experiment, on a given trial, both comparison stimuli were associated with samples to which the same sample–specific behavior was required.   For example, a vertical–line sample in the presence of which 10 pecks were required was associated with a horizontal–line comparison (responses to a red comparison were incorrect), but a red sample that required the same 10–peck sample response was associated with a red comparison (responses to a horizontal–line comparison were incorrect).   Similarly, a horizontal–line sample that required a different sample response (two pecks with at least a 3 sec interresponse time) was associated with a red comparison, but a green sample that required the same slow, two–peck sample response was associated with a horizontal–line comparison.   Thus correct choices were not predictable from the sample–response requirements.

Although the differential sample response could not provide a cue for correct comparison choice, it did facilitate acquisition, relative to controls, by helping the birds discriminate between vertical and horizontal samples and between red and green samples.   Thus, the facilitation of performance produced by differential sample behavior may not be attributable to facilitated associations between sample events and comparison stimuli, but rather to facilitated discriminability of the samples.

Empirically, one can distinguish among the three hypotheses described above by presenting pigeons with a matching–to–sample task in which the sample–specific behavior and the visual sample provide redundant, relevant cues for correct comparison choice.   Then, following acquisition, test trials can separate visual–sample effects from sample–specific behavior effects. If the sample–specific behavior provides an additional independent cue, then one should find additive control of performance level by both the visual sample and the sample–specific behavior.   If, on the other hand, sample–specific behavior increases the discriminability of visual samples, then the visual sample should control performance level.   Finally, if the visual samples and the sample–specific behavior form a configural cue, then one would expect neither cue by itself to control a high level of performance.

To distinguish among these hypotheses we trained pigeons on two independent matching–to–sample tasks in which sample hue and required sample response provided redundant, relevant cues for correct choice (Hogan, Zentall, & Pace, 1983).   On trials that involved red and yellow hues as comparison stimuli, the pigeon had to peck the sample ten times to illuminate the comparisons when the sample was red, and they had to peck the sample twice with at least 3 sec between responses when the sample was yellow.   On trials that involved green and blue hues as comparison stimuli, completion of the 10–peck response was required when the sample was blue, and completion of the slow, two–peck response was required when the sample was green.   During training, red and yellow comparisons never appeared with either a blue or a green sample, and blue and green comparisons never appeared with either a red or a yellow sample.   The birds were then tested on reconstructed trials formed by replacing one of the stimuli (i.e., the sample or either comparison) with a hue from the other task.   In this way we were able to present trials on which either the visual sample or the sample–specific behavior was the only relevant cue.   We found that the sample–specific behavior exerted primary control over performance.   Significant control by the visual cue was also found, however, and the level of performance (above chance) on control trials (i.e., the same as matching trials from training) corresponded to the sum of above chance performance on sample–specific–behavior trials and above chance performance on visual–sample trials (see Figure 22.3).   Thus, sample–specific behavior facilitates acquisition of conditional discrimination tasks by providing a salient, additional, independent cue that combines additively with the visual cue to determine level of performance.

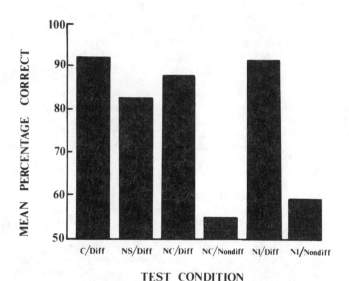

FIG.   22.3.   Test performance.   Control trials, New Correct comparison trials, New Incorrect comparison trials.   Sample–specific behavior provided either a differential cue (Diff) for correct choices or a nondifferential cue (Nondiff) for correct choices (from Hogan, Zentall, & Pace, 1983).

Why the response produced cues are more effective than the visual cues is not clear, especially since the pigeon must be able to detect which visual sample has been presented before being able to make the appropriate sample response.   Thus, in spite of the fact that on each trial the visual cue must be used to determine the appropriate sample response requirement, the visual cue plays very little role in comparison choice.   The data suggest that once the appropriate sample–specific response is made there is an active stimulus selection process, such that the pigeon not only ignores but virtually forgets the visual sample.

V.   OUTCOME FACTORS

Conditional discrimination learning has been characterized as the acquisition of a series of independent, sample/correct-comparison associations or chains. (Cumming & Berryman, 1965) The more easily discriminated are the chains from one another, the faster learning should be. As mentioned earlier, Carter and Eckerman (1975) showed that sample discriminability and comparison discriminability both affect rate of acquisition. In the preceding sections we have shown that other factors can affect acquisition rate as well. The relation between sample and comparison can affect acquisition and transfer of conditional discrimination learning by means of an identity concept, and differential sample responding can facilitate acquisition of conditional discriminations by serving as an additional relevant cue for appropriate comparison choice.

In the present section we will examine experiments that have assessed the effect of differential outcome on acquisition of conditional discriminations. Differential outcome means that correct comparison responses following presentation of one sample result in one outcome (e.g., mixed grain), whereas correct comparison responses following presentation of the other sample result in a different outcome (e.g., water). Peterson and his colleagues (Brodigan & Peterson, 1976; Peterson, this volume, Chapter 8; Peterson, Wheeler, & Armstrong, 1978) showed that such a differential outcome procedure can facilitate the acquisition of conditional discriminations.

We asked if pigeons trained with a differential outcome procedure would show significant positive transfer when tested with new stimulus pairings for which the outcome associated with the sample and the correct comparison stimulus was the only cue that could be used (Edwards, Jagielo, Zentall, & Hogan, 1982). For the experimental group, Phase 1 involved training on a shape matching-to-sample task with differential outcomes associated with the different correct comparison choices (e.g., correct "circle" matches were followed by peas; correct "plus" matches were followed by wheat). During Phase 2, the birds were trained on a color matching-to-sample task, again with differential outcomes (e.g., correct red matches were followed by peas; correct green matches were followed by wheat). During Phase 3, the pigeons were transferred to symbolic-matching trials involving (a) shape samples and color comparisons, and (b) color samples and shape comparisons. For half of the experimental birds, the positive transfer group, the correct comparison was defined as that stimulus which during training had been followed by the same outcome as the sample (e.g., red comparisons were correct in the presence of circle samples and green comparisons were correct in the presence of plus samples). For the remaining experimental birds, the negative transfer group, the correct comparison was defined as that stimulus which during training has been followed by a different outcome (e.g., red comparisons were correct in the presence of plus samples and green comparisons were correct in the presence of circle samples).

Control pigeons were trained with nondifferential outcome during Phases 1 and 2 (i.e., outcome, peas or wheat, was uncorrelated with sample shape or sample color). Differential outcomes were in effect for all groups during Phase 3.

The Phase 3 transfer data indicated that the positive transfer group performed at a significantly higher level than the control group and the negative transfer group performed at a significantly lower level than the control group (see Figure 22.4).

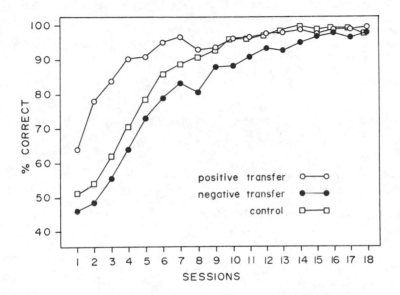

FIG. 22.4. Acquisition of a symbolic matching-to-sample initial transfer task. For positive transfer pigeons the sample and correct comparison had each been associated with the same outcome during training. For negative transfer pigeons the sample and correct comparison had each been associated with a different outcome during training. For control pigeons the sample and correct comparison had each been associated with both outcomes during training (from Edwards, Jagielo, Zentall, & Hogan, 1982).

In a second experiment, a mixed-grain control group, for which each correct response was followed by a mixture of peas and wheat (the typical outcome procedure for conditional discrimination tasks), transferred from shape and color matching to symbolic-matching trials at about the same level as an uncorrelated group, for which half the correct responses were followed by peas and the remaining correct responses were followed by wheat. Furthermore, both groups performed reliably below a correlated group.

Thus, differential outcomes appear to be able to mediate the sample/comparison association in conditional discrimination learning. What makes this finding particularly interesting is that the outcome can become associated not just with the correct comparison, but also retroactively with the sample.   Presentation of the sample apparently elicits a learned outcome expectancy (Brodigan & Peterson, 1976), and the outcome expectancy can be used as a cue for comparison choice.

In the case of earlier research (e.g., Brodigan & Peterson, 1976), it could be argued that overt, outcome-elicited behavior was responsible for the facilitated acquisition.   Jenkins and Moore (1973), for example, showed that pigeons pecked stimuli that were followed by grain differently from stimuli that were followed by water.   In Brodigan and Peterson's experiment pigeons may have pecked samples associated with grain differently from samples associated with water, and as we have seen earlier, differential sample responding can provide an additional discriminative cue.   To reduce the likelihood of differential, outcome-elicited, sample pecking we used two grains (peas and wheat) that were judged to produce highly similar consummatory behavior.   Furthermore, these were grains for which our pigeons showed no measurable differential preference (see Edwards *et al.*, 1982).   Thus, outcome expectancy was demonstrated under conditions where differential overt sample behavior was unlikely.

VI.   SOME PERSISTENT ISSUES

A. LEVEL OF TRANSFER PERFORMANCE

In concept learning with humans, development of a concept is confirmed by the maintenance of a high level of performance when novel exemplars of the concept are presented.   This "conservative" criterion is used because (a) it is assumed that concept learning is all-or-none, i.e., there are only two interpretable levels of performance, very high (indicating that the concept has been learned) and chance (indicating that the concept has not been learned), and (b) only a small number of transfer trials are generally presented so performance significantly above chance is often synonymous with very high levels of performance.

The wide spread use of a conservative criterion with human subjects may have resulted in the overly cautious application of such a criterion to other organisms.   Furthermore, such a conservative criterion may leave one in the awkward position of finding uninterpretable levels of concept transfer (i.e., performance levels significantly better than chance, but significantly below end-of-training performance levels).   If one assumes that concept learning may not be all-or-none, or that choice responding on transfer trials may be controlled by more than one factor, then it may be more reasonable to use chance as a reference level of performance against which to assess concept learning.

B. APPROPRIATE CONTROL GROUP

In our concept transfer research we have typically used a negative transfer control to assess the effects of concept learning on acquisition of the transfer task. In some of our earlier work (Zentall & Hogan, 1976), initial performance of the transfer task was at chance and a difference in the rate of acquisition of the transfer task was the only evidence for concept learning. Thus, the inclusion of a comparison group was particularly important.

Some have argued that a proper control group is one that is trained with the transfer task alone (Carter & Eckerman, 1976), but such a control group may perform significantly worse than an experimental (concept learning) group due to nonspecific transfer effects attributable to learning-to-learn (i.e., reduced position and stimulus preferences, adaptation to the test chamber, learning about the sequence of events on a trial, etc.). We are not convinced that there is a "neutral" control group that would control for all nonspecific transfer effects and yet would not provide the birds with some bias for or against matching or oddity. We have found the negative transfer control group to be ideal, for although it may magnify the measured concept learning effect, it also provides the most certain control for all factors other than concept learning. Furthermore, it provides an extended opportunity to observe concept transfer effects because both groups can be exposed to the transfer task until it is learned. Those who argue that concept transfer must be demonstrated on initial transfer tests (e.g., Carter & Eckerman, 1976) have an obligation to explain differential transfer-task acquisition functions without referring to some type of categorical conceptual behavior. Finally, if one can find relatively highly levels of initial transfer performance, as we have in some of our more recent work (Zentall & Hogan, 1978), the need for a comparison group is reduced.

C. CODING THE SAMPLE STIMULUS

It has been proposed that pigeons "code" or transform conditional stimuli such that they are quite different from their physical representations (Carter & Werner, 1978). According to his position, pigeons do not simply remember, for example, the color of the sample, but they label or transform that sample into a nonisomorphic representation. This conclusion was based in part on findings by Cumming and Berryman (1965) that pigeons (a) show little tendency to generalize from a sample stimulus of one color to a comparison stimulus of the same color, and (b) will sometimes treat a novel sample (but not a novel comparison) as if it were a stimulus with which the pigeon was familiar.

The notion of sample coding has been further explored by Roitblat (1980; 1982; this volume, Chapter 5). Roitblat argued that if the sample stimulus were coded in terms of the correct comparison stimulus in a symbolic matching task, the errors caused by confusion among similar comparison stimuli should increase as the retention interval between sample and comparison stimuli increased. If, on the other hand, the samples were not coded in terms of the correct comparison stimulus, errors caused by confusion among similar comparison stimuli should remain constant as the retention interval increased. Roitblat found that as the retention interval increased, errors that involved more similar comparison stimuli increased faster than errors that involved less similar comparison stimuli. Thus, under these conditions samples appear to be coded, at least in part, in terms of the correct comparison stimulus. But, it pigeons code the sample stimulus in terms of the correct comparison stimulus, then an identity relation between the sample and correct comparison stimulus would appear to be of little value to the pigeon.

It is, of course, possible that sample coding of the type found by Roitblat occurs only in a delayed symbolic matching task. In a matching task the sample stimulus may already be in a form suitable for identifying the correct comparison. It is also possible that sample coding does not occur when the pigeon can compare sample and comparison stimuli directly (i.e., as in the case with the simultaneous matching and the simultaneous oddity tasks we have used). If coding the sample is the pigeons' means of better remembering over a retention interval, then one might not expect an identity concept to develop when training involved a delayed matching task. On the other hand, identity learning and sample coding may not be mutually exclusive. The differential outcome data reported earlier (Edwards et al., 1982) indicate that under conditions which have been shown to promote the development of identity learning (i.e., simultaneous matching-to-sample with shape stimuli), pigeons can also code the samples in terms of differential outcomes. Thus, it is quite likely that both veridical stimulus matching and transformation or coding of the sample can occur at the same time.

D. NEGATIVE INSTANCES OF THE CONCEPT

When a pigeon learns a matching task it appears that very little is learned about the nonmatching stimulus (Zentall et al., 1981). It may be that in a matching task the degree of concept transfer to novel stimuli depends on the extent to which the pigeon learns to avoid the nonmatching comparison stimulus. To test this hypothesis we required pigeons to learn to avoid the nonmatching comparison by including trials on which both comparison stimuli were nonmatching, e.g., a red sample was presented with two green comparisons (Zentall & Hogan, 1978). On these "negative instance" trials, trial-termination was contingent on the omission of pecking for 3 sec. The pigeons readily learned not to peck on the negative instance trials and, relative to controls, showed improved concept transfer when novel stimuli were introduced. Although the process by which negative instance training facilitates concept transfer is not clear, the possibility exists that counter examples of the concept serve to better define the limiting conditions or boundaries of the concept. We are currently trying to identify the process that underlies the enhanced transfer effects found following negative instance training.

VII. SUMMARY

The data presented here indicate that stimulus–stimulus, stimulus–response, and stimulus–outcome relations can all affect conditional discrimination learning.

That the relation between sample and comparison plays a role in what is learned, suggests that identity is a meaningful concept for pigeons. Pigeons learn more than arbitrary S–R chains during acquisition of a matching–to–sample task. Similarly, that differential outcome can serve as a relevant cue for correct comparison choice suggests that expectancies can have stimulus properties not unlike those of external stimuli.

The relation between the sample stimulus and responses to the sample, also appears to be under cognitive control. Although the data can be described in terms of the simple additive effects of processing the visual stimulus and processing the feedback from sample–induced behavior, the fact that on each trial the visual stimulus must be processed sufficiently to identify the appropriate response suggests that birds are actively ignoring (and forgetting) the visual stimulus prior to the comparison response.

Finally, research we have described suggests that cognitive processing of stimuli during the acquisition of conditional discrimination learning tasks can also occur in other ways. The extent to which pigeons can use codes and can better define concepts when exposed to negative instance training further reflect their cognitive ability.

VIII. CONCLUSION

Research we have presented here supports the notion that cognitive processes are involved in pigeons' conditional discrimination learning. We have also identified a number of variables that might determine the extent to which pigeons will develop or demonstrate concept learning when training involves matching or oddity tasks. The goal of future research will be to identify the underlying cognitive processes involved, and to relate these findings to the development of conceptual behavior in other organisms, including humans.

ACKNOWLEDGMENTS

We thank Joyce A. Jagielo for her contribution to the research presented, W. K. Honig for his helpful comments on an earlier draft, and Shirley Jacobs and Julie McKinzie for their help in preparing the manuscript. The research presented was supported by NIMH Grants MH 19757, MH 24092, and MH 35378.

REFERENCES

Bolles, R. C. Avoidance and escape learning: Simultaneous acquisition of different responses. *Journal of Comparative and Physiological Psychology*, 1969, *68*, 355–358.

Brodigan, D. L., & Peterson, G. B. Two–choice conditional discrimination performance of pigeons as a function of reward expectancy, prechoice delay, and domesticity. *Animal Learning and Behavior*, 1976, *4*, 121–124.

Carter, D. E. & Eckerman, D. A. Symbolic matching by pigeons: Rate of learning complex discriminations predicted from simple discriminations. *Science*, 1975, *187*, 662–664.

Carter, D. E., & Eckerman, D. A. Reply to Zentall and Hogan. *Science*, 1976, *191*, 409.

Carter, D. E., & Werner, T. J. Complex learning and information processing by pigeons: A critical analysis. *Journal of the Experimental Analysis of Behavior*, 1978, *29*, 565–601.

Cohen, L. R., Brady, J., & Lowry, M. The role of differential responding in matching–to–sample and delayed matching performance. In M. L. Commons & J. A. Nevin (Eds.), *Quantitative analyses of behavior: Vol. I, Discriminative properties of reinforcement schedules*. Cambridge, Mass.: Ballinger, 1981.

Cohen, L. R., Looney, T. A., Brady, J. H., & Aucella, A. F. Differential sample response schedules in the acquisition of conditional discriminations by pigeons. *Journal of the Experimental Analysis of Behavior*, 1976, *26*, 301–314.

Cumming, W. W., & Berryman, R. The complex discriminated operant: Studies of matching–to–sample and related problems. In D. I. Mostofsky (Ed.), *Stimulus generalization*. Stanford: Stanford University Press, 1965.

Dobrzecka, C., Szwejkowska, G., & Konorski, J. Qualitative versus directional cues in two forms of differentiation. *Science*, 1966, *153*, 87–89.

Edwards, C. A., Jagielo, J. A., Zentall, T. R., & Hogan, D. E. Acquired equivalence and distinctiveness in matching to sample by pigeons: Mediation by reinforcer–specific expectancies. *Journal of Experimental Psychology: Animal Behavior Processes*, 1982, *8*, 244–259.

Garcia, J., & Koelling, R. A. Relation of cue to consequence in avoidance learning. *Psychonomic Science*, 1966, *4*, 123–124.

Hinde, D. E., & Stevenson–Hinde, J. (Eds.) *Constraints on learning*. London: Academic Press, 1973.

Hogan, D. E., Edwards, C. A., & Zentall, T. R. The role of identity in the learning and memory of a matching–to–sample problem by pigeons. *Bird Behaviour*, 1981, *3*, 27–36.

Hogan, D. E., Zentall, T. R., & Pace, G. M. Control of pigeons' matching–to–sample performance by differential sample response requirements. *American Journal of Psychology*, 1983, *96*, 37–49.

Hull, C. L. *Principles of behavior.* New York: Appleton–Century–Crofts, 1943.

Jenkins, H. M., & Moore, B. R. The form of the autoshaped response with food or water reinforcers. *Journal of the Experimental Analysis of Behavior,* 1973, *20,* 163–181.

Lyderson, T., & Perkins, D. Effects of response–produced stimuli upon conditional discrimination performance. *Journal of the Experimental Analysis of Behavior,* 1974, *21,* 307–314.

Peterson, G. B., Wheeler, R. L., & Armstrong, G. D. Expectancies as mediators in the differential–reward conditional discrimination performance of pigeons. *Animal Learning and Behavior,* 1978, *6,* 279–285.

Roitblat, H. L. Codes and coding processes in pigeon short–term memory. *Animal Learning and Behavior,* 1980, *8,* 341–351.

Roitblat, H. L. The meaning of representation in animal memory. *The Behavioral and Brain Sciences,* 1982, *5,* 353–406.

Sacks, R. A., Kamil, A. C., & Mack, R. The effects of fixed–ratio sample requirements on matching–to–sample in the pigeon. *Psychonomic Science,* 1972, *26,* 291–293.

Seligman, M. E. P. On the generality of the laws of learning. *Psychological Review,* 1970, *77,* 406–418.

Seligman, M. E. P., & Hager, J. L. (Eds.) *Biological boundaries of learning.* New York: Appleton–Century–Crofts, 1972.

Skinner, B. F. Are theories of learning necessary? *Psychological Review,* 1950, *57,* 193–216.

Tolman, E. C. *Purposive behavior in animals and men.* New York: Appleton–Century–Crofts, 1932.

Urcuioli, P. J., & Honig, W. K. Control of choice in conditional discriminations by sample–specific behaviors. *Journal of Experimental Psychology: Animal Behavior Processes,* 1980, *6,* 251–277.

Zentall, T. R., & Hogan, D. E. Memory in the pigeon: Proactive inhibition in a delayed matching task. *Bulletin of the Psychonomic Society,* 1974, *4,* 109–112.

Zentall, T. R., & Hogan, D. E. Abstract concept learning in the pigeon. *Journal of Experimental Psychology,* 1974, *102,* 393–398.

Zentall, T. R., & Hogan, D. E. Pigeons can learn identity, or difference, or both. *Science,* 1976, *191,* 408–409.

Zentall, T. R., & Hogan, D. E. Same/different concept learning in the pigeon: The effect of negative instances and prior adaptation to the transfer stimuli. *Journal of the Experimental Analysis of Behavior,* 1978, *30,* 177–186.

Zentall, T. R., Edwards, C. A., Moore, B. S., & Hogan, D. E. Identity: The basis for both matching and oddity learning in pigeons. *Journal of Experimental Psychology: Animal Behavior Processes,* 1981, *7,* 70–86.

Zentall, T. R., Hogan, D. E., Howard, M. M., & Moore, B. S. Delayed matching in the pigeon: Effect on performance of sample–specific observing responses and differential delay behavior. *Learning and Motivation,* 1978, *9,* 202–218.

VI. SPACE, TIME, AND NUMBER

23 TESTING THE GEOMETRIC POWER OF AN ANIMAL'S SPATIAL REPRESENTATION

Ken Cheng
C. R. Gallistel
University of Pennsylvania

I. INTRODUCTION

Animals move from one place to another in pursuit of various goals. Sometimes they may do this by homing on a recognized object or place, directed by a light or sound or odor emanating from the goal (beacon navigation). At other times, they may execute a motor routine that has brought them from one place to the other in the past (motor program navigation). In either of these cases, the process underlying the behavior makes no use of a representation of the spatial characteristics of the environment. Some feats of navigation. however, seem to imply the use of a spatial representation. By spatial representation, we mean a record that reflects certain spatial facts about an environment the animal has experienced, but which, unlike a motor program, does not specify any particular performance. The information in a spatial representation, unlike a motor program, may be used to generate a performance that is not foreseen at the time the representation is established and is not specified by the representation itself.

By considering the kinds of navigational tasks animals are capable of performing, and the kinds of transformations on their physical space that they are sensitive to, we may be able to make inferences as to the kinds of properties that they have encoded. Some animals may only encode the linear order of the points that they see (hear, feel, etc.). These animals would not be able to solve a navigational problem that required them to remember information about distances or angular separations. Their spatial representations, while useful in solving many kinds of problems, would be too weak to solve these problems.

Consider for example the digger wasp. The female digger wasp digs a hole in the ground in which it lays its eggs. It then covers the hole over carefully, takes a reconnaissance flight around it, and flies off in search of food for the yet-to-hatch larvae. This presents the wasp with the navigational problem of getting back to its nest. The nest is invisible at a distance of greater than at most one meter and gives off no sound, odor, or other stimulus that the returning mother wasp can home on from any distance away (Tinbergen & Kruyt, 1938). Nonetheless, the wasp that has found and captured suitable prey makes its way back overland to the burrow from tens of meters away. If an experimenter traps the wasp during the return journey and carries it inside an opaque box to an arbitrarily chosen location within about 100 meters of the burrow, the wasp still makes its way from the release site to the burrow (Thorpe, 1950; see Figure 23.1).   Navigation by beacon or by the path-specific "programs for getting from one point to another" proposed by Lieblich and Arbib (1982, p. 629) does not seem possible.   Thus, we are led to assume that the wasp has encoded some spatial relations holding between the location of the nest and the surrounding environment.   The wasp always makes a series of loops above its newly dug burrow before departing on its first hunting flight.   It seems likely that the record is made in the course of this reconnaissance flight.

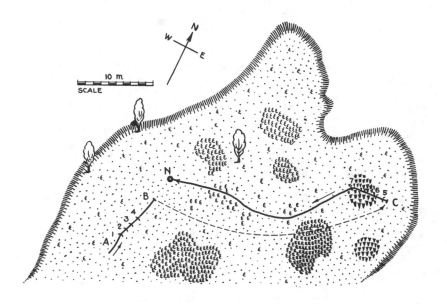

FIG. 23.1. Diagram to show the route taken during detour and displacement experiments by an individual of A. pubescens engaged in dragging its prey towards the nest. Heavy line indicates course of insect. A = point at which first observed.   Numbers 1–4 indicate points at which metal screens were placed in its path for detour tests.   B = point at which insect is captured.   Broken line indicates transfer of insect in box to release point C. Numbers 5–7 indicate further detour experiments. N = nest. The shading indicates a slight depression in an area of gravelly heath land with small patches and scattered plants of Erica and Calluna indicated by the symbol 'E' and with small birch trees (Betula) about 4–6 feet high indicated by the conventional tree symbol. The remaining symbol indicates tussocks of juncus, etc. Conditions: Bright sunshine. Noon, 27 July, 1947.   Eversley Heath.   Berkshire.   Time taken by insect approximately 15 minutes. (From Thorpe, 1950, by permission of E.J. Brill, publisher, and the author.)

One's intuition is that the wasp's representation of space might be in some way more primitive than our own. This leads one to ask what kinds of spatial facts must at a minimum be recorded in order for the wasp to be capable of using the record to find the way home from an arbitrary locus within the vicinity.   One navigational strategy requiring only a relatively weak spatial representation is the following: On its reconnaissance flight, the wasp encodes the locus of the nest as a point lying at the intersection of several straight lines, each such straight line being defined by two landmarks lying on the horizon to either side of the nest.   With that kind of spatial representation, the wasp can solve its navigational problem.   Wherever it winds up, it looks for corresponding pairs of landmarks and steers for the point of intersection of two or more of the straight lines defined by these pairs of contra-positioned horizon landmarks.

The proposed solution to the wasp's task requires a spatial representation.   Certain spatial relationships between the burrow and other points in the world must be recorded in the brain of the wasp during its survey flight and then employed later on to control the wasp's march home.   The proposed solution presupposes that the animal's nervous system can record that: (1) the burrow lies on the point that is the intersection of several straight lines; (2) each straight line has two other distinct points on it -- the landmarks; and (3) the burrow lies between these landmarks.   The properties the wasp must encode. (incidence & betweenness relations) correspond to what are called order properties in formal geometry.   Other spatial properties and relations, however, play no role in our hypothetical solution, and need not be represented by the wasp. These are the metric properties of angles and distances between points and lines.   It is thus conceivable (though we do not have the relevant experimental evidence yet) that the digger wasp possesses a spatial representation that encodes the order properties of the spatial environment but not its metric properties.   In which case, we would say that the wasp's spatial representation had the power of order geometry.

The digger wasp example illustrates the kind of question we want to address in our work on the spatial representations underlying animal navigation.   The task we have set ourselves is to examine the animal's navigational performance under various experimentally contrived circumstances and ask *what kind of geometry would be needed in order to be able to formulate a model of the animal's spatial representation powerful enough to explain these results*. We are not trying to construct the model itself; rather, we are trying to determine how powerful a language (how powerful a geometry) one would have to use in making a model.   If certain results are only explicable by models that incorporate distance information, then an adequate model of the animal's spatial representation could not be formulated within a geometry that did not deal with distance.   In other words, we want to use experimental results to determine however powerful a geometry we must use in describing the spatial representation that underlies a given animal's navigational performance.   The geometry assigned to an animal's spatial representation is the geometry that captures the set of properties the experiments show must be preserved in the animal's representation.   The assignment of a geometry says nothing about the *way* in which these properties are represented, merely that they are somehow represented.

II. THE POWER OF SPATIAL REPRESENTATION

To pursue the question we have posed, one must have a more formal notion of what kinds of spatial representations there are that tasks may require. We have hinted that the specification may be described by formal geometries, for example order geometry. Geometries are the fruits of hundreds of years of effort to capture precisely the various aspects of our conception of what space is or could be. Geometries provide precise languages for talking about representations of space. It would be ill-advised for psychologists theorizing about animals' spatial representations to ignore these formal languages. It seems likely that any geometry (language) that might be required to describe the spatial representations of animals has already been developed. The task, as we see it, is to select amongst the great number of geometries those that seem reasonable candidates to use in describing animals' spatial representations.

A. PRINCIPLES TO RESTRICT THE RANGE OF GEOMETRIES TO BE CONSIDERED

To select theoretical candidates from all geometries that have been investigated mathematically would prove futile since the field is too large. We believe, however, that with the aid of two plausible principles, the range of geometries to be considered can be drastically reduced. We state these principles briefly, as lack of space prevents full discussion.

1. Principle 1.   Animals may have weaker representations of space, but they are unlikely to have systematically wrong ones.

The justification for this principle rests on evolutionary grounds. A radically wrong representation of space is likely to prove harmful in that it would systematically mislead the animal as to the true facts about the world the animal had to navigate. Thus, geometries that are fundamentally wrong in the way in which they represent the world we actually live in (e.g. geometries with Minkowski metrics) should be selected against. By contrast, geometries that do not represent the Euclidean world we experience wrongly, but merely underrepresent it can be useful, as the hypothetical explanation of digger wasp navigation shows. Since these geometries represent correctly those aspects of experienced space that they represent at all, they cannot do harm by misleading the animal. Hence, they seem biologically plausible. (When we say that they represent correctly, we mean that they treat, for example, angles and distances, in a way that corresponds to the way angles and distances work in our Euclidean world, not that they necessarily get the value for a specific distance right.) Principle 1 rules out of court *a priori* any geometry that represents the world's metric or order properties in a radically wrong fashion.

2. Principle 2.   Euclidean geometry is taken to be the most powerful description of local space we have. Should the geometric language required to describe an animal's representation of space be weaker than Euclidean geometry, then we assume that the language describing the representation will be weaker in a formally natural way, that is, in a way in which the structure of Euclidean geometry may be sensibly weakened on formal grounds.

In promulgating this principle we are evincing faith in a correspondence between formal naturalness and biological naturalness. We can consider geometries of differing power as characterizing different sets of spatial properties. Principle 2 can be taken to claim that those sets of spatial properties that have natural formal descriptions will be the sets that may be contained in the spatial representation of animals.

Both these principles are empirical claims subject to the usual inductive uncertainties. We believe, however, that they are both plausible and that they give us a conceptual basis for conducting research.

By applying the above two principles, we arrive at a small field of candidate geometries and the corresponding sets of spatial properties they characterize. The geometries we arrive at are those that may be derived from Euclidean geometry solely by the deletion of one or more natural sets of properties possessed by Euclidean geometry. Of particular importance are order properties, which we have discussed above, and affine properties. The latter properties, which include all the order properties but do not include the metric properties of angles and distances, are particularly powerful for navigational purposes. An affine representation of space thus is a serious challenge to a hypothesis of a Euclidean representation.

III. EXPERIMENTALLY TESTING THE POWER OF AN ANIMAL'S SPATIAL REPRESENTATION

We have thought of two lines of evidence that might be gathered to test the power of an animal's spatial representation. One approach is to present carefully chosen navigational problems to an animal, problems chosen for their geometric implications. Success at the task would mean that the animal must have represented at least the spatial properties required in principle to do the task. For reasons to be described in a moment, we have abandoned this approach. We now prefer to use a second approach -- a transformational approach. In this approach we have the animal perform a certain navigational task, and then we transform the experimental space in systematic ways. Should the animal behave in a significantly different fashion (usually indicated by a deterioration in performance), we may conclude that the animal was relying on a representation of some of the spatial properties that were changed by the transformation. The transformations of the environment that do or do not affect the animal's pattern of navigational behavior within that environment reveal what properties are and are not preserved in the animal's representation of the environment.

The principal problem with the first approach lies in the difficulty of determining the theoretical requirements of actual as opposed to idealized experimental tasks. Consider for example, a task that a four-year-old congenitally blind child can do. In an experiment by Landau, Gleitman, and Spelke (1981) such a child was led from starting landmark, A, to another landmark, B, and then back to A. She was then turned through the appropriate angle and led to another landmark, C. Her task then was to go directly from C to B, that is, to complete the triangle. The child was reasonably good at setting off in the right direction, and seemed also to have an idea of how far to go. Maier (1929) did somewhat similar experiments with rats. He forced them to take detours that led away from the direct route to their goal and showed that they could find their goal from the end of the detour. They could do this in total darkness, so that beacon cues were unlikely. We have done some pilot experiments with rats and found (using infra-red videotaping) that they set fairly direct courses from the end of the detour to their goal, as Maier's results implied (though he did not actually observe the paths taken by the rats).

Intuitively, these triangle completion tasks seem to require computations over angles and distances for their solution. Setting a course to the unperceived goal seems to require computations based on the metric relations between the other two legs of the triangle (e.g., for Landau et al. 's subject, the distances AB and AC and the angle BAC). A more formal analysis, however, reveals that this is true only if the landmarks at the ends of the paths (A, B, C in the case of Landau et al.'s subject) are considered single points. If they are thought of as clusters of points, as seems more realistically the case, then a solution requiring only computations over the affine properties of space is possible. In general, tasks requiring computations over metric properties require that the animals have available as perceivable landmarks only those that can be considered as points. It is difficult to create these and difficult to induce animals to use such minimal landmarks. Furthermore, we can think of no satisfactory way to test whether they really constitute points in the animal's representation. For all these resons, we have turned instead to the transformational approach.

A. THE TRANSFORMATIONAL APPROACH

An example of an experiment using the transformational approach is work done by Suzuki, Augerinos, and Black (1980) with rats in an eight-arm radial maze (Olton, 1978). The maze was surrounded by black curtains and had a black ceiling above and a black floor below. On the curtain over the end of each arm hung a salient stimulus cue, serving to define the place for the rat. The rat was allowed to travel down three predetermined arms at the start of each trial. It was then confined to the center platform, while the cues were rearranged. It was then allowed to choose freely among all the arms to collect the remaining baits.

In a control condition, nothing was altered. Here, in accord with much previous research (Olton, 1978; Olton & Samuelson, 1976; Roberts, this volume, Chapter 24), the rats went predominantly to new arms (arms they had not previously visited) following the delay. In a rotation condition, every cue was moved 180° about the center to hang over the opposite arm. The baits were also moved in this fashion. Performance in this condition tests whether the configuration of the cues truly defines the space for the rat. If it does, the rats should go to the "new" places, the places whose spatial relation to the cues is the same as before. The rats in fact did this. When we count the number of new arms they go to in the first five choices following delay, they were not significantly different from the control condition.

We can contrast their performance in the purely rotational control transformation with the performance in a third condition, involving a permutation transformation. Here, the cues were randomly switched about. However, one and only one cue still hung over the end of each arm after the permutation of the cues. Baits were moved with the cues. The rats did not go to new places, as defined by the permuted cues. Their performance following permutation of the cues was unrelated to their pre-delay choices; they did not avoid those arms (cues) that they had visited before the delay.

Suzuki *et al.* interpreted their results as showing that the animals were sensitive to the configurational properties of their spatial layout. We would like to be more explicit and ask what configurational properties the rats are sensitive to. The basic experimental logic of a transformation experiment is the following: if the transformation destroys some spatial properties to which the animal is sensitive, then the animal will change its behavior. The change in this case, as in most cases, involves the deterioration of a previously successful performance. Ironically, animals with weaker spatial representations should carry on their previous successful performance. Applying this experimental logic to the study just described, we can say that the permutation has changed some spatial properties to which the rat is sensitive, that is, properties that it captures in its representation of the maze. It is not clear, however, what these properties are. The permutation not only alters the metric properties of angles and distances, it changes affine, order, and topological properties as well. It leaves only graph-theoretic properties intact (which were the only properties represented in Deutsch's (1960) model of map based navigation). For example, pairs of cues that lined up with the center of the platform before the transformation in general do not do so afterwards (a change in order properties). It is not clear which of the formally natural sets of spatial properties the rat was sensitive to in the Suzuki *et al.* study. The experiment is thus not a strong case for arguing that the rat represents metric properties. It is, however, a strong reason for rejecting the Deutsch (1960) graph-theoretic model, which like all association-based models, preserves very little truly spatial information.

B. AFFINE TRANSFORMATION AND AFFINE GEOMETRY

To test whether Euclidean properties are represented, it is especially important to test how the animal reacts to affine transformations. These transformations leave the affine properties of the space unaltered. Affine properties form an important set of properties because they are the largest formally natural subset of Euclidean properties that do not include any metric properties. An affine transformation is by definition a transformation which leaves collinear points in a Euclidean space or plane collinear.

The set of properties such transformations leave invariant in a Euclidean space constitutes affine geometry. Any affine transformation on a Euclidean plane can be characterized (Modenov & Parkhomenko, 1965) as a uniform squash or stretch along one dimension followed by a (possibly different) squash or stretch along an orthogonal dimension, followed by any Euclidean transformation (a transformation that preserves all metric properties in the space). A uniform squash or stretch does just what we intuitively think it would: it expands or contracts the entire plane uniformly along the axis of expansion or contraction. In general, such squashes and stretches change distances and angles.

Affine geometry plays a central role in our undertaking because it is what you have when you remove the metric properties and only the metric properties from Euclidean geometry. It includes all of the other properties -- the properties that enable one to deal with straight lines, to deal with the order in which points occur along a straight line, to deal with the intersection of lines, and so on. Affine geometry is powerful enough to make possible the solution to many navigational problems. For example, the hypothetical solution to the wasp's problem discussed above does not exploit the full power of affine geometry. Yet, the removal of the metric properties weakens the language enough to make impossible a number of things that are readily done in a geometry that recognizes metric properties. For example, there is no basis within affine geometry for distinguishing one triangle from another. An animal with only an affine representation of a triangular space could not tell if one altered that space from, say, an equilateral triangle to a triangle with an obtuse angle and no side equal to another. The animal with only an affine representation of previously experienced space would have no way of recognizing the change, *provided that it was not sensitive to a concomitant alteration in some other, non-spatial characteristics*, provided, for example, that the animal did not respond to the fact that one of the corners in the new triangle did not feel tight when the animal squeezed into it, as all of the corners of an equilateral triangle would. This proviso is a troublesome one; the desire to avoid all non-spatial changes dictates the particular kind of affine transformation that we have employed in our experiment.

C. EXPERIMENTS WITH AFFINE AND REFLECTION TRANSFORMATIONS

In the experiment we now review, we used a variant of the differential-baiting paradigm of Hulse and O'Leary (1982) and Olton, Shapiro, and Hulse (Chapter 10, this volume). The maze is illustrated in Figure 23.2. It was an elongated X situated inside a rectangular enclosure, which, like the X, was twice as long as it was wide. Four movable boards with simple patterns on them served as landmarks. On every trial, one arm was baited with 18 45-mg Noyes pellets, one with 6, one with 1, and one with none. Which arm had which bait was constant for a given rat throughout the experiment ("which arm" being determined by the stimulus pattern at the end of the arm). This was done in the hope (successful) of inducing the rats to choose the arms in the order of the magnitudes of their baits, thereby improving the experiment's ability to reveal a perturbation in choice behavior following transformations.

The stimulus patterns in the corners were symmetrical about the vertical axis, like the letters used to represent these landmarks in Figure

23.2. The boards containing the patterns were inserted into the corners at an angle such that each board formed the hypotenuse of a $1:2:5^{1/2}$ right triangle, the shorter side of which was a portion of the shorter wall of the enclosure. Their symmetry and the angle at which they were set into the corners of the rectangle were necessary for carrying out an accurate affine transformation without changing the landmarks themselves nor their distance from the rat's choice point. The maze and the wall of the rectangle enclosing it were painted black to discourage the animal's relying on cues from them. The rectangle was covered with a masonite board into the center of which was set a rectangular (2:1) half silvered mirror, through which the experimenter could observe the rat, without allowing the rat a view of anything outside the enclosure. In other words, every effort was made to isolate the rat from any spatial information coming from outside the enclosure.

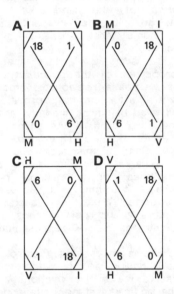

FIG. 23.2. The experimental arrangement used to test the rat's reaction to various theoretically motivated transformations of the environment. A. The training setup. The letters written just outside the rectangle represent the landmarks, which are in fact mounted on the diagonals in the corners of the rectangular enclosure. The numbers beside each arm of the X maze indicate the number of pellets with which that arm is baited. B. An affine transformation of the training setup. C. A Euclidean transformation (a rotation). D. A reflection.

In this experimental environment, one can produce an affine transformation of the landmarks without altering the landmarks in any way and without changing the conditions under which the animal views them, at least from the center of the X, where it stands when choosing an arm.

The transformation is to move all the landmarks into the neighboring corner (Figure 23.2b). This manipulation effects the following affine transformation: Squash the rectangular space along its long axis by a factor of two, stretch it along the other axis by a factor of two, then rotate the space by 90° about its center. The result is that the two pairs of landmarks that straddled the two short ends of the rectangle (I & V and M & H) now straddle the long sides and vice versa. The distances from one landmark to the other have been altered and so have the angles between the center of the maze and any pair of landmarks, but each landmark looks exactly the same as it did before and is still located at the end of an arm projecting into the corner of the rectangular enclosure. As in the Suzuki et al. experiment, neither the maze nor the landmarks have been altered in any way, and the distance of the landmarks from the center of the X configuration is the same as before. In other words, the only basis for recognizing the transformation has occurred is the change in the *metric configuration* of the landmarks. In making such a transformation, the baits go with the landmarks, so that the large bait remains under the same landmark that it has always been found under, and likewise for the lesser baits.

Unless the animal is making use of a spatial representation that preserves the metric properties of the space, it should be blind to this affine transformation of the spatial relationship amongst the landmarks. It should perform just as well after the transformation as before: it should choose the arm with the large bait first, then the arm with the next largest bait, and so on. On the other hand, if the animal is using a metric map, when it tries to make its map of the maze correspond with the transformed maze, it will find a mismatch due to the altered metric relations amongst the landmarks. Insofar as the animal relies on these mismatching metric properties for navigational purposes (that is, for heading to the larger baits), the map will be useless and the animal's performance should deteriorate. Thus, the weaker spatial representation in this paradigm leads to more robust performance than does the stronger representation, because, with the weaker representation, the transformation must go undetected.

Our experiments include as an essential control a rotation condition, a condition that moves the landmarks around with respect to the maze but in such a way as to preserve the metric relations amongst them. This is done by interchanging the landmarks at opposite ends of the same diagonal (Figure 23.2c). Thus, the pair that formerly straddled one end of the rectangle now straddle the other end and vice versa. This condition tests the extent to which the landmarks and only the landmarks define the space for the animal. It is essential that the animal's performance be unperturbed by this pure rotation transformation (as was the case in the Suzuki et al. experiment). For a system that is relying entirely upon the configuration of the maze and its enclosure plus the landmarks in the corners, nothing has been changed by moving the landmarks 180° about the center of the maze. The performance of such a system could not be perturbed by this transformation, because the system would have no way of knowing that anything had been done. Any perturbation in performance following this transformation implies that the animal is relying upon more than simply the configuration of the X maze and its rectangular enclosure plus the landmarks. It must be picking up spatial cues from such things as a knot hole on the wall or sounds from outside the enclosure, in which case results from the affine transformation will be uninterpretable.

We have also included another transformation, designed to detect whether a rat's spatial representation preserves another important property of space known as sense. Sense is the distinction between right and left. The preservation of sense is tested by a reflection transformation (Figure 23.2d), which can be done by interchanging pairs of stimuli straddling the long or short walls of the rectangle. All metric properties are preserved by this transformation, but right is turned into left. It is possible to have spatial representations that capture the metric properties of space (and, therefore, *necessarily* the lower properties such as the order of points along a straight line) but do not capture the distinction between right and left. If the rat's spatial representation does not preserve sense, then the rat's performance should be undisturbed by the reflection transformation, but if the representation does preserve sense, then again the rat should be unable to make its map of the pretransformation environment coincide with the environment it now confronts and its performance should deteriorate.

Four adult male Sprague–Dawley rats served as subjects. They were given 13 days of training followed by 18 days of experimental conditions. During training, all the rats were always confronted with the same experimental space: The X–maze, the rectangular enclosure around it, the roof above the apparatus, and the boards in the corner were always in the same configuration. The experimental space as a whole (enclosure, maze, landmarks, and roof all together) was moved about the experimental chamber from trial to trial, in order to discourage the animals from using any cues that, despite our precautions, they might be able to pick up from outside the enclosure. The magnitude of the bait under each landmark remained identical from trail to trail for each rat. Each rat, however, had a different assignment of baits to landmarks. The animals each ran five trials per day throughout the entire experiment.

In the experimental phase, control trials alternated with experimental trials. In the control trials, conditions were exactly like the training phase. In the experimental trials, a transformation of the experimental space took place before the trial. In effecting the transformation, only the boards in the corners were moved. *In the rotation condition* each board was moved to the diagonally opposite corner. *In the affine condition,* each board was moved one corner over in the counterclockwise direction. *In the reflection condition,* each board exchanged places with its neighbor along the long wall. In all transformations, the baits went with the transformations, that is, baits were rotated, affine transformed, or reflected in the same manner as the boards. This means that the pattern on each board continued to be associated with its usual bait size for each rat throughout all the transformations, that is, baits were rotated, affine transformed, or reflected in the same manner as the boards. After training, the rats first had six days of testing in which rotation trials were mixed with control trials, then six consecutive days in which affine trials were mixed with control trials, and another six days in which reflection trials were mixed. with control trials. The order of the affine and reflection conditions was counterbalanced across rats. The entire experimental space continued to be moved about the experimental chamber from trial to trial. As the experimental trials alternated with control trials, this gives a total of six experimental conditions: the three transformation conditions of rotation, affine transformation, and reflection, along with the control trials given in alternation with each of these.

On each trial, we noted the order in which the rat chose the arms. We calculated a quantitative index of how optimal the order of choices was, optimal order being to choose the largest first, the next largest next, and so on. This was done by rank correlating the order of the rat's choices with the order of bait magnitudes. Optimal choice sequences have a value of one, and sequences chosen at random have an average rank correlation value near zero. It is this dependent variable which appears in Figure 23.3 for each condition.

The results show that in the rotation condition, the animals chose sequences in as optimal a fashion as in the control conditions or the last three days of acquisition. This means that the boards defined the places for the rats, rather than any peculiarities of the arms of the maze, the walls, or roof. Under both the affine and reflection conditions, however, performance dropped drastically, average values of rank correlation with the optimal order being close to zero. The results differed significantly across conditions for each individual rat by a nonparametric Kruskal–Wallis one-way ANOVA ($p < 0.005$ in each case). The results show that the rats were repeatedly and consistently disoriented by the transformations even though the baits remained at the same landmarks and even though the rearrangement changed only sense in the reflection condition and only the metric relations amongst the landmarks in the affine condition. The rats treated the transformed space in each case as if it were a new space. This in turn implies that they must have encoded in their representations of space both the metric properties and sense. We take the results to show that rats use a sense–preserving Euclidean representation of space for navigational purposes.

IV. DISCUSSION

This program of formally characterizing the power of spatial representations is in some ways analogous to Chomsky's attempts at formally characterizing natural languages. Chomsky (1956) attempted to show, for example, that the structure of natural languages is formally too complex to be captured by the class of regular languages (finite state automata). Likewise, we are starting off by characterizing which of several natural classes of spatial properties in a formal hierarchy are needed to describe adequately the power of an animal's spatial representation.

Our choice of the spatial faculty as a topic of study is motivated by a modular view of cognition similar to the view that Chomsky (1980) has urged in connection with his choice of human language as a faculty to be studied. In this view, the faculty for forming spatial representations is like the faculty for dealing with language in that it has an extensive set of characteristics that are peculiar to it and reflect the particular character of the exigencies that have shaped the evolution of this particular faculty. Chomsky (1980) views the function of the brain as best characterized in

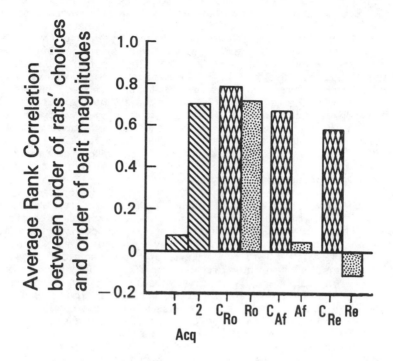

FIG. 23.3. The average rank correlation between the order of the rat's choices and the order of bait magnitudes for all four rats in the first three days of acquisition (Acq 1), the last three days of acquisition (Acq 2), the control trials mixed with the rotation (C_{Ro}), the affine transformation (C_{Af}), and the reflection (C_{Re}), and the experimental conditions of rotation (Ro), affine transformation (Af), and reflection (Re).

terms of a small number of distinct but (to various degrees) interacting "mental organs" or "modules." Each such module resembles a specialized organ such as the kidney in that it has functional properties unlike those of other modules. The functional properties of the kidney are not those of the liver. Likewise, we assume that the functional properties of the module that deals with the representation of space are not those of the module that deals with language -- a module that appears to be absent in all but humans. This view contrasts with the more prevalent view that the laws of learning and memory are "uniform across domains. . .homogeneous and undifferentiated (Chomsky, 1980, p. 40)."

The question of whether the mind of any animal is best characterized as one all encompassing structure or a number of interacting specialized modules is a theoretical issue to be settled on empirical grounds. Insofar as we can show that an animal's capacity to represent the space it must move through depends on the ability to code peculiarly spatial properties such as distance, angle, collinearity, and the like, then one-faculty accounts of animal cognition become less plausible. It is not clear how accounts invoking classical conditioning, operant conditioning, or the formation of associative nets can come to grips with the existence of a metric representation of experienced environments. It is difficult to embed the language of Euclidean geometry in the language of associationism.

In one important respect, however, our subject of study differs from Chomsky's. While the language faculty is unique to the human species, the spatial faculty is found in diverse animals, from digger wasps to human beings. This is hardly surprising. Most animals move about, and many make permanent or long lasting homes to which they regularly return after outings. Any accurate representation of space can help in this regard and improve chances for survival. One central interest of ours is to compare the power of spatial representations across species. At the level of comparison that we seek, in terms of the kinds of spatial properties which are encoded and used, the task seems sensible and feasible. While animals' differing nervous systems may encode spatial properties in different ways, the kinds of properties that they encode will likely come from the few formally natural sets we have posited. This presumption, along with the fact that very similar kinds of experiments can be adapted to different species, make interspecific comparisons both easy and deeply interesting. Transformations can be performed in any kind of environment over which the experimenter has control. Thus, for example, the willingness of nutcrackers to store seeds in an experimental room and later retrieve them (see Balda & Turek, Chapter 28, this volume) allows the possibility for doing transformation experiments with that animal.

One interesting question is whether any animal possesses a spatial representation that contains less that the Euclidean properties. We end this paper with some more speculations about the digger wasp. Van Beusekom (1948) built various configurations around digger wasps while they were digging their nests in the ground. (The wasps use the local configuration to orient by when they return to the nest.) In a series of experiments, van Beusekom built a circle of pine cones around the wasp while it was digging its nest. While it was away hunting for prey, he built a second configuration (of various shapes in different experiments) and waited to see which configuration the wasp chose to land at when it came back. The wasp preferred the circle over an incomplete arc of a circle, a heap of cones, or a straight line of cones. These discriminations contrast with tests using other closed figures formed with cones. Van Beusekom tried the square, the triangle, and the ellipse. The wasps preferred the circle over the square (51 to 29), and over the triangle (71 to 40), but not over the ellipse (12 to 10 in favor of the ellipse). Van Beusekom could find no reason for these differing patterns of discriminations. We can, however, offer this speculation. The ellipse is an affine transformation of the circle, while the square and the triangle are not and neither are straight lines, open arcs, or heaps. It is possible that the wasp is not sensitive to those transformations of space that preserve all the affine properties, and hence fails to differentiate a circle from an ellipse. Van Beusekom's results are at least consistent with this hypothesis. Does this mean that the digger wasp does possess a weaker spatial representation than the rat or human? The evidence is too meager and subject to many alternative interpretations. Future research with more systematic transformation experiments can tell us.

ACKNOWLEDGMENTS

This research was supported by Biomedical Research Support Grant RR 07083-16 Sub 19 awarded by the University of Pennsylvania with funds from NIH. The authors benefitted from extensive discussions with Elizabeth Spelke and Barbara Landau.

REFERENCES

Chomsky, N. Three models for the description of languages. *IRE Transactions on Information Theory*, 1956, *2*, 113-124.

Chomsky, N. *Rules and representations.* New York: Columbia University Press, 1980.

Deutsch, J. A. *The structural basis of behavior.* Chicago: University of Chicago Press, 1960.

Hulse, S. H., & O'Leary, D. K. Serial pattern learning: Teaching an alphabet to rats. *Journal of Experimental Psychology: Animal Behavior Processes*, 1982, *8*, 260-273.

Landau, B., Gleitman, H., & Spelke, E. Spatial knowledge and geometric representation in a child blind from birth. *Science*, 1981, *213*, 1275-78.

Lieblich, I., & Arbib, M. A. Multiple representations of space underlying behavior. *The Behavioral and Brain Sciences*, 1982, *5*, 627-659.

Maier, N. R. F. Reasoning in white rats. *Comparative Psychology Monographs*, 1929, *6*, 1-93.

Modenov, P. S., & Parkhomenko, A. S. *Geometric transformations.* New York: Academic Press, 1965.

Olton, D. S. Characteristics of spatial memory. In S. H. Hulse, H. Fowler, & W. K. Honig (Eds.), *Cognitive processes in animal behavior.* Hillsdale, N.J.: Erlbaum, 1978.

Olton, D., & Samuelson, R. J. Remembrance of places passed: Spatial memory in rats. *Journal of Experimental Psychology: Animal Behavior Processes*, 1976, *2*, 97-116.

Suzuki, S., Augerinos, G., & Black, A. H. Stimulus control of spatial behavior on the eight-arm maze in rats. *Learning and Motivation*, 1980, *11*, 1-18.

Thorpe, W. H. A note on detour behaviour with *Ammophila pubescens* Curt. *Behaviour*, 1950, *2*, 257-264.

Tinbergen, N., & Kruyt, W. Uber die orientierung des Bienenwolfes (*philanthus triangulum* Fabr.). III. Die bvorzugung bestimmer wegmarken. *Zeitschrift fur vergleichende Phisiologie*, 1938, *25*, 292-334.

van Beusekom, G. Some experiments on the optical orientation in *Philanthus triangulum* Fabr. *Behavior*, 1948, *1*, 195-225.

24 SOME ISSUES IN ANIMAL SPATIAL MEMORY

William A. Roberts
The University of Western Ontario

I. INTRODUCTION

In 1976, Olton and Samuelson published a paper titled "Remembrance of places passed: Spatial memory in rats," in which they introduced a new piece of apparatus, the radial maze, and reported some surprising findings about rats' ability to traverse this maze. The radial maze consisted of a central elevated octagonal platform, from which eight elevated arms radiated outward for about a meter. Rats initially were trained to find food at the end of each arm and then were tested on repeated trials in which food was placed at the end of each arm and the rat was allowed to start at the center platform and roam the maze freely until all eight pieces of food were collected. The primary finding of these experiments was that rats learned to collect all available food pellets, with a very low frequency of repeating entrances into alleys. After 10 to 20 days of practice, it was common for a rat to enter eight different alleys on its first eight choices. Although rats did occasionally repeat alley entrances (defined as an error), the average number of different alleys entered on the first eight choices was between 7.5 and 8.0.

Most researchers, when initially confronted with these findings, suspect that rats' success is based on some more basic process than memory. For example, rats may be using an algorithmic strategy or fixed order of alley entrances, such as always entering the alley adjacent to the one just exited and proceeding around the maze in a clockwise or counterclockwise direction. Olton and Samuelson (1976) argued that this was not the case, since their rats chose alleys in random sequences. Later work has shown that, under certain circumstances, rats will choose adjacent alleys (Roberts & Dale, 1981). However, the important point is that rats do not need to use algorithms in order to choose accurately. Numerous experiments now have shown that rats initially forced to enter a randomly chosen subset of the alleys on a radial maze and then allowed to choose freely will accurately pick out the unentered alleys. Another initially entertained explanation of the rat's prowess on the radial maze was the use of odor cues. Rats might smell the presence or absence of food at the end of arms, or they might lay down an odor trail on an entered alley which would serve as a cue not to reenter that alley. Numerous control experiments have ruled out these possibilities. When alleys are rebaited after entrance, rats still choose the unentered alleys (Olton & Samuelson, 1976; Roberts, 1979), and experiments which have either rearranged alleys in the middle of a trial (Olton & Samuelson, 1976) or directly interfered with olfaction (Maki, Brokofsky, & Berg, 1979; Olton & Samuelson, 1976; Zoladek & Roberts, 1978) uniformly have failed to support the notion that animals avoid an odor trail. These experiments and others do suggest strongly, however, that rats are using memories of arms previously visited and that these memories involve extramaze visual cues.

Summaries of the initial findings from research with the radial maze can be found in several papers by Olton (1977, 1978, 1979). Of particular interest here is a model of spatial memory tentatively put forth by Olton (1978). Three aspects of this model are of particular concern to this paper. First, it was suggested that spatial memory could be conceived of as a working memory in which a limited number of items or places visited could be stored. Second, it was held that information placed in working memory would not decay or be forgotten over the few minutes required for a test. However, once a trial was completed, a reset mechanism allowed the animal to delete the contents of working memory and thereby leave it empty for recording the events of a subsequent trial. Finally, the format of storage was held to be that of a list in which discrete representations of each place visited were held. Each of these properties of Olton's model may be debated.

It will be argued in this chapter that spatial memory is not limited in the structural sense that there are a fixed number of slots in which memories of places visited can be stored. Evidence will be offered to show that spatial memory is susceptible to both proactive and retroactive interference and that spatial memory cannot be reset by deleting the contents of working memory at the end of a trial. Finally, findings from a number of experiments will be marshalled to argue that the representational format of spatial memory is map-like rather than list-like.

II. THE CAPACITY OF SPATIAL MEMORY

The notion of a limited capacity storage system arises from Miller's (1956) seminal paper, in which it was suggested that humans can simultaneously process only 7 ± 2 units of information. This idea subsequently was incorporated into two-store models of human memory, the most well-known being that of Atkinson and Shiffrin (1968). Within this model, it was held that short-term storage could hold only about seven items at one time. Evidence for this assumption came from such observations as the fact that memory span or people's ability to perfectly recite a list of numbers or letters deteriorates rapidly beyond seven to nine items and that the recency effect seen in the serial-position curve of free recall reaches asymptote around six or seven items from the end of the list. In other words, a rather sharp break in performance is seen at the point which defines limited capacity.

If the rat brain is built with a limited capacity for spatial information, we might expect that this limit would be equivalent to or even less than the human brain's limited capacity for short-term storage of verbal information. No such equivalency seems to be found. We know that rats can perform almost errorlessly on an eight-arm maze, and we might expect that accuracy would decline sharply if more arms were added. Olton, Collison and Werz (1977) tested rats on a 17–arm radial maze and found that rats learned to enter between 14 and 15 different alleys in the first 17 choices.   Of particular interest is Figure 24.1, taken from the Olton *et al.* article, which plots probability of correct response as a function of choices 2 through 17.   The notion that rat spatial memory is limited to about eight items leads to the prediction that rats would respond accurately over the first eight choices, but as the number of places which had to be kept in memory accumulated beyond the eighth choice, a sharp drop in the curve should appear.   Contrary to this prediction, no discontinuity in the function is found.   Accuracy of response declines at a slow but continuous rate over choices 2 through 17 but is still substantially above chance at the 17th choice.   The authors suggest that if this curve were extrapolated beyond 17 choices, the curve would not reach chance until between 25 to 30 arms.   If we assume that the rat's capacity for spatial memory is between 25 and 30 items, this would extend it beyond the postulated capacity of human short–term memory by a factor of about four.

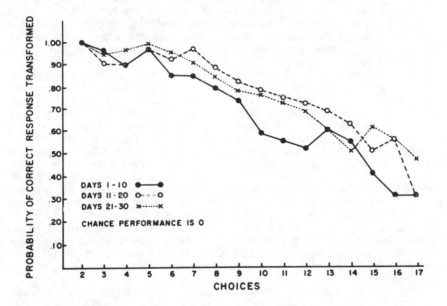

FIG. 24.1. Transformed probability of correct response plotted as a function of choices for rats on the 17–arm maze.   Different curves plot performance for different stages of training.   (From Olton, Collison, & Werz, 1977.   Copyright 1977 by Academic Press.   Reprinted by permission.)

It could be argued that environmental demands placed upon rat evolution have led to the development of a spatial memory with a large capacity and therefore we should not expect rat spatial memory to be limited to a relatively small number of items, as is the case with human verbal memory. However, if capacity is limited, we would still expect that at some critical point, performance would decline sharply. Some relevant experiments carried out by Roberts (1979) have made use of a hierarchical radial maze. The hierarchical maze is shown in Figure 24.2. It consists of the standard eight elevated arms radiating from a central platform (called primary alleys) and three secondary alleys radiating from the end of each primary alley. Three groups of rats were tested on the hierarchical maze , one group with only one secondary alley open, a second group with two secondary alleys, and a third group with three secondary alleys. Animals initially were allowed to learn the maze with a food pellet placed at the end of each secondary alley. After about 10 days of practice, the rats in each group had learned to run down primary alleys and very efficiently collect the pellets in all available secondary alleys. In subsequent experiments, memory for secondary alleys entered, as well as primary alleys, was tested by using trials which consisted of two runs on the maze. On the first run, selected secondary alleys were blocked and others were open and baited with a food pellet. After collecting all of the available food pellets on this initial run, the blocks were removed and a second run was given with food available only in the previously blocked alleys. Subjects were able to select the previously blocked alleys at a level of accuracy far above chance. This was the case regardless of variation in the number and position of the blocked secondary alleys from one primary alley to another.

Two aspects of these findings are important for the question of capacity of spatial memory. For one, in the group with three secondary alleys at the end of each primary alley, animals were required to keep track of the status (entered or unentered) of 8 x 3 = 24 secondary alleys. If we add the eight primary alleys to be remembered, rats were dealing effectively with 32 different choices. It might be argued that the number of places rats can remember should be raised from the 17 arms used by Olton *et al.* (1977) to the 32 primary and secondary alleys used in this study. A second observation is that performance on primary alleys was unaffected by the number of secondary alleys which had to be retained. Errors on primary alleys did not vary between the groups tested with one, two or three secondary alleys at the end of each primary alley. If raising the total number of secondary alleys from 8 to 16 to 24 added increasing numbers of items to be remembered to a limited capacity storage system, we would expect that an overloading effect would result in which performance on primary alleys would decrease as number of secondary alleys increased.

The number of spatial alternatives available to a rat may thus be raised from 17 to 32 and still excellent spatial memory is found. Evidence for spatial memory for a large number of places is not limited to rodents. Menzel (1973, 1978) has shown that chimpanzees remember very well 18 different locations where food was hidden in an open field, and Shettleworth and Krebs (1982) have found that marsh tits show good spatial memory for the location of seeds hidden in as many as 30–35 different places. Similar performance has also been observed in Clark's nutcrackers by Balda and Turek (see Chapter 28, this volume).

At this point, my preference is to assume that spatial memory is not

FIG. 24.2. A diagram of the hierarchical maze; alleys leading away from the central platform are called primary alleys, and the three alleys branching off the end of each primary alley are called secondary alleys. (From Roberts, 1979. Copyright 1979 by Academic Press. Reprinted by permission.)

limited in the structural sense that human short-term memory has been assumed to be limited. Rather, the evidence suggests to me that spatial memory in animals may be unlimited in capacity or limited only in the sense that animals can remember as many discriminable spatial locations as nature or an experimenter can arrange. Earlier in this century, many researchers attempted to define the intelligence of different species of animals by finding the maximal time for which an animal could delay before performance dropped to chance accuracy within a delayed response task. The effort proved futile because no fixed delay time could be found for any species. Each time a "limit" had been established, a more clever experiment would reveal a still longer delay. Similarly, tertiary alleys might be placed on the end of secondary alleys on the hierarchical maze, and evidence of still a higher capacity for spatial memory might be revealed. My guess is that a search for the absolute capacity of spatial memory in an animal would be equally as unrewarding as the search for a fixed delay interval.

III.   FORGETTING AND RESETTING

Several lines of evidence lead to the conclusion that spatial memory is very robust and resistant to forgetting.   Olton and Samuelson (1976) allowed rats to make four free choices on the eight-arm radial maze and them confined subjects to the center platform for a period of at least 2 minutes.   When animals were allowed to resume responding on the maze, they chose the remaining four alleys at a level of accuracy equivalent to that of animals with no delay interpolated between choices.   Far more impressive is a recent experiment by Beatty and Shavalia (1980a), in which rats were removed from the maze between the fourth and fifth choices and returned to the home cage for retention intervals as long as 24 hours. Performance on choices 5-8 was over 90% accurate at retention intervals as long as 4 hours.   Other evidence suggests that spatial memory may be impervious to interference.   Olton (1977, 1978; Olton & Samuelson, 1976) has concluded that rats are not susceptible to proactive interference (PI) on the radial maze, based on experiments in which animals were tested on repeated massed trials within days.   Evidence that spatial memory may not demonstrate retroactive interference (RI) comes from experiments by Maki, Brokofsky, and Berg (1979) and Beatty and Shavalia (1980b). In these experiments, events or activities interpolated between the fourth and fifth choices on the eight-arm maze caused no loss of retention relative to control tests.

Although impressive evidence thus can be marshalled for the resistance of spatial memory to forgetting, some recent experiments indicate that it is not totally resistant to interference and that proactive and retroactive interference can be demonstrated.   The case for proactive interference will be examined first.   Olton and Samuelson (1976; Experiment 6) tested rats repeatedly on eight successive trials within days.   After entering all eight alleys on the radial maze on one trial, an animal was detained on the center platform for only about the 1-minute period required to rebait alleys before another trial was initiated.   Curves which plotted probability of correct choice against choices 2-8 showed that the curves for each trial declined from choice 2 to choice 8 but always returned to errorless performance on the initial choices of the next trial (Olton, 1977, 1978). On the basis of this return to errorless performance, Olton (1977, 1978) concluded that experiences on early trials did not interfere with retention on later trials, because rats reset working memory at the end of each trial. Specifically, the process of resetting was defined as deleting the contents of working memory.

Examination of Olton's curves lead Dale and me (Roberts & Dale, 1981) to a different interpretation.   We were impressed by the observation that although each trial's curve started at perfect performance on initial choices, the rate at which performance declined over choices increased as trials progressed.   This pattern of findings seemed very similar to proactive interference curves found in studies of human short-term memory.

Keppel and Underwood (1962) tested people's retention of verbal material over successive trials; at a short retention interval, retention was equally good on early and late trials, but as the retention interval was lengthened, retention curves dropped faster for late trials than for early trials.   On the radial maze, the initial choices of a trial are made with a short retention interval and only a few preceding visits to remember.   At the end of the trial, however, the retention interval since initial choices is longer and the memory load is greater.   The human data suggest that proactive interference should be observed on later choices, and this is exactly the point at which performance becomes worse across trials.

Dale and I attempted to replicate Olton's findings by testing rats on five massed trials per day over a period of 5 days (Roberts & Dale, 1981; Experiment 1).   An interesting aspect of this experiment was that animals developed a strong tendency to enter adjacent alleys in either a clockwise or counterclockwise direction, so that by the fifth day of testing all ten of the rats were using this strategy on almost all choices.   This appears to be a unique observation, since in no other studies that I know of have all subjects adopted the adjacent alleys algorithm.   It is tempting to assume that rats adopted this strategy in the face of interference with retention from massed trials, since choosing adjacent alleys requires retention of only the most recently entered alley.   On the first two days of testing, however, the tendency to enter adjacent alleys was not strong, and sufficient errors were made to examine curves which plotted percent correct responses as a function of choices 2–8.   These curves are shown in Figure 24.3.   Accuracy dropped from trial 1 to trial 2 on both days and did not vary significantly from trial 2 to trial 5.   Separate curves therefore are plotted for trial 1 and the combined data of trials 2–5.   These curves appear to show basically the same results as those of Olton.   Both curves begin at errorless performance on choice 2, but the curve for trial 1 remains errorless through choice 5, whereas the curve for trials 2–5 drops much sooner to a lower level of accuracy.   In keeping with a proactive interference analysis, retention dropped faster on the later trials of a day than on the initial trial.

In a second experiment (Roberts & Dale, 1981, Experiment 2), a procedure was used which prevented animals from choosing correctly by using an adjacent alleys pattern of responding.   Three trials were run in succession on each day but each trial consisted of initial forced entry into four randomly chosen alleys followed by a free choice retention test among all eight alleys.   A variable of critical interest was the delay or retention interval between the forced choices and the retention test; this interval was set at zero sec on half the days and at 60 sec on the other days.   In Figure 24.4, percentage of correct choices is shown as a function of trials, with separate curves for 0-sec and 60-sec delays.   As in the first experiment, accuracy drops from trial 1 to trial 2, but the loss in retention is far more marked with a 60-sec retention interval than with a 0-sec retention interval.   These findings again suggest that the initial trial or trials proactively interfere with retention on subsequent trials and that this interference is most obvious when information must be retained over a relatively long interval.

The discovery of proactive interference in rat spatial memory argues against the use of a resetting mechanism.   If the contents of working memory were deleted after each trial, no proactive interference should be observed.   One other analysis of the data from our first experiment further reinforces this conclusion.   On each day that rats received five successive trials, the final alleys entered on trial $n - 1$ were compared

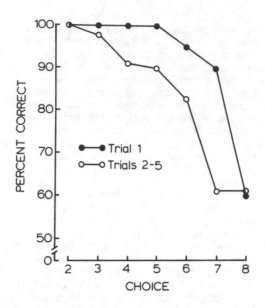

FIG. 24.3. Curves showing the drop in performance across choices for Trial 1 and Trials 2-5 in a sequence of five massed trials. Proactive interference is revealed by faster decline of the curve for Trials 2-5, than of the curve for Trial 1. (From Roberts & Dale, 1981. Copyright 1981 by Academic Press. Reprinted by permission.)

with the initial alleys entered on trial n. It was found that animals clearly tended to avoid the final alleys entered on trial $n - 1$ on their initial choices of trial n. In other words, the tendency to enter different alleys, seen within trials, also is found between trials. This observation suggests further that rats remember the events of a preceding trial and cannot simply erase this information from working memory in preparation for the next trial.

Turning to the question of retroactive interference in spatial memory, Maki *et al.* (1979) allowed rats to make four forced choices on an eight-arm maze and then, in different experiments, subjected rats to interpolated events such as lights, sounds, a distinctive odor, feeding on the maze, and running a four-arm radial maze in a different room. The retention test consisted of placing animals back on the eight-arm maze and determining how accurately they could choose previously unentered alleys on choices 5-8. None of these interpolated experiences reduced retention, relative to control tests on which animals remained idle during the retention interval. The finding that interpolated experience on another maze produces no retroactive interference is particularly impressive. Beatty and Shavalia (1980b) extended this experiment by requiring rats to enter four arms on a different eight-arm maze at varying temporal points within a 4-hour retention interval. In one experiment, testing ended after a retention test on the initially experienced eight-arm maze; in another experiment, rats

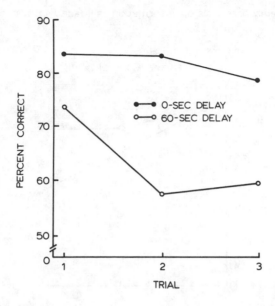

FIG. 24.4. Percentage of correct responses on free choices at each of three consecutive daily trials. More proactive interference is seen with a 60-sec delay between forced and free choices than with a 0-sec delay. (From Roberts & Dale, 1981. Copyright 1981 by Academic Press. Reprinted by permission.)

were subsequently tested for retention of forced choices on the interpolated maze. Under none of these conditions was any evidence found for retroactive interference.

In some experiments carried out in my laboratory, we pursued the question of retroactive interference further (Roberts, 1981). Some of the early work on retroactive interference in maze learning (Marx, 1944) indicated that retroactive interference might be found if more than one interpolated maze were used. We carried out an experiment in which rats were given four forced choices on either one, two or three interpolated mazes, each in a different room. Animals then were returned to the first room for a retention test on the initial maze and they were tested for retention of choices on each of the interpolated mazes. On control tests for each interpolated condition, animals ran the first maze and then were confined for a retention interval equal to the length of time taken to run interpolated mazes. In Figure 24.5, performance is plotted as a function of the number of interpolated mazes. When one interpolated maze was used, the results replicate those of Maki, et al. (1979) and Beatty and Shavalia (1980b), in that experimental and control conditions do not differ. However, when two interpolated mazes are required, the curves begin to separate (not a significant effect), and, at three interpolated mazes, a significant retroactive interference effect is apparent. These findings suggest that spatial memory is susceptible to retroactive interference, provided sufficient interpolated information is loaded into memory.

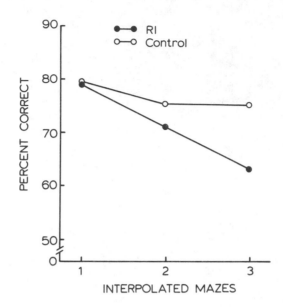

FIG. 24.5. Retention scores for retroactive interference conditions in which forced choices on 1, 2, or 3 mazes were interpolated between forced choice and a retention test on a target maze. In the control conditions, animals remained idle during the period of time required to run the interpolated mazes. (From Roberts, 1981. Copyright 1981 by the Psychonomic Society. Reprinted by permission.)

It could be argued that the demonstration of retroactive interference with three interpolated mazes contradicts the previous suggestion that spatial memory is unlimited in capacity. The argument here is that enough interpolated information exceeds the capacity of the memory store and crowds out memories of the initial forced choices. This argument follows if one assumes that forgetting arises only from displacement of old information by new information. As an alternative, however, forgetting could arise from confusion between memories. Since the rooms in which initial and interpolated choices were given were similar, animals may have confused memories established in different rooms. Forgetting may have arisen then not from limited capacity but from failure to discriminate between memories established in different contexts.

On the other hand, the fact that retroactive interference was not found with one and two interpolated mazes suggests that rats possess the ability to discriminate memories established on different mazes. It may be that with one or two interpolated mazes, differences in room cues were sufficient to allow memories established in these different contexts to be distinguished. In a further set of experiments (Roberts, 1981, Experiments

2a-2c), we attempted to establish retroactive interference with only one interpolated maze by placing the test maze and the interpolated maze within the same room. It was reasoned that alleys on mazes placed close together within the same context would share many cues and should therefore provide greater confusion between spatial memories. In a first experiment, the mazes were placed side by side, with only 30 cm separating them, and corresponding arms of each maze pointed in the same compass directions. After four forced choices on one maze, an animal was placed on the other (interpolated) maze and given four forced choices on that maze. The subject then was returned to the first maze for a free choice retention test among all eight arms. an experimental design was used which involved several interpolated conditions, one of which attempted to maximize conditions for retroactive interference by forcing animals to enter arms on the interpolated maze which pointed in exactly those directions the animal had not traveled on the initial maze. The experiment revealed no evidence of retroactive interference. In a second experiment, the mazes were displaced vertically instead of horizontally, by placing one maze on top of the other. The experiment was performed again, and once more no difference between interference and control conditions was found. In a final experiment, only a single maze was used; after four initial forced choices, interpolated conditions consisted of placing animals on four arms of the maze in succession. The retention test now revealed significant interference in that condition in which animals had been placed on those arms which had not been entered on the initial forced choices (i.e., the correct arms on the retention test).

Taken as a whole, these experiments indicate both that retroactive interference can be created under certain circumstances and that spatial memory is surprisingly resistant to retroactive interference. It appears that rats can easily discriminate memories of alley entrances made on mazes only a short distance apart within the same context. Apparently, differences in the distance and perspective from which extramaze cues are viewed on two mazes separated by less than a meter allow the rat to encode alleys entered on different mazes as clearly different points in space. Interference appears to arise only when (a) spatial memory is loaded with a large amount of interpolated information, i.e., three interpolated mazes in the first experiment, or (b) as in the final experiment, the interpolated experience involves placement at exactly those places where an animal must go in order to demonstrate retention.

IV. ENCODING FORMAT IN SPATIAL MEMORY

Two formats for encoding and representation of spatial information in animal memory have been suggested, each based on a different analogy. In Olton's (1978) model, visits to points in space on the radial maze were conceived of in analogy to a list of verbal items often used in tests of human memory. Each place visited would have a set of cues which would be entered into a slot in working memory. The other format is that of a cognitive map (Menzel, 1978; O'Keefe & Nadel, 1978; Suzuki, Augerinos, & Black, 1980). The concept of a cognitive map in animal learning originated with Tolman (1948) and suggests that animals retain not just sets of isolated cues but a representation which preserves the relationships between places in an environment. The analogy on which a cognitive map is based is the pictorial visual image of human phenomenology, although the extent to which visual images are pictorial has been challenged (Pylyshyn, 1973).

Tolman (1948) supported his concept of a cognitive map with evidence from several types of experiments, including demonstrations of latent learning and the use of a shortcut in spatial orientation experiments. Olton (1979) has criticized some of this earlier work and suggested that it does not provide totally convincing evidence for the use of cognitive maps. With regard to the radial maze, either a map or a list format would allow a rat to accurately keep track of alleys entered. However, several recent lines of evidence, either with different apparatus or with modifications of the radial maze situation, argue in favor of a map-like representation in spatial memory.

It has been suggested that cognitive mapping involves animals learning about the locations of places in an environment by observing the relations of places to various other landmarks in the environment. It should be possible, then, for an animal to find its way efficiently to a given place on the map from any other place on the map. New paths and shortcuts should be taken (Tolman, Ritchie, & Kalish, 1946a). In a recent experiment, Walker and Olton (Olton, 1979) trained rats on an apparatus which contained a central goal box and four start boxes on the north, east, south and west sides of the goal box. From each start box, rats could go directly to the goal box or to one of three other boxes. Animals initially were trained to approach the goal box from the east, north, and west sides. On test trials, animals were placed in the south start box for the first time; although the subjects had never approached the goal from this direction before, the majority of animals (80%) went directly to the goal.

A similar example comes from an experiment carried out by Ellen, Parko, Wages, Doherty, and Herrmann (1982, Experiment 1). The Maier three-table problem was used in this experiment. Three tables were interconnected with one another by three alleys which formed either a Y or a triangular configuration. In the Y configuration, the three alleys intersected at a common center point, and, in the triangular configuration, pairs of tables were connected with common alleys to form a triangular perimeter. The procedure involved an initial period of exploration of the tables and alleys, direct feeding on one of the tables, and a final test phase in which the animal was placed on one of the two other tables and allowed to find its way to either the table on which it had been fed (correct response) or to the table on which it had not been fed (error). Two sets of control animals were tested with the same configuration used during exploration and testing, one set with the Y and one set with the triangle. Two experimental groups differed from the controls by being given exploration and testing on different configurations. One group explored the Y and then was tested on the triangle, and the other group explored the triangle and was tested on the Y. While control animals performed at about 80% accuracy over the 24 trials of testing, experimental animals performed only at chance (50%) on the initial trials. By the final trials of testing, however experimental animals were scoring as well as control animals. These results indicate that rats learned to locate the goal (feeding table) and travel to it over a route they had never traveled during exploration.

A third example in the same vein is found in a clever experiment by Morris (1981, Experiment 2). Rats were tested in a circular tank, filled with an opaque mixture of milk and water. When placed in the tank, rats were required to swim about until they located a submerged platform, upon which they could stand and receive relief from the exhaustion of swimming. Three groups of animals initially were trained with the platform located in the northeast quadrant of the pool and the starting position on the west side of the pool. After 15 escape training trials, all animals swam directly to the platform. On transfer tests, the procedure now was changed for two groups of six subjects each. The same-place group now had the platform still in the northeast quadrant, but two animals were started from the north side of the tank, two from the east, and two from the south. Animals in the new-place group also started from different points but had the position of the platform moved also; the angular location of the platform always was the same as it had been in training, so that swimming in a leftward direction would lead to the platform. Finally, three control animals continued to swim from the west side of the tank to the platform in the northeast quadrant. The paths taken by animals in locating the platform are shown in Figure 24.6. Most of the new-place animals encountered considerable difficulty in finding the platform, whereas control and same-place animals swam directly to the platform. Although same-place rats had never swum from the north, east, or south sides of the pool before, they located the point in space of the submerged platform from those positions and took very direct routes to it. In agreement with earlier arguments made by Tolman (Tolman, Ritchie, & Kalish, 1946b), rats appeared to learn the place where the platform was located rather than a fixed swimming response. However, the representation of the safe place could not be based on local cues (Restle, 1957), since the platform was submerged. The rat's ability to locate the platform must have been based upon the relationships between several sources of environmental cues external to the swimming tank.

One final example of the use of shortcuts in spatial memory is seen in Menzel's experiments with chimpanzees (Menzel, 1973, 1978). Menzel tested six wild-born chimpanzees and used an open outdoor field as his laboratory. One animal was taken from the enclosure and carried around the field by an experimenter, who periodically hid 18 pieces of fruit in randomly selected places. The subject then was returned to its enclosure. Two min later, the subject was released, along with five other chimpanzees. These latter animals served as controls, for the possible location of food through sight or odor. Over a number of trials, the subjects shown the locations of food collected 12.5 pieces per trial, while the control animals combined, found only 0.21 pieces per trial. Equally as impressive as the chimpanzee's ability to locate hidden food were the paths taken during food collection. Animals collected food according to a *least-effort principle*. That is, regardless of the order in which food was hidden, the subject collected all of the food in one area before moving on to another area. After considering alternative accounts of these findings, Menzel (1978) concluded that only a concept of cognitive mapping does

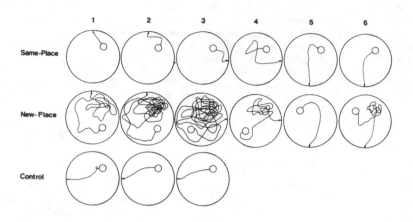

FIG. 24.6. Swimming paths taken by rats in locating the submerged platform. Rats in the same-place and control groups had the platform placed in the same place as it had been on preceding trials, whereas the position of the platform was changed for animals in the new-place group. (From Morris, 1981. Copyright 1981 by Academic Press. Reprinted by permission.)

them justice.

A mapping theory suggests that animals should be at a loss to find known locations when the environment no longer matches their map. Suzuki *et al.* (1980) investigated spatial memory in rats by using a highly controlled environment, a cylindrical chamber, around an eight-arm maze, with specific stimuli placed at the end of each arm. Prominent stimuli were used, such as Christmas tree lights, a fan with tinsel, and a wooden pyramid. On test trials, rats were allowed to make three forced choices and then were confined to the center platform for 2.5 min before being allowed to choose freely among the eight arms. One of three different manipulations was carried out during the retention interval. In a control condition, the walls of the cylinder, on which the stimuli were mounted, were rotated varying distances but always were returned to the original positions of the cues. In two other conditions, transpose and rotate, the positions of the stimuli were changed. The rotate condition involved rotating the walls and stimuli 180º; although the compass positions of the stimuli had been changed, the pattern or relative positions of the cues had been maintained. In the transpose condition, both the positions and the configuration of the cues were altered by randomly interchanging the stimuli. Performance on the free choices was scored in terms of how accurately animals approached stimuli which had not been approached on the forced choices. It was found that subjects performed at a high level of accuracy on control and rotate tests but were very inaccurate on the transpose tests. As long as the configuration of stimuli was preserved, in the control and rotate conditions, animals responded as if no change in cues had occurred. When the configuration was altered, however, animals behaved largely as if they were in an unfamiliar setting. Such results are exactly what we would expect if the environment now no longer matches the map which the animal initially formed of it. A list theory which holds that each alley entrance was encoded as just the stimulus at the end of the alley predicts that performance should be high in all three conditions, since the same list members are available and should guide accurate choice under all three conditions.

If animals form cognitive maps, it would seem that maps would be best formed when an animal has full access to an environment. If perception of the environment is hindered in some way, only an incomplete map may be formed. On the other hand, restricted observation of the environment should be perfectly effective in forming a list of cues to be approached or avoided. Mazmanian and Roberts (1983) examined this question by using a procedure divised by Walker and Olton (1979). Walker and Olton found that rats did not have to run a radial maze in order to remember where they had been. They simply placed rats on the end of one, two, or three arms of a four-arm radial maze and then returned the subjects to the center for a test. Far above chance, animals chose the arm(s) on which they had not been placed.

In the Mazmanian and Roberts experiments, rats were placed on the ends of two randomly chosen arms of a four-arm maze, with the extent of spatial and temporal observation of the environment varied. Three spatial viewing conditions were used, tunnel, 180°, and 360°. In the tunnel condition, animals looked down a narrow alley at only the cues directly in line with the arm. In the 180° condition, a straight partition was placed across the arm to allow the subject to see only 180° of the environment. The third viewing condition placed no spatial restriction on an animal's perception of the environment and was called the 360° condition. As a variable manipulated orthogonal to the spatial viewing condition, the time that animals were allowed to view the environment from the end of an arm was set at 2 or 20 sec. It was found that accuracy improved as a direct function of both the spatial and temporal extent of view. In the tunnel condition, animals performed little above chance, and the length of observation had no effect. Animals performed progressively better at the 180° and 360° conditions, and 20 sec of viewing time was better than 2 sec at both of these conditions of spatial viewing. Spatial memory appears to be sensitive to both the scope of view taken in from a given position in space and the length of time an animal is allowed to look. In agreement with a cognitive mapping position, an animal's ability to locate its position in space becomes worse as the degree of the environment surrounding it is progressively reduced. If only a list of the cues at the end of each alley was required for good spatial memory on the radial maze, we would expect equally high scores under each spatial viewing condition.

V.   CONCLUSION

Olton's (1978) model viewed spatial memory in the rat as a limited number of slots, with information about the cues seen at the end of each arm on the radial maze stored in each slot. When a trial had ended (all rewards had been collected), the list of cues could be flushed, by resetting working memory. My review of issues raised by this model suggests that it may not accurately depict the mechanisms of spatial memory in animals. While it cannot be proved that spatial memory does not have a limited capacity, recent discoveries of good spatial memory for an increasingly larger number of places indicates that the discovery of such a limit may not be in the offing. The notion that rats reset working memory at the end of a trial seems unlikely now, since recent findings indicate that rats are susceptible to proactive interference within a sequence of massed trials and that memories of final choices on trial $n - 1$ lead to avoidance of those choices on the initial selections of trial n. Finally, a variety of evidence seems to have cumulated, which suggests that spatial memory in animals has the characteristics of a map and not those of a list.

While no formal alternative theory of spatial memory will be presented here, some features of an alternative theory can be mentioned. Such a theory might hold that animals form map-like representations of familiar environments and that these representations are held in a long-term or reference memory (Roitblat, 1982). Places visited within such a mapped environment would be noted or marked on the map as a recently visited place. There would be no limit on the number of places on the map that could be so marked; however, performance may begin to suffer when nearby places are marked as visited, since spatial discrimination of similar areas may become difficult.

Proactive and retroactive interference were found to arise under similar circumstances, when rats were required to make trips to identical points in space. In proactive interference experiments, rats showed forgetting when required to return to alleys entered on immediately preceding trials, and retroactive interference was found when rats were placed directly on the correct alleys before a retention test. When mazes were placed a short distance apart, no retroactive interference was found. If rats record a trip to an arm as a marker placed on that part of the map, exploratory tendencies and past learning may lead the animal to avoid returning to that place. However, when repeated trips are made to the same places, more than one marker will be placed at the same position on the reference map. Coding along a temporal dimension would seem to be required here in order to deal with repeated visits (Dale & Staddon, 1982).

In a proactive interference experiment, the rat must be able to discriminate recent markers from earlier ones. This becomes a problem in temporal discrimination, and if a rat confuses entry markers on trial n and trial $n - 1$, unentered alleys will be avoided and entered alleys will be reentered. Similarly, in the RI experiment, direct placements may establish entry markers which are difficult to discriminate temporally from those established by the initial forced entrances. It should be emphasized that rats performed well above chance level in both proactive interference and retroactive interference experiments. Temporal discrimination between markers in memory may not be perfect, thus giving rise to interference effects, but it is far better than total confusion.

In summary, I feel that current guidelines for a general conception of spatial memory in the rat are (a) a map-like representation of space, (b) unlimited capacity for visits to places represented on the map, or at least capacity limited by the number of discriminable points on the map, and (c) interference effects caused by failure to discriminate between points in time at which markers depicting visits to places were established. A number of questions remain unanswered. Primary among these is the specific characteristics of spatial memory representation. If encoding is map-like, is there one map or a number of maps? If there are multiple maps of a given environment, how many are there and how integrated are they? What is the relationship between the environment and the map or maps? Is the map pictorial, or are only certain particularly salient aspects of the environment represented? I feel that these are among the most interesting questions now facing the field of spatial memory, and I think they will be answered by transfer experiments like that of Suzuki *et al.* (1980), in which the environment is systematically manipulated between initial input and a retention test.

ACKNOWLEDGMENTS

Preparation of this paper and research reported in it were supported by Grant A7894 from the Natural Sciences and Engineering Research Council of Canada.

REFERENCES

Atkinson, R. C., & Shiffrin, R. M. Human memory: A proposed system and its control processes. In K. W. Spence & J. T. Spence (Eds.), *The psychology of learning and motivation.* (Vol. 2.) New York: Academic Press, 1968.

Beatty, W. W., & Shavalia, D. A. Spatial memory in rats: Time course of working memory and effect of anesthetics. *Behavioral Biology,* 1980, *28,* 454–462. (a)

Beatty, W. W., & Shavalia, D. A. Rat spatial memory: Resistance to retroactive interference at long retention intervals. *Animal Learning and Behavior,* 1980, *8,* 550–552. (b)

Dale, R. H. I., & Staddon, J. E. R. A theory of spatial memory. Manuscript submitted for publication.

Ellen, P., Parko, E. M., Wages, C., Doherty, D., & Herrmann, T. Spatial problem solving by rats: Exploration and cognitive maps. *Learning and Motivation,* 1982, *13,* 81–94.

Keppel, G., & Underwood, B. J. Proactive inhibition in short-term retention of single items. *Journal of Verbal Learning and Verbal Behavior,* 1962, *1,* 153–161.

Maki, W. S., Brokofsky, S., & Berg, B. Spatial memory in rats: Resistance to retroactive interference. *Animal Learning and Behavior*, 1979, *7*, 25–30.

Marx, M. H. The effects of cumulative training upon retroactive inhibition and transfer. *Comparative Psychology Monographs*, 1944, *18*, 1–62.

Mazmanian, D. S., & Roberts, W. A. Spatial memory in rats under restricted viewing conditions. *Learning and Motivation*, 1983, *12*, 261–281.

Menzel, E. W. Chimpanzee spatial memory organization. *Science*, 1973, *182*, 943–945.

Menzel, E. W. Cognitive mapping in chimpanzees. In S. H. Hulse, H. Fowler, & W. K. Honig (Eds.), *Cognitive processes in animal behavior*. Hillsdale, N.J.: Erlbaum, 1978.

Miller, G. A. The magical number seven, plus or minus two: Some limits on our capacity for processing information. *Psychological Review*, 1956, *63*, 81–97.

Morris, R. G. M. Spatial localization does not require the presence of local cues. *Learning and Motivation*, 1981, *12*, 239–260.

O'Keefe, J., & Nadel, L. *The hippocampus as a cognitive map*. Oxford: Clarendon Press, 1978.

Olton, D. S. Spatial memory. *Scientific American*, 1977, *236*, 82–98.

Olton, D. S. Characteristics of spatial memory. In S. H. Hulse, H. Fowler, & W. K. Honig (Eds.), *Cognitive processes in animal behavior*. Hillsdale, N.J.: Erlbaum, 1978.

Olton, D. S. Mazes, maps, and memory. *American Psychologist*, 1979, *34*, 583–596.

Olton, D., & Samuelson, R. J. Remembrance of places passed: Spatial memory in rats. *Journal of Experimental Psychology: Animal Behavior Processes*, 1976, *2*, 97–116.

Olton, D. S., Collison, C., & Werz, M. A. Spatial memory and radial arm maze performance of rats. *Learning and Motivation*, 1977, *8*, 289–314.

Pylyshyn, Z. W. What the mind's eye tells the mind's brain: A critique of mental imagery. *Psychological Bulletin*, 1973, *80*, 1–24.

Restle, F. Discrimination of cues in mazes: A resolution of the "Place-vs.-Response" question. *Psychological Review*, 1957, *64*, 217–228.

Roberts, W. A. Spatial memory in the rat on a hierarchical maze. *Learning and Motivation*, 1979, *10*, 117–140.

Roberts, W. A. Retroactive inhibition in rat spatial memory. *Animal Learning and Behavior*, 1981, *9*, 566–574.

Roberts, W. A., & Dale, R. H. I. Remembrance of places lasts: Proactive inhibition and patterns of choice in rat spatial memory. *Learning and Motivation*, 1981, *12*, 261–281.

Roitblat, H. L. The meaning of representation in animal memory. *The Behavioral and Brain Sciences*, 1982, *5*, 353–406.

Shettleworth, S. J., & Krebs, J. R. How marsh tits find their hoards: The roles of site preference and spatial memory. *Journal of Experimental Psychology: Animal Behavior Processes*, 1982, *8*, 354–375.

Suzuki, S., Augerinos, G., & Black, A. H. Stimulus control of spatial behavior on the eight-arm maze in rats. *Learning and Motivation*, 1980, *11*, 1–18.

Tolman, E. C. Cognitive maps in rats and men. *Psychological Review*, 1948, *55*, 189–208.

Tolman, E. C., Ritchie, B. F., & Kalish, D. Studies in spatial learning. I. Orientation and the short-cut. *Journal of Experimental Psychology*, 1946, *36*, 13–24. (a)

Tolman, E. C., Ritchie, B. F., & Kalish, D. Studies in spatial learning. II. Place learning versus response learning. *Journal of Experimental Psychology*, 1946, *36*, 221–229. (b)

Walker, J. A., & Olton, D. S. The role of response and reward in spatial memory. *Learning and Motivation*, 1979, *10*, 73–84.

Zoladek, L., & Roberts, W. A. The sensory basis of spatial memory in the rat. *Animal Learning and Behavior*, 1978, *6*, 77–81.

25 THE NUMERICAL ATTRIBUTE OF STIMULI

Russell M. Church
Warren H. Meck
Brown University

I. INTRODUCTION

The present chapter concerns the attribute of number. In it we make the following points:

1. The number of successive events (stimuli or responses) can serve as an effective stimulus for behavior, even when all temporal cues are counterbalanced or held constant,

2. There is cross-modal transfer of the attribute of number, and

3. The same internal mechanism is used for duration, number, and rate discrimination.

II. IDENTIFICATION OF "NUMBER" AS A STIMULUS ATTRIBUTE

The first problem is to determine whether or not the number of successive events (stimuli or responses) can serve as an effective discriminative cue for behavior. The number of events is usually confounded with other variables, especially temporal variables, and the task is to separate any influence of the number of events from such confounding variables. Hobson and Newman (1981) have provided a good review of studies of discrimination and differentiation based upon the number of responses. They provide some evidence to suggest that the number of responses, independently of the duration of responding, can serve as a discriminative cue. Even if this is the case, there is a potential problem of interpretation of the results of any response counting experiment. Suppose that each response duration is roughly constant and that an animal sums these durations. Then the animal would be timing rather than counting. It is reasonable to assume that rats can sum the duration of discrete stimuli because that is the essential finding of the "gap" experiments in which rats ran their clocks during the stimuli but stopped them during the gaps (e.g., Meck, Church, & Olton, in press; Roberts & Church, 1978). Thus, it is possible that they could sum the durations of fixed-duration responses. Although it may be difficult to discredit this idea in the case of counting responses, tests are straightforward in the case of counting of stimuli.

Davis and Memmott (1982) have prepared a critical evaluation of counting behavior in animals. Most of the evidence concerns the number of responses as the discriminative stimulus, but there are a few demonstration experiments showing that the number of successive stimuli can serve as a discriminative stimulus. For example, with an autocontingency procedure, rats can discriminate three successive events when time between events is a random variable (Davis, Memmott, & Hurwitz, 1975). In that experiment, the animals were trained to press a lever on a 30-sec variable interval schedule of reinforcement. In one condition they received a maximum of three fixed-duration shocks in a session, but at irregular times. As a result, the response rate of the rats increased dramatically after the third shock. This suggests that they may have been able to count the number of shocks delivered in a session, at least up to three. Alternatively, the rats may have summed the durations of the fixed-duration shocks.

Fernandes and Church (1982) have shown that rats can discriminate the number of sequentially presented events, even when all temporal cues are counterbalanced or held constant. They presented 10 rats with a sequence of short sounds. Then two levers were inserted into the box. If a sequence of two sounds had been presented, a right lever response ("few") was reinforced; if a sequence of four sounds had been presented, a right lever response ("many") was reinforced. (Each sound in a sequence will be called a "stimulus.") The sequences could be described in terms of the number of stimuli (n), the duration of each stimulus (a), and the interstimulus interval (b) which is defined as the interval between the termination of one stimulus and the onset of the next. These determine the cycle duration ($c=a+b$), the total sequence duration ($d=na+[n-1]b$), the total stimulus-on duration (na), the stimulus rate (n/d), and other characteristics of the sequence. In various phases, each of the time intervals was excluded as the major relevant variable on which the discrimination was being made. (See Table 25.1.) Thus Fernandes and Church concluded that the controlling variable in that experiment was the number of sequential stimuli (2 or 4).

One method to demonstrate that the number of successive stimuli is an effective cue for behavior is to train animals to respond differentially to the number of stimuli when each of the time intervals is varied. If the behavior is determined primarily by the number of stimuli and it is relatively unaffected by the time intervals, one may conclude that the discrimination is based primarily on the number of stimuli. This was the method used by Fernandes and Church (1982). An alternative method is to confound completely the number of stimuli and the duration of a sequence during training. Then, during testing, the number of stimuli can be held constant while the duration of a sequence is varied to assess control by duration, and the duration of a sequence can be held constant while the number of stimuli is varied to assess control by number. This is the method used by Meck and Church (1983). The results showed that the choice responses were controlled by the number of stimuli, even though the duration of the sequence was held constant. These results are discussed in more detail in the section on "Equivalence of the bisection function for duration and number."

TABLE 25.1
Median Percentage and Interquartile Range
of a Left or "Many" Response
as a Function of Signal Conditions

a	b	c	d	na	n/d	Reinforced Response	Left Response %	Range
				Number of stimuli = 2				
.2	.8	1.0	1.2	.4	1.67	Right	14	9
.2	1.3	1.5	1.7	.4	1.18	-----	17	23
.2	1.8	2.0	2.2	.4	.91	-----	28	21
.2	2.3	2.5	2.7	.4	.74	-----	20	20
.2	2.8	3.0	3.2	.4	.63	Right	20	19
.4	.8	1.2	1.6	.8	1.25	-----	23	16
.4	2.8	3.2	3.6	.8	.56	-----	32	21
				Number of stimuli = 4				
.4	.8	1.2	3.2	.8	1.25	Left	72	20
.4	1.3	1.7	4.7	.8	.85	-----	70	22
.4	1.8	2.2	6.2	.8	.65	-----	80	25
.4	2.3	2.7	7.7	.8	.52	-----	67	30
.4	2.8	3.2	9.2	.8	.43	-----	61	36
.1	.8	.9	2.8	.4	1.43	-----	82	18
.1	2.8	2.9	8.8	.4	.45	-----	94	6

Note-- a = Duration of each stimulus, b = interstimulus interval, c = cycle duration, d = total duration, na = total stimulus-on duration, n/d = stimulus rate (stimuli per sec), range = interquartile range of left (i.e., "many") responses. All durations are in sec. Data from Fernandes & Church, 1982.

III.   CROSS-MODAL TRANSFER

A signal can be described in terms of modality, e.g., vision or audition. It can also be described in terms of amodal attributes, e.g., duration, number, location, and intensity. An animal can classify a signal according to its modality or according to its amodal attributes or both. For example, a rat can classify a signal according to its duration: If trained to discriminate between a 2-sec light and an 8-sec light, it can make the same discrimination immediately when presented with a 2-sec sound or an 8-sec sound (Meck & Church, 1982a, b; Roberts, 1982). Apparently the rats classified the signals as 2-sec or 8-sec amodal events; there is some psychological reality to the abstract concept of "duration" independent of any specific stimulus energy.

In the experiments on "number" as a stimulus attribute, were the rats attending to the number of sounds or were they attending to the abstract number of events, regardless of the nature of the events? The straightforward test is to observe their reaction to a sequence in another modality, e.g., vision. The 10 rats from the Fernandes and Church experiment were given five sessions of retraining with sequences of either two or four sounds. The sounds were 1.0 sec in duration and the interval between sounds was either 0.8 or 4.4 sec. If one of the two 2-sound sequences was presented, a right ("few") response was reinforced; if one of the two 4-sound sequences was presented, a left ("many") response was reinforced. Table 25.2 shows that the rats discriminated between the 2-sound and 4-sound sequences. On the sixth session a light stimulus was used instead of a sound stimulus. The response rule was reversed for half of the rats and it was unchanged for the other half of the rats. Table 25.2 shows that rats in the nonreversal group continued to make a "many" response following a sequence of 4-lights and a "few" response to a sequence of 2-lights, although not to the extent that they did for sound. Rats in the reversal group had a tendency to respond "many" following sequences of two lights. The effect of the number of light signals on the response choice was significantly different in the nonreversal and reversal groups, $F(1,6) = 25.3$, $p < 0.01$. Apparently during training on the sound sequences the animals learned to discriminate by number, and this transferred to a new modality.

TABLE 25.2
Median Percentage and Interquartile Range
of a Left or "Many" Response
as a Function of Signal Conditions

a	b	c	d	%	Range	%	Nonreversal Range	%	Reversal Range
					Sound		Light		

Number of stimuli = 2

a	b	c	d	%	Range	%	Range	%	Range
1.0	.8	1.8	2.8	10	7	18	22	68	18
1.0	4.4	5.4	6.4	8	10	26	8	64	12

Number of stimuli = 4

a	b	c	d	%	Range	%	Range	%	Range
1.0	.8	1.8	6.4	86	23	60	17	44	28
1.0	4.4	5.4	17.2	88	18	70	16	42	28

Note-- a = Duration of each stimulus, b = interstimulus interval, c = cycle duration, d = total duration, range = interquartile range of left (i.e., "many") responses. All durations are in sec.

In this chapter an animal is said to be "counting" if the number of events is a discriminative stimulus. Others use the term in a more restricted sense (e.g., Davis & Memmott, 1982). Our model of how the animal estimates the number of events has some of the following features: The mode switch has the discontinuity property: It closes some integer number of times that corresponds to the number of events. The representation in the accumulator of each number of events is distinct, and it is ordered in the same way as the number of events. Our model of animal counting does not exclude the possibility that response errors could occur, since there are several potential sources of variance (see Gibbon & Church, Chapter 26, this volume). Other evidence of cross-modal transfer of number has been provided by Meck & Church. (1983)

IV. THE INTERNAL MECHANISM FOR DURATION, NUMBER, AND RATE DISCRIMINATION

An animal is said to be timing if the duration of an event, independently of the number of events, is a discriminative stimulus; an animal is said to be counting if the number of events, independently of their duration, is a discriminative stimulus. How may timing and counting be related? First, it is possible that different and unrelated mechanisms are used for timing and counting. Second, it is possible that one of these functions is an artifact of the other. Although we have reviewed some evidence in the section on "Identification of 'number' as a stimulus attribute" that an animal that appears to be counting is not timing some interval, perhaps an animal that is timing is counting some repeating event (e.g., Rilling, 1967). Finally, timing and counting may use the same internal mechanism. This final proposal is the one supported by the evidence in this chapter.

Gibbon and Church (Chapter 26, this volume) have presented an outline of an information-processing model of timing. It may also apply to counting. The only difference between counting and timing may be in the action of the "switch" (see Figure 26.3 of Gibbon & Church, Chapter 26). Figure 25.1 of the present chapter shows three modes of operation of such a switch. The top line shows a stimulus sequence. The next three lines show how the animal might respond to the stimulus: In "Run" mode the initial stimulus starts an accumulation process that continues until the end of the trial; in "Stop" mode the process occurs whenever the stimulus occurs; in "Event" mode the onset of the stimulus produces a relatively fixed-duration of the process. This is a mechanism that can be used either for timing (the "Run" and "Stop" modes) or for counting (the "Event" mode). When the number of responses is the discriminative stimulus it is normally impossible to distinguish between the "Stop" and "Event" modes since the responses are usually brief and relatively constant in duration. When the number of experimenter-controlled stimuli is the discriminative stimulus, however, it is possible to distinguish between the "Stop" and "Event" modes since the duration of each stimulus can be varied.

The mode switch may correspond to a physiological mechanism with gating properties. There is some evidence that the same neural net can be activated in either the run, stop, or event mode. (Swigert, 1970) These neural modes may have some relationship to the behavioral evidence for an

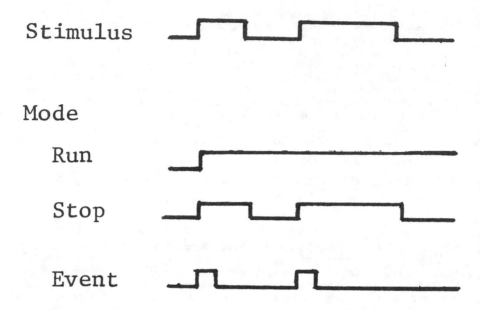

FIG. 25.1. Diagram of three modes of operation of the accumulation process.

internal mechanism that can be operated in several modes.   Regardless of any physiological evidence, this one-mechanism proposal is simpler than a two-mechanism proposal with a separate clock and counter, so it is the preferred proposal unless inconsistent data are found.   There is also substantial evidence in support of the notion that the same mechanism is used for timing and counting.   This is revised in the next three sections.

A. EQUIVALENCE OF THE BISECTION FUNCTION FOR DURATION AND NUMBER

Meck and Church (in press-a) found that the psychophysical functions for duration and number were indistinguishable when the ratio of the extremes was constant.   Either the same mechanism was used for timing and counting or the two mechanisms, by coincidence, had the same sensitivity.

In one experiment duration and number were both relevant.   During original training a left response was reinforced following a 2-cycle signal lasting 2-sec; a right response was reinforced following an 8-cycle signal lasting 8-sec.   During testing for duration discrimination, number was held constant at 4 cycles and sequence duration was varied; during testing for number, duration was held constant at 4 sec and the number of cycles was varied.   The logic of holding one of the signal dimensions constant at an intermediate value while varying the other was to neutralize that dimension, thereby showing the degree of control by the other dimension.

The stimulus sequences used for testing in this experiment are shown in Figure 25.2. One set of sequences held the number of cycles constant at 4 while varying total signal duration between 2 and 8 sec (2, 3, 4, 5, 6, and 8 sec). The other set of sequences held total signal duration constant at 4 sec while varying the number of cycles between 2 and 8 cycles (2, 3, 4, 5, 6, and 8 cycles). There was no reinforcement for a response to any of the intermediate sequences during this single test session, i.e., disregard the column labelled "Reinforced Response" at this point. To maintain the discrimination, half of the trials were the same as during original training. At the end of a stimulus sequence both levers were inserted into the box. If the animal made the correct response, a pellet of food was delivered; if it made the incorrect response, no food was delivered. Then there was an intertrial interval of random duration with a mean of 40 sec.

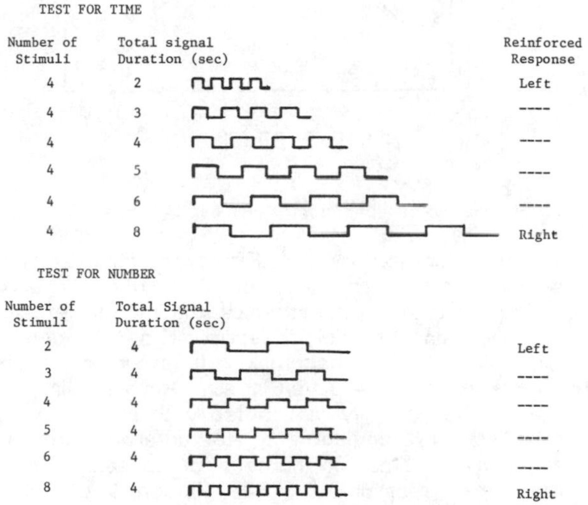

TEST FOR TIME

Number of Stimuli	Total signal Duration (sec)		Reinforced Response
4	2		Left
4	3		----
4	4		----
4	5		----
4	6		----
4	8		Right

TEST FOR NUMBER

Number of Stimuli	Total Signal Duration (sec)		
2	4		Left
3	4		----
4	4		----
5	4		----
6	4		----
8	4		Right

FIG. 25.2. Diagram of the stimulus sequences.

The results of the test session are shown in Figure 25.3. Open circles represent signal variation along the duration dimension and closed circles represent signal variation along the number dimension. The psychophysical function relating the proportion of right response ("many") to the number of stimuli was very similar to the function relating the proportion of right responses ("long") to the signal duration. In both cases the point of indifference was near the geometric mean of the reinforced extremes ($[2 \times 8]^{1/2} = 4$).

The smooth function drawn near the data points is based on a version of scalar timing theory that Gibbon (1981) refers to as "sample known

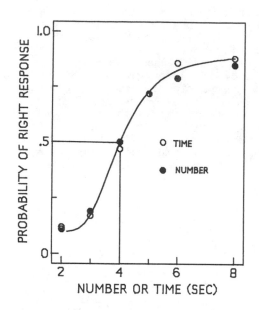

FIG. 25.3. Psychophysical function for number and duration.

exactly with similarity decision rule."   The quantitative application of this model to these data has been described elsewhere (Meck & Church, 1983). The description of the application of the model here is in terms of the information—processing model described by Gibbon and Church (Chapter 26, this volume).   The pacemaker is assumed to emit pulses at some interpulse interval on any given trial.   Across trials the interpulse interval is assumed to be normally distributed with some mean and standard deviation.   As the standard deviation of the interpulse interval increases, the animal's sensitivity to time or number decreases.   This is the only source of variance that is assumed.   (See Gibbon & Church, Chapter 26, this volume, for other plausible sources of variance.)

From the results of the experiment just described, we concluded that the rat has a representation of the sequence duration and of the number of stimuli on each trial. This can be accomplished with the switch for one accumulator set in run mode (for timing) and the switch for another accumulator set in event mode (for counting). When the switch is set in run mode, the pacemaker—switch—accumulator mechanism may be called a clock; when the switch is set in event mode this mechanism may be called a counter. The values in the accumulators are available in working memory to be compared to the memory of reinforced values stored in reference memory.   A similarity response rule is used.   The animal makes a left response if the value in working memory is closer to the value in reference memory of a reinforced left response; it makes the right response if the value in working memory is closer to the value in reference memory of a reinforced right response.   The measure of distance is a ratio of the two values being compared, the larger divided by the smaller. This is the basis for choice on trials on which the animal is paying attention to the stimulus sequence. On some trials the animal may not pay attention to the stimulus sequence and the process described above cannot apply. On these trials we assume that the animal has some constant probability of making a particular response.

The equations for estimating the parameters of this model are adapted for this application by Meck and Church (1983) from the analysis of Gibbon (1981). The values of the parameters were found by exhaustive search of the parameter space (with a step size of 0.05). The coefficient of variation in this version is simply the ratio of the standard deviation of the interpulse interval of the pacemaker to its mean interpulse interval. This was estimated to be 0.25, a value similar to that obtained in other experiments. The mean interpulse interval of the pacemaker would be expected to be the same during original training and testing, and the estimate of the ratio of those values was 1.0 The probability of attending to duration (or number) was 0.8; there was no bias, i.e., the probability of a long (or many) response given inattention to duration (or number) was 0.5. For both duration and number discrimination, the model accounted for a satisfactory percentage of the variance, $\omega^2 > 0.99$ and there were no systematic deviations from theory.

The predicted function shown in Figure 25.3 was calculated from the equations derived by Gibbon (1981) and used by Meck and Church (1983), but the process can also be simulated.

B. EQUIVALENCE OF THE EFFECT OF METHAMPHETAMINE ON TIMING AND COUNTING

Previous research has shown that 1.5 mg/kg of methamphetamine, administered ip, can produce a leftward shift in the psychophysical time estimation function of about 10% (Maricq, Roberts, & Church, 1981; Meck, 1983a); that is, durations were judged to be 10% longer than on control trials without the drug. If the same mechanism that is used in Run Mode for timing is used in Event mode for counting, a variable that affects the psychophysical time estimation function should affect the psychophysical count estimation function in the same manner. Thus, if methamphetamine produces a leftward shift in the function relating probability of a "long" response to signal duration, it should produce a similar leftward shift in the function relating probability of a "many" response to the number of stimuli. This was found to be the case (Meck & Church, 1983) as shown in Figure 25.4. The smooth function near the data points is based on the model described previously. The coefficient of variation remained at 0.25 and again there was no bias for one response or the other, i.e., the probability of a long (or many) response given inattention to duration (or number) remained at 0.5. With the additional training the probability of attention increased to 0.9. Most importantly, methamphetamine decreased the mean interpulse interval of the pacemaker by about 10%, and the same decrease was found for duration and number. This decrease in the mean interpulse interval translates into an increased rate of pulse accumulation. The fits again were satisfactory [$\omega^2 = .99$] and there were no systematic deviations.

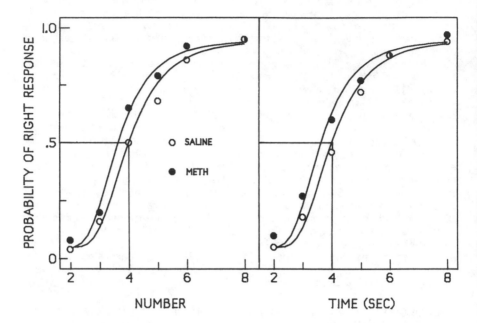

FIG. 25.4. Psychophysical function for number and duration with methamphetamine and saline.

Such leftward shifts of the psychophysical function are not obtained on all dimensions. For example, pigeons trained on a wavelength discrimination do not show a horizontal shift of the psychophysical function when amphetamine is administered (Hayes, 1981). We assume that such leftward shifts would not occur on all intensive dimensions, such as stimulus intensity, but this must be tested. The fact that a similar leftward shift occurred for both timing and counting suggests that there is a fundamental similarity between these two processes.

C. OTHER POSSIBLE SIMILARITIES BETWEEN TIMING AND COUNTING

Probably our most exciting empirical findings about timing have concerned the control operations that animals can do with time (Church, 1978; Meck, in press-b; Roberts & Church, 1978). The first results came from a fixed-interval procedure with a gap test in which an animal trained to time the duration of some stimulus (e.g., a light) was presented with a stimulus that was on for some time (a), off for some gap (b), and then on again for some time (c). The animals initially "stopped their clock" during the gap and reported the time as a+c, although with training they could learn to stop, run, or reset their clocks, i.e., they could report the time intervals, a+c, *a*+*b*+*c*, or *c*.

The next result came from the shift test in which an animal was trained with one temporal criterion with one stimulus (e.g., a light) and a different temporal criterion with another stimulus (e.g., a sound). When, on some test trials, a stimulus began in one modality and changed to the other modality, an animal shifted its criterion and reported the time as the sum of the durations of the two stimuli (Church, 1978; Roberts & Church, 1978). This suggested that the rats used the same clock to time lights and sounds, that they timed up rather than down to a fixed criterion of zero, and that they timed in an absolute rather than a proportional manner. Roberts (1981) has replicated both the gap and shift results with the peak procedure.

As another example of control, rats can time two stimuli simultaneously (Meck, 1983b; Meck & Church, in press). In one demonstration, rats were trained on a 50-sec peak procedure with food reinforcement. Then, at random times during some of these trials, a 10-sec sound signal followed by electric shock was added. The rats showed good temporal discrimination of both intervals, and the timing of the 10-sec shock interval did not interfere with the accurate timing of the 50-sec food interval. In addition, the timing of the 50-sec interval could be temporarily stopped while the animals continued to time the 10-sec interval, indicating that multiple accumulators are used to process the temporal attributes of simultaneous events.

There are no comparable experiments on the control of counting, but if the single-mechanism hypothesis is correct, one can speculate about the operations on numbers that animals can do. With a gap test, perhaps an animal can count on command, stop counting for a period of time, and then continue counting from where it left off. With a shift test, perhaps an animal initially counts toward the first criterion and, when the signal is shifted, changes the criterion without losing count.

Can an animal simultaneously count the number of signals from two modalities? For example, could it report either the number of light signals or the number of sound signals? Again, if the analogy to time is correct, we should expect to find (as did Meck, 1983b, and Meck & Church, in press, for time) that animals can be trained to simultaneously count two signal sources. This suggests that any model involving a single counting source (pacemaker) without multiple accumulators is too simple to account for the performance of animals.

V.   OTHER TYPES OF EVIDENCE FOR A SINGLE MECHANISM FOR COUNTING AND TIMING

The evidence that the same mechanism is responsible for timing and counting has been based upon two similarities of the input-output relations for timing and counting: (a) the psychophysical functions relating probability of a right response to duration and number were indistinguishable such that the same parameter values of a scalar timing theory applied equally well to both functions, and (b) the two functions were equally affected by the same dose of methamphetamine. How can we know that mechanisms for timing and counting are not distinct, but have the same properties? An increase in the number of correspondences between duration and number discrimination will not resolve this problem. There are, however, two promising approaches.

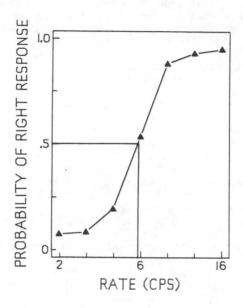

FIG. 25.5. Psychophysical function for stimulus rate.

A. THE "SHIFT" EXPERIMENT

Rats can be trained in the presence of one stimulus that reinforcement occurs after one time interval, and in the presence of another stimulus that reinforcement occurs after another time interval. For example, food can be delivered following the first response after 30 sec in the presence of light and after 60 sec in the presence of sound. In the shift experiment occasional critical trials began with the 30-sec stimulus but later shifted to the 60-sec stimulus (Roberts, 1981; Roberts & Church, 1978). As described above, the response rate was a function of the time since the trial began and the current stimulus; it did not matter whether the interval was begun with the light or a sound stimulus. This suggests that rats used the same internal clock to time light and sound but adjusted the remembered time of reinforcement to the value appropriate for the current stimulus.

A similar experiment can be done to determine if the same mechanism is used to time one signal and count another one. Suppose rats were trained to count the number of sound bursts and to time the duration of a light. Then, on some critical trials a sequence of sound bursts is first presented and then a continuous light is presented. Would the sound bursts affect the expected time of the reinforcement of the light? If the same accumulator was used to count the sound bursts and to time light, then filling the accumulator with pacemaker pulses from the sound bursts would reduce the time the light needs to be on before the time criterion would be reached. The magnitude of the reduction would serve as a measure of the equivalence between an increment of one count and a unit of time.

This experiment has not been done, and it could fail even if animals use the same mechanism for timing and counting. We now recognize that an animal can time two independent, asynchronous events without interference (see above). Such simultaneous temporal processing demonstrates that rats have multiple accumulators and so they would be able to use one of them for accumulating times and others for accumulating counts. The simplest explanation of the results of the shift experiment with times, is that rats apparently do not use multiple accumulators; if they also use the same accumulator in the shift experiment from timing to counting, then the interference experiment described here could be successful.

B. THE "COMPOUND STIMULUS" EXPERIMENT

A second approach to testing the equivalence of the mechanism for timing and counting that does not depend upon the observation of similarities involves tests with compound signals. If the same mechanism is used for timing and counting, these tests will produce different results for timing and counting. First, consider timing elementary stimuli and their compounds. In a 50-sec peak procedure, a signal occurs for some period of time on each trial. On some trials, food is available after 50 sec and delivered immediately following the next response; on other trials no food is available and the trial lasts for 130 sec. As a consequence of this procedure, response rate increases to a maximum near the time that reinforcement is made available (e.g., 50 sec) and then declines in a fairly symmetrical fashion (e.g., Roberts, 1981). If the stimulus on half the trials is light and the stimulus on the other half of the trials is noise, the performance is similar in the presence of the two signals. What happens when a compound signal of light and sound is presented? A rational expectation is that the function would be similar to that produced by the light and noise alone and the maximum response rate would be in a similar location, i.e., at 50 sec. Unpublished observations of Roberts showed that the maximum was less than the time that reinforcement is made available. Meck & Church (in press-b) replicated this observation and found that the scalar timing model they had used to account for over 99% of the variance in the peak functions for light and sound could account for the peak function for 99% of the variance of the compound signal. All parameters were the same for compound and elements, except the attention to time was reduced for the novel, unreinforced compound trials. The basic assumption was that in compound signals the animal took a sample from the distribution of remembered time of reinforcement following a light stimulus and a sample of the remembered time of reinforcement following a sound stimulus and used the shorter of these two values. This not only accounted for a leftward shift in the function relating response rate to time since the trial began; it accounted for the magnitude of the leftward shift. In any case, the peak of the compound was significantly earlier than the peak of either element, and it certainly was much earlier than the sum of the two. In terms of the information-processing model we assume that either light or sound can close the mode switch for the duration of the stimulus so that impulses travel from the pacemaker to the accumulator continuously as long as the signal is present (Run mode). Thus, when both light and sound are presented in compound, the mode switch is set to connect the pacemaker to an accumulator by one of the stimuli and the other stimulus has no effect on this accumulator, although it may be timed simultaneously by a second clock.

The same information-processing model has a rather different implication for counting elementary stimuli and then compounds. If an animal is trained to respond "few" to a sequence of two sound stimuli or two light stimuli and to respond "many" to a sequence of four sound stimuli or four light stimuli, how will the animal respond to the initial presentations of two light-sound compound signals? In terms of the information-processing model we assume that either the onset of light or sound will briefly close then open the mode switch (Event mode). In a compound stimulus the sound and light presumably do not have their effects at precisely the same instant, so a compound stimulus is able to operate the same mode switch twice. Presumably, the same accumulator in Event mode is increased a relatively fixed amount with each stimulus onset, whether stimuli are simultaneously or successively presented. This implies that animals may treat two light-sound compound as four signals, and respond "many."

Five rats were trained to make a right response following a 2-sound sequence or a 2-light sequence and to make a left response following a 4-sound sequence or a 4-light sequence. Each stimulus was on for 1 sec and the interstimulus interval was either 0.8 or 4.4 sec. Then on a single test day half the trials were the same as during training and half were unreinforced trials with compound stimuli. The compound-stimulus trials consisted of one or two sound-light stimuli, and when there were two stimuli, the interstimulus interval was 0.8 or 4.4 sec. The mean proportion of "many" responses was calculated for two and four elementary stimuli and for one and two compound stimuli. A one-factor analysis of variance demonstrated that conditions made a difference, $F(3, 12) = 22.1$, $p < .001$, and the Tukey HSD test indicated that one compound stimulus did not have significantly different effect than two elementary stimuli and two compound stimuli did not have significantly different effect than four elementary stimuli, but all other pairwise differences were significantly different ($p < .01$). Thus each compound stimulus was equivalent to two elementary stimuli. There is no reason to believe that this analysis could not be extended to compounds of more than two elements.

If an animal is trained to respond "few" to a sequence of two light-sound compounds and to respond "many" to a sequence of four light-sound compounds, how will it respond to the initial presentations of four elementary stimuli (lights or sounds)?   By the argument in the preceding paragraph, it is possible that during original training the animal would have learned to respond "few" to four events (two compound stimuli) and to respond "many" to eight events (four compound stimuli).   Thus, it is possible that animals that responded "few" to four elementary events on unreinforced test trials would respond "many" to four compound events, although the result would be surprising.

C. SUMMARY

The proposal that the same mechanism is used for timing and counting is reasonable.   A one-mechanism model is a simpler solution than a two-mechanism model; it accounts for the similarity of the psychophysical function for timing and counting when the ratio of the extremes is equal; it accounts for the quantitative equivalence of the effect of methamphetamine on the psychophysical function for timing and counting; and it accounts for the difference in the combination rules for duration and number in the presence of a compound stimulus.   The one-mechanism hypothesis, however, needs further testing.   Hobson and Newman (1981) challenge the assumption that ratio-based and time-based schedules reflect the same behavioral process.   They describe four differences in the performance under ratio-based and time-based schedules (p. 200).   For example, they show that the Weber fraction decreases with ratio size, but it remains relatively constant over a wide range of time-based schedules. Fetterman, Stubbs, and Dreyfus (1981) found that pigeons bisected time-based schedules at the geometric mean, but bisected response-based schedules at or below the harmonic mean.   These investigators did not try to rule out the possibility that some time interval, rather than number of responses, was the controlling stimulus.   Any attempt to integrate timing and counting must be able to deal with such findings.

VI.   DISCRIMINATION OF STIMULUS RATE

Stimulus rate is defined as the number of stimuli per unit time.   Rats are able to discriminate between stimulus rates (Sidman, 1960), but the relationship between rate, duration, and number discrimination has not been worked out.   In this section we describe some reasons for believing that rate discrimination involves counting for a relatively fixed period of time.

   In one experiment, four stimulus sequences were used in training (Meck, Church, & Olton, in press). They were 2 cycles/sec for 2 or 8 sec and 16 cycles/sec for 2 or 8 sec.   Ten rats were trained to make a left response following a 2 cycle/sec sequence of either duration and a right response following a 16 cycle/sec sequence of either duration.   Then signals of intermediate rates were introduced.   They were 3, 4, 6, 8, or 11 cycles/sec.   These sequences varied in duration (2, 2.6, 3.2, 4.0, 5.0, 6.4, and 8.0 sec) and they were never reinforced.   The psychophysical function for signal rate, averaged across signal durations, is shown in Figure 25.5.   The probability of a right response increased as a function of signal rate. Although the animals could not achieve a high performance on these sequences with a pure counting strategy, they could use a timing strategy. The duration of each stimulus and the cycle duration were both perfectly correlated with the reported rate.

In previous research with duration and number discrimination, we have found that the point of subjective equality of the bisection function is at the geometric mean. Thus, if the rats were using the duration of each stimulus as the effective cue, the point of subjective equality should be when the duration of each stimulus was about $[(1/4)(1/32)]^{1/2} = 0.09$ sec, i.e., a rate of about 11.3 cycles/sec. If the rats were using the cycle duration as the effective cue, the point of subjective equality should be about $[(1/2)(1/16)]^{1/2} = 0.18$ sec, i.e., a rate of about 5.7 cycles/sec. In fact, the point of subjective equality was 5.9, indistinguishable from the geometric mean of the cycle duration of the extreme sequences. It should be noted, however, that the geometric mean between the two extreme rates, 2 and 16 cycles per second, is equivalently near the observed point of subjective equality.

The psychophysical function relating choice to signal duration is symmetrical when signal duration is scaled in logarithmic units (Church & Deluty, 1977). The psychophysical function relating choice to both signal duration and number shown in Figure 25.3 becomes symmetrical when signal duration and number is scaled in logarithmic units. This symmetry on a logarithmic scale is a consequence of the scalar timing theory. Figure 25.5 shows that the psychophysical function relating choice to log cycles per sec was also symmetrical. Unfortunately, this means that the function relating choice to log cycle duration would also be symmetrical, so this cannot serve as a basis for distinguishing between a rate and duration discrimination as the controlling variable.

Initially, it appeared impossible to distinguish between a rate and duration discrimination since one is the reciprocal of the other. There are, however, ways to distinguish between the variables that are described in the rest of this section.

When signal rate is confounded with signal duration, one can determine whether a signal with many cycles per sec (i.e., with a short cycle duration) is treated more like a long or short duration signal. In one experiment a white–noise signal was presented to rats at a rate of 2 cycles per sec or 16 cycles per sec for a duration of 2 sec or 8 sec (Meck, Church, & Olton, in press). A left response was reinforced following a 2–sec signal of either rate and a right response was reinforced following an 8–sec signal of either rate. In training, choice accuracy was significantly higher for the 2–sec signal when the rate was slow (2 cycles per sec) and for the 8–sec signal when the rate was fast (16 cycles per sec). Thus, a short signal duration is like a slow rate. This result suggests that animals are counting the number of stimulus onsets for a relatively fixed period of time rather than timing the duration between stimulus onsets.

Some drugs have effects that are best interpreted as affecting the rate of the pacemaker (Maricq, Roberts, & Church, 1981; Maricq & Church, 1983; Meck, 1983a); others have effects that are best interpreted as affecting the remembered time of reinforcement (Meck, 1983a). It now appears that pacemaker rate is positively related to the effective level of dopamine and that the remembered time of reinforcement is inversely related to the effective level of acetylcholine. Whether or not these neurotransmitter explanations survive further tests of their consequences, there are pharmacological manipulations for altering pacemaker rate and time of a remembered reinforcement. Lesions may also affect the remembered time of reinforcement, possibly also through modification in the effective level of acetylcholine in relevant parts of the brain. Frontal lesions in rats produce a permanent rightward shift in the psychophysical function consistent with an increase in the remembered time of reinforcement (Maricq, 1978); a fimbria fornix lesion produces a permanent leftward shift in the psychophysical function consistent with a decrease in the remembered time of reinforcement (Meck, Church, & Olton, in press).

A fimbria–fornix lesion also produces a permanent leftward shift in the psychophysical function relating the probability of a right response as a function of stimulus rate (Meck, Church, & Olton, in press). This is a critical fact. Suppose the rats were making a duration discrimination. If the cycle duration is remembered as shorter, the rate is remembered as faster. Thus, if the animals were discriminating on the basis of cycle duration, one would expect a fimbria–fornix lesion to produce a rightward shift in the psychophysical function relating the probability of a right response to stimulus rate. The opposite was observed. If on the other hand, the rats based their discrimination on the number of stimuli in a relatively fixed subjective time period, a leftward shift would be expected. This would occur if the actual number of stimuli in the subjective time period was unchanged by the lesion, so the remembered number of pulses at reinforcement would be reduced. This is equivalent to a reduction in the perceived stimulus rate at reinforcement, or a leftward shift in the function.

Meck (1983a) has shown the selective influence of various drugs on timing functions. Further tests of the assumption that rats are discriminating the number of cycles in a fixed time period rather than the cycle duration can employ such drugs. Some drugs appear to decrease the remembered time of reinforcement (e.g., vasopressin, oxytocin, physostigmine) or increase the remembered time of reinforcement (e.g., atropine). Other drugs appear to increase clock speed (e.g., methamphetamine) or to decrease clock speed (e.g., haloperidol). If stimulus rate is the controlling variable, leftward shifts of the psychophysical function relating the probability of a right (high rate) response to stimulus rate would be expected following drugs that decrease the remembered time of reinforcement or increase clock speed; if cycle duration is the controlling variable, rightward shifts would be expected. The present evidence suggests that rate is the controlling variable.

VII. FUNCTIONAL SIGNIFICANCE OF COUNTING

The evolutionary significance of counting skills is uncertain, but the fact that such skills are easily observed suggests that they may serve some function. There are some reasons to suspect that counting abilities may have evolved in animals since they could contribute to reproductive success and fitness.   For example, according to optimal foraging theory animals will forage preferentially in areas where prey density is highest (Krebs & Davies, 1978).   (Prey can refer to plants as well as animals.) Although density estimates are usually thought to occur with time estimates (a combination of inter-prey intervals), they might also occur with count estimates (the number of prey) or rate estimates (the number of prey in a fixed time interval).   An excellent set of chapters and references on optimal foraging theory is found in Kamil and Sargent (1981).

Of course, it is possible that animals in a laboratory situation will provide evidence of capacities which have not been used as yet in nature. Such "prospective" evolution may result from (1) the natural selection of a general-purpose device that is likely to be successful in varied and changing environmental circumstances, (2) a concomitant development dependent upon the natural selection of some other specific capacity, or (3) the random variation in the evolution of cognitive abilities through speciation (e.g., Gould, 1980; Gould & Eldredge, 1977). This is in contrast to the natural selection of special-purpose devices that are proven successful in particular environments.   For timing, counting, and rate discrimination, a single general-purpose accumulator may have evolved that can be operated in several modes (e.g., Run, Stop, and Event) rather than separate devices sensitive to temporal, numerical, and density attributes of a stimulus.

ACKNOWLEDGMENTS

This research was supported by NSF Research grants BNS 79-04792 and BNS 82-09834.

REFERENCES

Church, R. M. The internal clock. In S. H. Hulse, H. Fowler, & W. K. Honig (Eds.), *Cognitive processes in animal behavior*. Hillsdale, N.J.: Erlbaum, 1978.

Church, R. M., & Deluty, M. Z. Bisection of temporal intervals. *Journal of Experimental Psychology: Animal Behavior Processes*, 1977, *3*, 216-228.

Davis, H., & Memmott, J. Counting behavior in animals: A critical evaluation. *Psychological Bulletin*, 1982, *92*, 547-571.

Davis, H., Memmott, J., & Hurwitz, M. B. Autocontingencies: A model for subtle behavioral control. *Journal of Experimental Psychology: General*, 1975, *104*, 169-188.

Fernandes, D. M., & Church, R. M. Discrimination of the number of sequential events by rats. *Animal Learning and Behavior*, 1982, *10*, 171-176.

Fetterman, J. G., Stubbs, D. A., & Dreyfus, L. R. Generalization of response and duration–based stimuli. Paper presented at a meeting of the Eastern Psychological Association, New York Society. April, 1981.

Gibbon, J. On the form and location of the psychometric bisection function for time. *Journal of Mathematical Psychology*, 1981, *24*, 58–87.

Gould, S. J. *The panda's thumb*. New York: Norton, 1980.

Gould, S. J., & Eldredge, N. Punctuated equilibria: The tempo and mode of evolution reconsidered. *Paleobiology*, 1977, *63*, 115–151.

Hayes, W. F. *Drug effects on wavelength discrimination in pigeons*. PhD Thesis, Brown University, 1981.

Hobson, S. L., & Newman, F. Fixed–ratio–counting schedules. In M. L. Commons & J. A. Nevin (Eds.), *Quantitative analyses of behavior: Vol. I, Discriminative properties of reinforcement schedules.* Cambridge, Mass.: Ballinger, 1981.

Kamil, A. C., & Sargent, T. D. *Foraging behavior: Ecological, ethological, and psychological approaches.* New York: Garland STPM Press, 1981.

Krebs, J. R., & Davies, N. B. *Behavior ecology: An evolutionary approach.* Oxford: Blackwell Scientific Publications, 1978.

Maricq, A. V. Some effects of lesions of the prefrontal cortex on timing behavior in the rat. Honors thesis, Brown University, 1978.

Maricq, A. V., & Church, R. W. The differential effects of haloperidol and methamphetamine of time estimation in the rat. *Psychopharmacology*, 1983, *79*, 10–15.

Maricq, A. V., Roberts, S., & Church, R. W. Methamphetamine and time estimation. *Journal of Experimental Psychology: Animal Behavior Processes*, 1981, *7*, 18–30.

Meck, W. H. Selective adjustment of the speed of internal clock and memory processes. *Journal of Experimental Psychology: Animal Behavior Processes*, 1983, *9*, 171–201. (a)

Meck, W. H. Simultaneous temporal processing: Isolation of multiple internal clocks. Manuscript in preparation, 1983. (b)

Meck, W. H., & Church, R. M. Abstraction of temporal attributes. *Journal of Experimental Psychology: Animal Behavior Processes*, 1982, *8*, 226–243. (a)

Meck, W. H., & Church, R. M. Discrimination of intertrial intervals in cross–modal transfer of duration. *Bulletin of the Psychonomic Society*, 1982, *19*, 234–236. (b)

Meck, W. H., & Church, R. M. A mode control model of counting and timing processes. *Journal of Experimental Psychology: Animal Behavior Processes*, 1983, *9*, 320–334.

Meck, W. H., & Church, R. M. Simultaneous temporal processing. *Journal of Experimental Psychology: Animal Behavior Processes*, in press.

Meck, W. H., Church, R. M., & Olton, D. S. Hippocampus, time, and memory. *Behavioral Neuroscience*, in press.

Rilling, M. Number of responses as a stimulus in fixed–interval and fixed ratio schedules. *Journal of Comparative and Physiological Psychology*, 1967, *63*, 60–65.

Roberts, S. Isolation of an internal clock. *Journal of Experimental Psychology: Animal Behavior Processes*, 1981, *7*, 242–268.

Roberts, S. Cross-modal use of an internal clock. *Journal of Experimental Psychology: Animal Behavior Processes*, 1982, *8*, 2–22.

Roberts, S., & Church, R. M. Control of an internal clock. *Journal of Experimental Psychology: Animal Behavior Processes*, 1978, *4*, 318–337.

Sidman, M. Stimulus generalization in an avoidance situation. *Journal of the Experimental Analysis of Behavior*, 1960, *4*, 157–169.

Swigert, C. J. A mode control model of a neuron's axon and dendrites. *Kybernetik*, 1970, *7*, 31–41.

26 SOURCES OF VARIANCE IN AN INFORMATION PROCESSING THEORY OF TIMING

John Gibbon
*N.Y.S. Psychiatric Institute
and Columbia University*
Russell M. Church
Brown University

I. INTRODUCTION

When a signal of some duration is presented to an animal, what does the animal perceive as its duration? It is common in sensory psychophysics to postulate a subjective scale with scale values and variance properties so that, in some sense, what the animal perceives is a position on the y axis of the graph relating subjective time to physical time. This is a value that may be temporarily stored in a working memory. If food is delivered for a response after a signal of one duration, and not others, what does the animal remember? Presumably, some representation of the value stored in working memory is transferred to a more permanent reference memory. After the presentation of a signal of some duration, how does the animal decide whether or not to respond? Presumably, a comparison is made between a current experience of a duration in working memory and a remembered value in reference memory. The purpose of this chapter is to develop an information processing model of timing along these lines that is compatible with experimental results. The development focuses particularly upon where, in the course of perceiving and discriminating time, variability arises.

The key features of our view of the processing of time are exemplified in a simple experimental procedure that will be the reference experiment for our analysis of sources of variance in the perception and memory of time. This reference experiment is a straightforward temporal generalization task with rats reported earlier (Church & Gibbon, 1982). The procedure is simple. The houselight in an operant chamber is turned off for a duration of time, T. Then a retractable lever is inserted into the chamber. After a 5 sec opportunity to respond, the lever is withdrawn and a 30 sec intertrial interval begins. If the light–off stimulus lasted the "correct" duration, say, 4 sec, a response is reinforced with a pellet of food. If it was not 4 sec, no reinforcement is forthcoming.

The typical result is that rats come to respond with high probability when the duration is 4 sec, and with decreasing probability for stimuli either longer or shorter than 4 sec. Figure 26.1 shows how this decline changes with changes in the size of the reinforced duration (S+). The

465

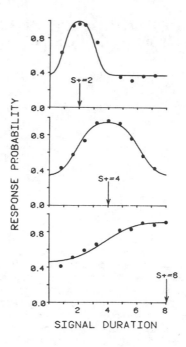

FIG. 26.1. Response probability gradients for three different placements of the positive signal. Data points represent median response probability values for groups of rats studied under each condition.

probability of responding on the lever is shown as a function of signal duration when the rewarded duration was either 2 sec (top panel), 4 sec (middle panel), or 8 sec (bottom panel). Two features of these data are important: The maximum response probability was near S+, and the spread of responsiveness around S+ increased considerably with increases in S+ value.

The way in which the spread is related to the size of S+ is shown in Figure 26.2 which plots the results of many experiments in this series including those of Figure 26.1. The data are plotted on a relative time scale in which signal durations are taken as proportions of the S+ signal. The data have been treated in another way as well, to remove the effects of inattention and overall responsiveness that are evident in the non-zero resting (operant) levels in Figure 26.1. The details of this transformation are available in Church and Gibbon (1982). For the present purposes it is important only to note that data from a variety of different groups in different experiments roughly superpose in this metric. In particular, the data from Figure 26.1 are not distinguishable from the rest when plotted on a relative time scale. This property, which we have called the scalar property, is ubiquitous in animal temporal discrimination work in the seconds-to-minutes range (cf. Gibbon, 1977). The scalar property in

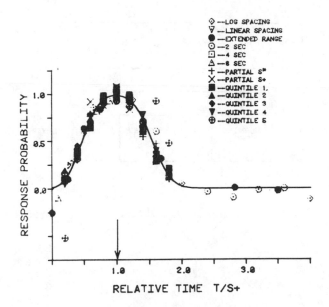

FIG. 26.2. Response probability gradients plotted in relative time (T/S+). Points are taken from six experiments reported in Church and Gibbon, 1982. Data from Figure 26.1 are included as open circles, squares, and triangles. Data have been treated to remove the effects of differing overall levels of responsiveness and differing degrees of attention to the task.

Figure 26.2, we will show below, is a strong constraint on admissible sources of variance and on admissible comparison rules for the decision whether or not to respond.

Our information processing model for this situation is shown in Figure 26.3. The top row shows the clock process that includes a pacemaker, switch, and accumulator. The pacemaker generates pulses at a mean rate (Λ) that we assume is high relative to the time values (seconds to minutes) that we use in these experiments. The switch, after appropriate training (instructions), gates pulses for a mean duration (D_T) to an accumulator when the timing signal is present. The accumulator records the number of pulses (mean of N_T). The second row shows the memory processes that includes a working and a reference memory. A representation proportional to the number of pulses in the accumulator is stored in working memory (mean of M_T). When, at the end of a given trial, a response is made and reinforced (T=S+), the time value recorded in working memory on that trial is stored in a reference memory for reinforced values (mean of M^*_{S+}). The third row shows the decision process. A response occurs when a comparator yields a judgment that the current record in working memory for this trial is "close enough" to the reference memory for the reinforced duration to warrant a response. If the response was reinforced, the current representation in working memory would be stored in reference

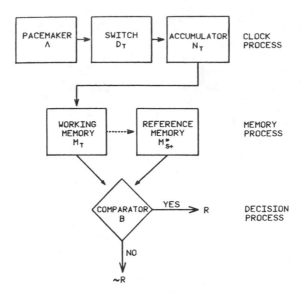

FIG. 26.3. Schematic diagram of information processing components in a temporal generalization task. The top row represents components of a clock process, the middle row represents working (short—term) and reference (long—term) memory. The decision process at the bottom involves a comparison between the contents of working memory and reference memory.

memory.

In Figure 26.4 we show the proposed clock process in more detail. Imagine a pacemaker generating pulses with interpulse interval, τ, with mean rate λ. In this figure, the pulses are evenly spaced indicating no variance. Later analyses allow variance in these parameters. The pulses are switched into the accumulator by the switch indicated in the middle box. The switch is assumed to have some latency to close (t_1) after the signal goes on, and some latency to open (t_2) after the signal goes off. Thus, the mean time during which pulses are gated into the accumulator is $D_T = T - T_0$, where T_0 is the expected difference between the latencies to close and open the switch.[1] In principle, T_0 may be negative as well as positive. The graph above the switch shows this linear relation between D_T and T. The intercept is the minimum signal duration below which counts do not register in the accumulator. For signal durations less than this minimum, the switch is opened before it has closed, so no pulses are switched into the accumulator. If the latency to re—open the switch

---

[1]We generally refer to random variables with lower case letters and their expectations with the corresponding upper case letter (e.g., $D_T = E(d_T)$). An exception is λ, which can play a dual role in the Poisson pacemaker.

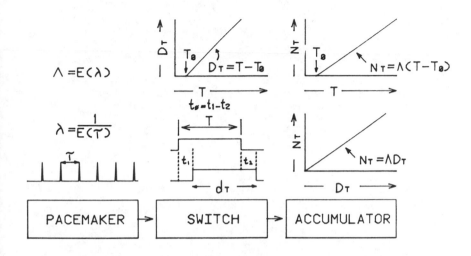

FIG. 26.4. Diagram of the clock process. The pacemaker generates impulses at intervals of τ and these pulses are gated into an accumulator by the switch. The switch has a latency to close (t_1) at the beginning of the stimulus and a latency to open (t_2) at the end of the stimulus. The accumulator simply records the number of impulses gated to it during the interval that the switch is closed.

exceeds the initial latency to close it ($t_2 > t_1$), even a very short signal duration suffices to allow counts to register during the t_2 latency. If t_1 and t_2 were precisely the same, the switch duration would mimic exactly the signal duration. In general, this situation is unlikely, however, and these two latencies constitute one of the sources of variance that we will consider later.

The accumulator in this scheme simply records the number of impulses gated to it. The mean accumulated number, N_T, therefore is just the impulse rate times the duration that the gate is closed, as shown in the lower graph directly above the accumulator. The upper graph shows the accumulation value as a function of signal duration. The switch introduces the non-zero intercept.

The memory system that we propose here is much simpler than is probably realistic for memory models (e.g., Heinemann, 1984a, b). However, it is quite complicated enough to analyze, as we will show, even when variability is restricted to one component at a time. We propose that the working memory directly reflects the accumulated count as shown in the proportional plot above working memory in Figure 26.5. When, for a given condition (series of trials), S+ is fixed at some value, and when on a particular trial S+ is presented (T = S+) and a response is made and

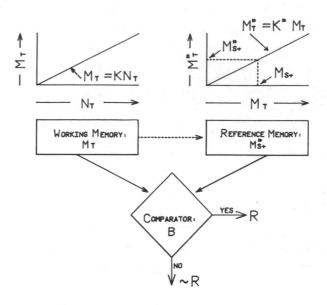

FIG. 26.5. Schematic of the memory system. The working memory holds a value proportional to the number of pulses in the accumulator associated with a particular T. A value proportional to this value in working memory is recorded in reference memory on reinforced trials.

reinforced, the value stored in reference memory, M_{S+}^*, is a proportional representation of the value on that trial recorded in working memory. The comparison, then, on subsequent trials, is between the current value in working memory, M_T, and a stored value in reference memory, M_{S+}^*. The judgment whether or not to respond is based on some comparison between these two values.

Several major features of this account are present in an early proposal of Treisman (1963) in which a pacemaker, counter, store (memory), and comparator occur. To our knowledge, that was the first model to use a form of scalar timing (Equation 11) explicitly in a timing system (but see also Werner & Mountcastle, 1963, for a data-based application of the scalar property in neuronal discharge). Our account parallels much of Treisman's analysis, particularly with respect to the role and properties of the pacemaker as the basic mechanism for measuring time.

A full account of the process outlined in Figure 26.3 would provide evidence for the distinctions between each component. Good evidence might consist of independent variables that affected one but not the other components in the system. Such an approach has been made by Roberts (1981) who identified some variables affecting the timing system (what we have called the clock process and memory process) independently of responding; and other variables affecting the decision to respond independently of the timing and memory processes. Gibbon (1972) used this approach to provide evidence for a dissociation between whether a response will be made or not, and when it will be made.

More recently, a number of studies suggest that clock speed may be changed differentially by some drugs, such as methamphetamine (Maricq & Church, 1983; Maricq, Roberts, & Church, 1981). Other studies suggest that the translation of a value in working memory to reference memory can be altered differentially by some drugs, such as physostigmine (Meck, 1983), and some lesions, such as a lesion of the fimbria fornix (Meck, Church, & Olton, in press).

Within the clock process, several behavioral studies suggest that subjects may be trained to operate the switch gating pulses into the accumulator in different modes. They may, for example, allow the switch to remain closed from the onset of a stimulus to the end of a trial or from the onset of a stimulus until the end of a stimulus (see Roberts & Church, 1978). They may also be induced to close the switch for a fixed, brief period of time with each stimulus (Church & Meck, Chapter 25, this volume, and Meck & Church, 1983).

Finally, the comparison rule may depend upon the task. The proximity rule that we describe below appears to be used in the temporal generalization task analyzed here (Church & Gibbon, 1982). However, different rules have been used for other timing tasks that involve a choice between responses (Gibbon, 1981; Gibbon & Church, 1981; Meck, 1983).

The above discussion is intended to motivate exploration of our information processing system rather than provide a definitive justification for it. These several lines of evidence suggest an inquiry into the formal properties of this system to reveal constraints on acceptable sources of variability and acceptable decision rules for temporal generalization.

We first examine the case in which there is no variance in the system. Here, the values in the accumulator and working memory are one-to-one with signal durations, and reference memory contains a single value corresponding to S+. We then introduce variance at different points in the system — in the pacemaker, switch, memory, and comparator. In each case, we examine an absolute and relative response rule. The consequence of these sources of variance and response rules is examined. Many of them will be seen to lead to predictions qualitatively different from the results shown in Figures 26.1 and 26.2.

II. NO VARIANCE

It is instructive to consider first the case in which the clock and the memory contain no variance and the comparison is based on fixed values that are perfectly correlated with T and S+. If the pacemaker generates pulses at a high steady rate, and if the switch gates these pulses into the accumulator with constant latencies to close and open, T_1, T_2, and if the accumulator records and stores the number of pulses accurately, then: the memory values (Figure 26.5) are linear in the duration of the signal, and the reference memory associated with reinforcement is a single value.

$$M_T = K\Lambda(T - T_0), \qquad\qquad T > T_0 \qquad\qquad\qquad (1a)$$

and for a particular S+ = S,

$$M_S^* = K^* K\Lambda(S - T_0) \qquad\qquad\qquad\qquad\qquad (1b)$$

The memory representations for the current trial duration and for the remembered reinforced value are linear functions of the corresponding physical durations.

A. ABSOLUTE RULE

Now these psychological values must be compared in such a way that a response is made when they are close and no response is made otherwise. We consider two response rules for proximity or "closeness." The first is an absolute rule, which simply takes the algebraic difference between the two values and dictates a response when the absolute difference is smaller than some threshold value, B.

$$|M_{S+}^* - M_T| < B. \qquad\qquad\qquad\qquad\qquad (2a)$$

Note that the reference memory and working memory values may differ even for T = S+, since we have allowed a proportional translation during storage in reference memory. The account presently assumes that the comparator extracts information both from working memory and reference memory independently, and performs some kind of closeness match such as that exemplified in Equation 2a. A bias during translation to reference memory, however, would introduce asymmetry in errors around S+. To allow for generality in translation between working and reference memory, the absolute rule would have to be altered to accommodate signed differences deviating not from 0 but from a biased set point. For the present purposes, however, this complication will be ignored, since our data (Figure 26.2) are roughly symmetric around S+. We will assume here that $K^* = 1$, or, equivalently, that the values in reference memory are brought back through an inverse transform to working memory before comparison.

In Figure 26.6 the top row shows the absolute rule for two values of S+ differing by a factor of 2 (like those in Figure 26.1). The left-most tic on the abscissa represents T_0, the minimal signal duration above which counts may accumulate in the accumulator and hence increment in memory. The second tic (at S) indicates the position of the reinforced time value for our first example. This might be appropriate to the 2 sec data in Figure 26.1. When reinforcement is available for the responding to S but not to other durations, subjects may adopt the rule indicated in Equation 2a, and respond when the absolute difference between the current time value and the remembered time value for this S+ is below threshold. The absolute difference between M_{S+}^* and M_T is shown decreasing to zero at S and increasing beyond it (solid line function). The positive diagonal hatching shows the acceptance region within which responses are required on this rule.

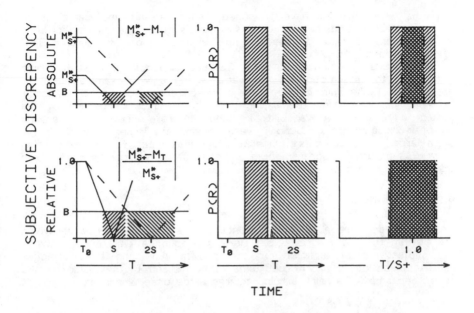

SUBJECTIVE DISCREPENCY

FIG. 26.6. Two comparison rules. The top row shows subjective discrepancy (left column) and its associated response probability (right columns) for an absolute response rule in which the absolute difference between working memory and reference memory forms the basis of the response decision. The bottom row shows subjective discrepancy and response probability for a relative discrepancy rule.

In the same panel, we examine what this rule would produce if the S+ value was doubled. This situation is depicted above the third tic at 2S on the right. There, the difference between working memory and the memory for 2S is shown as the dashed function decreasing to zero at T = 2S and increasing again beyond. The region below threshold, the acceptance window around 2S, is indicated by negative diagonal hatching.

In the middle panel in the top row, the response consequences are plotted against real time. In the first case, when S+ = S, a response is made whenever T is within an absolute distance of S. A little algebra on Equation 2a shows this distance to be $B/K\Lambda$. The same rule on an S+ = 2S task produces the same spread around the reinforced value, that is, responding occurs within this same fixed window around S+. The result is that when the 2S function is plotted in relative time in the far right panel, comparable to Figure 26.2, the efficiency of the system has been vastly improved. The spread around 1.0 in this plot for S+ = 2S is approximately half that around 1.0 for S+ = S.

This is inconsistent with the results shown in Figure 26.2. Thus, it is unlikely that an absolute comparison rule is tenable with our linear scale for time. Of course, this conclusion depends on the linearity of subjective time and possibly also on the no-variance assumption. We will not discuss in depth the issue of the linearity of subjective time, except to note that elsewhere we have proposed theoretical and empirical evidence from at least two different sources which support the assumption that subjective time is linear or nearly so in real time (Gibbon & Church, 1981; Church & Gibbon, 1982).

The greater efficiency as S+ = 2S also depends, at least to some extent, on the assumption of no variance. We will examine later the absolute proximity rule in the context of different sources of variance. Possibly, this rule might be salvaged if the variance properties of the system were such that variability in either the reference memory around S+ or working memory around T were considerably larger when S+ and T were larger. If much more variance were introduced for S+ = 2S than S+ = S, then greater spread would be induced for this case around 1.0 in the upper right panel.

B. RELATIVE RULE

An alternative to the absolute difference rule is the relative rule shown in the bottom row of Figure 26.6. Here, the absolute difference between the subjective representation of the current time and the subjective representation of the reinforced time is normalized by the value of the reinforced time. The relative rule requires a response whenever:

$$\left| \frac{M^*_{S+} - M_T}{M^*_{S+}} \right| < B. \tag{2b}$$

The subjective discrepancy is taken as a proportion of the reference memory value for reinforcement. In the panel in the lower left, the solid line discrepancy function for S+ = S is shown decreasing from 1.0 down to 0 and back up for T greater than S, as in the panel above. Now, however, the threshold, B, is a proportion. When the discrepancy is less than this proportion of M^*_{S+}, responding is dictated. Again, the acceptance region is indicated by positive diagonal hatching.

The dashed function shows the subjective discrepancy when S+ = 2S. Now when S+ is doubled, the window size nearly doubles too. The negative diagonally hatched area is increased (almost) proportionally.

In the middle panel in the bottom row, the consequences for responding are shown in real time and the broader acceptance region for 2S is clear. In the right-most panel, the two acceptance regions are nearly equivalent when plotted in relative time. They are not precisely equivalent because of the T_0 intercept in the accumulation of subjective time. In this example, T_0 has been chosen rather large, equal to 1/4 of S, to show that even when this value is large the relative comparison rule produces near superposition.

This analysis of the no-variance case argues for a relative comparison rule to dictate an acceptance region around the remembered positive value. Such a ratio rule is consonant with previous theorizing of ours about how subjective times are discriminated and is consonant with the ubiquitous Weber's Law finding in animal timing in several varied contexts (cf. Gibbon, 1977). In our previous theorizing on this matter we have spoken about this finding as being entailed by the scalar property in the underlying memory variance. In the analysis here the superposition is evident (or not) as a function of the kind of comparison rule employed, quite independently of variance sources. Even when there is no variance, the response rules constrain the appearance of superposition.

The data of Figures 26.1 and 26.2, of course, require variability. The response probability gradients are not square waves. Subjects are uncertain about whether to respond or not when the time values deviate from the "correct" one. The result is the smooth shoulders observed in the data around S+. We consider next a variety of sources for this variability. We will restrict the examination here to different sources acting alone. That is, we will consider what the predicted response would be, using both the absolute and relative rules shown in Figure 26.6, when there is but one source of variance at some position in the chain of Figure 26.3.

III. PACEMAKER VARIANCE

A. POISSON

Returning to the beginning of our system, one source of variance that has received classical attention is variability in the interpulse interval of the pacemaker. From a variety of considerations, variance here ought to follow the Poisson law (Creelman, 1962; McGill, 1967; Green & Luce, 1974; Treisman, 1963). In this construction τ is a random variable with an expected value of $1/\Lambda$ so that the intensity parameter, Λ, describes the mean rate of pulse generation. Interpulse intervals are exponentially distributed in a steadily varying stream. The accumulator is then a Poisson counter. Mechanisms of this kind have been studied by several workers in the time domain. Creelman (1962) proposed such a pacemaker for temporal duration discrimination. The fact that his duration discrimination model required considerable (and rather unwieldy) modification to fit his data presages the convergence in current thinking on a linear scale for time and Weber's Law properties in the appropriate ranges (cf. Getty, 1975; Gibbon & Church, 1981; Allan, 1979). Nevertheless, for our analysis we wish to pursue this possibility to see what modifications of a Poisson neural count generator might be appropriate here.

In the Poisson system, variance increases directly with the mean so that the system is more efficient, i.e., the ratio of standard deviation to mean is lower at long times than at short times, much as with the absolute response rule. In Appendix B, response probability functions are derived for the Poisson case, for both response rules. The results are shown in Figure 26.7. The top row shows the absolute response rule, and the bottom row the relative response rule. In the left column, absolute time for the S+ = S and S+ = 2S conditions are shown. In the right column the same gradients are plotted in relative time (T/S+).

On any trial the pacemaker generating counts for D_T units of time produces a count, n_T, which varies from trial to trial in the Poisson fashion. After many trials then, at any single value of signal duration, T,

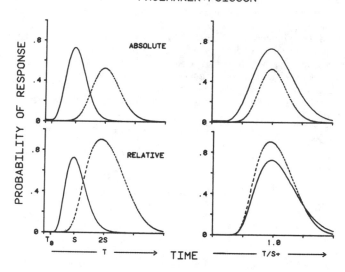

FIG. 26.7. Response probability functions associated with the absolute discrepancy (top row) and relative discrepancy (bottom row) rules when the only source of variance is the Poisson source in the pacemaker.

the accumulator, and consequently the memory system, has registered a distribution of counts with a mean linearly related to T, and a variance directly related to T. The results shown here assume that the mean count is large so that a Gaussian (normal) form is appropriate for the random variable, n_T.

The results of these calculations show that with no variance in the switch, the memory or the comparison stages, Poisson variance in the pacemaker results in nearly symmetric response functions around S+ which differ both in level and spread depending on which kind of rule is used. In the upper left panel of Figure 26.7 when S+ = S the function form is narrower than when S+ = 2S. When these two are plotted in relative time (upper right panel) it is clear that the increase in variability from S+ = S to S+ = 2S is not as great as required by the data in Figure 26.2. Relatively speaking, discrimination of 2S is more efficient. Moreover, when the absolute difference rule is used, what might be thought of as the "accuracy" of the discrimination, indexed by the height of the peak at S+, drops as S+ is increased. Hence, superposition fails in two ways here.

The relative rule shown in the bottom row overcompensates for these violations of superposition. Here, as S+ increases, accuracy increases also (lower left panel), so that when these forms are plotted in relative time (lower right) greater accuracy occurs when S+ = 2S. Thus, the introduction of Poisson variance has produced a violation of the superposition seen under the no-variance case for this rule, in the direction of increasing accuracy -- that is, too much responding at the larger S+.

The efficiency of the Poisson system for the relative rule is about right for superposition in the extreme wings of the gradient. However, there is a sizable discrepancy between the two curves near the peak. These results, we feel, rule out Poisson variance acting alone with either rule.[2]

B. SCALAR

An alternative source of pacemaker variance is a drifting rate. Imagine that the time between pulses, τ, is fixed on any trial, but that from trial to trial, the pulse rate, $\lambda = 1/\tau$, varies normally around a mean, Λ. A more realistic version might allow rapid variation of local pulse rate both within and between trials, but for our present purposes, it is simplest to think of a locally constant rate which varies from trial to trial. In Appendix C we show that assuming a normal form for local rate, λ, we arrive at a system which is linear in real time with the scalar property: variance in the accumulator, and therefore the memory, increases approximately as the square of the mean. This is the property that we have previously used in scalar timing models and it is the keystone of our thinking about Weber's Law. Here, it arises because of Equation 1. The linear property applies to the mean number of counts, and since the rate of accumulation enters as a multiplier, the result is scalar variance.

A graphic representation of the scalar property for the accumulated number of counts, n_T, is shown in Figure 26.8. The description is similar to that in the right panel of Figure 26.4. The solid line function rising from T_0 represents the expected increment in counts associated with the mean pulse rate, Λ. The two dashed rays above and below it correspond to plus and minus one standard deviation in pulse rate.

The lower distribution on the ordinate for S+ = S represents the variability in accumulated counts associated with T = S. Now when S+ is doubled, the distribution of accumulated counts is almost a scale transform of the distribution associated with S. The solid function distribution for T = 2S is generated by our information processing system. The dashed function that is nearly identical to it is the simple scale transform of the S distribution which would result from strict proportionality, an axis multiplier of 2. The small discrepancy reflects the role of T_0, which in this example is S/4. Again, this rather large value of T_0 does not result in substantial differences between the strict scale transform version of scalar timing and the linear transform version required by switch latency.

The scalar variance generated by the varying pacemaker rate is analyzed in Appendix C and the results are shown in Figure 26.9. Again, the top and bottom rows show the absolute and relative response rules and the left and right columns show the response gradients plotted in real

---

[2]In a later paper (Gibbon, Church, & Meck, in press), we treat the interesting question of whether Poisson variance may yet be present but contribute to the behavioral end result only in shorter time ranges.

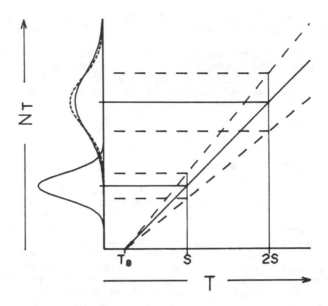

FIG. 26.8. Distributions of accumulator values associated with different S+ durations when the only source of variance is a scalar source consisting of fluctuations in the rate of the pacemaker.

time and relative time. In the upper left, the spread around S+ = 2S is greatly increased and accuracy decreased over that for S+ = S with the absolute discrepancy rule. A comparison with the Poisson system (Figure 26.7) shows the same trend but less extreme. The discrepancies between the two curves remain when plotted in relative time in the upper right panel. In particular, the height difference, reflecting accuracy at S+, is large. Thus, even the scalar property on variance, which does have the right spread, cannot salvage the simpler absolute discrepancy comparison rule because of the decrease in accuracy with increasing S+.

The relative rule in the bottom row shows that both accuracy and spread are properly constrained with the relative comparison rule. In the lower left panel, accuracy is about the same for S and 2S. In the lower right panel, the two functions approximately superpose when plotted in relative time. Note again, that the superposition is not exact due to the presence of T_0, but it is certainly within the range of data variability evident in our generalization gradients (Figure 26.2). Moreover, there is some slight skew in these functions which is also a feature of the data.

The scalar source of variance in the pacemaker, namely its rate, reflects three key properties of the response gradients: accuracy at S+ remains about the same with increasing S+, spread increases with increasing S+ in such a way that superposition is obtained when the gradients are plotted in relative time, and the functions have a slight positive skew.

PACEMAKER : SCALAR

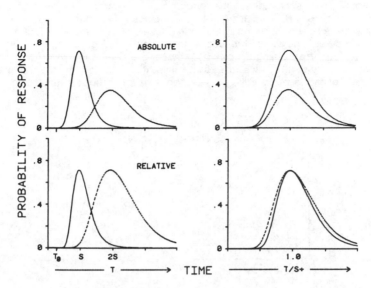

FIG. 26.9. Response probability functions associated with the absolute and relative rules when the only source of variance is the scalar source produced by fluctuations in pacemaker rate.

IV.   SWITCH VARIANCE

A. CONSTANT

Moving through our information processing system in Figure 26.3, we might realistically expect that variability in the appreciation of these stimulus durations would be induced by variation in the latency to close and open switch. Latency variance would induce variance in the counts even if the pacemaker were generating pulses at a constant rate. This situation is analyzed in Appendix D assuming t_1 and t_2 are independent and have the normal form.[3]

Essentially what variance here implies is that the x–intercept of the accumulator function (Figures 26.4 and 26.8) is now a random variable with a mean equal to T_0. This assumption induces a constant increment in

---

[3]The assumption of normality is probably not realistic as latencies tend to be exponentially distributed and so their difference would be distributed as a Laplace variate with a high discontinuous peak. However, these features of a more realistic description do not alter the conclusions we wish to draw which depend upon qualitative discrepancies with the data.

variance at all values of T. The result for the absolute discrepancy comparison rule is response gradients which are simply shifted on the time axis for different S+ values. Since variance is constant in real time, a constant perturbation around S+ is introduced when the absolute response rule is used, as shown in the upper left panel of Figure 26.10. Note that accuracy does not change, so when these gradients are plotted in relative time (upper right) they are the same height. But, as in the no-variance case, the spread around S+ = 2S is too small for superposition in relative time.

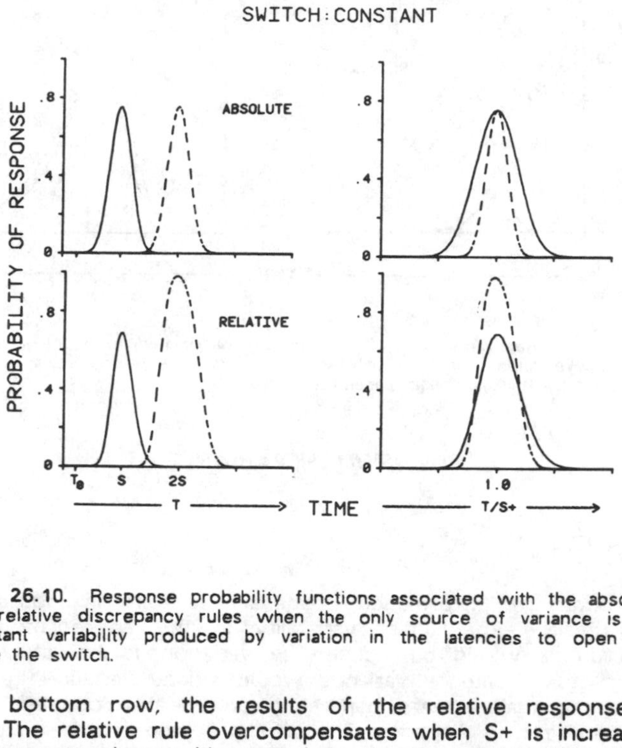

FIG. 26.10. Response probability functions associated with the absolute and relative discrepancy rules when the only source of variance is the constant variability produced by variation in the latencies to open and close the switch.

In the bottom row, the results of the relative response rule are displayed. The relative rule overcompensates when S+ is increased. Here accuracy increases along with spread so that in relative time (lower right) the gradient associated with the longer S+ is considerably too high. This is an extreme form of the violation of superposition noted with the Poisson variance case. Poisson variance does not increase fast enough for superposition, while constant variance does not increase at all.

V. MEMORY VARIANCE

A. SCALAR

The memory system we describe translates accumulator values with at most a multiplicative transform into working memory, and translates working memory values with at most a multiplicative transform into reference memory on reinforced trials. Either of these transformations might involve variability in the multiplier and therefore induce variance in response output. These sources are not analyzed separately here because they are similar to the pacemaker rate variance analysis, since a multiplicative random variable is assumed. Thus multiplicative sources of variance in the translations either to working memory or reference memory or both, result in similar qualitative implications to those discussed previously for pacemaker rate variance. Of course, the variance components would be interpreted differently.

VI. COMPARATOR VARIANCE

A. CONSTANT AND SCALAR

Still another source of variance is realistic for this system. This is variance in the proximity of T to S+ that subjects on a given trial deem "close enough." One can readily imagine fluctuations in this level induced, for example, by momentary motivational fluctuations which are not under experimental control. If the threshold varies so that for any given trial a sample from a distribution of thresholds is used, then the resulting response gradient will show very different properties for the two differing comparison rules. With the absolute response rule, threshold variance produces results similar to the switch variance case. With the relative rule, threshold variance produces results similar to the pacemaker rate variance case. Intuitively, the absolute rule with a varying threshold (Equation 2a) shows the threshold value as an additive component and hence this component may be expected to add a constant variance at all S+ values. In contrast, if we multiply the relative discrepancy rule (Equation 2b) by the norming value, threshold variance enters as a multiplier of memory for S+. It may be thus expected to introduce scalar variance in a similar manner to the pacemaker rate variance case.

In Appendix E the derivation of these properties is detailed and the predictions for response probability derived. The results are displayed in Figure 26.11 for the absolute and relative discrepancy rules in the top and bottom rows respectively. In the top row the constant accuracy property for the absolute rule, and the failure to increase spread with increasing S+, is quite similar to the top row of Figure 26.10 for switch variance. Accuracy at S+ is constant, but superposition for other T values is violated.

The bottom row of Figure 26.11 also shows constant accuracy at S+,

THRESHOLD

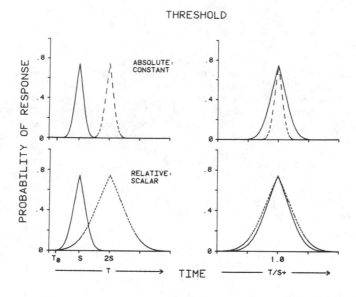

FIG. 26.11. Response probability functions for the absolute and relative response rules produced by variation in the threshold of the comparator. The absolute discrepancy rule results in constant variance similarly to the switch case, while the relative discrepancy rule results in scalar variance similarly to the pacemaker rate case.

but increasing spread, so that in the lower right panel near superposition is obtained, similarly to the lower right panel of Figure 26.9, for scalar pacemaker rate variance.

In both threshold cases, however, the function forms are different. They are sharply peaked and, in fact, discontinuous at T = S+. The discontinuity stems from the possibility in theory that occasional negative threshold values may preclude responding for any discrepancy. If the account is altered to constrain all thresholds positive, then an alternative difficulty arises. Accuracy at T = S+ is always perfect. Unless further modified, these features make it unlikely that the smooth shoulders and less than perfect accuracy we see in Figure 26.2 may be reconciled with threshold variance alone.

VII. SUMMARY

Our conclusion from this analysis is that an information processing model that satisfies the proportionality or superposition finding may be obtained from an appropriate selection of a response rule and an adroit introduction of variance in the structures we describe.   An absolute response rule failed to describe the data in any of the cases examined. A relative response rule can lead to superposition; in fact, it produces superposition even without any variance in the system. Some variance is required, however, to account for the smooth form of the observed response gradients and less than complete accuracy at T = S+.

A. TYPE OF VARIANCE

Introduction of variance into the system with a relative response rule violates the superposition property unless the source itself is scalar. Thus, we believe that there is at least one source of scalar variance. The inclusion of at least one scalar source of variance, of course, does not preclude contributions from other sources of variance. It only precludes them if they dominate the scalar source, and this is a problem yet to be analyzed.

Within the class of multiplicative variance sources which induce the scalar property, we have identified four, of which three are not distinguishable in this setting. These are the pacemaker rate, and working memory or reference memory translation constants. Variation in any or all of these has a similar form.

This form is somewhat different than that for the fourth source, threshold variance with the relative rule. This may be seen by comparing the strict symmetry in the response probability functions for the threshold scalar case with the slight asymmetry for the pacemaker scalar case (Figure 26.9). The asymmetry here arises as a generic property of a variable pacemaker system. Long time values are associated with greater variability in their representation in memory (Figure 26.8). This means that in comparing T with S+, more variability should occur in the representation of T when it is greater than S+ than when it is less than S+, hence more generalization with S+ on the high side, and positive skew.

Threshold variance with the relative discrepancy rule, does not induce asymmetry in the response probability functions. There is no variance associated with T in this case, so comparisons a given distance above and below S+ are equally likely to fall below the varying threshold. This source of variance achieves the scalar property because the threshold multiplies the representation of S+. Note that with the absolute discrepancy rule, the variance introduced is constant since here threshold variance is additive.

B. LOCUS OF VARIANCE

Of these four candidates for a scalar source of variance, threshold variance alone, for the reasons described previously, is unlikely. On the other hand, a scalar source in the pacemaker rate or memory storage is unlikely to be operating alone either. The positive skew in the theoretical response gradients is nearly always greater than the slight positive skew in the data. The data are asymmetrical, but usually not quite asymmetrical enough to be readily fit by an account which uses one of these scalar variance sources acting alone.

A combination of variance in pacemaker rate for memory and in threshold, however, may be satisfactory. The solid curves fit to the data in Figures 26.1 and 26.2 were derived assuming both sources (see Church & Gibbon, 1982 for details). Thus we believe that at least these two sources of variability are present in our timing tasks. Fluctuation in the speed of the clock or in the translation to memory, and fluctuation in the proximity subjects deem sufficient for responding, may both be introduced without violating the requirements of constant accuracy and near superposition in relative time that are the hallmarks of our analysis of temporal generalization.

VIII. APPENDIX

A. GENERAL CASE: $P(R|T)$

Let $S+ = S$. For variability in any structure except the comparator, the absolute decision rule is

$$-B < m_S^* - m_T < B,$$ (A2a)

and the relative decision rule is

$$1 - B < \frac{m_T}{m_S^*} < 1 + B,$$ (A2b)

where m_T, m_S^* are random variables with means given by Equations 1.

$$E(m_T) \equiv M_T = K\Lambda(T - T_0),$$ (A1a)

$$E(m_S^*) \equiv M_S^* = K^*K\Lambda(S - T_0), \qquad T, S > T_0.$$ (A1b)

For cases symmetric around $S+ = S$, $K^* = 1$ and the m^* are a subset of m_T, $m_S^* = m_S$.

The variance associated with the memory quantities differ depending on their source. However, once determined we may define

$$\sigma_T^2 \equiv \text{Var}(m_T),$$

and

$$\sigma_S^2 \equiv \text{Var}(m_S^*) = \text{Var}(m_S)$$

1. *Absolute discrepancy.*   Let $V_i = (-1)^i B$, $i = 1,2$.   Then the absolute rule (A2a) may be written,

$$V_1 < y < V_2,$$ (A3)

where

$$y = m_S^* - m_T.$$

We assume that whatever the source of variance, on each trial subjects sample independently from reference memory and compare that sample, m_S^*, with the current working memory sample, m_T.   Hence for $K^* = 1$,

$$E(y) = M_S^* - M_T = K\Lambda(S-T),$$ (A4)

and

$$Var(y) = \sigma_S^2 + \sigma_T^2.$$ (A5)

Defining

$$Y_i = \frac{V_i - M_S^* + M_T}{[\sigma_S^2 + \sigma_T^2]^{1/2}}, \quad i = 1, 2,$$ (A6a)

for m normal we have, from (A3),

$$P(R|T) = \Phi(Y_2) - \Phi(Y_1),$$ (A6b)

where Φ is the unit normal distribution function.

2. *Relative discrepancy.*   Let $W_i = 1 + (-1)^i B$, $i = 1, 2$. The relative rule (A2b) is then

$$W_1 < \frac{m_T}{m_S^*} < W_2.$$ (A7)

Assuming $m_S^* > 0$, the upper inequality is satisfied when

$$z = m_T - W_2 m_S^* < 0.$$ (A8)

The random variable z has mean and variance,

$$E(z) = M_T - W_2 M_S^*,$$

and

$$Var(z) = \sigma_T^2 + W_2^2 \sigma_S^2.$$

Hence defining

$$z_i = \frac{W_i M_S^* - M_T}{[\sigma_T^2 + W_i^2 \sigma_S^2]^{1/2}}, \quad i = 1, 2$$ (A9a)

analogously to the Absolute Discrepancy case, we have

$$P(R|T) = \Phi(z_2) - \Phi(z_1).$$ (A9b)

Now we require expressions for the standard deviations σ_T, σ_S, which depend on the source in the information processing chain.

B. PACEMAKER: POISSON

Let τ be exponentially distributed with mean $1/\lambda$. If $\lambda = \Lambda$ is constant, then the number of counts, n_T, accumulated in a fixed time has the Poisson distribution with

$$E(n_T) \equiv N_T = Var(n_T).$$ (A10)

But since $m_T = K n_T$, and if $K^* = 1$, we have immediately

$$\sigma_T = K[N_T]^{1/2} = K[\Lambda(T - T_0)]^{1/2}, \quad T > T_0.$$ (A11)

Assuming a large Λ, the m_T are approximately normal, and substituting (A11) in (A6) and (A9) gives the results used in Figure 26.7.

C. PACEMAKER: SCALAR

Let the intensity parameter, λ, reflect a fixed interpulse interval, τ, within a trial, which varies from trial to trial with mean and variance $E(\lambda) = \Lambda$, $\sigma_\lambda^2 = (\gamma\Lambda)^2$. Assuming λ is normal, $m_T = Kn_T = K\lambda D_T$ is too, and

$$\sigma_T = K\gamma\Lambda(T - T_0), \quad T > T_0. \tag{A12}$$

Note that for variance in k only, a similar argument yields

$$\sigma_T = \sigma_k\Lambda(T - T_0).$$

D. SWITCH: CONSTANT

Let the t_i be independent normal variates (T_i, σ_i), so that $t_0 = t_1 - t_2$ is also normal with

$$E(t_0) \equiv T_0 = T_1 - T_2, \tag{A13a}$$

and

$$Var(t_0) \equiv \sigma_0^2 = \sigma_1^2 + \sigma_2^2. \tag{A13b}$$

Since $m_T = K\Lambda(T - t_0)$, m_T is also normal with

$$\sigma_T = K\Lambda\sigma_0, \tag{A14}$$

a value which does not depend on T.

E. COMPARATOR: CONSTANT AND SCALAR

If the only variance in the system lies in the threshold, then response probability is simply a family of rescaled distribution functions of the threshold. Let b be a random variable, normal (B, σ_b).

1. Absolute discrepancy rule. For $T < S$ $(> S)$, $M_T < M_S^*$ $(> M_S^*)$ so this rule requires

$$M_S^* - M_T < b$$

or

$$K\Lambda(S - T) < b, \qquad T < S, \tag{A15a}$$

and

$$M_T - M_S^* < b$$

or

$$K\Lambda(T - S) < b, \qquad T > S. \tag{A15b}$$

Setting

$$A_i = \frac{1}{\sigma_b}[B + (-1)^i K\Lambda(S - T)], \text{ for } i=1,2 \qquad (A16)$$

as $T < S$ or $T > S$ respectively, the rule (A15) becomes $z < A_i$ where z is unit normal. Thus

$$P(R|T) = \Phi(A_i), \text{ for } i = 1,2 \qquad (A17)$$

as $T < S$ or $T > S$, respectively, for the Absolute Rule.

2. Relative discrepancy rule. For this rule a similar analysis gives

$$(-1)^i \left(\frac{T - S}{S - T_0}\right) < b, \text{ for } i = 1,2 \qquad (A18)$$

as $T < S$ or $T > S$, respectively.
Setting

$$R_i = \frac{1}{\sigma_b}[B + (-1)^i \left(\frac{S - T}{S - T_0}\right)], \qquad (A19)$$

the rule (A18) becomes $z < R_i$ and

$$P(R|T) = \Phi(R_i), \qquad i = 1,2,$$

as $T < S$ or $T > S$ respectively. Note that as $T \longrightarrow S$, the criteria A_i, $R_i \longrightarrow B/\sigma_b$, which implies high accuracy only if b is several sigma units above zero.

ACKNOWLEDGMENTS

We are indebted to Steven Fairhurst and Amy Waring for implementation of the computer analysis displayed in the figures. Work was supported by Grants MH37528-01 and BNS-81-19748 to Gibbon and Grants BNS-79-04792 and BNS-82-09834 to Church.

REFERENCES

Allan, L. G. The perception of time. *Perception & Psychophysics*, 1979, *26*, 340-354.

Church, R. M., & Gibbon, J. Temporal generalization. *Journal of Experimental Psychology: Animal Behavior Processes*, 1982, *8*, 165-186.

Creelman, C. D. Human discrimination of auditory duration. *Journal of the Acoustical Society of America*, 1962, *34*, 582-593.

Getty, D. J. Discrimination of short temporal intervals: A comparison of two models. *Perception & Psychophysics*, 1975, *18*, 1-8.

Gibbon, J. Timing and discrimination of shock density in avoidance. *Psychological Review*, 1972, *79*, 68-92.

Gibbon, J. Scalar expectancy theory and Weber's law in animal timing. *Psychological Review*, 1977, *84*, 279–325.

Gibbon, J. On the form and location of the psychometric bisection function for time. *Journal of Mathematical Psychology*, 1981, *24*, 58–87.

Gibbon, J., & Church, R. M. Time left: Linear versus logarithmic subjective time. *Journal of Experimental Psychology: Animal Behavior Processes*, 1981, *7*, 87–108.

Green, D. M., & Luce, R. D. Counting and timing mechanisms in auditory discrimination and reaction time. In D. H. Krantz, R. C. Atkinson, R. D. Luce, & P. Suppes (Eds.), *Contemporary developments in mathematical psychology: Measurement, psychophysics, and neural information processing.* (Vol. 2.) San Francisco, Calif.: Freeman, 1974.

Heinemann, E. G. A memory model for decision processes in pigeons. In M. L. Commons, R. J. Herrnstein, & A. R. Wagner (Eds.), *Quantitative analyses of behavior: Vol. III, Acquisition.* Cambridge, Mass.: Ballinger, 1984. In press. (a)

Heinemann, E. G. The presolution period and the detection of statistical associations. In M. L. Commons, R. J. Herrnstein, & A. R. Wagner (Eds.), *Quantitative analyses of behavior: Vol. IV, Discrimination processes.* Cambridge, Mass.: Ballinger, 1984. In press. (b)

Maricq, A. V., & Church, R. W. The differential effects of haloperidol and methamphetamine of time estimation in the rat. *Psychopharmacology*, 1983, *79*, 10–15.

Maricq, A. V., Roberts, S., & Church, R. W. Methamphetamine and time estimation. *Journal of Experimental Psychology: Animal Behavior Processes*, 1981, *7*, 18–30.

McGill, W. J. Neural counting mechanisms and energy detection in audition. *Journal of Mathematical Psychology*, 1967, *4*, 351–376.

Meck, W. H. Selective adjustment of the speed of internal clock and memory processes. *Journal of Experimental Psychology: Animal Behavior Processes*, 1983, *9*, 171–201. (a)

Meck, W. H., & Church, R. M. A mode control model of counting and timing processes. *Journal of Experimental Psychology: Animal Behavior Processes*, 1983, *9*, 320–334.

Meck, W. H., Church, R. M., & Olton, D. S. Hippocampus, time, and memory. *Behavioral Neuroscience*, in press.

Roberts, S. Isolation of an internal clock. *Journal of Experimental Psychology: Animal Behavior Processes*, 1981, *7*, 242–268.

Roberts, S., & Church, R. M. Control of an internal clock. *Journal of Experimental Psychology: Animal Behavior Processes*, 1978, *4*, 318–337.

Treisman, M. Temporal discrimination and the indifference interval: Implications for a model of the "Internal Clock". *Psychological Monographs*, 1963, *77*, 13.

Werner, G., & Mountcastle, V. B. The variabiity of central neural activity in a sensory system and its implications for the central reflection of sensory events. *Journal of Neurophysiology*, 1963, *26*, 958–974.

VII.   EVOLUTION AND DEVELOPMENT

27 THE ECOLOGY AND BRAIN OF TWO-HANDED BIPEDALISM: AN ANALYTIC, COGNITIVE, AND EVOLUTIONARY ASSESSMENT

Alexander Marshack
Harvard University

I. THE ADAPTIVE "POTENTIAL VARIABLE CAPACITY"

The papers at this conference on animal cognition explore selected, measurable or observable aspects of species memory, problem–solving capacity and neurological function. To that extent they are inquiries into the range and variety of species potential capacity. The capacities and behaviors being studied and measured, however, do not represent the range of adaptive capacities and behaviors used in the wild or that range of potential capacities available to a species, of which only a part can be used in any temporal context. In biological evolution, natural selection probably occurs as much for the range of "potential variable capacity" as for the structures of morphology and the patterned structures of species behavior.

Cognition itself represents a particular aspect of species adaptive capacity, that aspect which is not dependent on genetically coded or programmed behaviors but on a variable, if constrained, neurological response to the phenomena, patterns and structures of the real world. Potentially variable adaptive capacities and systems exist, of course, at the simplest evolutionary levels but, with development of the chordates, vertebrates, mammals, and primates, these become increasingly complex, variable and specialized. Given sufficient functional data, a species can probably be described and defined as much by the nature and range of its potentially variable capacity and behavior as by the more traditional categories of morphology and observed patterns of behavior in the wild. The problem is crucial to the following discussion of hominization as an aspect of natural selection for an increase in the specialized "potentially variable capacity" of an early pongid.

II. POTENTIAL VARIABLE CAPACITY IN HOMINIZATION

If the process of hominization, beginning some five or more million years ago, represented the branching of a potentially new species and the incipient development of new modes or levels of adaptive behavior, then the range of cognitive capacities involved must have been derived and

491

selected from a range extant on the pongid line. That range, however, is not documented in anthropological observations of primate behavior in the wild or in laboratory studies of certain selected aspects of primate capacity. Laboratory and field studies always represent those aspects of a problem which are of interest to a group of specialists historically, that is, at a particular time. These change decade by decade.

In contemporary discussions of hominization, two aspects of the hominid adaptation have often been left out, the nature and range of the evolving hominid capacity and neurology and the nature of the *cultural* contents and contexts that were probably always needed for hominid adaptations and for the initiation of natural selection for the cognitive capacities involved. We now know that adaptive cultural traditions exist among regional troops of chimpanzees, baboons, and macaques. It is likely that hominization may have begun as an aspect of cultural adaptation to new ecological conditions and of natural selection for the cognitive and neurological capacities involved, long before there was measurable morphological change.

There is one body of archaeological evidence that deals more or less directly with the full spectrum of problems relating to the evolution of the hominid cognitive capacity and complexity. It involves species-specific capacities that are qualitatively somewhat different from the many discussed at the present conference.

The earliest available evidence of a complex cultural symbolic tradition appears in the archaeological record, c. 30,000 B.C., with *Homo sapiens sapiens*, or "Cro Magnon" as he is popularly called, during the period of the European Upper Paleolithic or last Ice Age. When I began research into the cognitive, symbolic contents of that early, widespread human culture, the archaeological consensus was that Cro-Magnon was a superior, more variable hunter than prior Neanderthal man *because* he had a more complex and efficient tool technology and set of subsistence strategies. It was largely assumed that Cro-Magnon had cognitive, manipulative, symbolic and linguistic capacities beyond that possible for Neanderthal man. It was also assumed that Cro-Magnon image, symbol and art were "primitive," the mere expression of "magic" and of the aesthetic and spiritual aspect of man. Art and symbol were "superstructure." Economic parameters were assumed to be the fundamental, determining factors of human adaptation. The stone tools, the larger animals hunted, the changing hominid skeleton and the increase in brain volume were considered to be the hard, material evidence for a linear evolution of human capacity and economic adaptation. This materialist-economic approach was accepted in the West as much as in avowedly "Marxist" countries.

Art and symbol, like the evolution of language, were assumed to have proceeded from the adaptive subsistence needs of hominid culture. It was therefore surprising that my research with early human symbol systems indicated that the internal cognitive complexities, the implied neurological complexity, and the implied cultural complexities of these Ice Age symbolic traditions were many orders higher than is implied in the tool technologies and the remains of meals with which they were contemporaneous. These symbol systems provided a far different level of data and set of references for discussing the nature and evolution of the hominid cognitive capacity (Marshack, 1972-1981) and for making comparisons with pongid cognitive and symbolic capacities.

III.   THE EARLY UPPER PALEOLITHIC DATA

From an early Cro-Magnon level (the Aurignacian-Perigordian Period) c. 30,000 - 28,000 B.C., a carved and engraved bone plaque was excavated in the early 20th century (Didon, 1911) at the rock shelter of Blanchard in the Dordogne region of southwest France.   It was referred to as a "decorated" polisher, presumably used to soften skins.   Microscopic analysis revealed that it was not a polisher and was not decorated (Figure 27.1).

The small plaque (10 cm), just large enough to be held in the hand, had high polish at the rear where it sat against the palm, but was broken at the front from persistent pressure against hard objects.   The artifact was, in fact, a pressure-flaker for the fine retouch of stone tools.   The polish and frontal breakage suggested that it had been used for a considerable period, of perhaps some months.   The engraving on the surface, within an area the size of a large wrist watch (4.4 cm), was shown to be a slowly accumulated series of 69 marks broken into 24 discrete sets of marks of from one to seven or eight units each.   Each set was made by a different tool or point and with a different type of stroke (Figure 27.2, 27.3) (Marshack, 1970b, 1972a, b).

The accumulation began in the center and proceeded in a "serpentine" manner, with two turns at left and two at right.   Clearly, the accumulation and the turning were intentional and had occurred over a period of time. This seemed to be some form of *notation*, made fully 20,000 years before the invention of a proto-writing system of record-keeping and accounting in the neolithic of the Middle East (Schmandt-Besserat, 1977, 1981) c. 8,000   B.C., and even longer before the true writing systems of c.   4,000 B.C. Cognitively, on a single artifact, we had two types of "tools," one practical and one conceptual or ideational, both of which functioned at different levels of neurological complexity, and in different ways.   Both, however, were part of a complex tool technology and required an equally complex understanding and use of "reality," that is, of diverse processes, materials and phenomena.

The published analyses have suggested that the Blanchard engraving represented a *non-arithmetic*, observational lunar notation covering a period of 2-1/4 months, with the turns representing the periods of "turning" in the lunar phases: the full moon periods at left, the periods of crescent moon and invisibility at right, and the half moons in mid-line. From the point of view of early cognitive and cultural complexity, the document was perhaps as important as the discovery of Ice Age art in the 19th century.   As with the early art, archaeologists were, for a long time, unwilling to accept the presence of notation at so early a date and did not know what to make of a symbolic and cultural tradition that did not fit into the categories of tool typology, subsistence strategy or ecological context.   It was only as the anthropological and ethnographic evidence began to accumulate in recent years for wide-spread traditions of non-arithmetical, astronomical observation and usage among historical pre-

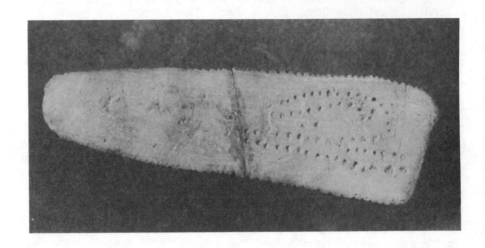

FIG. 27.1. Bone pressure flaker from the early Aurignacian period of
the European Upper Paleolithic. c. 28,000 B.C., engraved with a sequential,
cumulative notation. From the Abri Blanchard, France.

literate peoples and as evidence for astronomical observation in prehistory
began to be documented by the new sub-discipline called
"archaeoastronomy" that the Ice Age notations began to be accepted as a
tradition that may have been used as a means of ecological and economic-
subsistence adaptation. The research and concept began to find its way
into college texts (Thomas, 1979) and into archaeological theoretical
discussion (Boehm, 1978; Edwards, 1978; Gamble, 1980; Raskin, 1981;
Binford, 1982).

The Blanchard notation is perhaps the most complex single cultural
artifact to come from this early period. In it we are dealing with an

FIG. 27.2. Detail of the engraved portion of the Blanchard plaque, indicating the sets of marks made by different tools and in different styles of strokes.

aspect of human cross-modal, associational and sequential capacity that Luria, the Soviet neuropsychologist, had earlier in this century termed the "higher mental processes" of the evolved neocortex, that aspect of human capacity which, like language or speech and more than tool use, differentiates man from the great apes.

It is probable that a cognitive and cultural system as complex as this notation, involving non-linguistic, two-handed manipulo-spatial acuity, visual observation of periodic phenomena in the sky and visual modes of abstraction and symbolizing, also required a use of spoken language to explain and maintain the tradition. But even with language and merely at

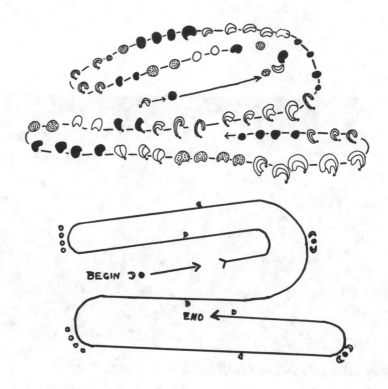

FIG. 27.3. Schematic representation of the serpentine mode of accumulating the sequential notation within a small area. The "turns" fall at phase points in the presumed observational lunar notation.

the level of visual/manipulo-spatial competence, association and abstraction, the notation documents a cognitive and cultural complexity far greater than for any specialized tool or tool kit of the period. If one adds the implication of possible language, it becomes a cultural artifact of a different order than the bone and stone tools. The manufacture and use of tools does not require a complex use of language and is, in fact, learned largely without a use of language.

It can be assumed that the constellation of capacities and skills evident in the notation could not have appeared and evolved "suddenly," that is, at the moment of their appearance in the archaeological record. The neurological complexity would seem to indicate that it is the end-product of hominid evolution and long cultural usage. I have indicated that certain aspects of the two-handed morphology and neurology, evidenced in the notation are derived from a pongid incipience (Marshack, 1979).

By 30,000 B.C. these are qualitatively and evolutionarily of a different order. In fact, nothing quite of this complexity and phenomenological reference is known from any other species. It is a *cognitive* artifact that cannot be derived from any of the genetically-programmed mammalian behaviors discussed in present day sociobiology. Yet it appears in the archaeological record full-blown, not only suggesting a long evolution of the competencies involved, but a long prior cultural use of relevant seasonal and lunar observation before its artifactual appearance. This would put the cultural use back at least into the prior Mousterian period of Neanderthal man. The fact is, the presence of the notation in the archaeological record is an artifact, largely, of a particular technology. It was the Cro-Magnon cultures that began the profuse use of bone and antler and developed the specialized stone tools to make them. The presence of the notation on a bone pressure flaker, a technical artifact of that new tool technology, may simply be an accident of that development. I therefore backtrack to an earlier, Mousterian example of the manipulo-spatial, two-handed competence and the Neanderthal use of symbol.

From a Mousterian level at Tata, Hungary (Marshack, 1979), c. 45,000 B.C., comes a small ovaloid plaque carved from a large compound mammoth molar. It is almost the same size and shape as the Blanchard plaque. It has traditionally been assumed in archaeology that bone work of this quality could not exist in the Mousterian and did not begin until the Cro-Magnon period. The supposition was based on the lack of a developed bone industry in the record and from a belief that the Neanderthals had a less developed cognitive capacity. Microscopic analysis of the plaque revealed that it had been carefully carved and shaped from difficult material. First a single section or lamelle had been separated from a compound molar. This had then been shaped, the soft rear had been beveled and, according to the hand polish around the edges, it had then been used for a considerable period. During that period it had also been painted with red ochre. The plaque, then, was a manufactured, non-utilitarian symbolic artifact that had had long use. It had been made by a form of modern man for which a manipulo-spatial capacity and acuity of this order, the presence of manufactured symbol, and even the presence of language, was being discussed as being "impossible."

I have recently published evidence for the use of red ochre in the Acheulian period within Central Europe, c. 250,000 B.C., made by a late member of *Homo erectus* (Marshack, 1981). We have evidence, therefore, of the non-utilitarian use of a symbolic technology and the manufacture of symbolic artifacts in the Neanderthal period and perhaps as far back as the Acheulian. If the analytic data are valid, the Blanchard plaque was no "sudden" invention but was instead an instance of the late use of slowly evolved cognitive, manipulo-spatial capacities and skills, and the elaboration of a cultural tradition involving observation of the passage of time when a precise determination of lunar periodicity had assumed adaptive significance. If so, then this "time-factoring" aspect of the Cro-Magnon cultures, in a sharply seasonal ecology occurring within the "temperate zone" latitude, was probably as important as the tool kits themselves. The tool kits, incidentally, may also have varied seasonally and contextually.

IV.   END OF THE ICE AGE: THE LATE DATA

In 1972, a fragment of bone was found at the Grotte du Tai (Drome), in east central France (Brochier & Brochier, 1973; Marshack, 1973a). The plaque , of roughly the same size and shape as the Blanchard bone (8.6 cm), came from a level at the very end of the Ice Age, the terminal Magdalenian, c. 9,000 – 8,500 B.C., where it lay in contact with the following post–Wurm Azilian culture (Figure 27.4, 27.5). It was, therefore, some 18-20,000 years later than the Blanchard notation and came from a period when the climate was warming and the great ice had begun to disappear.   This was the period when incipient farming was beginning in the Middle East and a thousand or so years before the first economic record-keeping appeared (Schmandt-Besseratt, 1977, 1978, 1981).

Analysis of the plaque proceeds at a number of levels, each representing a different aspect of problem-solving and abstraction.   To the eye it seems that there are nine or ten horizontal lines or rows (A to I). At the top is a form that looks like a broken, right-angle meander. Microscopic analysis revealed an entirely different type of complexity.   It documented that the engraver had engaged in *ad hoc* forms of problem-solving within a well-known notational tradition.   Uncovering these strategies helped reveal the direction of engraving and the sequential system involved.

The plaque had apparently been a tool, perhaps a retoucher like the Blanchard plaque, but it had broken in use and had then been cut at far right and snapped to make a small portable surface or slate to carry a cumulative engraving.   At the far right, near the break, are two right-angle descents among four of the horizontal rows (E-F and G-H). Each of these descending sets is connected at the bottom by a horizontal bar.   It seems that the engraver ran into that difficulty indicated in a popular American sign that hangs in offices, admonishing one to "THINK AHE$_{AD}$" but where the spacing of the letters had not been adequately planned.   Apparently, the engraver had not planned his spacing for the number of unit marks he required on the horizontal row (E).   He needed more space to complete an intended or expected sequence.

The first descent comes where the breakage had created an absence of space.   Having added a length of line and having marked this with units, the engraver went upward on the next line, leading to row F. The engraver then found he had created the same problem for lines G-H, which now also had to dip and turn up.   The descending sequence of horizontal lines and their marks, were therefore clearly made in a "boustrephedon" mode. The whole was a continuous descending sequence that formed a "serpentine" notation.

The microscope then revealed that the horizontal rows were broken or sub-divided into short sub-sections, one appended to the next, though it would have been far easier to incise a single line across the small surface. Each of these sub-sections was engraved with its own set of marks, made with a different tool and rhythm of marking, suggesting that each set was intended to be encompassed by that sub-section acting as a "containing" line.   On the Blanchard plaque the "containing" line was never engraved; it was conceptual, and the sets of marks were shorter.   The

FIG. 27.4. Bone plaque from the terminal Magdalenian period of the European Upper Paleolithic, c. 9,000 B.C., engraved with a sequential, boustrephedon notation. From the Grotte du Tai, France.

turns on the Grotte du Tai plaque came at the end of far longer periods. The cognitive strategies and the notational mode, including the serpentine sequencing of a continuous notation within a limited space, were the same, however, as on the Blanchard plaque of 18–20,000 years earlier.

The full analysis of the Grotte du Tai notation, including the notation on the reverse face, has been prepared for publication (Marshack, in press–a). It indicates that each sub–section contains the marks, roughly, for one observational lunar month, that each horizontal made up of appended sub–sections contains approximately six months.   The turnings at right apparently came in the month of the summer solstice and at left in the

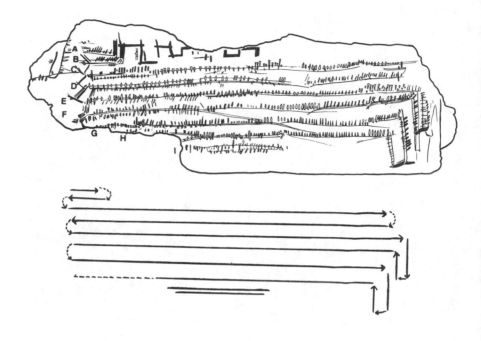

FIG. 27.5. Schematic rendition of all the engraved marks on the Grotte du Tai plaque, indicating the strategy of extending the horizontal containing line to take additional sets of marks. The continuous notation, based on a non-arithmetic lunar/solar observation, is made in a serpentine or boustrephedon mode.

month of the winter solstice. The "year" apparently began at left with the winter solstice and "turned" at the summer solstice, when the sun had reached its farthest point north on the horizon and then began to rise and set each day further to the south, and when the days now began to grow shorter. The plaque noted an observational 3-1/2 years of lunar-solar observation on this face, as opposed to the earlier and simpler 2-1/4 lunar months on the Blanchard notation.

In almost every respect -- the scale of engraving and the hand/eye acuity, the observation of periodic time and phenomena in the sky, the cultural use of symbolized time, and the problem-solving strategies involved in sequencing a continuous, periodic notation in a small space -- the two notations are similar. They represent the same evolved cognitive capacities and the same cultural tradition, though the Grotte du Tai notation is the more evolved. What is significant is that the two-handed, visuo-manipulo-spatial competence and the sequencing strategies are the same as would be used later, with linguistic contents, in the early writing systems. The "boustrephedon" mode, in fact, was used in many of the early writing systems that appeared in different parts of the world.   The lateralized, neurological, vision-oriented sequencing and abstracting capacities and processes evidenced in these early notations would be used later in other symbol systems: arithmetic, astronomy and geometry.   They are similar to the neurological sequencing and abstracting capacities found in human language though speech is performed in the auditory/vocal mode.

The right-angle meander at the top of the plaque seemingly represents a near-iconic abstraction derived from the boustrephedon mode, an image and symbol of the "turning" of periodic time visualized linearally.   It seems here to serve as an introductory sign suggesting that "we have just completed a sequence of turning years and now begin another." Throughout the Upper Paleolithic the closely related images of the meander, boustrephedon, spiral and serpentine appear as near-iconic images of periodic time (Marshack, in press-a). They appear also in the following Mesolithic period (Marshack, 1983, 1984a) and in the later farming cultures and civilizations.   The derivation, uses, development and variation of these near-iconic signs are discussed and documented in the volume in press (1984a).

V.   TWO-HANDED HOMINID CAPACITY

The Ice Age notations (Marshack, 1970a, 1972a, b, c) represent one set of cultural end-products in the unique biological, neurological and morphological evolution of *Homo sapiens*, involving the use of the asymmetric, lateralized eye/hand system to deal with certain aspects of the phenomenological real world and the interpersonal, relational, social reality. In this context, it is not the traditional stone tool that was important, since the tool was merely an aspect and product of this more generalized developing competence.   Language itself probably evolved as part of the constellation of neurological, cognitive and cultural processes involved in and sustained by the developing two-handed competence.   I can perhaps clarify this point by discussing the Blanchard plaque as an aspect of the stone tool culture and tradition.

Tool-making by hammering crude flakes from a pebble with another stone goes back at least two million years to *Homo habilis*.   These pebble tools were probably cultural end-products of still earlier hominid tool-making traditions.   A more sophisticated tradition of tool-making by use of a hammer stone began with *Homo erectus*, c. 1,500,000   B.C. The use of a "soft" hammer of bone for the flaking and retouching of stone tools developed in the late Acheulian, c.   200,000 B.C.; this was followed by a Levallois tradition of striking flakes from a previously prepared core, then touching up the resulting flakes, c.   100,000 B.C. Finally, there came that apparent explosion of bone and stone technology in the Cro-Magnon period, c.   40-35,000 B.C., with the fine retouch of specialized stone tools by prepared bones such as the Blanchard plaque.   In the early archaeological record only evidence of the developing tool technologies remains.

The Blanchard plaque as a bone retoucher, therefore, represents the end-product of one tradition – some 2-1/2 million years of developing tool technology, concomitant with the development of a two-handed manipulative competence and a two-handed understanding of materials. The notation, far more complex neurologically and culturally, probably represents an equally long tradition involving an increasing hunter-gatherer observation, knowledge and use of seasonal periodicity and variability. This is a totally different and "higher" aspect of hominid cognitive capacity, yet one that would have been intimately related to the many functions of a two-handed adaptive capacity operating in a variable, mosaic ecology. On the single plaque we have aspects of what might be described as the operational and conceptual bases of the successful hominid adaptation. Neither was simple. As part of the stone tool tradition, the plaque involved a knowledge of which bone parts from which animals were appropriate, and what types and quality of stone could best be worked by such a bone retoucher. The tool tradition, in addition, represents the culmination of at least a four to five million year development of two-handed manipulative acuity and asymmetric, lateralized competence.

In two-handed action, the primary hand performs the specifying action; the secondary hand holds, turns and orients the stone or bone being worked. In such action there is knowledge not only of materials but, increasingly, of different classes of tools to be made for different purposes. At the simplest visual, manipulo-spatial level, such tool-making and use is performed at the Piagetian level of operational competence found in the human adolescent. Though two-dimensional writing and drawing are achieved earlier, the manipulative capacity required for manufacturing three-dimensional utilitarian artifacts, involving a knowledge both of materials and cultural usage, is not achieved until adolescence. Even skill in the making and use of such tools as the burin, bow and adze are not achieved until adolescence. The making of tools by the adolescent, in fact, becomes part of the play preparation for adulthood. However, it is that other and "non-material" component of the notational plaques, the cultural and conceptual skills involved, that is fully adult in the examples we have examined and which function at levels far beyond those possible at Piaget's childhood stages of cognitive development. In earlier tool-based theories of hominization, it was tool-making and use that was considered the important subsistence achievement of hominization. As described not long ago these skills were "the result of selective pressures which favored a more effective use of hands and tools to better extract food from the environment. . . The command of a better tool kit was the major factor in the spread and successful ecological adaptation. . .during the mid and late parts of the lower Pleistocene. . ." (Bordaz, 1970).

More recent archaeological theory has begun to discuss technology as an aspect of economic-social context: "I would favor models involving concurrent development with mutual reinforcement of adaptive advantage by matching changes in all components, and from this stance I would argue that hunting, food-sharing, division of labor, pair bonding and operation from a home base or camp, form a functional complex, the components of which are more likely to have developed in concert than in succession. It is easy to see that tools, language and social cooperation would fit into the functional complex. . . (Isaac, 1972)"

Though not explicitly stated, this social approach represents an essentially "economic" model involving the more efficient use of the ecology and resources of the hominids (Peters, 1979; Harding & Teleki, 1981). It leaves out one of the most important developments in hominization, the capacity to function at new levels of adaptation in a variable "time-factored," seasonal reality. There are other problems with such models. The behavioral, social approach is not based on available archaeological evidence but on interdisciplinary reference and comparison to field studies of primate behavior and the behavior of historical hunting-gathering peoples (Chevalier-Skolnikoff & Poirer, 1977; Harding & Teleki, 1981). In such models, sets of behavioral strategies are hypothesized to meet sets of economic, subsistence needs and pressures. Adaptation is assumed to occur by essentially mammalian behaviors, including changing strategies of reproduction and parenting or technologically by cultural "inventions" and discoveries. Left out of such models is the fact that during hominization, a new species-specific reality or realm was being created and it was being dealt with, in large part, by a developing two-handed, vision-oriented neurology and the ability to use this variable capacity differentially in space and time.

During hominization the increasing potential for open and variable behaviors was probably the crucial and significant adaptive mode being developed neurologically and morphologically to interface with a seasonally variable mosaic ecology. The hominids had taken an evolutionary path, incipient among the apes, that was not dependent on patterned ethological mammalian or primate behaviors and, one might add, not dependent on tools. The changing neurology, probably before there was significant change in brain size, had begun mediating a rather complex biological constellation: bipedalism, changing subsistence strategies and intraspecific behaviors, and the uses of smaller teeth. One cannot go developmentally from current ethological, behavioral and social models and theories of hominization to the contents, contexts, variability and complexity of human cultures. One can, however, do so if one assumes a long, slow evolution and natural selection for a two-handed, vision-oriented neurology and cognitive capacity with a developing competence for functioning in an increasingly variable phenomenological and social reality. The Blanchard and Grotte du Tai plaques interface with both of these levels of hominid function.

The Blanchard plaque, as a retoucher, represents that hominid skill involved in day-to-day problem-solving and tool-mediated economic, subsistence activity; but as notation it represents a "higher" conceptual, abstracting skill, involving an understanding of the temporal periods and processes which structured the sequence of social relations and subsistence strategies.

The tool-making (McGrew, 1977; Kitahara-Frisch, 1980) and the conceptual skills (Boehm, 1978; Edwards, 1978) derive, in a Darwinian sense, from earlier pongid capacity and incipience. It is commonly acknowledged that specialized seasonal behaviors and activities, including changing cognitive, problem-solving activities and interpersonal behaviors and strategies, occur in most species. At the pongid level, clearly seasonal tool-making and tool-use occurs as a regional "cultural" activity at Gombe National Park, Tanzania, when chimpanzees fish for termites. It is probable that such seasonal activity in the chimpanzee is initiated by sensory cues emanating from the termite mound. On the Blanchard plaque we are neurologically, cognitively, and culturally far beyond such direct sensory cueing and pongid incipience.

A paper at the present conference described the capacity for species-specific "cognitive-mapping" and memory, evolved as a seasonal, ecological adaptation by Clark's nutcracker (Balda & Turek, Chapter 28, this volume, and Shettleworth, 1983). The nutcracker has a memory for locating hundreds of hidden and dispersed caches of pinyon nut. This capacity is based on an avian neurology, avian flight and vision and a specialized beak. The same seasonal, ecological problem was solved by man in the same region, in the American southwest and Great Basin, by a more elaborate set of human strategies.

When the climate changed and dried at the end of the last Ice Age, the lakes and rivers dried, the lowland forests thinned and animals and vegetation became scarce. The human hunter-gatherers living in the Great Basin adapted by scheduling sequences of annual, seasonal movements and variable sets of strategies within their territory, each requiring different types of two-handed skills and tools and different sets of interpersonal relations and forms of cooperation. They developed a seasonal round that moved periodically to the area of best resources. In the late summer or fall, as the winter approached, they went into the hills to collect huge quantities of the same pinyon nuts cached by the nutcracker. These nuts were the same winter food for both species. Each had adapted to the same seasonal ecology using different species-specific modes, morphologies and forms of "cognitive mapping" and memory. The nutcracker had devised a relatively complex but limited and genetically constrained adaptive strategy, one that could conceivably be shifted to another form of nut or seed if the ecology again changed, while man was able to construct a series of variable strategies, by use of an extraordinarily generalized set of hands and two-handed cognitive skills. Man could shift his subsistence strategies radically within a few years without requiring genetic, biologic changes by natural selection. These are the capacities evidenced in the Blanchard plaque and notation. It represents a totally different order of seasonal, ecological adaptation than is found in the genetically coded and sensorily-cued behaviors of other species; yet, in a Darwinian sense, it serves comparable adaptive functions.

In the European Ice Age, the concept of sequential, periodic time, determined by observation of the sky and other processes in nature, would have encompassed a series of relevant activities: the preparation of animal skins for winter, a preparation accomplished biologically by other species; the preparation of oils and fats for lamps and tanning; the accumulation of wood for fires; or bone and antler for carving; of nuts and perhaps smoked and dried foods; of mineral colors and paints for winter ritual and ceremonies; and the construction of windbreaks for shelter. These periodic cultural activities, involving the two-handed capacity, were made possible as well by the *concept* of repetitive, functionally specialized, periodic sequences of time. This is an aspect of normal "higher" human cognition and capacity that is not usually studied in psychology and neuropsychology, though failure of that capacity is noted in neurological insult and injury. It is an aspect of human capacity that

A. Luria (1966) had discussed as part of the "integration of programs of behavior," a process mediated by the frontal lobes and motivated and marked by the limbic system, a mediation that is evolutionarily and hierarchically above those capacities and productive skills encoded and mediated in the more rearward temporal and parietal areas of the neocortex.

At every level of analysis, then, the plaques are beyond the great ape capacity for tool-making or sensory and cognitive response to the seasons. From an evolutionary perspective, we face a problem. The temporal distance from a pongid incipience to the Ice Age notations is only some five or so million years, according to chromosomal and molecular genetic comparison. The neurological and cognitive differences, however, are far greater than is suggested by the small chromosomal difference; also human interpersonal, behavioral variability, capacity and complexity is greater than among the pongids. One must, therefore, discuss the nature and evolution of that difference in terms of the mediating role and capacity of the evolving brain and neurology. These are involved with the processes of development, maturation, experience and cognition, among the "potentially variable" aspects of biological function. In this paper I am discussing a single aspect of that neurological, cognitive difference.

VI. EVOLUTION OF HOMINID POTENTIAL VARIABLE CAPACITY

The constellation of capacities under discussion suggests that what was evolving on the hominid line was an "information-processing" capacity of extraordinary complexity, with language and speech probably evolving at least in part to mark the referential complexity. The nature of both the human competence and the pongid incipience suggests that the process probably began at the time of hominid divergence as part of the cognitive and manipulative adaptation to an inherently variable, seasonal, mosaic ecology, and what was in essence a more complex physical reality. Diverse hominid groups that later began to specialize in certain ecologically stable vegetational niches, such as the australopithecine *africanus, bosei* and *robustus*, would eventually become separated from those hominids which continued to become more competent two-handed problem-solvers in wider ranges of a variable ecology. These hominids established the lines of *Homo habilis, erectus* and *sapiens*. Hominid communication as it evolved would probably have referred more to the contexts and contents of the relevant and variable processual and material complexities than to such presumed problems as the manufacture of tools, cooperative hunting, food-sharing, pointing, naming or affective communication, none of which require any great semantic or syntactical complexity and all of which occur at some level among the great apes.

In contemporary models of hominization (Johanson & White, 1979; Lovejoy, 1981) the soft evolution of neurological adaptation to species-specific phenomenological realms and conditions, material and time-factored, is inadequately considered. For one, these processes do not appear in the archaeological record. How, for instance, can one posit the cognitive mapping capacity of Clark's nutcracker from the morphology of the beak? Or the human adaptation in the Great Basin from the size of the brain or the tool typology? The problem is largely avoided in paleontology and sociobiology by assuming, in large measure, the presence of structured, "morphological" behaviors, that is, coded programs of adaptive behaviors essentially closed sets of mammalian behaviors to accompany closed sets of skeletal, morphological changes.

What probably did occur in hominization was, in fact, the opposite of what the paleontological model assumes. The major anatomical changes occurring during the process of achieving full bipedalism (in posture, hip, leg and foot) were accompanied by a *lessening* of the manipulative, grasping, orienting capacity of the foot, though the foot is a major grasping organ of the primate and pongid. This was accompanied by a lessening of the generalized functional capacity *and* neurological complexity of the foot. By contrast, the hand, which did *not* undergo as large physical change, except for increasing the manipulability of the thumb, increased its general manipulability, sensitivity and acuity of fine action, its *two*-handed associational, asymmetric capacity and neurological, brain-mediated complexity. The process established something new in biological adaptation, an increasingly specialized but "open-ended," innovative adaptive realm. I should note that paws, beaks and wings all become part of the variable repertoires of other species, often in secondary roles not devoted to locomotion or feeding. None of such acquired secondary roles, however, has ever achieved the functional complexity and potential variability that is found in the asymmetrically lateralized hands of the hominids.

The crucial selective changes probably began quite early in hominization, *before* the morphological changes of a developing bipedalism would have become apparent, and *before* many of the intraspecific sexual and reproductive changes theorized by Lovejoy, for instance, would have been genetically installed. These selective changes would have involved the adaptive, probably cultural, use of and natural selection for an increase in the already extant pongid "potential variable capacity," making it possible for an incipient hominid to function more successfully in a complex seasonal, mosaic ecology. In such a process, we would have selection for an increase in neurologically-based "information-processing" capacity. This should not surprise us; it maintains one line and direction in general pongid evolution. The outstanding functional characteristic of the great apes is not morphology or brain size, but brain capacity, the basis of "potential variable capacity."

In the first stages of "hominization," natural selection would have occurred for an increase in particular neurological capacities, selected from within the range of variation existing in a regional genetic pool. Such selection would probably involve the morphology in a feedback relationship, since an individual more able to use a bipedal stance and gait would probably also learn to use his hands and brain with greater efficiency. Selection for one would work in tandem for selection of both. But at either level, the process would involve a selection that was in essence "determined" and constrained by the nature of the phenomenological, real-world, material and ecological reality, *and* the evolutionary history of the species. The "soft," information-processing changes involved in these initial stages of selection would not show up morphologically until much later, since in evolution morphology as often follows successful behavioral adaptation as it adds to it.

As noted, a new level of "information-processing" complexity was being instituted, essentially a two-handed, vision-oriented, lateralized adaptation to an "open-ended" ecology and material, phenomenological reality. Unless this process is understood, hominization and the complex end-products seen in *sapiens* cultures cannot be understood. We are not a product of bipedalism, a larger birth canal, a bigger brain, or of altered mammalian sexual and parenting behaviors, but the product of a qualitatively and quantitatively different neurological adaptive mode. Because of the neurological complexity and concomitant morphological changes in the feedback process, selective evolution probably occurred slowly at first rather than by saltational leaps or major random mutations. It would have occurred as selection for an increase in the pongid "potential variable capacity," occurring from within that pool of extant genetic variability which touched on these parameters.

The presence of such pongid incipience is partially evidenced by the range of handed capacities and functions that have been observed and tested for among the great apes, the structural asymmetry already present in the pongid brain and the tendency for preferential handedness in complex two-handed problem-solving among the great apes, a tendency that increases as the sequence of handed actions becomes more complex. Natural selection and ecological pressure for more efficient two-handed behavior, particularly in times of pressure or crisis, would have tended to favor a neurology making such behavior possible. The great apes indicate the nature of that incipience.

The chimpanzee makes and uses primitive tools in simple acts of problem-solving: termite fishing, sponging water from a tree notch, wiping its bottom with a leaf, hurling objects. These are essentially *one*-handed actions, with the secondary hand used, as in the manufacture of a "fishing" tool, to hold or grasp the twig while the primary hand performs the intended action of stripping the leaves. It is not possible practically, let alone neurologically, for both hands to simultaneously hold and strip the leaves. In large measure, it is the dichotic nature of the material, physical problem that creates the impossibility. Leaf-stripping by the chimpanzee is, therefore, already asymmetrically two-handed. Termite "fishing" itself is essentially a *one*-handed action, much as the use of the beak by Darwin's finches to hold a twig for grub "fishing" is a one organ probing action. Among marsupials and mammals (shrews, raccoons, rodents, squirrels, monkeys, apes) simple acts of holding are symmetrically two-handed, with the mouth performing the "specifying" action. In complex manipulative sequences, use of the mouth would be impractical.

Handedness, or preferential right or left hand use, occurs in random distribution among the chimpanzees and, therefore, does not seem to have been genetically selected for the requirements of complex two-handed action. However, the fact that preferential handedness and neocortical asymmetry does exist suggests that, as two-handed problem solving became more complex, there would begin to be an advantage for those who could more efficiently perform lateralized two-handed actions. An interesting neurological and cognitive process would have been created, based on the nature of the three-dimensional physical reality and the two-handed, vision-oriented capacity.

In two-handed action, the primary hand is involved in the sequence of specifying actions, while the secondary hand assumes its own sequence and range of supporting actions. The actions and information processing sequences of each hand/eye set differ when working wood, stone, bone, vines, twigs, nuts, skins or animal carcasses, or when collecting, carrying, nest or shelter-building. Therefore, chimpanzee hands and tools already perform actions that involve a complex range of *materials* and *processes*. The hands are also used *affectively* for interpersonal grooming, for begging and comforting, hugging and tickling. This capacity for *variable* processes and behaviors *in terms of* handed action would have begun to be selected for during the process of increasing bipedality and the freeing of hands, as an aspect of neurology. We are dealing with a two-handed neurological competence that is functional with or *without* tools, and selection for this competence would have begun before there was major brain enlargement.

In the proposed model, bipedalism was not adaptive because it made possible scanning for carnivores over the savannah grass, or made possible the "carrying" of food or infants, as has been theorized. These concomitants of bipedalism occur in various forms among other species. Nor was bipedalism crucial because it opened the birth canal to a larger fetal brain, since there are other biological solutions for acquiring a larger brain, including changing rates of post-natal growth and development. Instead, it was adaptive because a bipedal hominid could "carry" a generalized and developing two-handed competence to a wider range of ecologically variable niches and problem-solving opportunities. Hominid tool-use, cooperative hunting and food portage were to become secondary functional aspects of that species-specific competence.

In such a model, there would develop a neurologically restructured way of "seeing" the world in terms of the two-handed competence, much as other species "see" and relate to their species-specific reality in terms of flying, swimming, brachiation, burrowing or running, involving specialized sensory and neurological capacities, limits and constraints. As a result of the hominizing process being proposed, an interesting and profound evolutionary development would have been instituted, different from anything assayed before, a process involved epistomologically and cognitively with the meaning of human "knowledge." The two-hands, with their increasingly learned and encoded skills, would have become differentially concerned with relevant aspects of the handed reality. The eyes would be increasingly capable of judging a range of objects, in terms of the evolving two-handed competence and potential. The non-genetic and experiential nature of the two-handed skills would tend to make this knowledge increasingly cultural. The tool industries, symbols and notations of *Homo sapiens* are end-products of this development.

Since I am discussing "animal cognition" in a particular evolutionary and ecological context, I would suggest that it was probably only after a two-handed, asymmetric competence and neurological reordering had taken place and was already adaptive within a small-brained hominid that the adaptive advantage of an increase in the "information processing" and encoding capacity of the neurological system, by the addition of relevant tissue in specialized areas and for specialized functions, would have become functionally apparent. Adding tissue to a chimpanzee brain, even if it rode on a bipedal frame, would not give us a hominid, no matter how the mammalian sexual, bonding and parenting behaviors were altered. What was evolving was a generalized, variable behavioral and cognitive capacity. Even those behaviors derived from animal ethology and proposed as genetically programmed adaptive behaviors during hominization were probably actually potentially *variable* behaviors. The essence of hominid and human adaptative behavior is this variability, a variability that is present among humans in the parameters of male/female bonding, parental investment, experiential maturation, subsistence technologies, sequences and programs, and *all* aspects of culture.

Chronologically, the point at which an already reorganized two-handed neurology may have found it advantageous and possible to increase its information-processing carrying capacity may have occurred in that million year period of transition from the small-brained australopithecine line of "Lucy" (Johanson & White, 1979) c. 4 million years ago, to the larger-brained line of *Homo habilis*, the maker of the pebble tools. The mosaic process would probably have involved changes in the early fetal and natal brain and changes in the rate and sequence of maturation of different sets of functional areas and capacities of the brain, particularly those involved in cross-modal association and inter- and intra-hemispheric exchange. Of particular importance, it would have involved installation of that hierarchical capacity for short-term evaluation and long-term programming of integrated cultural behaviors, which mature so slowly during ontogeny. That is both a hierarchical associational process and a *motivational* process that includes cortical, frontal lobe encoding, and mediation and sub-cortical limbic marking.

This brings us to the problem of the origins of language and speech as a corollary cognitive development in the pattern of adaptive evolution being modeled. As noted, language and speech serve to mark, differentiate, and refer to objects, processes and relations of the relevant and increasingly complex material and cultural reality. Since speech occurs in the vocal/auditory mode, it requires a different sensory network. The co-evolution of language as a corollary referential and marking system of the increasingly complex reality and the developing, lateralized hominid neurology, can only be discussed in a separate paper. Like the imaging and symbolling skills discussed in the early notations and art, or the skills of two-handed problem-solving just discussed, language is a specialized cognitive capacity and system of the evolving brain. As such, it evaluates and produces sequences and equations of relevance in the one modality. But it is a product of the same brain, evolution, adaptation and ontogeny as the other cognitive skills. Language as a specialized cognitive mode and aspect of the evolved hominid "potential variable capacity" will be discussed in a separate paper.

ACKNOWLEDGMENTS

The research for this paper was supported in part by grants from the Harry Frank Guggenheim Foundation and The National Endowment for the Humanities. Thanks are given to S. Chevalier-Skolnikoff, N. Geschwind, M. Kinsbourne, J. Kitahara-Frisch, P. Tobias, C. Trevarthen and others, whose suggestions, advice and papers, sent in response to a first draft of this paper, helped in the revision.

REFERENCES

Binford, L. R. Comment on R. White, "Rethinking the Middle/Upper Paleolithic transition". *Current Anthropology*, 1982, *23*, 178-179.

Boehm, C. Rational preselection from Hamadryas to *Homo sapiens*: The place of decisions in adaptive process. *American Anthropology*, 1978, *80*, 265-296.

Bordaz, J. *Tools of the old and new Stone Age*. New York: Natural History Press, 1970.

Brochier, J. F., & Brochier, J. L. L'art mobilier de deux nouveaux gisements magadeleniens a Saint-Nazaire-en-Royans (Drome). *Etudes Prehistorique*, 1973, *4*, 1-12.

Chevalier-Skolnikoff, S., & Poirer, F. E. *Primate bio-social development: Biological, social, and ecological determinants*. New York: Garland, 1977.

Didon, L. L'Abri Blanchard des Roches. *Bull. de la Soc. Hist. et Archaeo. du Perigord*, 1911, *38*, 321-345.

Edwards, S. W. Nonutilitarian activities in the Lower Paleolithic: A look at two kinds of evidence. *Current Anthropology*, 1978, *19*, 135-137.

Gamble, C. Information exchange in the paleolithic. *Nature*, 1980, *283*, 522-523.

Luria, A. R. *Higher cortical function in Man*. London: Tavistock, 1966.

Harding, R. S. O., & Teleki, G. (Eds.) *Omnivorous primates: Gathering and hunting in human evolution*. New York: Columbia University Press, 1981.

Isaac, G. Ll. Early phases of human behavior: Models in Lower Paleolithic archaeology. In D. L. Clark (Ed.), *Models in archaeology*. London: Methuen, 1972.

Johanson, D. C., & White, T. D. A systematic assessment of early African hominids. *Science*, 1979, *203*, 321-330.

Kitahara-Frisch, J. Apes and the making of stone tools. *Current Anthropology*, 1980, *21*, 359.

Lovejoy, C. O. The origin of man. *Science*, 1981, *211*, 341-350.

Marshack, A. Le baton de commandement de Montgaudier (Charente). Reexamen au microscope et interpretation nouvelle. *l'Anthropologie*, 1970, *74*, 321-352. (a)

Marshack, A. *Notations Dans Les Gravures du Palelithique Superieur*. Bordeaux, France: Institute de Prehistoire, Universie de Bordeaux, 1970. (b)

Marshack, A. *The roots of civilization*. New York: McGraw-Hill, 1972. (a)

Marshack, A. Cognitive aspects of Upper Paleolithic engraving. *Current Anthropology*, 1972, *13*, 445–477. (b)

Marshack, A. Preliminary analysis of an azilian notational engraving from the Grotte du Tai (St. Nazaire-en-Royans, Drome). *Etudes Prehistoriques*, 1973, *4*, 13–17.

Marshack, A. Upper Paleolithic notation and symbol. *Science*, 1973, *178*, 817–828. (b)

Marshack, A. Proper data for a theory of language origins. *The Behavioral and Brain Sciences*, 1979, *2*, 394–396. Commentary on Parker & Gibson article.

Marshack, A. Paleolithique ochre and the early uses of color and symbol. *Current Anthropology*, 1981, *23*, 188–191.

Marshack, A. *Image of man*. New York: Harcourt Brace Jovanovich, 1984. In press. (a)

Marshack, A. Epipaleoithic–Early Neolithic iconography: a cognitive, comparative analysis of the Lepenski Vir iconography & symbolism, its roots and influence. Paper at Int'l Symposium: The culture of Lepenski Vir and the problems of the formation of the neolithic cultures in Southeastern and Central Europe. Cologne. *Journal of Mediterranean Anthropology and Archaeology*, 1983–84, in press. Zeitschrift. Band 2. (b)

McGrew, W. C. Socialization and object manipulation of wild chimpanzees. In R. S. O. Harding & G. Teleki (Eds.), *Omnivorous primates: Gathering and hunting in human evolution*. New York: Columbia University Press, 1977.

Peters, C. R. Toward an ecological model of African Plio–Pleistocene hominid adaptations. *American Anthropologist*, 1979, *81*, 261–278.

Raskin, S. R. Archaeology of mind: An analysis of some Paleolithic notations. Paper presented at the 80th annual meeting of the American Anthropologists Association, Los Angeles Society. 1981.

Schmandt-Besseratt, D. An archaic recording system and the origins of writing. *Syro-Mesopotamian Studies 1-2*. Monographic Journal of the Near East, 1977.

Schmandt-Besseratt, D. The earliest precursor of writing. *Scientific American*, 1978, *238*, 50–59.

Schmandt-Besseratt, D. From tokens to tablets: A re-evaluation of the so-called 'Numerical Tablets'. *Visible Language*, 1981, *14*, 321–344.

Shettleworth, S. J. Memory in food-hoarding birds. *Scientific American*, 1983, *248*, 102–110.

Thomas, D. H. *Archaeology*. New York: Holt, Rinehart, Winston, 1979.

28 THE CACHE-RECOVERY SYSTEM AS AN EXAMPLE OF MEMORY CAPABILITIES IN CLARK'S NUTCRACKER

Russell P. Balda
Richard J. Turek
Northern Arizona University

I. INTRODUCTION

The results of natural history studies on vertebrate animals often suggest that these animals are more than simple stimulus–response creatures making appropriate responses to environmental stimuli. In some species, foraging, mating, territory selection, and predator avoidance behaviors, appear to require a complex set of decisions that suggest some higher level of cognition than normally accepted. Yet, naturalists who study and understand these behaviors find it virtually impossible to design experiments or make appropriate observations to critically test their assumptions and hypotheses about animal cognition. These assumptions and hypotheses are concentrated on the theme that animals seem to be making decisions that have, as a result, a probabilistic set of outcomes.

Two examples from our recent long term work with color banded wild populations of birds may suffice to illustrate this point:

1. Male Pinyon Jays (*Gymnorhinus cyanocephalus*) appear to select mates based on a complex constellation of female characteristics including size, age, and past breeding experience. Based on this set of criteria, which vary for each male based on his morphology and age, males pick the "best" available female, i.e., the one that best matches his criteria. When the pool of available females is sparse, selection criteria are relaxed but not discarded. Reproductive success is reflected in how well these criteria are adhered to (Marzluff and Balda, in prep.).

2. Pygmy Nuthatches (*Sitta pygamea*) band together during the cold winter months into groups numbering from six to 40+ birds and collectively defend a territory. These territories contain numerous tree cavities originally excavated by woodpeckers. These cavities serve as roost–sites for the groups of nuthatches throughout the winter. Cavities vary greatly in size, shape, thermal quality and microenvironment. Different cavities are used by birds on different nights. It appears that the decision to use a particular cavity depends

on the size of the group, consensus of the group, and nighttime temperature and precipitation. Wise cavity selection results in an energetic savings to the birds which can have direct implications for their survival (Guntert, Hay, and Balda, in prep.).

In both of the above examples the birds appear to be presented with a highly complex set of interactive factors. By integrating these factors, a best possible solution (best mate, best roosting cavity) can be achieved. Some of the factors require long-term memory (past success of female, location of cavity) and most likely, the final decision is reached by a complex set of mental activities. The decision made will have a direct effect on the fitness of the birds and therefore, this type of decision making will be influenced by natural selection.

Studies on animal cognition that have been undertaken often employ subjects that do not match the test procedures or vice versa, so that the results bear little relationship to the experiences and problems animals face in the real world. The transfer of cognitive abilities from an animal's natural environment to artificial laboratory situations has never been satisfactorily demonstrated.

The appropriate procedures and subject organisms should match the ecological setting in which the organism has experienced selective pressures. Often studies on learning, memory, and cognition have been completely divorced from natural conditions and the normal functions these mental processes may play in the life of an organism. One such natural history tactic that seems to be ideal for studies on cognition is the food caching behavior of members of the family Corvidae (Turcek & Kelso, 1968; Vander Wall & Balda, 1981).

Some species of corvids take advantage of the large crops of conifer seeds produced in some years by harvesting seeds from cones and storing them in subterranean caches. Each autumn these birds spend considerable amounts of time and energy performing these behaviors. For example a single Clark's nutcracker (*Nucifraga columbiana*) in a year of a large seed crop may cache up to 33,000 seeds (Tomback, 1978; Vander Wall & Balda, 1977). These birds live in high, cold alpine environments where other sources of energy and nutrients are in short supply during the long, harsh winter. In order to survive, a Clark's nutcracker must recover about 2,500 caches (Tomback, 1978; Vander Wall & Balda, 1977) each winter.

The function of caching behavior is to remove seeds from conspicuous locations (cones) where they are available for exploitation by competitors (other birds, ground squirrels, etc.) and place them in inconspicuous locations. The seeds are thus placed under the control of the cacher. These seeds serve as a ready supply of food when other food items are not available (Roberts, 1979).

In order for this to occur it is necessary for the bird to somehow locate these caches up to 11 months later (Vander Wall & Hutchins, personal communication). A constellation of behavioral and morphological adaptations (Bock, Balda, & Vander Wall, 1973; Vander Wall & Balda, 1981) for caching seeds should then be matched by a neurological system capable of providing the bird with the information for recovering them.

The adaptations for caching seeds would, in fact, be severely selected against, according to evolutionary theory, if the critical adaptations for recovery of theses seeds were not present.

During the caching period, birds invest time and energy and expose themselves to predators. There must be a "pay-off," in a selective sense, for performing in such a manner. These "pay-offs" can only be realized if the bird can locate these seeds at a later date. The pay-offs, summarized below are discussed in detail elsewhere (Vander Wall & Balda, 1981):

1. Winter Diet: The winter diet of nutcrackers consists of between 80 and 100% conifer seeds (Guintoli & Mewaldt, 1978). These seeds most likely come from seed caches as cone crops are usually depleted by mid-November.

2. Timing of the Breeding Season: Nutcrackers are among the earliest nesting passerine birds in North America, often breeding in alpine environments in March and April. Cached seeds may well provide the energy for nest building, egg laying, and incubation.

3. Nestling Diet: Nestling nutcrackers are fed on a diet consisting almost entirely of seeds, an unusual occurrence in birds. These seeds are retrieved from caches made months earlier.

Thus, the birds have a somewhat predictable source of food in the winter which results in a higher probability of survival than without this stored food. The birds can nest very early, giving the young-of-the-year additional time to mature before the onset of the harsh winter.

Field evidence suggests that nutcrackers can find specific caches made many months earlier (Mattes, 1978; Swanberg, 1951; Tomback, 1980).

Recent laboratory experiments also demonstrate a large capacity memory system used by nutcrackers to find seeds (Balda, 1980; Vander Wall, 1982). Thus, the nutcracker-conifer seed relationship may offer scientists a unique opportunity to examine the limits of memory and the type of information encoded about specific cache sites.

The research described herein has the following objectives:

1. To explore some characteristics of cache making and cache recovery in Clark's nutcracker. Do nutcrackers use sight, smell, random search, memory, or a combination of the above to find caches?

2. To determine the capacity a nutcracker has for storage of information about cache sites made at an earlier time.

3. To determine the configuration of the memorized information (if present) about the location of specific cache sites. What kind of information does the nutcracker have about specific caches?

II.   METHODS AND TRIAL ENVIRONMENTS

The five trials to be described herein were all conducted in the same 3.5 x 3.9 m room. The floor was covered with 6 cm of fine sand which was brushed to a smooth surface between and within each trial to obliterate all signs of cache making and cache presence. Conspicuous objects such as feces, feathers, and seed hulls were also carefully removed from the sandy substrate. The north wall contained a window in the east corner that was covered during each trial. The east and west walls each contained a wall socket and brackets for the attachment of three 3.5 m long perches. The south wall contained a sink, a small bench, a door, and a light fixture 0.3 m from the ceiling. These features of the room remained constant for each trial.

For all five trials a 15 cm square seed tray was placed near the center of the room. It was not moved between the caching and recovering phases of each trial. In Trials 1 and 2 the seed tray was on the floor while in Trials 3, 4, and 5 it was 1 m above the floor. A water dish was always located next to the seed tray, but remained on the floor.

All trials were conducted with a single semi-tamed Clark's nutcracker that had been in captivity for 7 months prior to the initiation of the trials. The sex and age of this bird were not known. Trials 1 through 3 were conducted between September 1980 and February 1981. Trials 4 and 5 were run in September and October 1981. The bird spent the intervening spring and summer of 1981 in a large outdoor flight cage. During this and other intervals, during which the bird was not in the above described room, it was fed insects, mice, a variety of seeds and a sparse supply of pinyon pine (*Pinus edulis*) seeds. The bird occasionally cached and recovered seeds in this situation.

All cache-recovery trials were conducted between 06:45 and 10:00 unless stated otherwise (Trial 4). Light and temperature conditions were the same between the cache-recovery phases of all trials.

During the caching and recovery phases of all trials, one of us (RPB) sat on the bench on the south wall and made direct observations in full view of the bird. In all trials RPB recorded the behavior of the bird and plotted the approximate location of each cache, on a map of the floor. Each cache was sequentially numbered as it was made. The time at which caching began and ended was recorded.

All trials followed a similar protocol. The nutcracker was put in the room seven days before each trial. During these seven days, the bird was fed hulled pinyon pine seeds, a mixture of various commercially available seeds, popcorn, and an occasional mouse. Each day the bird was fed 10 to 15 intact pinyon pine seeds in our presence. Intact pinyon pine seeds were never cached and only rarely did we find other food in caches. Eighteen hours before the initiation of a caching session the bird was deprived of all food.

At the beginning of a trial 210 large, intact, pinyon pine seeds were placed in the feeding tray. The bird typically ate four to six seeds before caching seeds in the substrate. After the bird ate or cached all 210 seeds, the lights were turned off, the bird was captured, and removed from the room. The bird was placed in a holding cage for the isolation period which was either 31 days (Trials 1, 2, 3) or 10 days (Trials 4 and 5). During this period, the bird was fed 10 to 20 pinyon pine seeds, sunflower seeds, and popcorn, each day.

After the bird was removed from the room all caches were located, dug up, and the number of seeds per cache recorded. Steel measuring tapes, and meter sticks were used to determine the exact location of each cache. These were then plotted on a map.

The area encompassed by all caches made during a single trial was measured by connecting a straight line through the outermost caches. The enclosed area was then measured.

Half of all caches were randomly selected and removed in Trials 1, 2, 3. The others were reburied at the proper depth and location. Efforts were made to insure that all reburied caches presented no visual clues as to their location.

After the prescribed isolation period the bird was deprived of food for 20 hrs and reintroduced into the room. This was done by bringing the bird into the darkened room in the isolation cage. After the light was switched on the cage door was opened and the bird could leave at will. During the recovery phase of each trial, RPB sat on the bench and made direct observations of what caches were found, the sequence in which they were found, the bird's search route, and where the bird probed. These data were recorded on the map containing accurate locations of all caches.

After the recovery phase, we removed all seeds from the remaining caches in the presence of the bird. The bird was then allowed to explore the room for seven days as described earlier.

III. DEFINITION AND DESCRIPTION OF BEHAVIORS

Each trial consisted of three phases: caching, isolation and recovery. In this section we describe the behavior of the bird during each phase and define terms where appropriate.

When offered pine seeds during the caching phase the bird generally began eating them within the first minute. Most seeds were carried up to a perch where they were opened and eaten. The seed hulls fell to the floor. Once the bird commenced caching, it made repeated trips from the seed tray to the floor and back, stopping only briefly to eat one or two seeds. When picking up seeds from the tray the bird occasionally "bill clicked" or "bill weighted" the seeds (Ligon & Martin, 1974) but few were discarded. "Bill clicking" and "bill weighing" are behaviors used to assess the quality of the pine seeds. "Bill clicking" is done by holding a seed mid-way from the tip of the bill and rapidly opening and closing the mandibles. "Bill weighing" is performed by holding a seed for a brief time mid-way in the bill. Edible, healthy seeds are then placed in a sublingual pouch (Bock et al. 1973) whereas spoiled or empty seed hulls are cast out of the bill.

Over 95% of the caches were made by the bird burying all the seeds it carried in its pouch in a single subterranean cache. When the bird landed on the floor, it would often stop, look about, carefully scan the floor, hop a number of times, again look around and at the floor, and then begin to stick the seeds into the sand with strong jabs of the bill. Sometimes the bird performed up to five hop-look sequences before making a cache. After the seeds were placed in the sand the bird used a sideways bill swipe to cover over the seed cache with sand. Often the bird performed 5 or 6 bill swipes when covering a cache. After completing the above task the bird flew directly back to the seed tray to pick up more seeds.

An edge cache is one made within 10 cm of a stationary object (wall, seed tray, rock, etc.). Caches located greater distances from objects are referred to as open caches.

During the period the bird was isolated from the room, it was held in an indoor research laboratory. Here the bird was exposed to human intrusions and noises normally associated with such an establishment. Because the bird was semi-tame it did not appear overly disturbed by these activities but they most certainly constituted an "unnatural" setting in which to spend the isolation phase.

The recovery phase of all trials was initiated by placing the bird back in the experimental room after the floor had been prepared. Normally the bird flew to a perch and acted rather disturbed. Jumping from perch to perch, flying from one end of the room to the other, and loud calling were typical behaviors. During the recovery phase, some postures and soft songs were given that we never heard under other conditions. Eventually the bird flew to the floor and hopped slowly about looking at the walls and sandy floor. On occasion the bird would stop, peer down at some spot on the floor, and then hop away. On other occasions, after peering at the floor, the bird would make either a forceful jab straight down into the substrate or a sideways bill swipe. In either case, if a cache was located the attempt was labeled a successful probe. Upon location of a cache the bird extracted some or all of the seeds and flew to some location where the seeds were opened and eaten. Most often all seeds were taken from a cache before the bird proceeded to search for another cache. If the bird probed or swiped at a cache-site from which we had removed the seeds the bird often vigorously repeated this behavior five or six times in an attempt to locate seeds. Sometimes the bird would then begin "area searching" by making extended bill swipes in areas adjacent to the correct location of the cache. These probes were not recorded as incorrect probes.

On occasion a bird would return to a cache site it had previously emptied. These were easy to locate because of the obvious probe hole present. Sometimes the bird would simply peer into the probe hole and then move on. In other instances the bird would probe vigorously, but unsuccessfully, at these sites. These probes were recorded and classified as "repeat" probes. They were not used to calculate the percent accuracy figures presented herein.

IV. RESULTS

A. TRIAL 1

In this trial the floor of the room was bare except for the centrally located seed and water trays. In this trial we hypothesized that the bird may have great difficulty locating its caches because of the lack of surface objects. We suggest that these objects may serve as orientational clues that are used by the bird when finding caches.

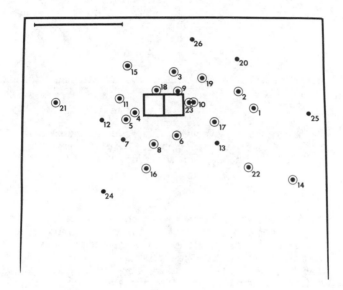

FIG. 28.1. Map of the distribution of caches in Trial 1. The numbers indicate the caching sequence. Dots with circles indicate caches located by the bird. Single line is 1 m in length.

The bird made 26 caches in an 5.7 m^2 area (Figure 28.1, Table 28.1). These caches were made in 22 min. Thirteen of these caches were then randomly removed by the experimenters.

After 31 days of isolation the bird recovered seeds from 18 caches, or 69% of all caches made. The bird probed at a total of 20 sites, only two of which were sites where no cache had been previously placed. Another eight probes were made at cache-sites from which the bird had recovered seeds earlier in the recovery phase of this trial. The bird located eight full caches and 10 caches from which the experimenters had removed the seeds.

The bird, thus, probed with an accuracy of 90% even though there

TABLE 28.1
Cache-Recovery Data for Five Trials

Trial	Cache Area	Caches	Caches Found		Probes Used	Random Probes Needed
			N	%		
1	5.7	26	18	69	21	1158
2	5.2	30	10	33	11	505
3	8.7	24	3	13	6	334
4	1.1	14	4	29	18	91
5	5.9	27	8	30	44	519

Note-- Cache area = area circumscribed by caches in square meters, N = number of caches found.

were few surface objects available to use as orientational clues. The bird searched at 10 sites that we had emptied so other direct clues (sight of the seed, odor, soil disturbance, etc.) were not present.

B. TRIAL 2

In addition to the seed and water trays, the floor contained three logs arranged in a perpendicular pattern with a rock in each quadrate. The symmetry of the room was maintained. Our hypothesis was that the more surface clues present, the better the accuracy of the bird. In this trial the nutcracker made 30 caches (Figure 28.2, Table 28.1) in 35 min. Fifteen of these caches were randomly removed by the experimenters.

After 31 days of isolation the bird recovered seeds from 10 caches which was 33% of the total caches made. The bird probed at 11 different sites and used a total of 13 probes. Two probes were made at cache-sites previously harvested by the bird. The bird probed with an accuracy of 91%. Four of the 10 caches located by the bird contained seeds whereas the other six had been emptied by the experimenters. Thus, the accuracy of the bird did not improve when we added surface features to the room.

C. TRIAL 3

In this trial the logs, rocks, and water dish were arranged as in the previous trial. The seed tray, however, was placed on a pedestal one m off the floor. The bird made 24 caches in 19 min. The distribution of caches in the room showed greater scatter than in the previous two trials when the food tray was on the floor (Figure 28.3). This occurred because

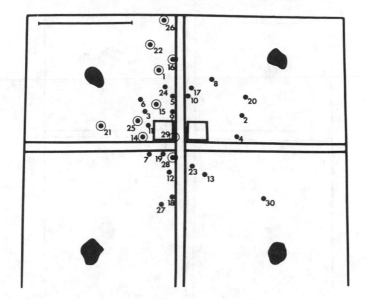

FIG. 28.2. Distribution of caches in Trial 2 when logs and rocks were added to the sand floor. Symbols same as in Figure 28.1.

the bird now had to glide from the seed tray to the floor.

During the 31 day isolation phase, we removed the rocks, logs, seed, and water trays. This was done to simulate conditions in which all surface objects would be unavailable to the recovering bird such as would occur after a snow storm or a forest fire.

When placed back in the caching room the bird acted uneasy and nervous. Its frequency of calling, flying, and hopping between perches was greater than in the first two trials. Also, the bird spent more time on the floor looking about, but not probing, than in the previous trials. After 30 min the bird began actively probing for caches. Ultimately the bird made six probes, locating three caches for an accuracy of 50%. All three located caches were ones in which we had left the seeds. These results support the idea that surface clues, although important, are not the single source of information the bird had stored about where its caches were located.

D. TRIAL 4

In this trial the logs, rocks, seed and water trays were arranged as in Trial 3. This trial was designed to test if the bird remembered those caches it had previously recovered. To do this we used the same procedures for the caching phase but we shortened the isolation phase to 10 days and had two recovery phases at 10 day intervals. Also, all caches were

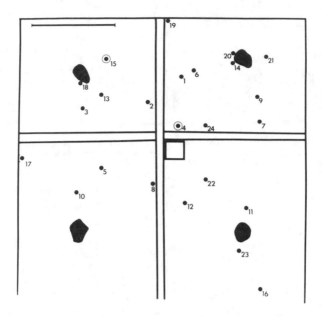

FIG. 28.3. Distribution of caches in Trial 3 before logs and rocks were removed from room.

undisturbed. For some reason the bird did not appear motivated to cache and made only 14 caches in 64 min. Many seeds were left in the seed tray.

When the bird was placed back in the caching room after 10 days of isolation it recovered four caches using 18 probes. Two of these, however, were repeat probes at sites where caches had previously been harvested. Thus, the bird had an accuracy of only 25%.

During this recovery phase, some of the seeds were recached rather than eaten. Four new caches were made.

During the second 10 day isolation phase, the floor was gently brushed to provide the bird no visual clues about its earlier recovery activities. The bird now had a total of 14 caches available to it. Ten of these were made during the caching phase and four during the first recovery session. After isolation, the bird found seven caches using 20 probes. Four of these were caches made during the first recovery session and three were made during the initial caching phase (Figure 28.4). The bird did not demonstrate a preference for either set of caches. Of the 20 probes made, eight were incorrect, two were at cache sites emptied during the first recovery phase and three at sites emptied earlier in this phase. Thus, the bird had an accuracy of 47%. The bird demonstrated no pattern as to which empty cache sites it revisited.

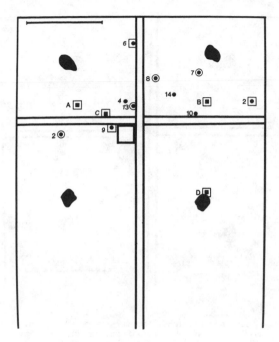

FIG. 28.4. Distribution of caches in Trial 4. Letters indicate caches made during first recovery phase and located during second recovery phase.

E. TRIAL 5

All objects remained in place as in the previous trial. This trial was designed to see if the bird could find its caches after all surface clues were moved as might occur during a mud slide, avalanche, or wind storm. The bird made 27 caches in 23 min.

During the 10 day isolation phase, we moved the logs, rocks, seed and water trays 40 cm to the north of their original locations. The sandy floor was then gently brushed.

During the recovery session, the bird spent an unusually long time on the floor hopping about and peering at the substrate. The bird made 44 probes, eight of which located a cache (Figure 28.5). Fifteen of these probes, however, were made at cache sites the bird had previously emptied in this recovery trial and 21 were incorrect probes. Thus, the bird demonstrated a probing accuracy of only 28%. Four of the 21 unsuccessful probes were, however, within 1 cm of being displaced exactly 40 cm north of where a cache was located. If these probes are counted as correct, then the accuracy of the bird would be 41%.

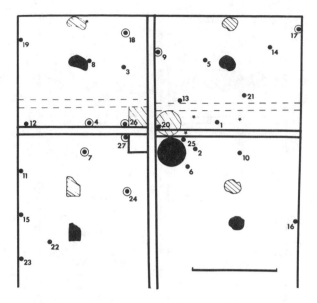

FIG. 28.5. Distribution of caches and surface objects during the caching phase in Trial 5. Dotted and outlined objects indicate their placement after being moved 40 cm north. Stars indicate probes that would have been correct if the objects had not been moved.

V. DISCUSSION

What types of clues does the nutcracker use for locating caches? Clues can be of two kinds, direct or indirect. A direct clue is one given by the cache itself, such as a portion of a seed visible on the substrate, an odor, or signs left from the placement of a cache, for example, footprints. An indirect clue is a set of circumstances or signs that point the way to a cache but are independent of the cache itself. These stimuli are present in the room whether or not a cache is made.

To distinguish the above two classes we attempted to obliterate some of the direct clues by ensuring that caches were buried under the sand, brushing the surface of the floor (all trials) and randomly removing caches (first three trials).

Trials 1 through 5 demonstrate that the bird did not use sight of the seeds to locate the caches and in two of the first three trials olfaction could not have been a factor. This is consistent with the findings of Bossema and Pot (1974) for the Jay (*Garrulus glandarius*), the Eurasian nutcracker (*Nucifraga caryocatactes*) (Balda, 1980), and also demonstrated by Vander Wall (1982) for Clark's nutcracker.

In the third trial, in which the seeds were removed from one-half of the caches and all surface clues were removed, the bird found three caches, all of which contained seeds. The probability of finding three or more full caches out of 12 caches is 0.11 using the hypergeometric distribution. In the birds' natural environment, they often cover caches with a leaf, small stone or a pine cone. These objects, in addition to footprints may also serve as direct clues, but they are easily moved (or removed) by wind, snow, and rain. (Vander Wall, 1982) Although direct clues are, in all likelihood, used by the bird, the fact that the accuracy of probing was better than 90% in the first two trials suggests that indirect clues play a much greater role in locating caches. In the terminology of Cheng and Gallistel (Chapter 23, this volume) the birds do not rely on beacon navigation but use some type of spatial representation.

Is the bird foraging randomly? The accuracy of probing on all five trials was far higher than one would expect from a random search pattern. To test for this we made the following calculations. The bird can expose 34 cm² of surface area with a bill swipe and a cache has a surface area of 3 cm². Using these figures we calculated (see Balda, 1980, for technique) the number of random probes needed to find the appropriate number of caches in each trial. This number ranged from a low of 91 to a high of 1158. In these two trials the bird used 18 and 21 probes, respectively. The most probes used on any trial was 44 (Table 28.1). Thus, random probing was not the technique used to locate caches. This is consistent with the findings of Balda (1980) with the Eurasian nutcracker.

We conclude from the above that nutcrackers do not need direct clues (beacon navigation) to find caches and also they do not use random probing when searching for caches. If the above are eliminated then some type of mental map or caching pattern may be important.

The simplest type of caching pattern would be to have typical sites that are used repeatedly for caches. If this were the case, then one should see an overlap in cache locations between trials. We compared maps of Trials 1, 2, and 3 which were all run in the same year, and looked for caches located within 10 cm of each other. Caches located 10 cm from each other between trials were classified as overlapping. Ten cm was used because a single bill swipe can cover that distance.

TABLE 28.2
Caches Overlapping Between Trials

Trial Pair	Number Overlapping	Binomial Probability
1 - 2	8	0.048
2 - 3	4	0.218
1 - 3	0	0.923

Only between Trials 1 and 2 did the bird make a significant number of overlapping caches. No caches from Trial 3 overlapped with Trial 1 caches (Table 28.2).   The fact that we moved the feeding tray off the floor may have influenced this pattern, but this argument is weakened by the fact that the bird, during Trial 5, placed no caches where ones were made dring Trial 4.   Between Trials 1 and 2 we placed the logs and rocks in the room so the bird may have used indirect clues from the room in general to make decisions about where to locate caches.   We conclude that the bird did not have a mental image of a set of cache-sites that it favored for filling on each trial.   This is consistent with the findings of Vander Wall (1982) for Clark's nutcracker but not with those of Balda (1980) for the Eurasian nutcracker.

In studies with the Eurasian nutcracker, Balda (1980) found a preference for caching within 10 cm of an object.   This preference was not followed by a preference to harvest caches from along "edge-lines." Edge-lines, nevertheless, may be important in providing indirect clues about the location of caches.   Using the five trials reported herein, the bird made between 12 and 56% of its caches within 10 cm of objects.   No preference, however, was evident for harvesting caches along edge-lines. We also looked for any pattern or sequence of recovering caches according to our open vs. edge categories and none existed.

Trial 3 provides some evidence about the type of information stored on the "mental map" the bird may have of its caches.   In Trial 3 all surface clues were removed from the floor and the bird subsequently recovered three of 24 caches using only six probes.   Accuracy in the first two trials had been better than 90%, thus, alteration of surface objects was probably responsible for this decline in accuracy.   But, even in Trial 3, the bird could not have achieved this level of accuracy using random probes alone (Table 28.1).   The bird must be integrating surface clues (rocks, logs, etc.) with larger objects in its environment such as the walls, lights, perches, etc. in the room.   By their structural nature, these objects are also more permanent and may be analogous to large boulders, tree trunks, and overhanging branches.   These permanent clues will remain in place and be visible after snow, rain, and high winds.   The surface objects may well be moved or covered by the above agents, and therefore of no use to the bird.   We suggest that the bird stores a number of pieces of information about each cache.   These pieces of information consist of the direction and distance from large, permanent objects and small ephemeral objects. Minimally, the bird should retain information from two large objects so that triangulation is possible (Cheng & Gallistel, Chapter 23, this volume). Smaller surface clues may simply be used (a) only when available, or (b) for improved precision because they are closer to the cache site (Vander Wall, 1982).   In some instances the bird may have been forced to include surface clues because more permanent clues were not available or appropriate.   Thus, the system of remembering indirect clues about each cache appears to have built-in redundancy.   The small surface clues are used when available and may provide high precision whereas the larger, more general clues are more predictable in their permanence but may not provide the same degree of precision.

The bird must undergo some process of selection about what clues to remember about each cache during the time the cache is being made.   In Trial 1, when few surface clues were available the bird was forced to, seemingly, incorporate information only about large permanent objects.

These objects were all present during the recovery phase of Trial 1 and the bird was very accurate when locating caches, indicating that these objects suffice for accurate cache location. In contrast, in Trial 3 when both surface objects and permanent objects were present during the caching phase but the former were absent during the recovery phase the bird's accuracy was low. Thus, the decision to store information about certain surface clues was, in this case detrimental to the bird. Because surface clues were left intact during Trials 1 and 2, the bird may have "anticipated" the same would happen in this trial.

In Trial 5 we attempted to duplicate what could happen to surface objects in a natural setting. After the caching phase all surface objects were moved 40 cm north of their original location. Now the bird probed with only 28% accuracy, locating eight caches and making 21 incorrect probes. More important, however, is the fact that four more probes were made at the proper locations if we had not moved the surface clues. We conclude that for some caches, surface clues may predominate. Certainly, in natural situations, these caches would never be found after heavy snows.

It is possible that the bird has a system of harvesting caches that is determined by the particular combination of surface and permanent clues used to remember each cache. For example, caches that are remembered by surface clues may be harvested early in the winter before these clues are obliterated, or in spring after snow melt. Caches located by more permanent clues may then be recovered in the winter. Mattes (1978) showed that probing accuracy did not decline between late fall and early spring in wild Eurasian nutcrackers. Thus, it is possible that both types of clues can result in accurate cache location. Our trials were of too short a duration to provide evidence for or against this hypothesis.

Regardless of what type of information is memorized about cache location the bird should have a system for efficiently exploiting these caches. For example, the bird may store information about the sequential pattern of making caches and then use this information when recovering caches. In all five trials, however, there was no significant correlation (Spearman rank correlation coefficient) between the order of making caches and their recovery. Thus, recency–primacy effects were not operating. This is consistent with findings on the Eurasian nutcracker (Balda, 1980).

A system was discovered in the Eurasian nutcracker (Balda, 1980) whereby the bird harvested nearest–neighbor caches first. In both of two trials, the bird first harvested caches that were close to other caches. Subsequent recoveries repeated this pattern so that those caches farthest from other caches were harvested last. This is consistent with optimal foraging theory (Pyke, Pulliam, & Charnov, 1977) and also strongly suggests that the bird integrates information about cache location so that each cache can be located independent of location or temporal sequence. Clark's nutcracker did not follow this nearest–neighbor harvesting pattern.

As the trials proceeded, the nutcracker did not maintain its initial high degree of accuracy. Earlier trials may have interfered with later ones, and/or the bird perceived that the ecological constraints normally present in its natural environment were not present in the caching room. This may have eliminated the necessity for particular patterns of caching and harvesting. If the bird gradually changed its perception of the ecological situation, it may also have changed its perception of the caching room. For example, the room may have been a novel place to cache seeds on the first trial but by the fifth trial the room may have been perceived as a familiar caching area.

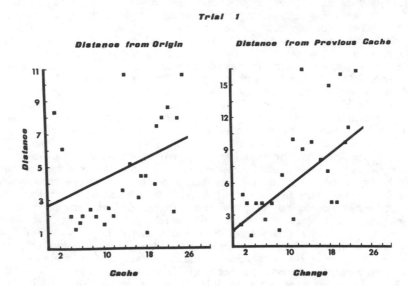

FIG. 28.6. Linear regression of distance of successive cache from the seed tray (left) and distance from previous cache (right) in Trial 1.

Evidence supporting a gradual perceptual change of the caching room comes from two different lines of analysis. For the five trials we measured the distances of successive caches from the seed tray. We also measured the distance between successive caches. For Trial 1 the distance from the seed tray to the cache sites increased linearly as did the distance between successive caches (Figure 28.6). Both regression lines were significant at the 5% level. In Trial 2 the distance of successive caches from the seed tray initially increased and then the effect damped (negative quadratic effect). The distance between successive caches followed a similar pattern (Figure 28.7). Obviously, the linear effect seen in Trial 1 could not continue indefinitely as the room was of finite size. Neither a significant linear nor quadratic equation could be generated. In Trial 3 the two measures showed a significant negative quadratic pattern as they did in Trial 5 (Trial 4 was not analysed because of small sample size) where the significance was even stronger (Figures 28.8 and 28.9). Thus, as the bird progressed through the five trials it gradually changed its pattern of distributing caches.

We anticipated that the bird would show some type of caching

Trial 2

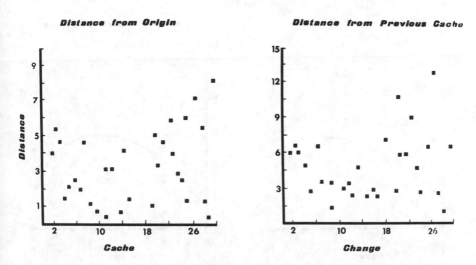

FIG. 28.7. Distribution of successive changes as a function of distance from seed tray (left) and distance between previous cache (right) in Trial 2.

pattern. We had no idea, however, about the form this pattern would take. The actual form it took in this series of trials may simply be an artifact of the experimental design (size of room, placement of food tray, arrangement of the perches, etc.) but, what is most important to us is the fact that a caching pattern did emerge. The pattern used by nutcrackers in their natural environment may be far different from that seen here, but nevertheless we feel we can make a logical argument for this one.

Which of the above patterns (linear or negative quadratic) would the bird be most likely to demonstrate in a natural setting? We can only speculate that the quadratic pattern could be more common for two reasons. First, by the fifth trial the bird was probably familiar with the caching room and a bird caching in a natural setting is most likely well acquainted with its caching areas. Although we can make a reasonable argument for caching in a negative quadratic pattern, we have no evidence from the field to support this argument (Vander Wall, personal communication). Second, we assume the centrally located seed tray was used as a focal point such as some large permanent object in nature (a large tree trunk, a branch, or a large rock) from which the bird could fly to the ground. When caching seeds the bird may use this permanent object as a clue. The permanent object may serve as an indicator of a patch full of caches. Next, the bird must distribute caches in such a manner so that the pattern is not easily interpreted by cache predators that exist in the bird's natural environment (Pivnik, 1960; Tomback, 1980). By

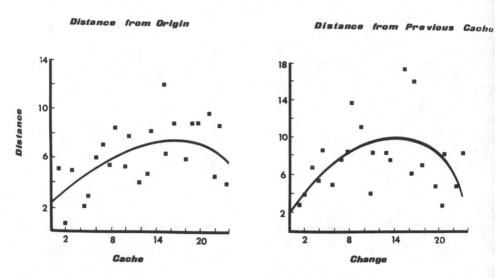

FIG. 28.8. Negative quadratic regression of the distance of successive caches from the seed tray (left) and distance between previous cache (right) in Trial 3.

gradually moving outward from the focal point the bird may inscribe a patch. The most efficient way to inscribe a patch around a centrally located focal point would be to gradually spiral outward from a starting point at or near the focal point. Our bird started at the focal point but did not put successive caches close to one another in a spiral pattern until the dampening effect occurred. Thus, as the caching area increased, the distance between caches did likewise. This means, if the bird first went north to make a cache, on its second caching attempt it went in some other direction or at least some greater distance than between the previous two caches. This seems to indicate that the bird, when familiar with the caching area, attempted to inscribe its boundaries as rapidly as possible and then later "fill in" open areas. A spiral pattern was not followed, therefore, cache predators would be deterred from using a fixed angle to move from one cache site to another, and the entire caching area would be, in effect, saturated with caches.

The focal point may then serve as a clue as to the location of many caches. If nearest—neighbor distances are important then caches within a patch will be efficiently harvested. Thus, the bird may store information about specific locations of dense patches (based on permanent objects) and specific caches within these patches. The above findings support this view as first proposed by Turcek (1966).

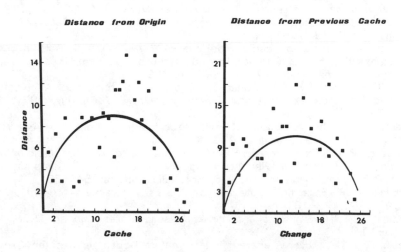

FIG. 28.9. Negative quadratic regression of the distance of successive caches from the seed tray (left) and distance between previous cache (right) in Trial 5.

ACKNOWLEDGMENTS

This research was supported by the Organized Research Committee of Northern Arizona University. Craig Benkman, Andreas Federschmidt and John Marzluff made substantial contributions in time and energy to carry out this project. Steve Vander Wall made many helpful suggestions on the manuscript. Jane Balda helped in all phases of the research. All of the above are gratefully thanked for their help.

REFERENCES

Balda, R. P. Recovery of cached seeds by a captive *Nucifraga caryocatactes. Zeitschrift fur Tierpsychologie,* 1980, *52,* 331–346.
Bock, W. J., Balda, R. P., & Vander Wall, S. B. Morphology of the sublingual pouch and tongue musculature in Clark's nutcracker. *Auk,* 1973, *90,* 491–519.
Bossema, I., & Pot, W. Het terugvinden van verstopt voedsel door de Vlaamse gaai (*Garrulus g. glandarius*). *De Levende Natuur,* 1974, *77,* 265–279.

Guintoli, M., & Mewaldt, L. R. Stomach contents of Clark's nutcrackers collected in western Montana. *Auk*, 1978, *95*, 595–598.

Ligon, J. D., & Martin, D. J. Pinon seed assessment by the pinon jay *(Gymnorhinus cyanocephalus)*. *Animal Behavior*, 1974, *22*, 421–429.

Mattes, H. Der tannenhaher im engadin. *Munstersche Geographische Arbeiten*, 1978, *3*, 87.

Pivnik, S. A. Renewal of cedar–pine stlannik *(Pinus pumila* Rgl.) in plant communities of the Cislenan Uplands (Yakutia). *Problemy Kedra, Trudy Po Lesnoe Khozyaistro Sibiri*, 1960, *6*, 129.

Pyke, G. H., Pulliam, H. R., & Charnov, E. L. Optimal foraging: A selective review of theories and tests. *Quarterly Review of Biology*, 1977, *52*, 137–154.

Roberts, R. C. The evolution of avian food–storing behavior. *American Naturalist*, 1979, *114*, 418–438.

Swanberg, P. O. *Food storage, territory, and song in the thick-billed nutcracker*. Technical Report, Proceedings of the Xth International Ornithological Congress, 1951. S. Horstadius, ed., pp. 545–554.

Tomback, D. F. Foraging strategies of Clark's nutcrackers. *Living Bird*, 1978, *16*, 123–161.

Tomback, D. F. How nutcrackers find their seed stores. *Condor*, 1980, *82*, 10–19.

Turcek, F. J. Uber das Wiederauffinden von im Boden versteckten Samen durch Tannen und Eichelhaher. *Waldhygiene*, 1966, *6*, 215–217.

Turcek, F. J., & Kelso, L. Ecological aspects of food transportation and storage in the Corvidae. *Communications of Behavioral Biology*, 1968, *1*, 277–297.

Vander Wall, S. B. An experimental analysis of cache recovery in Clark's nutcracker. *Animal Behavior*, 1982, *30*, 84–94.

Vander Wall, S. B., & Balda, R. P. Coadaptations of the Clark's nutcracker and the pinon pine for efficient seed harvest and dispersal. *Ecological Monographs*, 1977, *47*, 89–111.

Vander Wall, S. B., & Balda, R. P. Ecology and evolution of food–storage behavior in conifer–seed–caching corvids. *Zeitschrift fur Tierpsychologie*, 1981, *56*, 217–242.

29   ADAPTATION AND COGNITION: KNOWING WHAT COMES NATURALLY

Alan C. Kamil
University of Massachusetts/Amherst

I. INTRODUCTION

During the last 10-15 years, psychology and ethology have developed parallel and complementary interests in the cognitive mechanisms of animals. Psychologists have focused upon cognitive mechanisms such as memory and selective attention in attempting to explain the results of various experiments in laboratory learning situations. Ethologists, particularly behavioral ecologists, have become increasingly concerned with the effects of such cognitive mechanisms under natural conditions. The theme of this chapter is that the research and theories of psychologists interested in animal cognition could be improved by incorporating ethological concepts and methods (for the other side of the coin, benefits to ethologists of using psychological concepts and methods, see Kamil, in press; Kamil & Yoerg, 1982).

There can be no doubt that psychologists are taking a more cognitive view of animal learning. The very success of the Columbia Conference demonstrates this development. But it is not always completely clear, at least to me, exactly what is meant by the word cognitive. I am unaware of any simple, straightforward criterion by which cognitive processes are to be distinguished from the noncognitive (but see Chapter 1, by Terrace, this volume). However, the meaning of the cognitive approach does seem quite clear in general terms. It views the organism as a decision maker, emphasizing dynamic processes such as selective attention and memory. And this certainly represents a radical departure from the predominant approach to animal learning in the psychology of the past forty years or so.

There has also been a radical change in the ethological view of animals and the processes contributing to their behavior. At one time ethologists emphasized relatively fixed, inflexible behavior patterns, as reflected in concepts such as innate releasing mechanisms, fixed action patterns and ritualized display behavior. In contrast, the contemporary view, most accurately labeled behavioral ecology, tends to view the animal as a decision maker, responding dynamically to certain characteristics of its environment.

An example may help clarify the nature of this change. The traditional ethological view of territorial behavior has been a static one. Animals generally were regarded as being either territorial or not territorial, although it was known that there was some variation across different populations of the same species, or across different times of the year. However, an economic analysis of territoriality suggests a more dynamic pattern. If there are costs and benefits associated with being territorial, and if these cost–benefit values vary as a function of some set of environmental parameters, then territorial behavior could depend upon these parameters. Gill and Wolf (1975) studied this problem in a population of color–marked Golden-winged Sunbirds (*Nectarinia reichenowi*) in Kenya. They measured the costs (energy invested in defense) and benefits (increased amounts of nectar per flower) of territoriality. They found that the net benefit of being territorial depended on the average levels of food availability, with a maximum benefit at moderate levels. When Gill and Wolf observed individual birds over a substantial period of time, they found that there were day–to–day changes in territorial behavior which were correlated with nectar availability. The sunbirds were territorial only when nectar was at moderate levels. It was as if the sunbirds assessed the environment on a daily basis and then decided whether or not to attempt to maintain a feeding territory.

This example demonstrates the two factors, one theoretical and the other methodological, which have contributed most to the development of a cognitive view within behavioral ecology. The theoretical factor has been the emergence of quantitative optimization theory as a predictive tool. Optimization theory is based upon the assumption that through natural selection, animals have evolved so as to maximize their biological fitness. The basic approach is to develop a mathematical or graphical model of a specific problem that an animal regularly faces in nature, such as whether or not to be territorial. The model is then solved for its optimal solution, often in terms of maximizing either net rate of intake in the case of foraging models, or net genetic contribution to the next generation in the case of models of social behavior. The solution is then taken as a prediction. In most cases, the optimal solution depends upon environmental parameters, as in the case of sunbird territorial behavior. Thus the models predict behavioral flexibility in response to environmental variation.

The methodological factor which has contributed to the development of a more cognitive view of animal behavior among ethologists is a dramatic increase in the number of field studies in which identifiable individual animals are studied over extended periods of time. This increase largely has been a consequence of optimization models which make predictions about changes in individual behavior with environmental change. In any event, once such data began to be collected, it became apparent that such behavioral flexibility was common. Many of the phenomena observed are almost certainly the result of learning and cognition on the part of the animals. These findings have reinforced the emergence of a view of animals as dynamic, efficient decision makers.

The emergence of similar views of animals as decision makers in both psychology and behavioral ecology offers a unique opportunity for the development of interdisciplinary research and theory. This might be difficult, but could prove essential for both disciplines, especially psychology. It could provide psychological research into animal learning and cognition with something it has labored long without -- a broad biological framework within which to understand and evaluate data. We have ignored an essential attribute of the phenomena in which we are interested: that the existence of cognitive capacities in animals represents an evolutionary fact as well as a psychological one. Until recently, this may have been a moot point, since very little was known about how learning might operate in the field. This is no longer the case. There is a sizable and rapidly growing literature in behavioral ecology which has many implications for animal learning and cognition.

During the Columbia Conference, John Krebs illustrated one aspect of the relationship between psychology and behavioral ecology with an analogy. Imagine, he said, that you were a Martian scientist, and that you were given a camera to analyze. You could probably figure out the mechanisms of the camera, how it worked, even if you had no idea of what it was used for. But the analysis would be much easier if its picture-taking function was known. The data and models of behavioral ecologists are beginning to reveal some of the functions of learning and cognitive processes. The psychological analysis of these processes will be facilitated if it is carried out with an increased awareness of their adaptive function.

In fact, there are distinct risks associated with ignoring the adaptive significance of animal cognition. Time and effort may be wasted on epiphenomena, phenomena that are more a product of our procedures and techniques than of the capabilities of the animals we seek to understand (see Johnston, 1981). For example, performance on concurrent variable interval schedules and the matching law have generated an enormous literature in recent years. However, many aspects of behavior on these schedules may be related to certain peculiarities of variable interval schedules, particularly the relationship between the amount of time spent on one schedule, and the probability of a reinforcement having set up on the other schedule (Kamil & Yoerg, 1982).

Another risk which is associated with attempting to study animal cognition without reference to adaptive function is that we may remain ignorant of many important cognitive phenomena. It is quite likely that at least some of the cognitive abilities of animals are most apparent under natural conditions (Kamil, Peters, & Lindstrom, 1982). An excellent example is provided by the literature on spatial memory in food-caching birds (Balda, 1980, Chapter 28, by Balda & Turek, this volume; Sherry, Krebs, & Cowie, 1981; Shettleworth & Krebs, 1982; Vander Wall, 1982). These studies, using laboratory techniques simulating natural food-caching and recovery, have demonstrated a remarkable ability to remember spatial locations.

The camera analogy suggested by Krebs is useful because historical processes are important to both the camera and cognition. In the case of the camera, clues to its operation can be gleaned from knowledge of its function because the camera was designed as an image maker. An analogous process is responsible for the existence of cognitive processes in animals, although the historical processes are different and the use of the word design metaphorical. If the cognitive abilities of animals serve biologically significant functions, then natural selection can act upon them, and the cognitive capacities of animals will be best understood only if research design and theory take adaptive considerations into account.

Shettleworth (1982) has pointed out that it is not easy to combine functional and mechanistic considerations. One of the major stumbling blocks is that natural selection selects on the basis of results, and only indirectly for mechanism. If there are two mechanisms which could serve the same function equally well, either could be selected for. There is no necessary one-to-one relationship between function and mechanism.

Nonetheless, knowledge of function will often have important implications about mechanism. For example, a bird recovering cached seeds will most reliably find seeds in the most recently made caches if there is significant "stealing" of caches by other animals. There is some evidence for a recency effect in marsh tits (Shettleworth & Krebs, 1982).

In the next two sections of this chapter, I will present two examples of recent research in which we have attempted to combine ecological and psychological approaches. The first example demonstrates the feasibility of using ecological information to make predictions about performance on cognitive tasks. In the second example, psychological techniques are used to begin to analyze an ecological problem with interesting implications for learning and cognition.

II. STAY AND SHIFT LEARNING

Many of the problems studied by behavioral ecologists that are of most interest to psychologists involve foraging behavior. For example, many animals forage for food that is located in small, discrete patches which are easily depleted, but slow to renew. These animals can increase their foraging efficiency by avoiding revisits to those locations they have depleted recently. One ecological system which is particularly well-suited to studying this problem is the nectarivorous bird maintaining a feeding territory. Such a system is ideal for field studies because the food resource is stationary and relatively easy to measure, and the birds spend most of their time on the territories, and can often be kept under continuous observation. In many cases, a flower is completely emptied of nectar in a single visit by a bird, but produces nectar only slowly.

Several experiments have demonstrated that nectarivorous birds do avoid revisits to flowers they have recently emptied (sunbirds, Gill & Wolf, 1977; Hawaiian honeycreeper or amakihi, Kamil, 1978; hummingbirds, Gass & Montgomerie, 1981). For example, in a study of amakihi (*Loxops virens*), I recorded the visits of color-banded, territorial birds to a set of mapped flowers. It was quite clear that the birds made exactly one visit per flower more frequently than would be expected by chance. In addition, I found that while the resident birds avoided revisits to flowers they had emptied, intruding birds did not avoid these flowers. This aspect of the data strongly implies that some sort of memory for emptied flowers is involved (see Kamil, 1978, for details). This result has several implications. Fine scale spatial memory should be particularly well-developed in these birds. Since in many cases flowers produce nectar at predictable rates, nectar feeding birds may be able to fine-tune the temporal patterning of their flower visitations. Since immediate returns to already visited flowers never produce reward, while visits to unvisited flowers often do, nectarivorous birds may find it easier to learn to shift among different positions than to learn to return to the same position.

We have tested this last implication in a spatial learning experiment with hummingbirds (Cole, Hainsworth, Kamil, Mercier, & Wolf, 1982). The procedure involved a two part trial. In the first part of the trial, the information stage, the bird was presented with a single artificial flower located on either the right or left side of a styrofoam feeding board. This flower contained a small amount of 0.5M sucrose solution, and its location was random from trial to trial. The animal was allowed to take this nectar. The feeding board was then removed, and the second, choice stage of the trial began 10-12 sec later. During the choice stage, the hummingbird was presented with two artificial flowers on the feeding board, only one of which contained sucrose solution. The bird could choose either to return to the location visited during the information stage, or to avoid that location.

During stay learning, return visits were rewarded; during shift learning, only visits to the position not visited during the information stage were rewarded. All 8 hummingbirds received both stay and shift learning, with order of training counterbalanced. Training continued until a criterion of 3 consecutive days above 80% correct was achieved. The results were quite dramatic. Shift learning was always much more rapid than stay learning, regardless of whether it came first or second. The fastest stay learning took 282 trials and 130 errors, the slowest shift learning took 180 trials and 96 errors. There were two major differences between performance during shift learning and performance during stay learning. At the outset of the experiment, all birds showed a pre-experimental basis towards shifting. This may have been innate, or a result of the previous experience of these wild-caught birds. However, the rate of shift learning (in percentage improvement per day) was also significantly higher than the rate of stay learning. This held even when each task was being learned after the other, and initial rates of performance were equal. These results suggest that nectar feeding birds have an evolved tendency to shift locations of flower visits, which is manifested in both starting performance and in differential rates of learning.

To be sure that there is no misunderstanding, I would like to make the reasoning behind this experiment very explicit. The logical structure of the stay and shift tasks was very similar. In each case, the only cue predicting the location of food during the choice stage was the location visited during the information stage. In each case, the correct response was a visit to a specific location. Therefore, the reason for the difference in learning rates can be conceived of as residing in the structure of the animal being tested rather than in the tasks themselves.

For most nectarivorous birds, the major source of energy is floral nectar located in small, easily depleted, slowly renewing specific locations. Return visits to recently emptied flowers will never be rewarded; visits to new flowers will frequently be rewarded. Over evolutionary time, natural selection will favor those birds who learn to shift flower locations, but not those who repeatedly return to the same location. This process could be responsible for the results of our spatial learning experiment, and also is consistent with the ecological view.

Several points need to be made about an ecological view of learning and cognition. First, it needs to be contrasted with the idea of "biological constraints on learning" (e.g., Seligman & Hager, 1972). The basic idea of the biological constraints approach is that learning represents an evolved ability. As Seligman and Hager (1972) put it:

> What an organism learns in the laboratory or in his natural habitat is the result not only of the contingencies which he faces and has faced in his past but also of the contingencies which his species faced before him —— its evolutionary history and the genetic outcome (p.1).

However, in practice, most biological constraints research has focused upon apparent anomalies in performance in psychological paradigms, such as taste aversion learning and auto-shaping. These anomalies are then interpreted in terms of the adaptations of the subject. But the subjects are usually rats or pigeons, about whom relatively little is known outside of the laboratory. The tasks upon which animals are tested in many of these experiments are defined by the history of psychology rather than knowledge of the natural environment of the species being tested. Therefore, the adaptive explanations tend to be very speculative, and weak. Indeed, they might best be classified as just-so stories (Gould & Lewontin, 1979).

The ecological perspective advocated in this paper is in basic agreement with the philosophical underpinnings of the biological constraints approach. It simply takes the biological constraints view further towards it logical conclusion, and looks to detailed ecological information to make predictions about the characteristics of learning and cognition. And as the hummingbird spatial learning experiment demonstrates, the analysis of learning and cognition can begin from detailed knowledge of the ecology of the species being tested.

A second issue concerns levels of explanation. Some might argue that the hummingbird results simply demonstrate a well-known phenomenon, spontaneous alternation. There are some problems with such an argument, particularly in view of the differences in percentage improvement per day, even when each task was learned after the other. But at a more fundamental level, this argument misses an important point. Spontaneous alternation is merely a label for an observed behavior pattern; it explains nothing and predicts relatively little. The ecological point of view suggests a reason for the behavior pattern, and makes some interesting predictions, especially comparative ones.

For example, it is possible that tendencies to shift responses between different locations can result from two different adaptations —— exploration of novel environments, and foraging on depleted resources. The exploration tendency should be most evident when the animal is in a novel or highly variable environment. The research of Glickman and Sroges (1966) on curiosity and manipulative behavior suggests that exploration may be most prevalent in predatory species. As a foraging adaptation, the tendency to shift should be long-lasting, even within a relatively stable environment, as in our hummingbird experiment. This tendency should also vary as a function of ecological parameters, particularly the depletability and renewal rates of the food sources upon which the species in question tends to feed.

The ecological view generates predictions subject to empirical test, provided enough is known about the species being tested. Many of these predictions are comparative, and so may stimulate psychologists to begin work with new species. However, history suggests there will be considerable resistance to such a development. Consider the series of articles over the years urging research with a broader selection of species (e.g., Beach, 1950; Lockhard, 1971), and the lack of response to these articles. The ecological approach offers a distinct strategy for selecting species for study on the basis of similarities and differences in ecology (Kamil & Yoerg, 1982).

III. SELECTION AND PATCH DEPLETION

The purpose of the above example was to demonstrate how the ecological approach could be applied to a situation involving learning and memory. There are many other concepts and observations of behavioral ecology which are of interest to psychologists interested in animal cognition. Many of these concepts surround the problems of patch selection and perseverance in the face of patch depletion. For example, consider the problem of a forager with two patches in which it can hunt. These patches can differ in value (prey density)but which one is the better cannot be predicted in advance. What is the best way for the animal to deal with this problem?

The model proposed by Krebs, Kacelnik, and Taylor (1978) visualizes the animal as an information gatherer. According to their model, during an initial sampling stage, the animal should utilize each patch equally often until sufficient information has been collected to allow an accurate decision, after which exploitation should occur. This is a very cognitive view of the analogous probability learning situation. The model of Krebs et al. (1978) makes several interesting predictions. According to the model, the more similar two patches are in prey density, the longer the duration of the sampling period. In psychological terms, this can be translated into the prediction that it should take longer to learn the difference between similar patches than between dissimilar patches. Although confirmed in the Krebs et al. (1978) experiment, this is hardly a novel prediction. However, the model also makes another prediction which is, I believe, entirely novel. The duration of the sampling period should vary directly as a function of the length of the foraging bout. Data supporting this prediction have been reported by Kacelnik (1979).

Another problem with cognitive implications in which behavioral ecologists have been interested is persistence in the face of depletion within a patch. Food is often distributed in discrete patches separated from each other by some distance. There are several prey per patch, and as the animal hunts the patch becomes depleted. How long should the forager stay in a particular patch before giving up on that patch and traveling to the next one? Optimal foraging models (e.g., Charnov, 1976) predict that animals faced with this problem should react dynamically to several environmental parameters such as the time or energy required to travel between patches and average prey density. These predictions have been supported by several experiments (e.g., Cowie, 1977; Krebs, Ryan, & Charnov, 1974). These models and experimental results raise several interesting psychological issues.   For example, if the forager adjusts its behavior as its achieved rate of intake changes, as implied by the models, over what time interval does it integrate information about intake rate?   Another question concerns what types of decision rules the forager may use in making its decisions to leave a patch.

Iwasa, Higashi, and Yamamura (1981) have developed a model that predicts that the best rule for leaving a patch will depend upon the statistical distribution of prey within the patch.   For example, if the prey distribution is binomial, then a fixed-time rule would be best, but if the distribution is Poisson, then leaving after a fixed number of prey have been found would be best.   This idea could be interpreted in two ways. Particular species might be adapted to specific modes of prey depletion. In this case, a given species would use a particular rule regardless of the prey distribution it faces.   Alternatively, a species could be sensitive to differences in prey depletion patterns and use different rules when faced with different types of depleting patches.   These are not mutually exclusive alternatives.   Some species could be more flexible than others.

Sonja Yoerg and I have recently completed an experiment in which we examined the ability of blue jays to adjust to different within-patch depletion rules.   We reasoned that since blue jays are well-known as extreme generalists, they might be sensitive to variations in patterns of depletion.   It should be noted that this reasoning is not as tight as that underlying the hummingbird experiment.   It is not based upon detailed naturalistic information about the distribution of food and blue jay's responses in the field.   We are not testing an hypothesis in this case. Rather, we are exploring a behavioral problem in learning suggested by the ecological literature.

We used a simultaneous choice procedure in which the jay could choose to hunt for a prey item in either of two spatially distinct patches on each trial.   The patch on the right was the nondepleting patch.   There was a constant probability of 0.25 that this patch would contain a prey item on each of the 36 trials of each daily session.   The patch on the left was the depleting patch.   At the beginning of each session there was a probability of 0.50 that the depleting patch contained prey, but this probability was reduced to 0 partway through the session.   For two of the jays, the probability of the presence of prey in the depleting patch went to 0 after a fixed number of prey had been found in that patch, analogous to Iwasa et al.'s (1981) binomial distribution.   The number of prey preceding depletion for these birds was 6, and these prey occurred within the first 10 to 14 trials.   For the other two jays, the drop in prey probability always occurred after 12 trials, with the number of prey varying from 5 to 7, analogous to the Poisson case.

By using this procedure of pitting a depleting against a nondepleting patch, we hoped to make the most efficient strategy quite simple: begin each session in the depleting patch and switch to the nondepleting patch after either the appropriate number of prey or of trials.   The blue jays did adjust their choice behavior in this direction, but were only moderately efficient, even after 70 sessions.   At the beginning of the experiment, the jays chose the two patches with approximately equal frequency throughout the session.   By the end of the experiment, they were choosing the depleting side most of the time early in each session and the nondepleting patch late in each session. For example, during the last 10 sessions of the experiment, the jays chose the depleting patch 89% of the time during the first 12 trials of each session, but chose the depleting patch only 13% of the time during the last 12 trials. There were no apparent differences between the two jays exposed to the binomial distribution of prey and the two jays exposed to the Poisson distribution.

However, there was an unexpected result which indicates that the jays use different rules when deciding to switch between patches with differing prey distributions. Different events preceded switches out of the two patches. Switches from the depleting to the nondepleting patch usually were preceded by a "run of bad luck," two or more trials in a row without prey. Switches from the nondepleting to the depleting patch, in contrast, were usually preceded by a trial in which a prey item was found.

This difference in the events preceding switches out of each of the two patches, together with the marked tendency to choose the depleting patch only in the first third of the session, demonstrates that the jays were quite sensitive to the patch differences in prey distribution. One question that remains is whether the relatively inefficient switching behavior is due to some particular feature of the experimental design. For example, if depletion occurred more rapidly, would switching behavior more closely approximate maximal efficiency? Other questions concern the difference in switch rules. For example, if we assume that jay switches when its subjective estimates of the probability of finding a prey item in each patch favor the other patch, then careful manipulation of conditional probabilities in the two patches would be most revealing. In fact, in this experiment, the conditional probability of finding prey on the trial immediately following prey was equal to the conditional probability of finding prey immediately after no prey within each patch.

IV. CONCLUSIONS

Psychologists interested in animal learning and cognition need to reconsider and re-evaluate their research and its potential long-range significance. In and of itself, the psychological study of animal learning is of little interest. Research in this area acquires importance only as it relates to broader issues. Historically, the most important reason for animal learning research in psychology has been the hope that the results would be capable of generalization to our own species. Animals have been used as convenient subjects in research intended to discover general principles of learning which could then be applied to humans. While this approach has had its successes, it largely has failed. The attempt to carry out research which is related directly to human cognitive research can be regarded, at least in part, as a response to this failure. Whether or not the increased cognitive emphasis will enable animal learning research to be more relevant to human cognition remains to be seen.

But there is another context in which research or animal learning and cognition could prove very significant. Instead of thinking of animals as proto-humans, it is possible, and worthwhile, to study animals as animals. Behavioral ecology is an important area to which animal cognition is relevant, and from which psychologists can obtain new ideas and directions. Behavioral ecologists start with the working assumption that behavior is well-adapted to the natural environment of the species. Many field and laboratory studies have demonstrated that animals can and do adjust their behavior to environmental variation in ways that are very functional. Often these adjustments appear to involve cognitive mechanisms. These results strongly imply that cognition plays an important and direct role in adaptation.

Since psychologists have long experience with the theoretical and methodological difficulties that are presented by the experimental study of learning and cognition, they could make significant contributions to the investigation of learning in an adaptive framework. One of the easiest and most useful ways in which psychologists can get involved in this area would be to begin to investigate these apparent instances of learning highlighted by behavioral ecology. There are many interesting instances which call for study, including patch selection (e.g., Krebs, Kacelnik, & Taylor, 1978), patch perseverance (e.g., Krebs *et al.*, 1974; Cowie, 1977), food-caching and memory (e.g., Balda, 1980; Shettleworth & Krebs, 1982; Vander Wall, 1982), selective attention and prey crypticity (e.g., Bond, 1983; Dawkins, 1971; Pietrewicz & Kamil, 1981), and the selection of food items (e.g., Goss-Custard, 1981; Krebs, Erichsen, Webber, & Charnov, 1977). It is quite possible that experiments in these areas will completely change our conceptions of the cognitive abilities of animals.

A final cautionary note is necessary. It is altogether too easy to be superficial in attempting to make use of a discipline other than one's own, and this can lead to serious misunderstandings (Lehrman, 1974). An evolutionary approach demands thorough understanding of the ecology of the animals studied, so that the tasks which are set for them accurately reflect the tasks nature has set for them. This will require careful reading of a literature much different from the psychological literature. If done carelessly, or without an understanding of the context in which ecological research is carried out, serious misunderstanding could develop. The focus of the ecological literature is often descriptive and much less concerned with mechanism than is the psychological literature. But if these differences are appreciated, the ecological literature on behavior could provide ideas and directions which could contribute to a revitalization of the psychological study of animal learning and cognition.

ACKNOWLEDGMENTS

Preparation of this manuscript was supported by NSF grant BNS-81-02335. I would like to thank Sonja Yoerg, Pamela Real, Herbert Roitblat, and Deborah Olson for their helpful comments on an earlier draft of this paper.

REFERENCES

Balda, R. P. Recovery of cached seeds by a captive *Nucifraga caryocatactes*. *Zeitschrift fur Tierpsychologie*, 1980, *52*, 331–346.

Beach, F. A. The snark was a boojum. *American Psychologist*, 1950, *5*, 115–124.

Bond, A. B. Visual search and selection of natural stimuli in the pigeon: The attention threshold hypothesis. *Journal of Experimental Psychology: Animal Behavior Processes*, 1983, *9*, 292–306.

Charnov, E. L. Optimal foraging: The marginal value theorem. *Theoretical Population Biology*, 1976, *9*, 129–136.

Cole, S., Hainsworth, F. R., Kamil, A. C., Mercier, T., & Wolf, L. L. Spatial learning as an adaptation in hummingbirds. *Science*, 1982, *217*, 655–657.

Cowie, R. J. Optimal foraging in Great Tits (*Parus major*). *Nature*, 1977, *268*, 137–139.

Dawkins, M. Shifts in "attention" in chicks during feeding. *Animal Behavior*, 1971, *19*, 575–582.

Gass, C. L., & Montgomerie, R. D. Hummingbird foraging behavior: Decision-making and energy regulation. In A. C. Kamil & T. D. Sargent (Eds.), *Foraging behavior: Ecological, ethological, and psychological approaches*. New York: Garland, 1981.

Gill, F. B., & Wolf, L. L. Foraging strategies and energetics of East African sunbirds at mistletoe flowers. *American Naturalist*, 1975, *109*, 491–510.

Gill, F. B., & Wolf, L. L. Nonrandom foraging by sunbirds in a patchy environment. *Ecology*, 1977, *58*, 1284–1296.

Glickman, S. E., & Sroges, R. W. Curiosity in zoo animals. *Behaviour*, 1966, *24*, 151–188.

Goss-Custard, J. D. Feeding behavior of redshank, *Tringa totanus*, and optimal foraging theory. In A. C. Kamil & T. D. Sargent (Eds.), *Foraging behavior: Ecological, ethological, and psychological approaches*. New York: Garland, 1981.

Gould, S. J., & Lewontin, R. C. The spandrels of San Marco and the Panglossian paradigm: A critique of the adaptationist programme. *Proceedings of the Royal Society of London*, 1979, *205*, 581–598.

Iwasa, Y., Higashi, M., & Yamamura, N. Prey distribution as a factor determining the choice of optimal foraging strategy. *American Naturalist*, 1981, *117*, 710–723.

Johnston, T. D. Contrasting approaches to a theory of learning. *The Behavioral and Brain Sciences*, 1981, *4*, 125–139.

Kacelnik, A. J. *Studies of foraging behavior and time budgeting in great tits (Parus major)*. PhD Thesis, University of Oxford, 1979.

Kamil, A. C. Systematic foraging by a nectar-feeding bird, the Amakihi (*Loxops virens*). *Journal of Comparative and Physiological Psychology*, 1978, *92*, 388–396.

Kamil, A. C. Optimal foraging theory, and the psychology of learning. *American Zoology*, in press.

Kamil, A. C., & Yoerg, S. I. Learning and foraging behavior. In P. G. Bateson & P. H. Klopfer (Eds.), *Perspectives on ethology*. (Vol. 5.) New York: Plenum, 1982.

Kamil, A. C., Peters, J., & Lindstrom, F. An ecological perspective on the study of the allocation of behavior. In M. L. Commons, R. J. Herrnstein, & H. Rachlin (Eds.), *Quantitative analyses of behavior, Vol. II: Matching and maximizing accounts*. Cambridge, Mass.: Ballinger, 1982.

Krebs, J. R., Kacelnik, A., & Taylor, P. Test of optimal sampling by foraging in Great Tits. *Nature*, 1978, *275*, 27–31.

Krebs, J. R., Ryan, J. C., & Charnov, E. L. Hunting by expectation or optimal foraging? A study of patch use by chickadees. *Animal Behavior*, 1974, *22*, 953–964.

Lehrman, D. S. Can psychiatrists use ethology? In White (Ed.), *Ethology and psychiatry*. Toronto: University of Toronto Press, 1974.

Lockhard, R. B. Reflections on the fall of comparative psychology: Is there a message for us all? *American Psychologist*, 1971, *26*, 168–179.

Pietrewicz, A. T., & Kamil, A. C. Search images and the detection of cryptic prey: An operant approach. In A. C. Kamil & T. D. Sargent (Eds.), *Foraging behavior: Ethological, ethological, and psychological approaches*. New York: Plenum Press, 1981.

Seligman, M. E. P., & Hager, J. L. (Eds.) *Biological boundaries of learning*. New York: Appleton–Century–Crofts, 1972.

Sherry, D. F., Krebs, J. R., & Cowie, R. J. Memory for the location of stored food in marsh tits. *Animal Behavior*, 1981, *29*, 1260–1266.

Shettleworth, S. J. Function and mechanism in learning. In M. Zeiler & P. Harzen (Eds.), *Advances in analysis of behavior: Biological factors in learning*. (Vol. 3.) New York: Wiley, 1982.

Shettleworth, S. J., & Krebs, J. R. How marsh tits find their hoards: The roles of site preference and spatial memory. *Journal of Experimental Psychology: Animal Behavior Processes*, 1982, *8*, 354–375.

Vander Wall, S. B. An experimental analysis of cache recovery in Clark's nutcracker. *Animal Behavior*, 1982, *30*, 84–94.

30 ONTOGENETIC DIFFERENCES IN THE PROCESSING OF MULTI-ELEMENT STIMULI

Norman E. Spear
David Kucharski
State University of New York

I. INTRODUCTION

We discuss in this chapter some consequences of processing multiple-event episodes for learning and memory. The special focus is on how these consequences differ for immature and mature animals and what these differences might imply about the ontogeny of learning and memory.

It has been observed that in some circumstances one element of a stimulus compound "overshadows" the conditioning of another, but in other circumstances that element "potentiates" conditioning of another. Until recently "overshadowing" has been the more dominant phenomenon, both empirically and theoretically. One need only consider an animal's limited capacity for processing information to appreciate that if the stimulus compound AB is to be associated with some consequence, the more attention or associative strength the animal allocates to A, the less might be allocated to B. "Overshadowing" is the lesser conditioning of B when presented in an AB compound than when presented alone. A common observation since Pavlov's early work, this effect has also been a central consideration of contemporary theories of conditioning (Kamin, 1969; Mackintosh, 1975; Rescorla & Wagner, 1972; Wagner, 1981).

From this perspective, "potentiation" is paradoxical. Even aside from specific theories, to find greater conditioning of B when it is presented in an AB compound than when presented alone seems contrary to a limited-capacity system for processing and to economical allocation of the resources in such a system. Yet this effect has been found, labeled "potentiation" and established firmly as one possible consequence of multiple-element conditioning (for reviews, see Domjan, in press; Lett, 1982; Rescorla, 1981; Rescorla & Durlach, 1981). What has not been decided is precisely when overshadowing may be expected and when potentiation.

II.   EXPLANATIONS OF POTENTIATION

A.   AN EFFECT OF FEEDING, TERMED "ASSOCIATIVE POTENTIATION"

Potentiation may be exclusively a phenomenon of the feeding system, and its general importance for learning and memory might therefore be quite limited (Lett, 1982; Rusiniak, Hankins, Garcia, & Brett, 1979). The core of this argument has been that the clearest instances of potentiation have taken, exclusively, a singular form: presence of a distinctive taste has enhanced the conditioning of another element. The latter typically has been an odor or another taste, but not always (e.g., Galef and Osborne, 1978, found that taste potentiated conditioning of a visual stimulus).

The idea that potentiation is exclusively a feeding phenomenon has provided the framework for a particularly thorough set of experiments by Lett (1982), who calls the phenomenon "associative potentiation." In this view, the strength of the association between a CS (conditional stimulus) element and an illness is said to be enhanced in an absolute sense, in and of itself, because of the presence of another element -- explicitly a taste -- in the CS compound.

Lett's explanation is based on the general notion that the association between gustatory and visceral consequences, between a taste and a sickness, involves a special learning system that is phylogenetically old and can be seen in essentially all animals. Potentiation is viewed as synonymous with absolute strengthening of an association between a sickness and a stimulus feature such as an odor (or color) that ordinarily would not be readily associated with sickness, due to the presence of a taste in compound with that feature. The presence of the taste potentiates this association by endowing the odor with properties of the taste. In this view also, potentiation of the learning of odor by the presence of taste is particularly adaptive because odor is a distal cue that controls a different response component of feeding than does taste, the more proximal cue. Presumably the economy of behavior is therefore enhanced because upon sensing the conditioned odor, the animal can terminate its foraging for a substance to which taste aversion has been conditioned. Similarly, if a preference is conditioned to a taste, the potential conditioning of odors paired with that taste will result in more rapid approach to the substance involved.

This explanation has been subjected to an experimental analysis by Lett (1982) that is both broad and thorough. Her results are impressive. Potentiation would seem quite well suited as a special adaptation for feeding. But it is uncertain whether potentiation has generality beyond feeding per se and whether the mechanism for potentiation is always, or ever, an absolute strengthening of the association between the potentiated CS element and the US (unconditional stimulus). We present experiments that address both of these uncertainties. Our data will question the exclusivity of feeding for potentiation by showing that potentiation can occur in conditioning circumstances where ingestion is not involved, and when ingestion is central to the conditioning, potentiation sometimes occurs but sometimes does not, depending on other factors. Our data will also question the notion of potentiation as an absolute strengthening of a less likely association by showing that the effect of the potentiated element of the CS on behavior depends on the animal's subsequent experience with the alternative element.

B. THE MULTIPLE ASSOCIATION THEORY

According to the multiple association explanation of Rescorla and his colleagues (e.g., Rescorla & Durlach, 1981), potentiation is a direct consequence of separate associations between A and the illness, B and the illness, and between A and B. Because A is associated with B, which itself is aversive due to conditioning, the net aversion to A is inflated beyond its association with illness.

Our experiments do not directly bear on this explanation. It is evaluated indirectly, however, in considering our finding that preweanling (18-day old) rats show greater potentiation than adults. Compared to adults the younger rats might form stronger within-compound associations between elements A and B; or, they might generally have a greater probability of forming multiple associations among all pairs of elements, or between elements of a particular kind, within an episode. For the present data at least, we can dismiss the possibility that the strengths of the associations between A and the illness or B and the illness are greater for the younger animals. This simply was not indicated by any of the direct tests. There are no data to discount the possibility of a stronger within-compound association between A and B for preweanlings than adults, although such a difference would not be predicted from what is known about the ontogeny of conditioning in general. Yet the facts might still be accommodated by this theory as we shall discuss later.

C. THE GESTALT EXPLANATION

This explanation, termed "Gestalt" for lack of a better term (Rescorla, 1981), focuses on perception. The basic notion is that an association between A and B is not actually formed when they are presented in compound; instead, A and B are perceived together as a whole during conditioning, presumably as an "integral" rather than a "separable" compound (see Garner, 1970, 1974, and Chapter 19, this volume, by Riley). This explanation of potentiation requires three features: (1) When A and B are perceived as configuration, integral, or "blob," the net saliency (or intensity) of the compound CS is greater than that of A alone; (2) Greater CS intensity leads to greater conditioning; (3) The subsequent test of A would take advantage of the greater conditioning to the AB compound only if in the latter case there is sufficient compensation for the generalization decrement due to the difference between the conditioning and testing stimuli.

This is a particularly appealing hypothesis ontogenetically because of the evidence that young children seem to treat, cognitively, certain compound stimuli as integrals that are treated as separables at later ages (e.g., Shepp, 1978; Tighe & Tighe, 1979). The empirical tests we report provide some support for this hypothesis, in the context of ontogenetic change in the learning exhibited by the rat.

We did not set out to test these alternative theories explicitly. Our original intention was to test some ideas about ontogenetic change in memory processing for multi-event episodes. We were in fact somewhat surprised to find potentiation emerge as it did in our experiments. Ultimately, however, we were led to consideration of potentiation as a quite general phenomenon for understanding learning and memory, and in particular, as a potential tool in answering a classical question about the ontogeny of cognitive processes. This emphasis on ontogenetic considerations requires some brief background.

III. COMPOUND-CONDITIONING FEATURES OF ONTOGENETIC INTEREST

Our ontogenetic consideration of the overshadowing-potentiation distinction was directed by three factors. The first is that in a general sense, learning involving multi-element stimuli provides an analyzable model of what is required of the developing organism in learning to adapt to its multi-element world. This model could address the ancient question of whether the newborn encounters a world of perceptibly discrete elements ready for synthesis or whether the newborn's first step must be analysis (differentiation) of a perceptual glob, i.e., James' "booming, buzzing confusion." In other words, this issue could be approached experimentally through tests of learning and memory with multi-element stimuli. The ultimate question of course applies not only for the newborn but to adaptation required throughout ontogeny, a general issue handled most thoroughly by Gibson (1969).

The second reason for interest in this topic emerged from our overriding interest in infantile amnesia, modified from Freud's usage to refer to the relatively rapid forgetting among more immature animals (Campbell & Spear, 1972). Stated simply: (1) In spite of equivalent learning of some features of an episode, the learning of other aspects may differ among immature and mature animals; (2) animals of different ages may learn different things about the same episode; and (3) what is learned can determine what is remembered, not merely in the trivial sense that one could hardly be expected to remember X if X had not been learned, but in the more interesting sense that having learned Y might later help the remembering of X (Spear, 1978, 1979).

Our third reason for studying this topic is our belief that the prevailing conventional wisdom that the associative capacities of infants are not different from those of adults (e.g., Simon, 1974), which itself contrasts with an earlier view of infantile deficiencies in learning and memory, requires a hard assessment. Our initial experimental tests required systematic variation in the number of discrete elements an animal could, should, or must learn for adaptive behavior. A simple, preliminary approach is in conditioning to compound stimuli. We cannot claim even a firm beginning in testing for quantitative limitations in the immature animal's capacity for processing information. Although the question of such limitations cannot readily be dismissed, it is technically untestable for now. But perhaps more pertinent is the lack of empirical evidence that infants do acquire fewer memory attributes to represent a multi-event episode; to the contrary, the existing evidence suggests the opposite conclusion, that younger animals process *more* of the redundant aspects of an episode than do adults (see Spear & Kucharski, 1984).

IV. EXPERIMENTAL TESTS OF ONTOGENETIC DIFFERENCES IN COMPOUND CONDITIONING

Although our experiments were intended primarily for assessment of ontogenetic differences in conditioning to elements of a stimulus compound, they also bear on general issues of animal cognition in the learning and retention of complex stimuli. With one exception, the eight new studies we report employed classical conditioning of a taste (or, "flavor") aversion due to its pairing with an illness induced by injection of LiCl and analytical focus was on the conditioning of one taste element of a two-taste compound. The exception involves conditioning of an aversion to an odor by its pairing with a footshock; for one study the odor was presented in compound with a different odor, for the other it was presented in compound with a particular location. We provide here only a brief description of our methods and procedures; details omitted here can be found in descriptions of our previous work of this kind (e.g., Infurna, 1981; Kessler & Spear, 1980; Steinert, Infurna, Jardula, & Spear, 1979; Steinert, Infurna, & Spear, 1980; Steinert, 1981; Caza, Steinert, & Spear, 1980).

For our first experiment preweanling (18 days old) and adult (60 days old) rats were made ill with LiCl after they had tasted a particular flavor (15% sucrose solution) that either was or was not accompanied by a different flavor (1.25% decaffeinated coffee). The choice of these stimuli and particular procedures was facilitated by the thorough Ph.D. thesis of Pamela Steinert (1981) which had established, for instance, that equal single-element conditioning could be seen with these concentrations of these flavors. We will focus on conditioning to the 15% sucrose solution because we have learned a good deal about environmental and ontogenetic determinants of aversive conditioning to this particular substance (see above references) and because we could apply procedures to permit equivalent baseline conditioning for infants and adults with this as a single-element conditioned stimulus. These circumstances permitted reasonably clear ontogenetic comparisons of the two types of stimulus compounds we tested -- "simultaneous," in which the two flavors were mixed in a common solution, and "successive," in which one of the flavors was presented immediately following the other. Evidence with adult rats had indicated that this slight variation in form of the compound yielded substantial differences in the interaction between its elements. Rescorla and Durlach (1981) reported greater aversion to drinking either banana or almond flavored water when it had been mixed in the same solution as saccharin during conditioning than when presented alone, but lesser aversion when either "odor" had preceded or followed the saccharin during conditioning than if presented alone; in other words, they observed potentiation with simultaneous pairing of CS elements but overshadowing with successive pairing. Also there had been indications of ontogenetic differences among very young rats (4 days versus 8 days of age) in the consequences of simultaneous versus successive presentations of odors that serve as conditioned stimuli (Cheatle, 1980; Cheatle & Rudy, 1979).

A third variable was delay of the unconditional stimulus; injection of LiCl was given either immediately or one hour after consumption of the flavors. This variation in CS–US interval was included because of the ontogenetic differences known to occur in similar circumstances as a consequence of such delays and to avoid ceiling effects on measurement that might occur with zero CS–US interval. Careful control over temporal parameters, flavor order, and amount consumed was achieved in this experiment (and in those to follow) by allowing the hand-held animals to drink small amounts at specified times from the blunt end of a syringe that contained the solution, a procedure for which the animals were previously trained.

The full experimental design also included, for each experimental condition, explicitly unpaired controls given a 24-hour interval between CS and US. And, to control for general experience with the flavors prior to test, animals in the "single flavor" condition were given the alternative flavor also, one hour before they consumed the CS. Finally, all animals were tested on both flavors in counterbalanced order. The first test also served as an extinction treatment for that flavor, the significance of which will become clear shortly.

Three interesting results were obtained. First, although the conditioned aversion to sucrose solution was about the same for preweanlings and adults whether the sucrose had been presented "alone" or presented in the successive compound with coffee (and tended to be weaker in the latter case), conditioning to the sucrose solution was decisively greater for the younger rats when the sucrose and coffee had been tasted simultaneously (see Figure 30.1). In the simultaneous case, both the preweanlings and the adults showed potentiation -- greater conditioning to sucrose presented in compound with coffee than when presented alone -- but that of the younger rats was significantly stronger. Second, extinguishing (i.e., testing without reinforcement) the coffee element prior to testing the aversion to the sucrose element markedly reduced -- in fact, eliminated -- the preweanlings' aversion to sucrose. This extinction to coffee had a lesser, although statistically significant, effect on the adult's aversion to sucrose (see Figure 30.1). The third major result merely re-emphasizes the dramatically greater sucrose aversion for infants than adults when the simultaneous compound had been the CS: In all previous experiments to our knowledge, preweanlings have never shown a greater conditioned flavor aversion than adults and have always shown a weaker aversion when a 1-hour interval intervened between the CS and the US. Yet in the present experiment, infants given the simultaneous CS had significantly greater sucrose aversion than adults given that CS, whether the CS–US interval was 0 or 1 hour (see Figure 30.2).

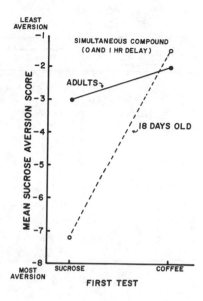

FIG. 30.1. The mean sucrose difference score for infants and adults conditioned with the compound and tested either before (no extinction) or after (extinction) the coffee consumption test. The coffee test is considered to result in some extinction of the coffee aversion. Because our definition of an aversion refers to a significant change from baseline preference due to conditioning, sucrose preference scores (sucrose intake/total intake) were further transformed into standardized difference scores [(individual experimental subject's preference score – mean of the respective control group's preference score) / standard deviation of respective control group's preference score].

A. GENERALITY OF THE MAJOR ONTOGENETIC DIFFERENCES

We conducted three studies to assess the generality of our finding of greater potentiation by preweanlings than adults, applying specific procedures somewhat different than those of the above experiment. The first included an entire replication of the major conditions of the above experiment except for a difference in the procedures for the critical "single–stimulus" condition; in this study presentation of the CS element was not preceded one hour earlier by the alternative element. The second study assessed potentiation of the conditioning of a taste aversion by the presence of an odor. The third study tested potentiation of conditioning to one novel odor by presence of a different odor, this time using an entirely different unconditioned stimulus, footshock.

The replication of the above experiment was relatively straightforward. It included the same orthogonal variation in age, CS–US interval, and type of CS, with explicitly unpaired control conditions accompanying each experimental condition. The only change from the previous experiment

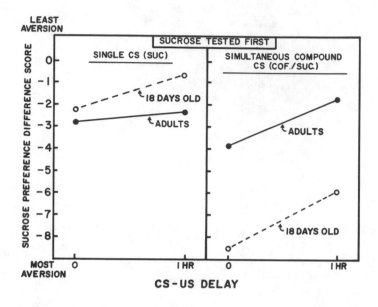

FIG. 30.2. The mean sucrose difference scores for infants and adults conditioned either with the simultaneous compound or the sucrose solution. The immediate CS–US groups were conditioned with no delay between the conditional and unconditional stimulus; 1–hr CS–US groups were conditioned with a 1 hour delay.

was in the procedure for presenting the sucrose solution alone as the CS. Whereas in the previous experiment the coffee flavor had been presented one hour prior to the sucrose flavor to equate taste experiences across conditions, the animals in the sucrose–only CS were not previously given coffee solution in the present experiment. The other type of CS tested was the simultaneous presentation of sucrose and coffee; the successive–CS condition was not tested in this replication experiment. Stated simply, the major results of the previous experiment were confirmed statistically (Figure 30.3).

A second study compared sucrose aversion among preweanlings and adults when it was the sole element of the CS in comparison to when it was one element of a multi–element compound. For some groups the latter consisted of a simultaneous compound of sucrose with lemon in solution plus an ambient novel odor, methyl salicylate (which smells like wintergreen), and for the other groups the compound consisted of only the sucrose solution plus the ambient odor. Still other experiments in this study included as one CS condition a simultaneous compound of sucrose and methyl salicylate tasted in solution plus methyl salicylate as an ambient odor. Generally speaking, potentiation was not found in all of these experiments; in some, overshadowing was found. But in each case, the preweanlings were less likely than the adults to be selective in their learning of the sucrose element, or in other words, the preweanlings were

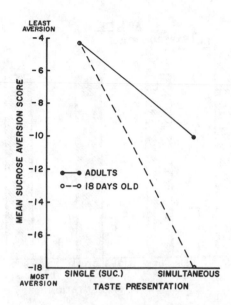

FIG. 30.3. The mean sucrose difference scores for infants and adults conditioned either with the compound or the sucrose solution.

more likely than adults to behave as if they responded to the alternative elements of the compound as well as the sucrose element. Potentiation was found among the preweanlings when sucrose was paired in a simultaneous compound with an ambient methyl salicylate odor; the preweanings' aversion to sucrose solution was greater in this case than occurred when sucrose alone was the CS. But such potentiation did not occur for adults in this experiment (see Figures 30.4 and 30.5). For the three other CS compounds tested in this study, however, the same ontogenetic pattern of results was observed, although at both ages there was more of a tendency for overshadowing and less for potentiation. Specifically, when the CS was a compound of a sucrose/lemon taste plus methyl salicylate odor, the adults showed overshadowing of the aversion to the sucrose but the infants did not. In other words, for the adults the aversion to sugar was less in magnitude when sucrose had appeared in any of these three stimulus compounds than when it had been the sole CS, but the infants' aversion to sucrose solution did not differ whether it had been presented in compound or had appeared alone as the CS.

It is difficult to interpret these ontogenetic differences in overshadowing. One view is that the greater overshadowing by adults indicates that their selection of the sucrose was impaired by competing selection of the other elements while the infants were relatively unaffected by the presence of the other elements. Yet, this conclusion would deviate from that based on the potentiation results and those showing that infants are more likely to respond to redundant contextual stimuli; each of these

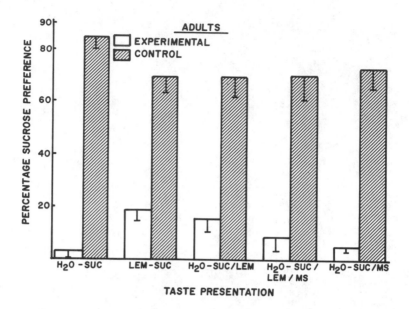

FIG. 30.4. The mean sucrose preference for adults given either water followed by sucrose (H$_2$O−suc), a lemon solution followed by sucrose (lem−suc), water followed by a simultaneous solution of lemon and sucrose presented with ambient methyl salicylate (H$_2$O−suc/lem/ms) or water followed by sucrose presented with methyl salicylate (H$_2$O−suc/ms): 24−hour delay controls, and immediate CS−US groups. Vertical lines in this and subsequent figures refer to standard errors of the mean.

results implies that the infants show less selection of the sucrose as the critical stimulus. But prior to further consideration of interpretation, the third replication study warrants brief mention.

Compared to the experiments we have discussed so far this third "replication" study included a quite different conditioning procedure. In this study odors were the CS and footshock the US (for characteristic details of the procedure, see Bryan, 1979; Kessler & Spear, 1980). For the basic potentiation condition the CS was a simultaneous compound of orange and lemon odors mixed together. When this CS predicted a footshock, a subsequent test for an aversion to either of the elements presented alone indicated significant potentiation for 18−day old rats; the aversions seen to either odor were greater than those found in other animals that had been conditioned with only the orange or only the lemon odor as the CS. In contrast, the adults showed overshadowing; conditioning to either of the odors was less when presented in compound than when they were presented alone as the CS. Control groups conditioned on only one odor but tested on the other confirmed that these effects were not artifacts of stimulus generalization between odors. This study therefore indicated that, once again, potentiation was greater for preweanlings than adults. In this experiment the adults actually had significant overshadowing, an effect

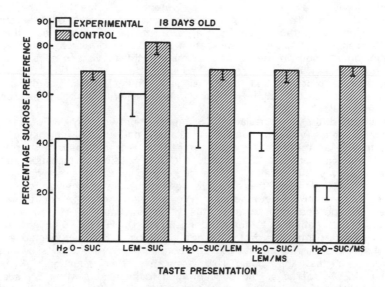

FIG. 30.5. The mean sucrose preference for 18-day olds given either water followed by sucrose (H_2O-suc), a lemon solution followed by sucrose (lem-suc), water followed by a simultaneous solution of lemon and sucrose presented with ambient methyl salicylate (H_2O-suc/lem/ms), or water followed by sucrose presented with ambient methyl salicylate (H_2O-suc/lem/ms): 24-hour delay controls, and immediate CS-US groups.

opposite to potentiation.

In summary, the results of four studies, each of which included internal replications and differed in the particular conditioning elements and conditioning procedures employed, agree that potentiation is greater for preweanlings than adults. There were also indications that when overshadowing occurred, this effect was greater for adults than preweanlings. We turn now to some studies that constitute our preliminary analysis of the basis for these ontogenetic differences in potentiation.

B. PRELIMINARY EXPERIMENTAL ANALYSIS TOWARD EXPLANATION OF THE ONTOGENETIC DIFFERENCES

We have focused initially on the possibility of interesting ontogenetic differences in responding to specific stimuli. These tests touch upon the alternative explanations of potentiation, but they were intended as only small analytic steps and not to be decisive. We briefly mention four of these studies that help move our story forward slightly.

The basic purpose of these studies was to determine the fit between age-related differences that actually occur with our procedures and those expected within explanations of our potentiation effects. The above discussion implies, for instance, that conditioning strength for the younger animals might show more gain from a more intense CS than would that for the older animals. A second implication is that in comparison with adults, conditioning strength in the younger animals might show less of a decrease due to generalization decrement caused by a difference between conditioning (compound) and testing (single element) stimuli, but more of a decrease due to extinction of any of the several associations formed during acquisition.

We tested the hypothesis of CS intensity in two ways. For the first experiment this variable was manipulated entirely in terms of the concentration of the decaffeinated coffee that was mixed with sucrose to provide the simultaneous CS compound. One possible explanation for the potentiation was that relative to the sucrose–only CS, the addition of the bitter coffee flavor had the effect of enhancing the perceived sweetness of the sucrose in the compound. We might therefore expect that the greater the coffee concentration the more intense, in effect, was the sucrose. Alternatively, a higher concentration of coffee might simply increase the net intensity of the compound "glob." Both suggested a need for systematic variation in coffee concentration in the simultaneous compound, a test that could also answer another question. We had selected 1.25% coffee as the concentration to be mixed with 15% sucrose because other of our data indicated that these concentrations of coffee and sucrose presented alone yielded the same conditioning strengths (Steinert, 1981). It was therefore possible that conditionability was the key metric for potentiation; that combining elements of equal conditionability yields potentiation, but that potentiation will be correspondingly less when conditionability of one element deviates from the other.

We therefore presented preweaning or adult rats with a simultaneous compound CS consisting of 15% sucrose mixed with one of seven concentrations of decaffeinated coffee, 0, 0.125, 0.25, 0.75, 1.25, 1.75, or 2.5%. For half the animals the LiCl injection immediately followed consumption of the CS and for the other half it was presented 24 hours later (explicitly unpaired control conditions). The results may be seen in Figures 30.6 and 30.7. It is quite clear that the preweanlings' degree of potentiation was an increasing function of coffee concentration, but this was not the case for the adults. Potentiation for the adults was statistically significant only with the 1.25 concentration; otherwise, conditioning of their aversion to sucrose was independent of the concentration of the coffee flavor mixed with it. These data conform to the expectation of the Gestalt interpretation by showing that both potentiation itself and the greater potentiation by preweanlings than adults can be more apparent the more intense the CS, when intensity is defined in terms of coffee concentration. That the adults did not show potentiation with more intense CSs while the preweanlings did, might be attributable to the adults' greater susceptibility to generalization decrement. We shall return to this point.

We next considered the influence of CS intensity in terms of sucrose concentration. Suppose, for instance, that the presence of the coffee in compound with sucrose inflated the perceived intensity of the sucrose concentration. Our previous studies had indicated that with no delay

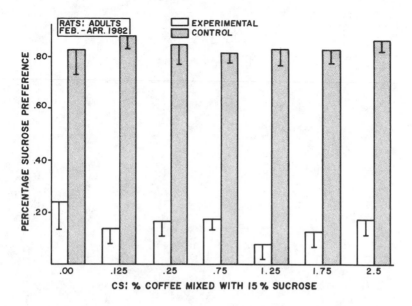

FIG. 30.6. The mean 15% sucrose preference for adults conditioned with 15% sucrose mixed with either 0.00%, 0.125%, 0.25%, 1.25%, 1.75%, or 2.5% coffee: 24 hour delay controls, and immediate CS–US delay groups.

between CS and US, higher sucrose concentrations as CS during *both* conditioning and testing led to poorer, not better, conditioning of an aversion to the sucrose solution (Steinert *et al.*, 1979), This inverse relationship between sucrose concentration and conditioning became apparent only by carefully accommodating the differences in intake among control animals when sucrose concentrations vary during testing. The poorer conditioning at higher concentrations of sucrose was significantly more evident for adults, however, than for 18–day old preweanlings. Especially when a 1–hour delay intervened between CS and US, the younger rats tended toward the opposite relationship, in terms of greater conditioning with the highest sucrose concentration (34%) than with 15% solution (Steinert *et al.*, 1979).

The Gestalt interpretation predicts better conditioning with the more intense conditional stimuli. This relationship tended to occur, within limits, for the preweanlings but not for the adults in the Steinert *et al.* (1979) study. Perhaps the Gestalt expectation would be confirmed more clearly within a more limited range; in the Steinert *et al.* experiment the highest concentration of sucrose solution, 34%, may simply have been too deviant from the next lower, 15%, to yield a positive relationship between intensity and conditioning for the adult animals. The Gestalt explanation also expects that a more intense CS during conditioning will result in greater conditioning strength even when tested with a less intense CS, or in other

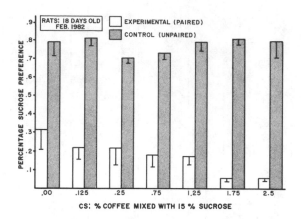

FIG. 30.7.   Data analogous to those in Fig. 30.6 from 18–day old rats.

words, in spite of the generalization decrement created by conditioning and testing with different stimuli. We tested these two expectations directly.

This experiment compared the generalization gradients resulting from conditioning with one of five sucrose concentrations (5, 10, 15, 20, or 30% sucrose solution) and testing always with the 15% concentration. Both 18–day old and adult rats were tested in this experiment and no delay intervened between the CS and the US during conditioning. The results tended to conform to the expectations of the Gestalt explanation of our age–related differences in potentiation (see Figure 30.8). For the 18–day old animals a relatively high concentration of sucrose solution during conditioning, 20%, yielded the greatest conditioning strength when tested with 15% solution. This confirmed the expectation that a higher CS intensity during conditioning can lead to greater conditioning strength for preweanling animals in spite of the generalization decrement that might occur from training with one concentration and testing with another. This relationship was not found for the adult subjects. The adults tended to show their greatest conditioning strength when they had been conditioned with the same 15% concentration used for testing. In comparison to this concentration the adults' conditioning strength was significantly less when conditioned with the two lower concentrations and was numerically less when conditioned with the two higher concentrations. Generally speaking, then, the results of this experiment were consistent with the expectation that conditioning strength for preweanlings can benefit from a high CS intensity during conditioning in spite of the generalization decrement that might be expected from testing with a lower sucrose concentration; this pattern of results was not seen with adults.

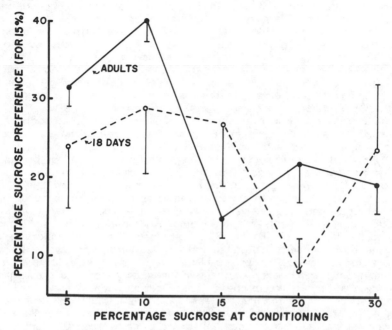

FIG. 30.8. The mean 15% sucrose preference for infants and adults conditioned with either a 5%, 10%, 15%, 20%, or 30% sucrose solution.

Taken together, the previous two experiments indicate that when an increase in the net intensity of a compound CS is created by increasing the concentrations of either of two kinds of elements (coffee or sucrose) mixed with a 15% sucrose solution, a stronger aversion to the 15% sucrose solution occurs; but, this relationship is seen only for preweanlings and not for adults. Given a testing stimulus that differed from the conditioning stimulus, the adults showed no indication of greater conditioning with more intense conditioning stimuli. In terms of the Gestalt explanation of potentiation, these results are consistent with the finding of greater potentiation in preweanling than adult animals. We caution, however, that while this confirmation is in accord with the Gestalt explanation, it would not be difficult to generate a multiple-associative explanation that would be equally consistent.

The final experiment in this series was to determine the extent to which adult and preweanling rats treat the simultaneous compound of sucrose and coffee as a configural (integral) stimulus. As an initial approach to this question, we conditioned and tested the animals with the simultaneous compound CS, while assessing the effects of extinction treatments consisting of exposure to one or more of the elements of the compound in the absence of a US.

Given the possibility that preweanling rats have a greater tendency to process the compound as an integral whole, we expected that among animals conditioned and tested with the sucrose–coffee compound, those given (interpolated) unreinforced presentations of one of the elements would show relatively little extinction to the simultaneous solution in comparison to those given unreinforced presentations of the compound. In effect, the preweanlings were expected to exhibit a substantial discrimination between the reinforced compound and the unreinforced elements.   Although preweanling animals given unreinforced presentations of an element show attenuated aversions to both elements (as seen in the previously described experiments and illustrated in Figure 30.1), these presentations may have little effect on their aversion to the compound.

Results of an experiment involving both age groups and exposure to one of four different extinction solutions (water, sucrose, coffee, or the same sucrose–coffee compound used as the CS) confirmed our prediction. Preweanling animals demonstrated significantly greater extinction to the compound CS when given unreinforced presentations of the same compound than when given unreinforced presentations of either of the elements presented separately, whereas for the adults, the effects of extinction to the elements and the compound did not differ significantly. This result, illustrated in Figure 30.9, provided one initial indication that while the adults treated the compound CS as merely the combination of two discrete elements, the preweanlings were more likely to treat it as different from a combination of the two elements.   Were the preweanlings behaving in a Gestalt–like manner?   These results alone clearly could not decide; one version of the multiple–association view could be, for instance, that the preweanlings extinguished on coffee still had their sucrose–LiCl association intact and so the compound might maintain more strength than after extinction on the compound, in which case both the coffee–LiCl and sucrose–LiCl associations would be degraded.

A further experiment involving only preweanling rats strengthened this finding and gave further support to the configuration notion.   Preweanling rats conditioned to the compound CS were given unreinforced presentations of both elements (order was counterbalanced) or the compound.   The multiple–association viewpoint would predict that, because in both cases the sucrose–LiCl and coffee–LiCl associations were extinguished, preference for the compound solution should be equally affected by these two extinction treatments.   According to the configural view, unreinforced presentations of the elements should result in little extinction of the compound. Results from this experiment supported the latter view; infant animals demonstrated a significantly greater preference for the compound if they had been given unreinforced presentations of the compound than if given unreinforced presentations of both elements. These results suggest that the infants viewed the simultaneous compound in a configural or integral manner, or in other words, as a different entity than the mere sum of two separate elements.   This would be consistent with the Gestalt theory.   The adults on the other hand tended to show nearly as much decrease in conditioning strength from extinction of either of the two elements as from extinction of the actual simultaneous compound itself, implying that an associative explanation of their potentiation (when it occurred) might be more appropriate.   It will be clear that these conclusions must be qualified because to the extent that degree

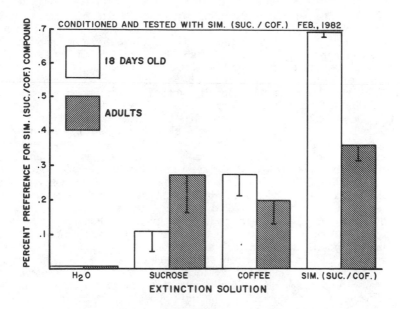

FIG. 30.9. The mean 15% sucrose preference for infants and adults that were conditioned on the compound and received reinforced presentations of water, sucrose, coffee or the compound.

of extinction differed for infants and adults. Our subsequent experiments have confirmed these same conclusions in identical tests following careful equation of extinction levels for rats of the two ages (Spear & Kucharski, 1984).

V. SUMMARY AND COMMENT

Our basic finding was that potentiation of conditioned aversion to a flavor was greater for preweanling and adult animals. Analysis focused on the enhanced conditioning to sucrose that resulted from presenting it in solution with coffee flavor. For preweanling animals this potentiation was sufficiently strong to override their otherwise ubiquitous deficiency, relative to adults, in conditioning with a delay between the CS and US. Potentiation was not seen for either infants or adults if the two flavors were presented successively, one immediately following the other, and especially for the infants, extinction of the association between coffee and LiCl reduced (in fact, eliminated) the potentiated conditioning of the sucrose solution.

To assess the generality of the basic results, we first replicated the original observations with a slight modification in certain control conditions. We determined also that the same ontogenetic difference occurred with each of two different combinations of flavors and in terms of potentiation of a flavor aversion by an ambient odor. Furthermore, this age difference also emerged in terms of potentiation of conditioning to one ambient odor by the presence of another, with footshock as the US. In most of these experiments the adults gave no indication of potentiation, and in some instances the opposite effect, overshadowing, was seen for these animals.

Collectively, these results suggested that the potentiation we observed was not due to the direct increase in associative strength suggested by Lett (1982) and others for potentiation effects that may be restricted to the feeding system. Potentiation was clearly reduced or eliminated by extinction of the alternative CS element. The finding that a novel ambient odor can potentiate a taste aversion is opposite the phenomena as characterized by Lett (namely, that taste potentiates conditioning of more distal stimuli). And finally, the circumstances of the experiments involving novel odors as the compound CS and footshock as the US, appear fairly more remote from those of feeding *per se*. Two other explanations seem more plausible, the "Gestalt" and "multiple-association" theories.

How would our finding of greater potentiation among more immature rats be reconciled in terms of the multiple-association theory? The preweanlings could have had stronger taste-illness associations or stronger within-compound associations than the adults; but as noted earlier, the former was not confirmed by our direct tests and there is no precedent for the latter among studies of the ontogeny of learning. Perhaps the more likely alternative is that in comparison to adults, preweanlings more readily form multiple associations within an episode, or in other words, are less selective about what is learned. This might seem not only inefficient, but also maladaptive in its tendency to produce unnecessary associative interference. Yet the latter could be avoided by another disposition of younger animals -- their special susceptibility to extinction (e.g., Spear, 1978, 1979). The present data suggest that following the pairing of an A/B simultaneous compound with an illness, extinction of associations between B and the illness drastically decreases the conditioning strength of A, suggesting that entire associative networks in the preweanlings might be relatively ephemeral and subject to change given the extinction of only one association within it. What emerges within this type of explanation is a sort of overproduction-and-selective-pruning model of learning and memory, reminiscent of the ontogenetic development of networks of neurons in the brain (cf. Jacobson, 1978).

The manner in which this overzealous learning occurs might not be associative at all, however, at least not in terms of the mechanism *initially* applied by the developing animal. What seems more likely is a form of "perceptual learning" (cf. Gibson, 1969). Such a mechanism would be consistent with what we have termed, for convenience, the "Gestalt" theory of potentiation.

Our final set of experiments assessed special assumptions of the Gestalt interpretation. We found that in accord with this theory, the conditioning of preweanlings was enhanced by either of two ways of increasing the intensity of the conditioned stimulus, an enhancement seen in spite of the generalization decrement that could be expected at the retention test. These effects were not found in adults. We also found that the preweanlings in particular tended to treat the simultaneous compound of two flavors as a configuration or integral stimulus that was differentiated from the simple addition of two separate flavor elements. This tendency toward configuration in preweanlings may be greater than that for adults; our data tend to confirm such a relationship.

A. COMMENT ON THE GENERALITY OF POTENTIATION.

Lett (1982) has provided a review of the potentiation literature and a description of several excellent new experiments on the topic.   She concludes that potentiation is exclusively a phenomenon of the feeding systems and has two primary characteristics: (1) The presence of a novel taste increases the conditioning observed to associated, nongustatory stimuli; (2) This enhancement of conditioning is due to a direct increase in the permanent associative strength between the latter stimulus and the unconditioned stimulus, which in all cases is toxicosis.   Our data suggest, however, that potentiation may be more general empirically than is indicated by Lett's analysis.   Alternatively, the instances of potentiation that seem to differ in their characteristics may represent fundamentally different effects and require different explanations.

Potentiation is a paradox, not only from a point of view that would expect overshadowing instead, but also in terms of its apparent generality. It seems on the one hand to be an elusive phenomenon, in the sense that some laboratories have seen no potentiation in circumstances that seem little different from those resulting in potentiation in other laboratories (e.g., Bouton and Whiting, 1982). Some of the experiments we report in the present chapter also indicate that for adults, seemingly minor changes in the nature of the compound flavors can result in quite different effects, potentiation in some instances but no potentiation or overshadowing in others.   And when circumstances do promote substantial potentiation, apparently "minor" factors such as whether the CS elements are presented in simultaneous or successive fashion can determine whether potentiation occurs (the above experiments; also see Rescorla & Durlach, 1981).

Yet there are, at least superficially, a variety of learning phenomena that are potentiation-like in their effects.   A contemporary example in basic conditioning is the "bridging effect," when the target element is separated temporally from the US and its conditioning is potentiated by the presence of a second CS element interpolated between the target CS and the US (for a review, see Kaplan and Hearst, 1982). An effect of substantial generality and broad importance, the basis of this enhanced conditioning of one element caused by the presence of another seems quite different from the present varieties of potentiation, but perhaps it is not.

Depending upon how one chooses to constitute the boundaries of the potentiation phenomenon, analogous effects can be found throughout a broad range of the literature in conditioning and learning. For instance, one might view as potentiation an effect seen in human infants by Fagan (e.g., 1981) in which recognition memory for a particular face or geometric design is enhanced if another version of that same face or design also is presented during the study phase of the experiment, an enhancement that is greater than if two identical versions of that face or design are presented. Generally speaking, one could identify potentiation with any instance in which a stimulus having a relatively large number of implicit or explicit associates more readily enters into an association with some other stimulus or response event. A variety of such cases are known in the field of human learning. Among the more striking are instances in which a verbal item is more likely recalled if in addition to a common context for learning and recall, a specific verbal associate is presented for that item during learning and recall (Spear, 1978; Tulving & Pearlstone, 1966; Tulving & Osler, 1968; Tulving & Thomson, 1973; Watkins, 1974). Taken in these terms we begin to make contact with what has been called the "associative paradox" (Underwood & Schultz, 1960). Arising most naturally out of the study of human learning -- where there have existed relatively convenient indices of the number and type of associations their subjects have to a particular stimulus element, i.e., "meaningfulness" -- the paradox is why stimuli with more associates acquired prior to learning are more apt to enter into new associations, in spite of what would appear to be correspondingly more sources of interference from prior associations to prevent formation of new ones.

This is not to suggest that potentiation is exclusively an associative phenomenon. Our evidence seems to indicate otherwise: at least for the present cases in which potentiation seems more prevalent in younger subjects, a perceptual–Gestalt interpretation might be more appropriate. The multiple–association theory remains an appealing and viable explanation, however, not eliminated conclusively by experiments to date. We are therefore not in a position for hard decisions on these theories. Instead, we assert two general points. First, a general definition of potentiation as any facilitation in the conditioning of stimulus X due to other associations held by stimulus X makes contact with a vast array of such effects, particularly in the field of human learning and cognition. Knowing this does not solve the associative paradox, which might be translated for basic conditioning into the question of, "Why potentiation instead of overshadowing (or vice versa)?" But it may reveal something of the scope of the problem and the nature of the solutions that have been pursued in the past. The second point, equally unprofound, is that with a phenomenon of such breadth, it should not be surprising that it could emerge for different reasons in different circumstances, or in other words, that there might be more than one local mechanism underlying different cases of potentiation; yet the possibility of subsuming them all under one general principle is real, and intriguing.

ACKNOWLEDGMENTS

This research was supported by grants from the National Science Foundation (BNS78-02360) and the National Institute of Mental Health (1RO1 MH35219) to Norman E. Spear. The authors wish to express their appreciation to Norman G. Richter, Sr. and Teri Tanenhaus.

REFERENCES

Bouton, M., & Whiting, M. R. *A comparison of odor-taste and taste-taste compounds in toxiphobia conditioning.* Paper presented at the 53rd annual meeting of the Eastern Psychological Association, Baltimore, Maryland, April, 1982.

Bryan, R. G. *Retention of odor-shock conditioning in neonatal rats: Effects of distribution of practice.* PhD Thesis, Rutgers, 1979.

Campbell, B. A., & Spear, N. E. Ontogeny of memory. *Psychological Review*, 1972, *79*, 215–236.

Caza, P. A., Steinert, P. A., & Spear, N. E. Comparison of circadian susceptibility to LiCl–induced taste aversion learning between preweanling and adult rats. *Physiology and Behavior*, 1980, *25*, 389–396.

Cheatle, M. D. *Ontogenetic differences in mechanisms of second order conditioning.* PhD Thesis, Princeton, 1980.

Cheatle, M. D., & Rudy, J. W. Ontogeny of second order odor–aversion conditioning in neonatal rats. *Journal of Experimental Psychology: Animal Behavior Processes*, 1979, *5*, 142–151.

Domjan, M. Biological constraints on learning 10 years later: Implications for general process learning theory. *Psychological Bulletin*, in press.

Fagan, J. F. III. Infant memory. In T. Field (Ed.), *Review in human development.* New York: Wiley, 1981.

Galef, B. G., Jr., & Osborne, B. Novel taste facilitation of the association of visual cues with toxicosis in rats. *Journal of Comparative and Physiological Psychology*, 1978, *92*, 906–916.

Garner, W. R. The stimulus in information processing. *American Psychologist*, 1970, *25*, 350–358.

Garner, W. R. *The processing of information and structure.* Hillsdale, N.J.: Erlbaum, 1974.

Gibson, E. J. *Principles of perceptual learning and development.* New York: Appleton–Century–Crofts, 1969.

Infurna, R. N. Daily biorhythmicity influences homing behavior, psychopharmacological responsiveness, learning, and retention of suckling rats. *Journal of Comparative and Physiological Psychology*, 1981, *95*, 896–914.

Jacobson, M. *Developmental neurobiology.* New York: Plenum, 1978. Second edition.

Kamin, L. J. Selective association and conditioning. In N. J. Mackintosh & W. K. Honig (Eds.), *Fundamental issues in associative learning.* Halifax: Dalhousie University Press, 1969.

Kaplan, P. S., & Hearst, E. Bridging temporal gaps between CS and US in autoshaping: Insertion of other stimuli before, during, and after CS. *Journal of Experimental Psychology: Animal Behavior Processes*, 1982, *8*, 187–203.

Kessler, P., & Spear, N. E. Neonatal thyroxine treatment enhances classical conditioning in the rat. *Hormones & Behavior*, 1980, *14*, 204–210.

Lett, B. T. Taste potentiation in poison avoidance learning. In R. Herrnstein (Ed.), *Harvard symposium on quantitative analysis of behavior.* (Vol. 4.) Hillsdale, N.J.: Erlbaum, 1982.

Mackintosh, N.J. A theory of attention: Variations in the associability of stimuli with reinforcement. *Psychological Review*, 1975, *82*, 276–298.

Rescorla, R. A. Simultaneous associations. In P. Harzem & M. D. Zeiler (Eds.), *Predictability, correlation, and contiguity.* New York: Wiley, 1981.

Rescorla, R. A., & Durlach, P. J. Within–event learning and Pavlovian conditioning. In N. E. Spear & R. R. Miller (Eds.), *Information processing in animals: Memory mechanisms.* Hillsdale, N.J.: Erlbaum, 1981.

Rescorla, R. A., & Wagner, A. R. A theory of Pavlovian conditioning: Variations in the effectiveness of reinforcement and nonreinforcement. In A. H. Black & W. F. Prokasy (Eds.), *Classical conditioning II: Current research and theory.* New York: Appleton–Century–Crofts, 1972.

Rusiniak, K. W., Hankins, W. G., Garcia, J., & Brett, L. P. Flavor–illness aversions: Potentiation of odor by taste in rats. *Behavioral and Neurobiology*, 1979, *25*, 1–17.

Shepp, B. E. From perceived similarity to dimensional structure: A new hypothesis about perspective development. In E. Rosch & B. B. Lloyd (Eds.), *Cognition and categorization.* Hillsdale, N.J.: Erlbaum, 1978.

Simon, H. A. How big is a chunk? *Science*, 1974, *183*, 482–488.

Spear, N. E. *The processing of memories: Forgetting and retention.* Hillsdale, N.J.: Erlbaum, 1978.

Spear, N. E. Memory storage factors in infantile amnesia. In G. Bower (Ed.), *The psychology of learning and motivation.* (Vol. 13.) New York: Academic Press, 1979.

Spear, N. E., & Kucharski, D. The ontogeny of stimulus selection: Developmental differences. In R. Kail & N. E. Spear (Eds.), *Memory development: Comparative perspectives.* Hillsdale, N.J.: Erlbaum, 1984. In press.

Steinert, P. A. *Stimulus selection among preweanling and adult rats as a function of CS amount and quality using a taste aversion paradigm.* PhD Thesis, State University of New York at Binghamton, 1981.

Steinert, P. A., Infurna, R. N., & Spear, N. E. Long–term retention of a conditioned taste aversion in preweanling and adult rats. *Animal Learning and Behavior*, 1980, *8*, 375–381.

Steinert, P. A., Infurna, R. N., Jardula, M. F., & Spear, N. E. Effects of CS concentration on long–delay taste aversion in preweanling and adult rats. *Behavioral and Neurobiology*, 1979, *27*, 487–502.

Tighe, T. J., & Tighe, L. S. A perceptual view of conceptual development. In R. D. Walk & H. L. Pick (Eds.), *Perception and experience.* New York: Plenum, 1978.

Tulving, E., & Osler, S. Effectiveness of retrieval cues in memory for words. *Journal of Experimental Psychology*, 1968, *77*, 593–601.

Tulving, E., & Pearlstone, Z. Availability versus accessibility of information in memory for words. *Journal of Verbal Learning and Verbal Behavior*, 1966, *5*, 381–391.

Tulving, E., & Thomson, D. M. Encoding specificity in retrieval processes in episodic memory. *Psychological Review*, 1973, *80*, 352–373.

Underwood, B. J., & Schultz, R. W. *Meaningfulness and verbal learning.* Philadelphia: Lippincott, 1960.

Wagner, A. R. SOP: A model of automatic memory processing in animal behavior. In N. E. Spear & R. R. Miller (Eds.), *Information processing in animals: Memory mechanisms.* Hillsdale, N.J.: Erlbaum, 1981.

Watkins, M. J. When is recall spectacularly higher than recognition? *Journal of Experimental Psychology*, 1974, *102*, 161-163.

31 THE EVOLUTION OF COGNITION IN PRIMATES: A COMPARATIVE PERSPECTIVE

Duane M. Rumbaugh
Georgia State University
Emory University
James L. Pate
Georgia State University

I. INTRODUCTION

Attempts to understand the complexities of animal and human behavior through the development of a *single* theoretical perspective by experimental psychologists and other behavioral scientists of this century have been relatively unsuccessful. Nothing has proven to be simple as questions regarding parameters and dynamics of behavior have been addressed. Though the operations necessary to obtain classical and instrumental conditioning seem simple enough, it has proven very difficult to define satisfactorily the processes which differentiate even these basic forms of learning (Rescorla & Holland, 1982).

Just how behavior should be viewed with respect to simple versus complex is itself controversial. It is maintained by the radical behaviorists, led by Skinner (1953, 1974), that, once a full understanding of basic operant behavior is achieved, the "complexities" of behavior will be found to be more apparent than real. By contrast, there are other psychologists who doubt that *all* of the complexities of creative problem solving by animals and humans are reducible to the operations which provide for the acquisition and emission of operants. They hold that an understanding of covert mediating mechanisms likely will be necessary to understand creative, innovative behaviors.

Mediating mechanisms are, by definition, not directly observable, not directly accessible for study by the psychologist. Their essence can only be inferred through the design of appropriate experiments. An important point to recognize is that neither the radical behaviorist, nor the psychologist who would include mediating mechanisms as a requisite for understanding of behavior, deny the private experiences of thought and feeling which humans universally admit having. The difference is only with whether private experiences are held to determine observable behavior. The radical behaviorist would insist that the mediating mechanisms are but another form of behavior and subject to the same contingencies as are observable behaviors (Skinner, 1953). Consequently, they are not viewed as causes of behavior. By contrast, those psychologists who posit the operations of mediating mechanisms would hold that, although they too, are controlled by natural laws, they do exert, by virtue of their operations, control over behavior (Kendler & Kendler, 1962; Miller, 1948; Osgood, 1953; Roitblat, 1982).

Thus, in the eyes of the present authors, the question is not whether or not a behavioristic approach is a requisite to the understanding of behavior but rather whether or not a behavioristic approach without the incorporation of mediating mechanisms will be sufficient to the task of achieving an understanding of the intricacies of behavior in all forms. There are behaviors, which are often referred to as *complex*, that are neither predictable from nor satisfactorily explained by basic conditioning principles. In some manner, mediating mechanisms, either simple or complex, are a requisite to the selection and execution of successful behaviors in various problem-solving situations. In these situations, behavioral adaptation is not to be achieved through responding in simple, straightforward ways to well defined stimuli. Rather, the relation between elements must be perceived, perhaps even on a *first* encounter, for successful behavioral adaptation to be achieved.

Perception of relationships between things, representations of things not present, and the covert manipulation of symbols are possible mediators and are believed to be requisites to creative behavioral adaptations. Just how well these processes are to be understood through empirical behavioral research is an open question but, nonetheless, is the one question on which cognitive psychology will probably stand or fall. If it proves to be superfluous to call upon mediating mechanisms to account for behavior, they should, of course, not be used. By way of summary, a necessary condition for any aspect of a cognition to be invoked is that there must be observable behavioral phenomena which otherwise cannot be satisfactorily accounted for or explained. This is not to deny, however, that psychologists, regardless of their orientation, are not at a loss to provide a *post hoc* explanation for any and all behavioral data once obtained. It is probably impossible to embarrass any perspective or school of psychology when it comes to giving an account for behavioral observations reported by others. The difference is whether or not the various perspectives or schools of psychology would *predict* that, given certain manipulations, interesting phenomena, marked with significant elements of surprise, are to be obtained. Further, the manipulations and phenomena that are judged to be important and significant differ for the various schools. This selection process, thus, directs the research program and the theoretical developments within a school.

What must be encountered, then, for any kind of cognitive process to be appealed to justifiably? It is the observation of behaviors, preferably on a first encounter with new situations, where the organism has no experience or learning from the past that bears in a straightforward manner upon the present situation, where there are new elements or new relationships between old elements that in some way must be perceived and capitalized upon for behavior to be at all competent, and where there is clearly an *integration* of past experiences and learning so as to organize the new behavioral pattern. The behavioral pattern must be more than that which can be accounted for in the simpler terms of stimulus generalization and response generalization. There must be a new topography in the pattern of behavior which compels an appeal by the psychologist to mediating mechanisms. None of this is to deny, however, that past experience is a requisite to the emergence of new, creative behaviors. But, creative behavior must involve more than simple training. On the other hand, neither animals nor humans can do that about which they know nothing! An appeal to mediating mechanisms, to cognitive processes of any kind, must never be construed to discount the importance of prior experience, prior learning. Although it is frequently alleged that Kohler (1925) appealed only to insight as an explanatory mechanism and discounted prior experience, the truth is really quite to the contrary. Kohler clearly appreciated the importance of prior experience, prior learning, for the manifestations of creative problem solving by his chimpanzees. For example, he explicitly stated that, if the difficulty of problems were escalated too rapidly, then the whole process of becoming a facile problem solver would be interrupted; this was indexed most obviously by the subject's balking. That Kohler did not weight or emphasize experience so much as he emphasized insight itself should not be construed to mean that he thought insight to be something independent of history of experience.

II. COGNITION AND INTELLIGENCE

The behavior of humans is so complex as to call for the introduction of concepts such as intelligence. Intelligence is said not only to differ among individuals, but its many dimensions and operations are dependent upon experience, the opportunity to learn information relevant to the tasks to be solved. Mediating mechanisms are certainly a requisite to the manifestation of intelligent behavior. To the degree that mediating mechanisms and cognition are to be equated, and we believe that they should be, cognition is requisite to intelligence.

Humans are members of the order Primates, which contains a very broad array of forms from the relatively primitive prosimians, to the monkeys of the Old and New Worlds, and to the lesser and great apes and humans themselves. Primates vary greatly in terms of their size, rates of development both prior to and after birth, probable life span, and brain development. Even more germane is the profound enhancement of the neocortex in relation to the rest of the brain as one moves from the most primitive primate forms, the prosimians, to the New and Old World monkeys, in turn, and to the lesser apes, great apes, and humans. The brain becomes more complex in its dendritic interlacings, its convolutions, and its gyrii. The cortex becomes disproportionately large relative to the rest of the brain.

Within the order *Primates*, then, there is an orderly progression of brain evolution. To the degree that the development of the complexities of the human brain are requisites to the operations of complex mediating mechanisms, the emergence of cognitions, and the potential for intelligent behavior, appropriate studies of a comparative nature within the order Primates might be of interest and significance. With evolution of the primate brain, it is reasonable to expect that there will be quantitative augmentation in the efficiency of operations. Whether there is concurrently the emergence of new processes which might be so different from prior levels of operations that the term qualitative should be introduced is a question that is closely related to the topic of this paper. It is, though, not crucial to the general case which will be made about learning abilities and their differences between species as revealed by empirical research.

The question is, how can we study indices of mediating mechanisms, cognition, and their possible relationship to what we call intelligence at the human level through comparative psychological studies with the diverse forms which comprise the order *Primates*?

III. LEARNING SET

Harlow (1949) defined the phenomenon known as learning set -- the formation of a facility to learn new problems of a given class as a function of experience in attempting to learn other problems from the same class. In its ultimate form, a learning set allows for one-trial learning of two-choice visual discrimination problems, with only the outcome of the choice on the first trial being necessary for the subject to perform errorlessly on subsequent trials, to know which object or stimulus must be selected subsequently if reward is to be obtained.

Learning set ability is, in part, a function of opportunities to learn problems of a given class, where each problem is presented for too few trials for mastery before the next is encountered. It also is a function of how well developed the brain of the learner is; at least this holds true, generally speaking, for species within the order *Primates*. (The fact that some birds are found to have facile learning set acquisition skills does not in any way conflict with the fact that there is an orderly relationship between brain development within the order *Primates* and ability to form a learning set. Rather, it is taken to mean that outside the order *Primates* there are life forms that have specialized learning skills that resemble at least basic learning set formation.) In Harlow's original view, it was with the formation of a learning set that the learner became freed of stimulus-response bondage to become a proficient, one-trial, insightful learner.

As stated, learning set has been found to be related, in general, to the development of the brain, and this makes learning set appear at first blush to be of obvious value for broad use as a comparative psychological tool. However, the diverse and ill defined differences in species' receptor-mechanisms and in their perceptual and motivational systems lead to questions regarding the wisdom of direct comparisons between species' learning set skills when their *absolute* performance levels are used as data.

In an attempt to improve Harlow's learning set measurement so as to achieve a defensible assay of primates' abilities to learn-how-to-learn, a variant of the discrimination-reversal learning set was developed to obtain what is called a Transfer Index (Rumbaugh, 1970; Rumbaugh & Gill, 1973).

Discrimination-reversal tasks differ from the basic learning set tasks in that at some point in training the initial cue values are reversed for the remainder of the problem so that the initially correct stimulus (S+) becomes incorrect (S-) and vice versa. The basic tenet of the Transfer Index is that the trials after reversal serve as a test of transfer of prereversal learning and that it is the *ability to transfer* that is of primary import in the understanding of basic mediating mechanisms.

IV. TRANSFER INDEX

Just as with learning set and basic discrimination-reversal tasks, Transfer Index methodology entails the administration of a long series of two-choice visual discrimination problems. The discriminanda generally are miscellaneous objects randomly paired, though they also might be two-dimensional patterns. Transfer Index methodology is distinct from conventional discrimination-reversal training in that the subject is retained in the initial discrimination with a given pair of objects which constitute a problem until a predetermined criterion of accuracy is achieved. Only upon the achievement of that criterion are the cue values reversed for purposes of testing for transfer-of-training. In Transfer Index methodology, the transfer test consists of ten trials on which the cues retain their values as reassigned on the trial immediately following the acquisition of the stated criterion. The prereversal training to a criterion serves at least three specific purposes: 1) it ensures that the subjects are attending to the task and are learning, 2) it ensures that the subjects of diverse species are equated on a *performance* criterion prior to the reversal of cues and the test for transfer-of-training abilities, and 3) it operationally provides control over the amount of information/learning which the subject has accrued at some specific point in discrimination learning. As the evidence to be discussed reveals strong species by prereversal criterial levels interaction, this technique is important for the purpose of achieving a comparative perspective of important processes.

The first of the above functions ensures that the learner is sufficiently motivated and is attending to the task at hand, e.g., the task and the learner are engaged. This is a particularly important consideration in dealing with diverse species where unique problems, motivation, and attention can result in "indifference" to the choices. The second of the above functions serves in effect to equate functionally (i.e., behaviorally) specimens of species that differ greatly in body size, weight, size of hands, strength, dexterity, appetites, and so on, any one of which, or in combination, might profoundly influence the absolute performance levels of the subjects in a learning task. The assumption is that if one gets diverse primate forms to function equivalently, as is held to be the case when they have achieved operationally the same criterion, the majority of effects due to differences in their receptor-effector mechanisms, etc., have been effectively attenuated.

Although any number of criterial levels for the prereversal phase of Transfer Index testing might be developed, to date only two have been employed -- the 67% and the 84% levels (see Table 31.1). Both are "under-learning" vs "over-learning" levels as they call for less than 100% accuracy. A limit of 60 trials is set for a subject to learn any single problem prior to reversal of cues for the test. Failure to learn within this limit results in discarding that problem. The 60 trial limit allows for transient shifts in motivation, attention, and the like, problems which might

TABLE 31.1

Responses correct required
for Transfer Index 67% and 84% criterial levels

Criterion	Trial no.	Required no. responses correct
67%	11	7 or 8
	14	9
	16	10
	19	12
	22-60	14 within last 21 trials
84%	11	9
	17	14
	21	17 or 18 (or alternate problems)
	22-60	17 or 18 (or alternate problems) within last 21 trials

Note-- First trial of training, in which chance prevails, is not included in the count. Hence, Trial 11 is twelfth trial of training. The table is to be read in the following manner. The 67% criterial level can be met by having 7 or 8 choices correct by the 11th trial, by having 9 choices correct by the 14th trial, or by having 10 choices correct by the 19th trial. Thereafter, the subject must have 14 choices correct within a span of 21 trials. If the subject exceeds these numbers of correct choices or if the subject fails to have 14 choices correct within a span of 21 trials by the 60th training trial, the problem is deleted. Interpretation of the 84% criterion is to be made in a similar manner.

be idiosyncratic to the individual subject or which might characterize a given species under study. Not more than 60 trials are allowed, however, for beyond that point a very strong negative correlation begins to emerge between total trials to criterion and the reversal-test performance. For fewer than 60 trials, the correlation is negligible and in the majority not significant for species studied to date.

The Transfer Index is calculated on the basis of ten problems per block. Percent correct responses from the reversal test trials, excluding the first reversal trial which is deleted because it serves to signal or inform the learner that the cues have been reversed, divided by the percentage for the prereversal trials (the criterial level) defines Transfer Index. Contingent upon the purposes for which the Transfer Index study is being conducted, any number of successive Transfer Index measurements might be obtained at one or more criterial levels of training. By way of example, for the calculation of a Transfer Index, assume that a subject having been trained to the 67% criterion, performs 76% correct on the reversal test trials on ten consecutive problems, where for each problem the subject has met the criterion as stated in Table 31.1. Dividing 76% by 67% yields the ratio of 1.13. The ratio indicates, regardless of the percentages involved, that the subject did relatively better on the reversal test trial than on the prereversal trials. On the other hand, assume that the subject was correct on only 40% of the reversal trials of ten consecutive Transfer Index problems, having been trained to the 67% level of correctness on each problem prior to test. The resultant Transfer Index would be 0.60. Again, regardless of the absolute percentages involved, the ratio tells us that relative to the prereversal criterion, the learner did poorly.

It should be clear then that Transfer Index values emphasize performance on the reversal test trials, which are viewed as transfer-of-training trials, relative to specific prereversal criteria. On these test trials, in a situation faced by the learner which is similar to but not identical to the original learning situation, the transfer may be negative (indicated by a Transfer Index less than 1.0), neutral (Transfer Index=1.0), or positive (Transfer Index greater than 1.0). The ability to transfer learning efficiently and to an advantage is certainly a hallmark of intelligence even though intelligence as we view it at the human level is certainly more than this ability. Thus, with respect to transfer-of-learning, we believe that an important perspective of a fundamental dimension of intelligent behavior has been obtained through comparative psychological studies with the Transfer Index paradigm.

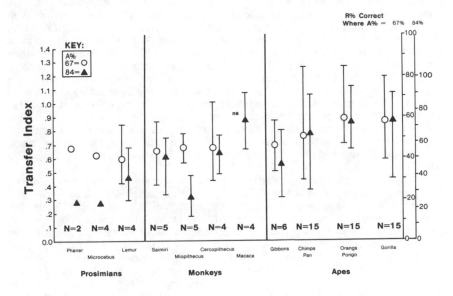

FIG. 31.1. Transfer Index values for the 67% and 84% criterion levels of learning. Responses correct on the reversal trials for each of the two levels can be accessed through use of the two axes at the right of the figure. Bars indicate the range of percent correct responses (Prosimian data were collected independently by Cooper, 1980, and ranges were not available.)

Figure 31.1 portrays Transfer Index data obtained in several studies for 79 primates of various species. Two successive Transfer Index values, each based on 10 problems, were obtained and averaged for each subject for each species, first at the 67% criterial level and then at the 84% criterial level. Where the Transfer Index remains constant regardless of the prereversal criterion value, it is clear that there were corresponding adjustments in the reversal performance levels. Thus, if the Transfer Index ratios are the same for a given species at both the 67% and 84% levels, the reversal performance increased proportionately as the training level was increased from 67% to 84%. On the other hand, where the Transfer Index value is lower for the 84% than it is for the 67% criterial level, there was no corresponding augmentation in the accuracy of reversal performance. In some instances there was, in fact, a profound drop in reversal performance. The reversal percent correct can be read from the two axes at the right of Figure 31.1.

The data indicate that as one moves from the prosimians through the monkeys to the great apes there is a general elevation of Transfer Index levels. The elevation is more pronounced for the 84% than for the 67% values. In general, for the great apes, the Transfer Index-67% and Transfer Index-84% values are comparable, indicating that as the prereversal criterion was elevated, there was a corresponding increment in reversal percent correct. This conclusion is corroborated by values to be obtained from the axes at the right of Figure 31.1. For example, the gorillas' reversal performance was 57% when trained in accordance with the Transfer Index-67% criterion schedule but was 73% when trained at the Transfer Index-84% criterion schedule. The 16% elevation in reversal percent correct approximates very closely the 17% difference between those two Transfer Index criterial levels. By contrast, the prosimians' reversal percent correct dropped sharply as the requirement was advanced from 67% to 84%. *Phaners'* and *Microcebus'* reversal percentages dropped 21% as the prereversal criterion was increased from 67% to 84%. This constitutes a drop greater than the increase in the Transfer Index criterion! Thus, whereas an increase in learning before reversal served to enhance the transfer test performances of the great apes (i.e., a strong trend toward *positive* transfer), it decimated the prosimians' (strong *negative* transfer).

It is unfortunate that Transfer Index-67% values are not available for the rhesus macaque (*Macaca*). The Transfer Index-84% values for that species are high and challenge both the averages and the upper limits for the great apes. The performance of the squirrel monkey (*Saimiri*) is perhaps unwarrantedly high due to the fact that they were five animals that had been collected across years for their high learning set proficiencies. At the time of that research, they were the only specimens of that species available to us. Talapoin (*Miopithecus*) performed as did the prosimians. Interestingly, the talapoin is viewed not infrequently as one of the most primitive Old World monkeys. If not the most primitive, it is certainly the smallest. Vervets (*Cercopithecus*) performed at intermediate levels, while the gibbons (*Hylobatidae*) performed more as monkeys than as the apes which they are.

Figure 31.2 portrays the change in reversal test performance as the prereversal criterion level was increased from 67% to 84%. The shift

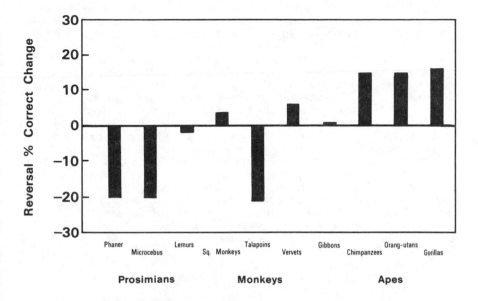

FIG. 31.2. Change in reversal percentages correct when the pre-reversal criterion was advanced from 67% to 84% correct. Strong negative transfer to the reversal trials is indicated for the prosimians and the talapoin monkeys; strong tendency to positive transfer is indicated for the great apes--a qualitative shift in effect.

from strong negative towards strong positive transfer is clear as one moves from the prosimians to the great apes. Although transfer-of-training can be thought of as a single dimension, few would quarrel with the assertion that a qualitative change has occurred as one moves from strong negative transfer to strong positive transfer. Just what that qualitative difference might be reflecting was explored in the following studies.

V. ARE THERE DIFFERENCES IN PRIMATES' LEARNING PROCESSES?

There is great reluctance on the part of some psychologists to admit to the possibility that there might be more than one kind of learning process; if there is more than one kind of learning, there are obvious complications for developing a generalizable set of laws that would apply across diverse species. In part, the reluctance is also a reflection of the robust character of present-day behaviorism with its emphasis upon environmental contingencies and schedules of reinforcement as they control the rates and patterns of operant responses (Skinner, 1953). It is well established that widely diverse species manifest very similar effects with each of a number of schedules of reinforcement.   Unfortunately, this observation has been taken to mean that there is but one set of learning or behavioral principles. Alternatively, it can be argued that the contingencies of operant behavior are so primitive, so basic to animal life that has survived to the present day, that they no longer serve to differentiate them behaviorally.

Despite this commonality, it should not be assumed that all principles are the same across species.   Obviously, there has been an evolution of the brain and physiological processes which suggests that the principles necessary to explain the behavior of more highly evolved forms *are not necessarily* the same as those necessary to explain the behavior of less highly evolved forms.   Even though the sets of principles may be different, the elementary principles of classical and operant conditioning are not invalidated.   Rather, new principles might be added to the more elementary ones. Thus we join those advocates of more than one kind of learning who have been sprinkled through the history of the twentieth century (see Kohler, 1925; Mowrer, 1947; Tolman, 1949; Yerkes, 1916; Yerkes & Yerkes, 1929). In the eyes of the present authors, the question remains an important one -- Are there qualitative differences in the learning processes of animals, notably primates?

TABLE 31.2
Associative vs. Mediational Learning Paradigm

Acquisition	Reversal Trial	Reversal Trials
Learn to 9/10 correct with	1	2-11
		A-B+ (control)
A+B-	A-B+	C-B+
		A-C+

Note-- A, B, C each refer to different stimuli.

In response to the data shown in Figure 31.2, the first author devised a paradigm intended to determine whether there were differences in learning process as diverse species achieved a given prereversal criterial level (Rumbaugh, 1971).   The paradigm, which was basically a variant of the Transfer Index methodology, is summarized in Table 31.2.   The paradigm was basically directed toward assessing the degree to which there was stimulus-response (S-R) associative learning in the achievement of the criterion as opposed to mediated learning, learning that would not clearly and obviously have the attributes commonly associated with associative learning. Hull's S-R associative learning implies that, in achieving criterion, an animal learns a positive habit of responding to whichever stimulus is correct and food rewarded and a negative habit of *not* responding to whichever stimulus was incorrect and *not* rewarded. These two habits, sH_R and sI_R, were, then, the opposing habits which would control the choice behavior of the learner. Mediational learning, by contrast, would entail the learning of rules that would provide a logical structure for the initial learning tasks *and* the transfer tests to be described.

The control condition for the paradigm consisted of conventional cue reversal, e.g., upon achievement of a 9/10 criterion, the cue values of both stimuli were exchanged and both stimuli were retained throughout the test trials (ten of them). By comparison, two other conditions were designed to assess the degree to which the achievement of the prereversal criterion had been achieved by S-R associative learning. In one condition, the initially correct stimulus (A+) was deleted *after* the first reversal trial, and a new stimulus (C-) was substituted in its stead. In the third and remaining condition, the initially *incorrect* stimulus (B-) was deleted after the first reversal trial, and in its stead a new stimulus (C+) was substituted. The deletion of A on the reversal test trials would preclude the necessity of extinguishing the tendency to choose it. Similarly, deletion of B would obviate the counter conditioning of the inhibition (sI_R) which had been built up to it as a function of nonreinforcement during criterial training. The *relative* differences between these two conditions and the reversal control condition are used as the bases for inferring something about relative strengths of A+ and B- in the prereversal learning. The absolute levels of accuracy are only of secondary interest.

To summarize, the paradigm consisted of a series of two-choice visual discrimination problems. Let us designate the correct and incorrect stimuli of a given problem pair as A+ and B-. The prereversal criterion was 9/10 responses correct. Upon achievement of criterion, the cue values became A- and B+ for the first reversal test trial. In the control condition, A- and B+ were sustained for the next ten trials. In the other two conditions, a new stimulus (C) was substituted for either A or B, resulting in C-B+ and A-C+ reversal test conditions. The relative difficulty of these two conditions would allow for the assessment of habit conditioned to approach and select A and of habit to avoid and not select B. To the degree that the criterion had been achieved through the formation of a strong habit to stimulus A, the C-B+ condition should yield high performance levels. Similarly, to the degree that the learner reached the initial criterion by avoiding B, developing inhibition, the C+A- condition should yield high performance levels.

But what can be expected if S-R associative learning is not occurring? What kind of performance would suggest that a mediating mechanism had become operational? We believe that when these three reversal test conditions are essentially equivalent a mediating mechanism is suggested. Generally, if one predicts "no difference" between conditions, it is a weak prediction. But, we argue to the contrary in this particular case, for these three reversal test conditions entail radical manipulations when compared with one another. If, despite those radical manipulations, a learner were to find them essentially one and the same with regard to difficulty, we believe that the introduction of a mediating process, probably of an inferential nature, would be mandatory. The three reversal test conditions, after trial one reversal, are alike in but one important regard -- each of them provides sufficient information for the subject to be correct on all subsequent trials. To be specific, these test-wise learners were given the A–B+ Trial 1 as potentially a signal that discrimination–reversal test trials were now to prevail. The subjects characteristically chose A on this trial and hence received no reward. Given that the experience A is now "–" was carried forward and paired with B, it could be inferred that B would be correct henceforth. Similarly, regardless of whether A– or B+ were deleted, whichever one was sustained could provide a basis for inference on the part of the learner as to what the cue value of the other member (C) was (cf. Chapter 22, by Zentall, Hogan, & Edwards, this volume). From a strictly behavioristic perspective, it might be argued that reinforcement or nonreinforcement on Trial 1 of the reversal becomes a cue as to which object should be selected on the next set of trials. However, what would allow a cue to serve that role unless one appealed to the response to the cue as providing some information to the subject? Too, such a behavioristic perspective would not likely inspire investigators to determine whether or not species differed in terms of their ability to use Trial 1 reversal reinforcement or its absence as a cue. On the other hand, if it is assumed that, with evolution of the brain, mediating processes are more likely to become operational (and the basic process of inference might be one of them), then the study becomes exciting enough to warrant the investment of time, energy, and effort. Twelve problems of each of the three types were collected, and the results are portrayed in Figure 31.3. The "bright apes" were a group selected for their superior performance on Transfer Index testing for the purpose of pursuing the question as to differences in learning between bright and dull apes (Gill & Rumbaugh, 1974).

Two main trends are to be emphasized. First, as one goes from the lemurs up through the group of retarded humans, there is a general elevation in overall accuracy of reversal performance, and second, the three curves become more and more similar. The first trend is basically consonant with the Transfer Index trend shown in Figure 31.1. Figure 31.4 portrays just A–B+ reversal test condition performances of Figure 31.3, the condition which most resembles the Transfer Index paradigm. If one accepts the order of brain evolution, ranked from relatively low to relatively high, as being lemur, talapoin, squirrel monkey, adult rhesus, gibbon, gorilla, and retarded adolescent humans, the Spearman rank order correlation of the groups ranked thus with their respective A–B+ averages is 0.93 ($p < 0.01$), with B–C+ averages is 0.54, ($p = 0.11$), and with the A–C+ averages is 0.86 ($p < 0.01$). These data not only corroborate the basic finding with the Transfer Index data, but they also serve to emphasize that it is primarily what learners do on the reversal test trials in response to the stimulus that was correct before the reversal which determines the proficiency of their performance level on the remaining reversal test trials. The overall level of responding is reflected in the intercepts of linear regression equations. With regard to the intercepts, the rank order correlation for the species as listed and the intercepts for their respective lines is significant for A–B+ (rho = 0.86, $p < 0.01$) but not for B+C– or A–C+. These data again corroborate the basic sensitivity and meaningfulness of the Transfer Index data as presented earlier.

On the strength of the foregoing analysis, the species were ranked

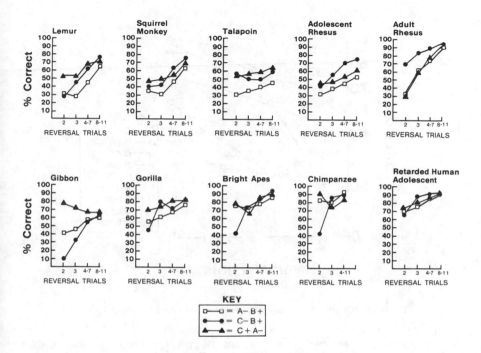

KEY

□——□ = A– B +
●——● = C– B +
▲——▲ = C + A–

FIG. 31.3. Reversal performances for each of the conditions in the associative–mediational assessment paradigm. (Data for adult rhesus are from Essock–Vitale, 1978, by permission. Data for retarded humans are from Meador and Rumbaugh, 1981, by permission.)

from low to high in accordance with their mean percentages correct on the A–B+ reversal conditions, which range from a low of 38 to a high of 81. Against these values are plotted the total deviations of the C+A– and the C–B+ conditions from the control condition, A–B+. The relationship is very strong and obvious as shown in Figure 31.5, with only the gorillas and the gibbons having any substantial departure from an otherwise pronounced rank ordering (Spearman rho = –0.90, p < 0.01 between absolute deviations and A–B+ means). The histogram portion of this figure supports the conclusion that there is a general enhancement of learning ability, defined by percent correct on the A–B+ reversal condition and a drawing together of the performance curves for the three reversal conditions.

The reduction in the deviations of the C+A– and the C–B+ curves from the control, A–B+, is not a simple ceiling effect. If it were but a ceiling effect (e.g., the curves becoming more and more alike because with the elevation of performance there was less opportunity for them to

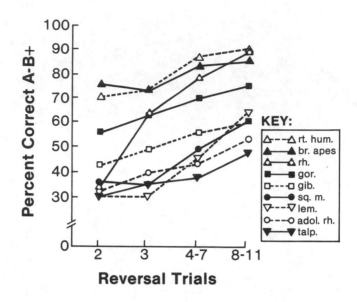

FIG. 31.4.   Percent responses correct for the A–B+ (control) reversal condition.

differ from one another), there would be no alterations in the relative difficulty of those three test conditions.   That there are such differences is clearly revealed in the top part of Figure 31.5.   There are, in fact, five different patterns among the groups: 1) talapoin and lemur, 2) adolescent rhesus, 3) gibbon, 4) adult rhesus, and 5) gorillas, bright apes, and retarded humans.   Although the squirrel monkeys' performances on the three reversal conditions did not differ significantly, which suggest mediation, they performed overall at a relatively low level of accuracy.   This group was comprised of five animals, three of which had histories of being above 90% correct on trial 2 over the span of 100 learning set problems. A randomly selected group of squirrel monkeys probably would not have given evidence of mediation.   The chimpanzee group marked "VC" was comprised of three animals, the performance of which was examined *post hoc*, as were the squirrel monkey performance, for correspondence with the data otherwise presented.

For talapoin and lemur, the C+A– and C–B+ curves were similar.   Both of these conditions were easier for these subjects than was the A–B+ control condition. This pattern seems to be an associative phenomenon with equivalent and large s_R and s_{IR}, the tendencies to approach A and avoid B. Deletion of either A or B equally facilitated reversal test performance.   This pattern is possibly the most primitive one for associative learning within the order *Primates*.

For the adolescent rhesus, a differential loading of the correct and incorrect stimuli is observed, with the correct stimulus being weighted

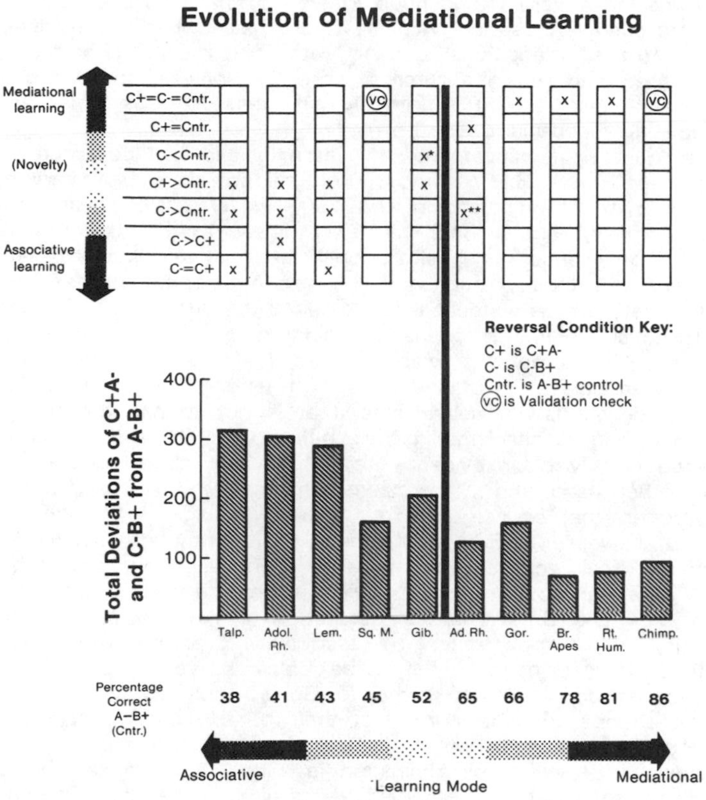

Evolution of Mediational Learning

Grid of learning conditions by species (reading approximate x-mark positions):

	Talp.	Adol. Rh.	Lem.	Sq. M.	Gib.	Ad. Rh.	Gor.	Br. Apes	Rt. Hum.	Chimp.
C+=C-=Cntr.					(VC)		x	x	x	(VC)
C+=Cntr.						x				
C-<Cntr.					x*					
C+>Cntr.	x	x	x		x					
C->Cntr.	x	x	x			x**				
C->C+	x									
C-=C+	x		x							

Left axis labels (top to bottom): Mediational learning · (Novelty) · Associative learning

Reversal Condition Key:
C+ is C+A-
C- is C-B+
Cntr. is A-B+ control
(VC) is Validation check

Vertical axis: Total Deviations of C+A- and C-B+ from A-B+ (400, 300, 200, 100)

Bar chart species (left to right): Talp. · Adol. Rh. · Lem. · Sq. M. · Gib. · Ad. Rh. · Gor. · Br. Apes · Rt. Hum. · Chimp.

Percentage Correct A-B+ (Cntr.)	38	41	43	45	52	65	66	78	81	86

Learning Mode: Associative ←——————→ Mediational

FIG. 31.5. The evolution of cognitive–mediational learning abilities. See text for explanation.

more with approach than is the incorrect stimulus weighted with conditioned inhibition. Performance was statistically superior in the C−B+ condition. Deletion of A, when incorrect, facilitated reversal performance.

A very strong novelty factor is suggested in the performance of the gibbon (*Hylobatidae*). Regardless of cue value, there was a strong tendency for the gibbons to select the new stimulus (C). An attractive, though very tentative, hypothesis is that the emergence of a strong novelty response is perhaps one form of evolutionary transition from primarily associative learning to operations that entail mediating mechanisms. In the three great ape groups, only a transient novelty effect is to be seen, that being in the suppressed trial two performances in the C−B+ condition (Figure 31.3).

The adult rhesus (*Macaca*) have a unique pattern; it suggests there was no loading whatever of the incorrect stimulus (B) with inhibition but that there remained a very pronounced loading of habit to the correct stimulus (A).   Interestingly, Essock–Vitale (1978) has published evidence which suggests *limited* mediational learning with adult rhesus. For the groups of gorillas, apes, and retarded humans, essential equivalence of the three test conditions is observed. It is primarily with these groups that a strong case for mediating mechanisms can be made.

The general enhancement of learning, as is documented in the histogram of Figure 31.5, is most likely that of an evolutionary continuum. That probability notwithstanding, the *differential patterns*, as noted in the top portion of Figure 31.5, are *not* likely to be the result of one continuum of change which alters only the relative distances between the curves and not their relationships.   We are strongly inclined to take the different patterns as evidence for *qualitative* differences in learning due to differential weightings of habit, inhibition, and novelty with an eventual introduction of something which serves, in effect, to discount specific stimuli.   Although it is argued here that among the great apes there is evidence for mediating mechanisms, that is not to say that all great apes utilize mediating mechanisms in their behaviors.   Gill and Rumbaugh (1974) reported equally clear evidence that there are *non*–mediating great apes. On the other hand, the chimpanzees Sherman and Austin, both language-trained over the course of four years (Savage–Rumbaugh, Rumbaugh, Smith, & Lawson, 1980), performed equivalently on all three reversal conditions.   Their accuracy on the A–B+ control problems was higher than for all other groups –– 88% correct (compare with values provided in Figures 31.4 and 31.5). In our estimation, in reversal tests they shed, in the majority, the constraints of associative learning and functioned as mediators of information.   Let it be clear, however, that if indeed this conclusion is correct, they came to do so only by virtue of extensive prior experience.   Furthermore, that their performance suggests high–level mediating proficiency is not to discount totally the role of S–R learning in the original prereversal discrimination learning.   It is to say, however, that in addition to whatever associative learning there was, some kind of higher–order mediating mechanism was operational.

In conclusion, it is argued that there is with evolution of the brain a continuum of enhancement for learning *and* attendant qualitative differences in learning.   The reasons for such differences are not at all obvious, and we do not now pretend to know what they are.   That limitation notwithstanding, the data do serve to caution strongly against the tenet so prevalent in psychology that principles of behavior and learning can be generalized in a straightforward, simplistic way across all species.   We believe that the data suggest very strongly that along with quantitative augmentations in learning there are qualitative and probably emergent characteristics of learning which are beyond simple associationistic learning processes.

VI. DISCUSSION

The data base reported in this paper is presented as very strong evidence in support of the hypothesis that quantitative increments in transfer-of-training among primates are coupled with qualitative alterations in learning and probably in the mechanisms for the transfer of that learning as well. With an enhancement of learning ability there is, first, a strong tendency to reduce equivalent and strong loadings of both approach and avoidance habits to the specific stimuli in prereversal criterial training. The first step toward this effect is achieved through a diminution of the inhibition accrued to the incorrect stimulus. Additional enhancement of learning ability next provides for a strong approach to novel stimuli regardless of their cue values, perhaps one way of breaking bonding to specific stimuli in the determination of choice behavior due to contingencies. Finally, further augmentation of learning ability provides for the equating of these three test conditions that are radically different in terms of their stimulus composition after the first reversal test trial, an observation reminiscent of Kluver (1933).

The data base of this study also strongly contradicts the conclusion that contingency effects in *discrimination* learning are one and the same across primate species. If one views each of the Transfer Index criterial levels (67% and 84%) as sets of contingencies, where each level prescribes a relationship between rewarded and unrewarded trials, the consequences are quite different for the prosimians at one extreme and for the great apes at the other. Whereas the prosimians exhibited very strong negative transfer on reversal test trials, the great apes manifested nearly equally strong positive transfer. Elevating the prereversal learning level for the prosimians proved to be a profound liability, whereas for the great apes it was a nearly equally profound asset. This difference between negative and positive transfer is in our view one of quality, despite the fact that one moves from negative to positive transfer or vice versa on a continuum.

There is in psychology today no theory of learning, no theory or set of behavioral principles which would have provided any logical basis for anticipating these differences. Our studies and analyses stop well short of articulating the details of the processes which provide for these differences; however, it is safe to anticipate that in part they reflect the relative importance of the positive and negative stimuli of the various groups' environmental challenges and the relative importance of certain primates' being able to respond to relationships between stimuli rather than just to stimuli as specific entities.

There is no denying that environmental contingencies are extremely powerful in the shaping and control of behavior. Notwithstanding that position, there are strong species by contingency interactions, ones which we believe are due to increased brain complexity as nicely availed to us within the order *Primates* A perspective attractive to us is that environmental contingencies in the more primitive forms, such as the prosimians, serve to regulate responses to specific stimuli whereas they serve to define information for the primates with the more highly evolved brains so that the higher-order processes of inference and cognition can equalize radically different transfer-of-training situations.

Perhaps there is a shift in how problem situations are perceived as a reflection of the changes in the primate brain. Perhaps the associative learners address situations by responding which is shaped rather specifically by contingencies. By comparison, mediational learners might perceive the same situations as *problem-solving* tasks or as *perceptual challenges*. With this presumed shift, elements of a problem are no longer just specific stimuli to which to respond. Rather, those elements and their relationships become informational units which can provide the basis for assessing probability of gain and loss for various options of responding available to the subject.

The one primary point of this paper is, then, that with evolution of the primate brain cognition somehow emerges from and supplants the more basic, primitive associative processes of learning.

ACKNOWLEDGMENTS

Preparation of this paper and a portion of the research reported was supported by a grant from the National Institute of Child Health and Development (HD-06016) and from the Division of Research Resources (RR-00165), National Institutes of Health.

REFERENCES

Essock-Vitale, S. M. Comparison of ape and monkey modes of problem solution. *Journal of Comparative and Physiological Psychology*, 1978, *92*, 942-957.

Gill, T. V., & Rumbaugh, D. M. Learning processes of bright and dull apes. *American Journal of Mental Deficiency*, 1974, *78*, 683-687.

Harlow, H. F. The formation of learning sets. *Psychological Review*, 1949, *56*, 51-65.

Kendler, H. H., & Kendler, T. S. Vertical and horizontal processes in problem solving. *Psychological Review*, 1962, *69*, 1-16.

Kluver, H. *Behavior mechanisms in monkeys*. Chicago: University of Chicago Press, 1933.

Kohler, W. *The mentality of apes*. New York: Kegan Paul, Trench, Trubner, 1925.

Miller, N. E. Studies of fear as an acquirable drive: I. Fear as motivation and fear-reduction as reinforcement in the learning of new responses. *Journal of Experimental Psychology*, 1948, *38*, 89-101.

Mowrer, O. H. On the dual nature of learning—a reinterpretation of "conditioning" and "problem-solving". *Harvard Educational Review*, 1947, *17*, 102-148.

Osgood, C. E. *Method and theory in experimental psychology*. New York: Oxford University Press, 1953.

Rescorla, R. A., & Holland, P. C. Behavioral studies of associative learning in animals. *Annual Review of Psychology*, 1982, *33*, 265-308.

Roitblat, H. L. The meaning of representation in animal memory. *The Behavioral and Brain Sciences*, 1982, *5*, 353–406.

Rumbaugh, D. M. Learning skills of anthropoids. In L. A. Rosenblum (Ed.), *Primate behavior: Developments in field and laboratory research*. (Vol. 1.) New York: Academic Press, 1970.

Rumbaugh, D. M. Evidence of qualitative differences in learning processes among primates. *Journal of Comparative and Physiological Psychology*, 1971, *76*, 250–255.

Rumbaugh, D. M., & Gill, T. V. The learning skills of great apes. *Journal of Human Evolution*, 1973, *2*, 171–179.

Savage–Rumbaugh, E. S., Rumbaugh, D. M., Smith, S. T., & Lawson, J. Reference: The linguistic essential. *Science*, 1980, *210*, 922–925.

Skinner, B. F. *Science and human behavior*. New York: Macmillan, 1953.

Skinner, B. F. *About behaviorism*. New York: Knopf, 1974.

Tolman, E. C. There is more than one kind of learning. *Psychological Review*, 1949, *56*, 144–155.

Yerkes, R. M. The mental life of monkeys and apes: A study in ideational behavior. *Behavior Monograph*, 1916, *3*, 1–145.

Yerkes, R. M., & Yerkes, A. W. *The great apes: A study of anthropoid life*. New Haven: Yale University Press, 1929.

VIII. NEUROPHYSIOLOGICAL APPROACHES

32 COMMON COMPONENTS OF INFORMATION PROCESSING UNDERLYING MEMORY DISORDERS IN HUMANS AND ANIMALS

John A. Walker
Good Samaritan Hospital and Medical Center
Portland, Oregon

I. INTRODUCTION

Throughout the course of an individual's lifetime, information is acquired, stored, and may be used again and again, often with continuing modifications. Behavior at any moment that seems to be influenced by preceding events can be said to be influenced by memory for those preceding events. Obviously, some kinds of information are more likely to be relatively unchanging over time, while others may vary greatly from moment to moment. The colors of the flag of the United States, for example, can be expected to be quite invariant over the life of an individual, while other kinds of information, such as the calendar date for yesterday, are constantly changing. Still other information, such as the name of the current president of the United States, changes at a rate intermediate to that of the two previous examples. It is likely, then, that memorability might be largely influenced by the relative stability or variability of the information to be remembered. What I wish to do in this paper is to review some selected work in animal and human memory deficits, to consider whether there are close links in the underlying functional and neural mechanisms studied in these two sometimes separate literatures, and to propose an approach which emphasizes how an appreciation of the relative stability and variability with which information is stored can be useful in advancing our understanding of the role of memory in information processing.

The study of memory deficits has a long history, but as recently as 30 years ago, Karl Lashley (1950) reported a marked lack of success in determining the locus of memory in the brain, concluding, in essence, that memory must be in many places simultaneously. Shortly afterwards, though, Scoville and Milner's (1957) important paper on patient H. M. appeared, stimulating the modern work in brain mechanisms of memory in both humans and nonhuman species. Whereas Lashley and others had primarily emphasized the role of cortical tissues, the limbic system involvement which seems to play a role in H. M.'s deficit led to the analysis of subcortical tissues which is so prominent in today's research. Although it is clear that the major progress of the past decades has been due to this emphasis on subcortical rather than cortical approaches, I will indicate later how the cortex is still likely to play a major role in an analysis of memory processes.

Most of the assertions made about the memory deficits found after brain damage are controversial to a greater or lesser degree, and await further work to resolve such controversies. My approach in this paper, though, will be to overlook many of these controversial details so as to focus on some of the larger issues that seem to emerge from this enormous body of research. In the context of a series of papers on animal cognition, I think it will be most useful to briefly review several of the prominent features of memory deficits in humans and animals, to a large extent focusing on those studies that I think are potentially useful for integrating the results from the different species. Although a number of authors, such as Terrace, (Chapter 1, this volume) are reluctant to explicitly link human and animal cognition, I think there are useful grounds for attempting such a link. I hope to be able to provide a unifying theme for understanding the role of memory in information processing in both brain-damaged and intact individuals. Because I wish to spend some time indicating how studies of human memory deficits can be especially useful for developing future ideas in animal cognition, I'll very briefly review the nature of memory deficits in animals, and spend a bit more time considering several important human studes in detail, so as to develop the implications of these studies for future studies in animal cognition.

II. STUDIES OF ANIMAL MEMORY DEFICITS

The wide variety of studies that have been done in animal memory deficits has been adequately reviewed elsewhere (Isaacson, 1976; Iverson, 1976; Weiskrantz, 1977; Olton, Becker, & Handelmann, 1979). Certainly one of the issues that we must constantly consider is that deficits in performance may be due to factors other than memory, such as decreased arousal or attention, changes in motivation, and changes in response biases. However, it seems clear that a wide variety of studies have shown deficits in performance after brain damage as due to a primary disorder of memory. In particular, I would like to concentrate on two subsets of the literature: discrimination tasks in primates, including alternation and reversal tasks, and spatial maze tasks in rodents. Most of these studies have involved various sorts of limbic system and other subcortical damage, and I will not review the many fine points that are crucial elements for understanding the neural substrates. Instead, I hope to summarize the general findings for this wide variety of neural structures in this paper.

In primates, tasks which require simple, repetitive discriminations such as a spatial discrimination, matching to sample, object discrimination, or oddity problems seem to show only mild deficits, if any, after damage to tissues such as the limbic system (Cordeau & Mahut, 1964; Correll & Scoville, 1965; Delacour, 1977; Drachman & Ommaya, 1964; Gaffan, 1974; Iverson, 1976; Mahut & Cordeau, 1963; Mishkin & Oubre, 1976). In contrast, much more dramatic deficits appear if these kinds of tasks require that the animal either reverse the original discrimination, or maintain on a trial-by-trial basis which of several different stimuli was most recently associated with the reward, as in alternation tasks (Brozoski,

Brown, Rosvold, & Goldman, 1979; Gaffan, 1974; Mahut, Moss, & Zola-Morgan, 1981; Moss, Mahut, & Zola-Morgan, 1981; Rehbein *et al.*, 1980).

That is, in tasks using similar stimulus materials and similar motor responses, far more pronounced deficits appear when the reward contingency association of a stimulus varies from trial to trial than when that association is constant from trial to trial, or if the previously stable association is suddenly altered, as in reversal. Although recent work with combined hippocampal–amygdala lesions or hippocampal–cortical lesions suggests some important qualifications to the preceding statements (Mahut *et al.*, 1981; Mishkin, 1978; Mishkin & Saunders, 1979), the general phenomena seem to hold, and I will indicate later how the Mahut and Mishkin results are actually supportive of a distinction based on stability versus variability of memory.

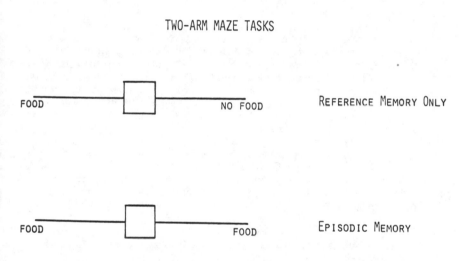

TWO-ARM MAZE TASKS

FOOD ——————————— NO FOOD   REFERENCE MEMORY ONLY

FOOD ——————————— FOOD   EPISODIC MEMORY

FIG. 32.1. A diagram of the apparatus used for testing Reference Memory or Episodic Memory. In the Reference Memory testing, the rat was required to choose the same, single arm on each presentation. In the Episodic Memory testing, the rat was forced to one arm, counterbalanced across trials. Upon returning to the center, the rat was required to choose the remaining arm.

A pattern of results similar to that just described for primates also seems to hold true for studies using rats as subjects, typically in maze tasks. That is, in tasks where the memory demand is discrimination (Winocur & Olds, 1978) or a consistent location or direction (Olton & Papas, 1979; Walker & Olton, in press; Walker, Skinner, Kosobud, Hennessy, & Black, in preparation; Winocur & Breckenridge, 1973), the brain-damaged animal seems to have a mild or transient impairment, at

best. In contrast, in those situations in which the correct response varies
from moment to moment, as in alternation, or is changed, as in reversals,
the brain-damaged animal is profoundly impaired (Oades, 1981; Olton &
Papas, 1979; Olton & Werz, 1978; Olton, Walker, & Gage, 1978;
Sinnamon, Freniere, & Kootz, 1978; Thomas, 1978; Thompson, 1981;
Walker & Olton, 1979; Walker & Olton, 1982; Walker et al., 1982;
Winocur & Breckenridge, 1973; Winocur & Olds, 1978; Winson, 1978).

The marked difference in performance after brain damage due to
stable versus variable memory demands can be illustrated even in the most
simple situations.   For example, my students and I (Walker, Skinner,
Kosobud, Hennessy, & Black, in preparation) conducted a study in which the
animals were tested in a two-arm maze, which consisted of a center
compartment and two arms which led from opposite ends of the center.
For one group of animals, both arms contained food at the begining of a
trial.   The animal was required to go to first one arm, then upon returning
to the center, to choose between the two arms.   The correct response
was to choose the unvisited arm (alternation task).   The second group of
rats was trained in the same apparatus, but in their task, only one arm
ever contained food, and they were required to choose the food baited
arm, and . avoid choices of the unbaited arm (consistent arm choice task).
Although the alternation task was far easier to learn than the consistent
arm choice task for normal rats, rats with brain damage had no difficulty
with the consistent arm choice task, measured by either acquisition· or
retention tests, as shown in Tables 32.1 and 32.2.   In contrast, other rats
with equivalent brain damage were completely unable to master the
alternation task, either in acquisition or retention tests, performing at
chance up to several months after surgery, even through several hundred
trials.   Thus, in the same apparatus, the brain-damaged rat was either as
competent as the control, or completely devastated, depending on whether
the task required only memory for a stable, invariant relationship between
places and reward, or demanded memory for constantly changing places
associated with reward.

TABLE 32.1
Two-arm maze task: Acquisition procedure

	Normals	Fornix Lesion	p values
Reference Memory	90%	90%	n.s.
Episodic Memory	90%	53%	<0.01

Note-- Scores = median group performance in last 30 trials.

The pattern of results described above for both primates and rodents
suggests to me that an important dimension that might underlie memory
deficits in animals might be the stability or variability with which
information must be retained and used.   Tasks that have a heavy

TABLE 32.2
Two-arm maze task: Retention procedure

	Normals	Fornix Lesion	p values
Reference Memory	95%	90%	n.s.
Episodic Memory	93%	62%	<0.01

Note-- Scores = median group performance in last 30 trials.

requirement for handling and manipulating information from one moment to the next seem to be much more debilitating for the brain-damaged animal than tasks that require simple, stable associations. It would be useful, then to see whether there is evidence of similar influences in the pattern of performance of humans with memory deficits.

III. STUDIES OF HUMAN MEMORY DEFICITS

Do humans have more difficulty with those kinds of information that change from moment to moment, compared to those kinds of information that are rather stable from moment to moment? For amnesic humans, this is generally true, at least anecdotally and in clinical practice. The amnesic, for example, may be completely unable to tell whether or not he had green beans for dinner last night, but can certainly tell if he ever ate green beans, or if he likes green beans or not, or if he prefers them to creamed corn. However, it is not sufficient support for the question of stability or variability of information in memory to rely on such clinical evidence, because most of the information humans deal with is not so easily categorized in terms of stability or variability, and moreover, some kinds of information, by their nature, are variable, while others are stable. This potential difference in the contents of what is usually considered stable and what is usually considered variable means that suitable experiments must be devised to study whether a difference in memorability is due to the stability-variability dimension. A variety of excellent reviews of the human amnesia literature exist (Butters & Cermak, 1980; Kinsbourne & Wood, 1975; Squire, 1982; Talland & Waugh, 1969; Warrington & Weiskrantz, 1973). Relatively few experiments have been designed to explicitly test whether the stability of information influences memorability in amnesics. However, there are several key lines of research that relate to the theme of stability versus variability in human memory deficits.

Warrington and Weiskrantz (1974) studied cueing effects on verbal memory in amnesics. The task was constructed so that the patient was asked to learn a list of common English words, then tested for recall by being cued with the first three letters of the word. Warrington and

Weiskrantz took advantage of the variability of such cues for specifying words in English, and manipulated this variability. For example, the cue ONI specifies only one word in English, ONION. In contrast, other cues, such as PRE, specify a large number of words. They found that amnesic subjects were as good as controls for those items that were uniquely specified by the cue, but were severely impaired relative to controls for those items that had many possibilities. A more important variation, for our purposes, was their use of cues that could specify only two possible words, such as MOA for MOAN and MOAT. Amnesic subjects were first taught a word list which contained only one of the two possible words, and their performance was indistinguishable from controls on cued recall. On the next trial, Warrington and Weiskrantz introduced a reversal procedure, such that the other member of the possible words specified by a cue was presented. With the introduction of the reversal, the amnesic patients' performance fell to chance, and remained there for several consecutive presentations of the reversal list. In contrast to controls, who showed a mild impairment upon the introduction of the reversal, but quickly returned to high levels of performance on repeated presentations of the reversal list. Note that the amnesic subjects did not show any sign of debilitation on this task until the link between the three letter cue was changed from a stable relationship (the first presentation) to a variable relationship (the introduction of the reversal). It also seems clear from this work that there might be a more long–lasting debilitating influence from variability than facilitation from stability in amnesics, in that one reversal presentation was sufficient to cause chance performance through repeated presentations of the reversal list. In many respects, this interesting report is similar to that described earlier for animals in reversal tasks, and reinforces the idea that the memory deficit is particularly acute for information which changes over time, but less severe, or absent, if the relationship is maintained with stability over time.

Intact perceptual learning seems to be an important preserved ability in amnesics. Severely amnesic patients have been tested with materials such as Gollin's (1960) series of fragmented drawings, or hidden figures designs. Performance on these tests has been shown to be useful in separating out autobiographical memory (when was this seen?) from perceptual memory (what is it, or where is the hidden figure?), in that amnesic patients are generally unable to tell that they have dealt with these specific materials before, yet show the benefits from prior exposure in subsequent identification that normal subjects show (Crovitz, Harvey, & McClanahan, 1981; Milner, Corkin, & Teuber, 1968; Warrington & Weiskrantz, 1968). Similar results for some perceptual–motor skills have also been reported (Brooks & Baddeley, 1976; Corkin, 1968). The extent to which autobiographical memory and perceptual learning can be dissociated in the amnesic is unknown at present, though recent advances in studying these issues with normal humans have been made (see Jacoby & Dallas, 1981, for a review), and it is likely that these kinds of approaches will have an impact on studies in amnesic patients. At least a large portion of such effects may be found to be mediated by factors such as the relative stability of the perceptual elements of these tasks, compared to more variability for the autobiographical aspects of the tasks.

Another set of tasks in which one sees preservation of some aspects of the tasks, but not others, in amnesics, is the recent work by Cohen and Squire. I think many aspects of this work can be readily accommodated by a consideration of the stability and variability of the various task demands. In a mirror-reading task (Cohen & Squire, 1980), subjects were required to read aloud words presented in mirror orientation. Normals, as well as amnesics, showed a decrease in reading time with repeated exposures of the same items that was in addition to the overall reduction of reading time that occurred as a consequence of familiarity with the task. This was in marked contrast to the amnesic subjects' inability to subsequently choose from a list of words those that had been seen in mirror orientation. Cohen and Squire suggested that this might be due to a separation between procedural and declarative methods of assessing memory, or a distinction between "knowing how" versus "knowing that." Though I am in general sympathy with their argument, it is clear from their own data that not all nonverbal, procedural skills are preserved. Cohen and Squire (1979) reported that the same amnesic subjects who show benefits in mirror-reading are profoundly impaired in a task involving delayed lever positioning. When the displacement was varied among several different possibilities, the amnesics were unable to perform accurately, and in fact showed a central tendency effect in their responses, which would seem to indicate that the subjects were aware of the possible range of movements (which would remain stable over the block of trials), but were unable to accurately remember the degree of displacement that had just occurred a few seconds earlier (which varied from trial to trial).

One of the crucial problems that is evident in analyzing the literature on human amnesia is the heavy emphasis on the contents of the memory, rather than the underlying processes. There appears to have been a relative lack of sensitivity to trying to capture the nature of the processing deficit, compared to a cataloging of the different kinds of materials and modalities that show impairment, although the Butters and Cermak (1980) studies of levels of encoding in Korsakoff patients are a notable exception. One can find a great number of studies in which the questions of interest concerned differences in materials, such as arithmetic, colors, high versus low frequency words, concreteness versus abstractness, and so forth. While these questions are certainly not uninteresting, they offer little potential for understanding the nature of memory deficits in species other than humans, and the emphasis on contents rather than processes may in the long run be harmful. Approaches which offer the potential for dissociating processes, independent of their specific contents, may be more useful.

Recently, I've conducted a series of experiments with human amnesics that was designed to be roughly comparable to the work I've done with brain-damaged animals, in terms of the processes studied. I was especially interested in manipulating the dimension of stability versus variability, but in keeping the contents equivalent. This word memory task was designed so that the subject was required to remember a target word which was chosen from a set of words, and he or she was asked to indicate that target when the entire set of words was then presented. The subject was seated in front of a video display, and the test materials were presented via computer. On each trial, the subject was first presented a set of four words, from which one word was then selected and presented as the target to be remembered. After a delay of a few seconds, the entire list of four words was re-presented, and the subject was asked to indicate which word had appeared as the target by pressing one of four keys corresponding to the spatial position of the target word on the screen. The lists of words were divided into two types: those that had stable targets, and those that had variable targets. For the stable target lists, only one of the four words was chosen at random, and was used as the target each time that list occurred. For the variable target lists, each of the words in the list occurred as a target within a block of trials. Thus, over a large block of trials, some word lists would have a target consistently paired with that list, while others would have targets that were inconsistent from trial to trial. I expected that amnesic subjects should do better for those items that were stable than for those items that were variable. Subjects were not informed of this difference in target type, but merely instructed to remember each target as it occurred, and to choose it from the list.

Because facilitation in this task might occur for stable targets simply because they might be more frequent, i.e., four occurrences in four presentations of a list, compared to one each in four presentations of the variable target lists, variable target lists were adjusted for frequency of target occurrence. The variable target lists were presented twice as often as the stable target lists. In practical terms, this meant that of eight presentations of a variable target list, one word was chosen and presented as the target on four occasions, one was presented as the target on two occasions and the other two each occurred as the target once. Thus, if there was any effect of frequency of occurrence as opposed to stability of occurrence, it could be sorted out in the data.

More than 20 amnesic subjects, of differing etiologies, have been tested, in addition to a number of normal subjects. Choice accuracy and reaction time were measured, during the list presentations, and after an interpolated task that lasted about 15 minutes, a recognition post-test was administered, which included words the subject had seen as well as distractor words. As might be expected, choice accuracy during the test itself was very high, averaging over 95% for amnesics, compared to over 98% for normals. No differences in accuracy were found for stable versus variable targets, but of course a ceiling effect had occurred. Reaction time, however, did show a significant decrease for stable targets compared to variable targets, which was completely independent of target frequency.

A similar pattern of decreased reaction time has been seen in normal subjects, as well. In contrast to the significant facilitation of reaction time due to target stability, amnesic subjects showed no differential effects due to target stability in the post-test retention task, essentially showing poor memory for both types of item. Normals also failed to show any effect in the post-test, but this was largely due to almost perfect performance on the post-test by normals. Although the facilitation predicted due to target stability emerged in only one of the measures, reaction time, the lack of effect in the other measures appeared to be due to asymptotically high or low levels of performance. It might be noted, though, that the pattern of facilitation in reaction time but not in the other measures is

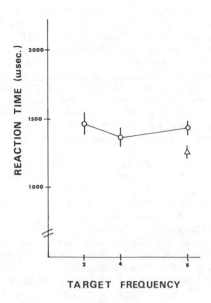

FIG. 32.2. Mean reaction times to correctly indicate the spatial position of a target word by human amnesics. The open circles represent targets chosen from lists where the target varied from trial to trial, while the triangle represents targets chosen from lists where the target was invariant from trail to trial.

maintained across different sets of items, reverses for specific items if the word lists are reversed in stable–variable dimension on different days with the same patient, and has even been observed in both languages when studied in bilinguals. Additional work is underway to determine the factors that influence such results, and to provide additional examples of materials that are useful in exploring the dimension of stability and variability in human amnesics.

IV. STABILITY AND VARIABILITY: APPLICATIONS TO MEMORY THEORIES

There should be little doubt that a wide variety of dimensions are important in determining what kinds of information are spared and what kinds are devastated in human and animal amnesias. However, I think there are sufficient similarities in the animal and human amnesic literatures to suggest some common underlying deficits in both humans and animals with brain damage. One important dimension, as I've indicated, seems to be based upon a differentiation of stability and variability. There is a whole family of somewhat related approaches to the separation of what is and is not preserved in amnesics, and presumably influential in normals' performance as well. Tulving's (1974) distinction between semantic and episodic memory, Olton's (1979; Olton et al., 1979) distinction between reference and working memory, and Cohen and Squire's (1980) distinction between procedural and declarative memory all have as a common attribute the idea that information which is relatively invariant is spared, while information that is variable is deficient. Of course, each theory applies its distinction to a somewhat different set of specific materials, and the competing theories sometimes make opposite predictions. However, I believe the next few years will see an increasing emphasis on such basic properties of information processing as the stability and variability of the information.

There remains one further point, which I believe will serve to reinforce the usefulness of a distinction based on stability and variability. In the work I have considered so far, I have emphasized the common defect in handling variable information that seems to be a chief feature of the deficit in both human and animal amnesias, with a common sparing of stable information processing. However, one might expect that the potential for a double dissociation exists, such that some brain damaged organisms might show deficits in the stable aspects of information processing, yet sparing in the variable aspects.

There are multiple examples of neurological patients that exhibit just such a pattern of deficits, though they are seldom labeled "amnesic." These kinds of patients include individuals with anomias and agnosias, and perhaps also other disorders such as agraphia, apraxia, and alexia (for good introductions and reviews of these disorders, see Hecaen & Albert, 1978; Heilman & Valenstein, 1979; and Kolb & Whishaw, 1980). These disorders can be characterized as involving the disruption of knowledge, but not of memory in the sense in which the patient labeled amnesic has memory difficulties. Consider the case of the agnosias, as perhaps the clearest case of a loss of semantic or stable information, but a sparing of episodic memory, or memory for past events which are variable. The patient with visual agnosia, usually as a consequence of posterior cortical damage, has an inability to adequately recognize even common objects. For example, if such a patient is presented a cup, or a comb, or a pen, he or she will be unable to demonstrate the use of the object, yet may be perfectly able to indicate that this particular item was presented earlier in the examination. It seems reasonable to consider the possibility that this individual has the reverse information processing problem of the amnesic patient, in that the agnosic patient can indicate that he's seen it before, but doesn't know what it is, while the amnesic patient knows what it is, but doesn't know if he's seen this particular one. The other types of disorders, such as anomia, apraxia, agraphia, and alexia, all of which may be characterized as a loss of stable but not necessarily variable information, all seem more likely after cortical damage than after subcortical damage. There might be different neural substrates that are involved in deficits in stable information processing compared to variable information processing. Individuals with large lesions might show both kinds of deficits. My current work is addressed to many of these issues in humans with neurological deficits.

A theory of memory which emphasizes a role for the stable side of

CONTINUUM

VARIABLE --------------------------- STABLE

AMNESIAS   AGNOSIAS

TEMPORAL-HIPPOCAMPAL   ANOMIAS
KORSAKOFF
CLOSED-HEAD INJURY   ALEXIAS
ETC.
                                      APRAXIAS

SUBCORTICAL/LIMBIC   CORTICAL

FIG. 32.3. A schematic illustration of how different human neurological deficits and presumed neural substrates might fall on a continuum of information stability.

information in addition to the variable side is, of course, primed to accept the cases of anomia or agnosias due to a breakdown at one end of a dimension, namely the stable end. The more traditional "amnesic" patient, of course, lies at the other end, having difficulty with the variable end. Whether there are two different disorders, or whether there are intermediate disorders on a continuum from stable to variable information processing remains for future investigation. However, it is quite likely that this kind of continuum approach is able to more adequately account for syndromes like Korsakoff amnesia, which seems to entail elements of both stable and variable information processing (Butters & Cermak, 1980; Squire, 1982).

There are already attempts in the literature to examine issues of stable information processing deficits due to cortical damage in nonhumans, most notably work by Boyd and Thomas (1977), Fuster and Jervey (1982) and Wilson and DeBauche (1981). Moreover, some of the controversial reports that have resulted from work with combined hippocampal-amygdala or hippocampal-cortical lesions (Mahut et al., 1981; Mishkin, 1978; Mishkin & Saunders, 1979) may be more easily reconciled if indeed these approaches represent damage to neural structures that are involved in both stable and variable information processing skills. Although existing tasks in the literature might be fruitfully reinterpreted to reflect on issues such as I've raised here, I think it will also be a very fruitful stimulus for those of us who are primarily involved with the assessment of capacities in normal animals to contribute techniques that will be useful to the neuroscientists, as well. Developing good tasks that tap into the stable side of information processing, such as categorization tasks, is certainly not impossible, and it will be increasingly useful to have animal analogs of tasks that demand the animal's ability to extract and form semantics.

V. CONCLUSION

In this paper, I have tried to emphasize how an analysis of memory deficits that emphasizes underlying processes, and in particular a dimension dealing with the stability with which information is stored, can be a useful way to not only incorporate the current literature, but to also expand the analysis into a number of different areas, especially having to do with semantic kinds of information processing, and their links to amnestic disorders. Baddeley (1976) has characterized the 1960's view of memory as "a store that had the sole function of holding information. . . ," while the views of the 1970's saw memory as "an integral part of other information-processing tasks, such as perception, pattern recognition, comprehension, and reasoning." I think the view for the 1980's will be to appreciate how factors such as the stability or variability of information influences a wide variety of tasks, some of which have traditionally been labeled memory and many of which have not.

REFERENCES

Baddeley, A. D. *The psychology of memory*. New York: Basic Books, 1976.
Boyd, M. G., & Thomas, R. K. Posterior association cortex lesions in rats: Mazes, pattern discrimination, and reversal learning. *Physiological Psychology*, 1977, *5*, 455–461.
Brooks, N. D., & Baddeley, A. D. What can the amnesic patient learn? *Neuropsychologia*, 1976, *14*, 111–122.
Brozoski, T. J., Brown, R. M., Rosvold, H. E., & Goldman, P. S. Cognitive deficit caused by regional depletion of dopamine in prefrontal cortex of rhesus monkey. *Science*, 1979, *205*, 929–931.
Butters, N., & Cermak, L. S. *Alcoholic Korsakoff's syndrome*. New York: Academic Press, 1980.
Cohen, N. J. & Squire, L. R. Skill acquisition in the amnesic syndrome. Paper presented at a meeting of the International Neuropsychological Society Society. 1979.
Cohen, N. J., & Squire, L. R. Preserved learning and retention of pattern-analyzing skill in amnesia: Dissociation of knowing how and knowing that. *Science*, 1980, *210*, 207–210.
Cordeau, J. P., & Mahut, H. Some long-term consequences of temporal resections on auditory and visual discrimination in monkeys. *Brain*, 1964, *87*, 177–188.

Corkin, S. Acquisition of motor skill after bilateral medial temporal–lobe excision. *Neuropsychologia*, 1968, *6*, 255–265.

Correll, R. E., & Scoville, W. B. Performance on delayed match following lesions of medial temporal lobe structures. *Journal of Comparative and Physiological Psychology*, 1965, *60*, 360–367.

Crovitz, H. F., Harvey, M. T., & McClanahan, S. Hidden memory: A rapid method for the study of amnesia using perceptual learning. *Cortex*, 1981, *17*, 273–278.

Delacour, J. Role of temporal lobe structures in visual short–term memory, using a new test. *Neuropsychologia*, 1977, *15*, 681–684.

Drachman, D. A., & Ommaya, A. E. Memory and hippocampal complex. *Archives of Neurology*, 1964, *10*, 411–425.

Fuster, J. M., & Jervey, J. P. Neuronal firing in the infero–temporal cortex of the monkey in a visual memory task. *Journal of Neuroscience*, 1982, *2*, 361–375.

Gaffan, D. Recognition impaired and association intact in the memory of the monkeys after transection of the fornix. *Journal of Comparative and Physiological Psychology*, 1974, *86*, 1100–1109.

Gollin, E. S. Developmental studies of visual recognition of incomplete pictures. *Perceptual and Motor Skills*, 1960, *11*, 289–298.

Hecaen, H. & Albert, M. L. *Human neuropsychology*. New York: Wiley, 1978.

Heilman, K. M. & Valenstein, E. *Clinical neuropsychology*. New York: Oxford University Press, 1979.

Isaacson, R. L. Experimental brain lesions and memory. In M. R. Rosenzweig & E. L. Bennet (Eds.), *Neural mechanisms of learning and memory*. Cambridge, Mass.: MIT Press, 1976.

Iverson, S. D. Do hippocampal lesions produce amnesia in animals? In C. C. Pfeiffer & J. R. Smythies (Eds.), *International review of neurobiology*. New York: Academic Press, 1976.

Jacoby, L. L., & Dallas, M. On the relationship between autobiographical memory and perceptual learning. *Journal of Experimental Psychology: General*, 1981, *110*, 306–340.

Kinsbourne, B. & Wood, F. Short–term memory processes and the amnesic syndrome. In D. Deutsch & J. A. Deutsch (Eds.), *Short-term memory processes*. New York: Academic Press, 1975.

Kolb, B. & Whishaw, I. Q. *Fundamentals of human neuropsychology*. San Francisco: Freeman, 1980.

Lashley, K. S. In search of the engram. *Symposia of the Society for Experimental Biology*, 1950, *4*, 454–482.

Mahut, H., & Cordeau, J. P. Spatial reversal deficit in monkeys with amygdala–hippocampal ablations. *Experimental Neurology*, 1963, *7*, 426–434.

Mahut, H., Moss, M., & Zola–Morgan, S. Retention deficits after combined amygdala–hippocampal and selective hippocampal resections in the monkey. *Neuropsychologia*, 1981, *19*, 201–225.

Milner, B., Corkin, S., & Teuber, H. L. Further analysis of hippocampal amnesic syndrome: 14 year follow-up study of H. M. *Neuropsychologia*, 1968, *6*, 215–234.

Mishkin, M. Memory in monkeys severely impaired by combined but not separate removal of amygdala and hippocampus. *Nature*, 1978, *273*, 297–298.

Mishkin, M. & Oubre, J. L. Dissociation of deficits on visual memory tasks after inferior temporal and amygdala lesions in monkeys. *Neuroscience Abstracts*, 1976, *2*, 1127.

Mishkin, M., & Saunders, R. C. Degree of memory impairment in monkeys related to amount of conjoint damage to amygdaloid and hippocampal systems. *Neuroscience Abstracts*, 1979, *5*, 1057.

Moss, M., Mahut, H., & Zola–Morgan, S. Concurrent discrimination learning of monkeys after hippocampal, entorhinal, or fornix lesions. *Journal of Neuroscience*, 1981, *1*, 227–240.

Oades, R. D. Types of memory or attention? Impairments after lesions of the hippocampus and limbic ventral tegmentum. *Brain Research Bulletin*, 1981, *7*, 221–226.

Olton, D. S. Mazes, maps, and memory. *American Psychologist*, 1979, *34*, 583–596.

Olton, D. S., & Papas, B. C. Spatial memory and hippocampal function. *Neuropsychologia*, 1979, *17*, 669–682.

Olton, D. S., & Werz, M. A. Hippocampal function and behavior: Spatial discrimination and response inhibition. *Psychology and Behavior*, 1978, *20*, 597–605.

Olton, D. S., Becker, J. T., & Handelmann, G. E. Hippocampus, space, and memory. *The Behavioral and Brain Sciences*, 1979, *2*, 313–322.

Olton, D. S., Walker, J. A., & Gage, F. H. Hippocampal connections and spatial discrimination. *Brain Research*, 1978, *139*, 295–308.

Rehbein, L., Zola–Morgan, S., Mahut, H., & Moss, M. Failure of sparing, or recovery, of early recognition memory after early hippocampal resection in the rhesus macaque. *Neuroscience Abstracts*, 1980, *6*, 88.

Scoville, W. B., & Milner, B. Loss of recent memory after bilateral hippocampal lesions. *Journal of Neurology, Neurosurgery, and Psychiatry*, 1957, *20*, 11–21.

Sinnamon, H. M., Freniere, S., & Kootz, J. Rat hippocampus and memory for places of changing significance. *Journal of Comparative and Physiological Psychology*, 1978, *92*, 142–155.

Squire, L. R. The neuropsychology of memory. *Annual Review of Neuroscience*, 1982, *5*, 241–273.

Talland, G. A. & Waugh, N. C. *The pathology of memory*. New York: Academic Press, 1969.

Thomas, G. Delayed alternation in rats after pre- or postcommissural fornicotomy. *Journal of Comparative and Physiological · Psychology*, 1978, *92*, 1128–1136.

Thompson, R. K. Rapid forgetting of a spatial habit in rats with hippocampal lesions. *Science*, 1981, *212*, 959–960.

Tulving, E. Episodic and semantic memory. In E. Tulving & W. D. Donaldson (Eds.), *Organization of memory*. New York: Academic Press, 1972.

Walker, J. A., & Olton, D. S. Spatial memory deficit following fimbria-fornix lesions: Independent of time for stimulus processing. *Physiology and Behavior*, 1979, *23*, 11–15.

Walker, J. A., & Olton, D. S. Fimbria–fornix lesions impair spatial working memory but not cognitive mapping. *Behavioral Neuroscience*, in press.

Walker, J. A., Skinner, L. E., Kosobud, A., Hennessy, C., & Black, J. Fimbria–fornix lesions impair both acquisition and retention of a working memory, but not a reference memory task. In preparation.

Warrington, E. K., & Weiskrantz, L. New method of testing long–term retention with special reference to amnesic patients. *Nature*, 1968, *217*, 972–974.

Warrington, E. K., & Weiskrantz, L. An analysis of short–term and long–term memory defects in man. In J. Deutsch (Ed.), *The physiological basis of memory*. New York: Academic Press, 1973.

Warrington, E. K., & Weiskrantz, L. The effect of prior learning on subsequent retention in amnesic patients. *Neuropsychologia*, 1974, *12*, 419–428.

Weiskrantz, L. Trying to bridge some neuropsychological gaps between monkey and man. *British Journal of Psychology*, 1977, *68*, 431–445.

Wilson, M., & DeBauche, B. A. Inferotemporal cortex lesions and categorical perception of visual stimuli by monkeys. *Neuropsychologia*, 1981, *19*, 29–41.

Winocur, G., & Breckenridge, C. B. Cue–dependent behavior of hippocampally damaged rats in a complex maze. *Journal of Comparative and Physiological Psychology*, 1973, *82*, 512–522.

Winocur, G., & Olds, J. Effects of context manipulation on memory and reversal learning in rats with hippocampal lesions. *Journal of Comparative and Physiological Psychology*, 1978, *92*, 312–321.

Winson, J. Loss of hippocampal theta rhythm results in spatial memory deficit in the rat. *Science*, 1978, *201*, 160–163.

33 THE HIPPOCAMPUS AS AN INTERFACE BETWEEN COGNITION AND EMOTION

Jeffrey A. Gray
Oxford University

I. INTRODUCTION

Psychology is replete with false dichotomies: consider, for example, the "biological" versus the "social" (but society is a biological invention), or "animal" versus "human" (but man is an animal). Two such dichotomies that are relevant to the theme of this chapter are: cognition versus emotion; and behaviorism versus its resuscitated and newly fashionable rival, cognitive science. No more than in my other examples are the terms of these dichotomies true polar opposites: thought rarely occurs without emotion, or emotion without thought; and behavioral and cognitive science, if either is to succeed, will need to be as intricately related as the controlled process (behavior) and controlling processes (cognition) that each seeks to specify.

My reflections on these issues will be presented in the context of a recently developed theory of the functions of the septo–hippocampal system and related structures in anxiety (Gray, 1982a, b).

The hippocampus has been the focus of several cognitive theories, especially in recent years (Hirsh, 1974; Kimble, 1975; O'Keefe & Nadel, 1978; Olton, Becker, & Handelmann, 1979; Winocur, 1982). It has figured in theories of the emotions less often, though this tradition began with panache in a famous paper by Papez (1937) and has a distinguished contemporary Soviet exponent in the person of P. V. Simonov (1974).

The new theory (Gray, 1982a) is to some extent an integration of these two traditions, since it treats the hippocampal formation and the Papez circuit, to which the hippocampus is so closely related, as an interface between cognitive and emotional processes.

The theory was developed in two main stages, and these correspond reasonably well to modes of "behaviorist" and "cognitivist" thinking, respectively. But they correspond also to a shift from reliance principally on psycho–pharmacological evidence as the data base for the theory to reliance principally on evidence from studies of the brain. Thus, the development of a concern with cognition has paralleled the need to theorise about the workings of the brain. This parallel should occasion no surprise, since I suppose we would all agree that the brain is the organ of cognition. Yet (with the honorable exception of those scientists cited above, who have offered cognitive theories of the functions of the hippocampus) contributors to the recent fluorescence of cognitive science have usually known little and cared less about the brain. This has been unfortunate, for, by tying a cognitive theory to the hard wiring of the brain, we gain real advantages in testability, an ingredient which many otherwise impressive cognitive theories sometimes rather conspicuously lack.

II. A THEORY OF ANXIETY

In its first stage of development the theory was largely deduced from observations in animals of the behavioral effects of anti-anxiety drugs (i.e., the benzodiazepines, barbiturates and ethanol).   Indeed, it was a simple inverse function of these observations, after they had been subjected to a certain amount of theoretical massage to produce two sufficiently simple but wide-ranging generalizations (Gray, 1975; Gray, 1977).   These generalizations are as follows. First, there is an equivalence between three types of stimuli, all of which produce essentially similar responses.   The stimuli concerned are those associated with punishment, those associated with frustrative nonreward, and novel stimuli (though the latter raise some special problems: Gray, 1982c).   The changes these stimuli produce in behavior are threefold: ongoing motor programs are brought to a halt; there is an increase in the level of arousal (in the sense that the next occurring behavior, whatever it is, is performed with especially great vigor); and there is an increase in attention to environmental (especially novel) stimuli.   Second, behavioral reactions of all these three kinds, and to all the relevant types of stimuli, are attenuated by all the anti-anxiety drugs.

If we now suppose that behavior antagonized by the anti-anxiety drugs is anxious behavior (a major assumption, but one that can be defended; Gray, 1982a), we may describe the types of stimuli listed above as "anxiogenic" and treat the reduction in anxiety produced by the anti-anxiety drugs as a reduction in the behavioral effects of anxiogenic stimuli.   These moves, so far, outline an esentially behaviorist theory of anxiety, though hard-line behaviorists would no doubt object to the language in which it is couched.   If we take one small further step we arrive at the hypothesis illustrated in Figure 33.1.   This step adds a hypothetical construct, the "behavioral inhibition system" (BIS), which is held to mediate the behavioral effects of anxiogenic stimuli, and upon which the anti-anxiety drugs act to attenuate those effects.

Hypothetical constructs already take us beyond the bound of behaviorist purity, though that is a purity that most behaviorists thankfully never accepted.   But we need more than a hypothetical construct or two before we can properly praise or abuse a theory as "cognitive."   To deserve this label the theory must specity the controlling processes which, inside the black box pictured in Figure 33.1, transmute input (anxiogenic stimuli) to output (anxious behavior).   The difficulty with developing such a specification is that there are a great many alternative models one can formulate, all of them compatible with the input-output relations which Figure 33.1 so crudely summarizes, and very difficult to distinguish by means of purely behavioral observations.

It is for this kind of reason that, even in human experimental psychology, where much finer-grained data bearing on cognitive models

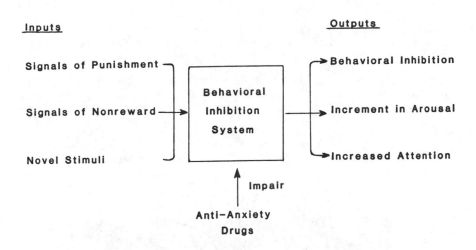

FIG. 33.1 The behavioral inhibition system.   (See text for further explanation.)

can be gathered, the choice between alternative models often remains a matter of personal taste.   But taste is unsuitable as an arbiter of the value of scientific theories; and, if better touchstones cannot be found, the current vogue for cognitive theorising will go the way of its forerunner a century ago.   To provide such a touchstone there is at present no alternative to the test of experiment, nor is there likely soon to be one. Thus we need a way of specifying the inside of the black box which will be open to experimental test; and if this cannot be done by behavioral observations alone, then these must be supplemented by other forms of observation.   It is in this context that we do well to remember that cognition is a product of the brain.   For, if we specify the processes which control behavior *and* the brain mechanisms which mediate these processes, it becomes possible to test the resulting theory by means of observations on the brain as well as on behavior; and, more powerfully still, by means of joint observations of both these kinds (for some examples, see Gray, 1982b).   It is such a joint specification of cognitive process and mediating brain mechanism that my theory of anxiety, in its second stage (Gray, 1982a; Gray, 1982b), attempts to construct.   I shall not try here to summarize the data bases (stemming from anatomical, physiological, pharmacological and behavioral experiments) on which this theory rests (see Gray, 1982a), but instead present it as a *fait accompli*.

III.   THE SEPTO-HIPPOCAMPAL SYSTEM

Figure 33.2 shows a photograph of the hippocampal formation and septal area excised from the brain of a rat. This piece of tissue seems to call forth culinary similes. O'Keefe and Nadel (1978) liken it to a sausage. I prefer to think of it as a pair of bananas joined at the front. The area where the bananas are joined is the septal area; *in vivo* it lies near the front of the brain, towards the animal's snout. The rest is the hippocampal formation. The septal area and the hippocampal formation are connected by two bundles of fibers: the fimbria, which sweeps along the outside edges of the two bananas, and the dorsal fornix, which keeps to the midline and courses in the hollow between the two bananas. The two bundles come together close to the septal area, where they are termed the "fimbria-fornix."

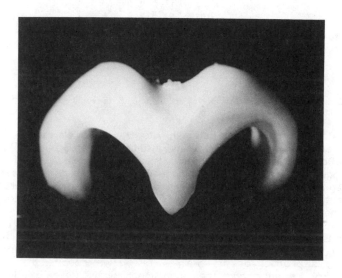

FIG. 33.2. The hippocampal formation and septal area of the rat (photograph by J. P. Broad).

The hippocampal formation has a beautifully ordered internal structure which has made it a favorite target for anatomists and physiologists alike. If you were to flatten the banana out to eliminate the complex curvature of its trajectory as it moves further and further away from the septal area towards the tip that is buried in the temporal lobe, and if you were then to cut it transverse to this "septo-temporal" axis, the section so revealed contains two major interlocking U-shaped rows of large cells which (to mix culinary metaphors) give the appearance of a slice of Swiss roll

(Figure 33.3). The upper U is the hippocampus proper (or "Ammon's horn"), the lower U the dentate gyrus or fascia dentata. On the basis of cell morphology the U of Ammon's horn is further subdivided. If we follow this U round, starting with the tip furthest from the fascia dentata, there is first a double row of medium—sized pyramidal cells (mostly belonging to the CA 3 group) running from the arch of the U to finish within the dentate gyrus; finally, the cells that make up the dentate U are granular rather than pyramidal in shape.

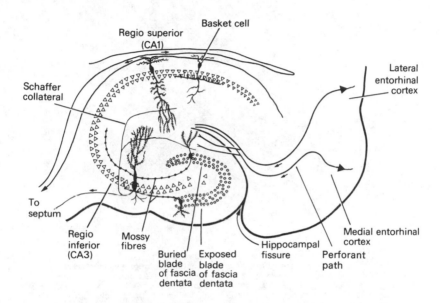

FIG. 33.3. Schematic diagram of the intra—hippocampal connections. Horizontal section through the right hippocampus; caudal is to the right of the figure (from O'Keefe and Nadel, 1978).

The interconnections between these different cell fields are as regular as their appearance. These interconnections are, for the most part, organized in strips or "lamellae" (Andersen, Bliss, & Skrede, 1971) transverse to the septo—temporal axis of the hippocampal formation (Figure 33.4). The starting point for the passage of information around a lamellar strip is a neuronal message that originates in the entorhinal cortex. This region of the neocortex itself receives sensory information in all modalities after extensive processing by association cortex. This information is passed along a fiber tract called the perforant path to the dentate granule cells. The granule cells in turn send impulses along their axons (known as the "mossy fibers") to the CA 3 pyramids. These pyramids have bifurcating axons, of which one branch exits from the hippocampus into the fimbria, destined for the lateral septal area, while the other (the "Schaffer collateral") synapses with the dendrites of the CA 1 pyramidal cells. Finally, the CA 1 pyramids send their axons out of Ammon's horn in the alveus, a sheet of fibers which covers the outside of the hippocampal banana like a gleaming white skin.

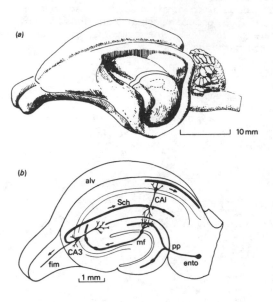

FIG. 33.4. Lamellar organization of the hippocampus. (a) Lateral view of the rabbit brain with the parietal and temporal neocortex removed to expose the hippocampal formation. The lamellar slice indicated has been presented separately in (b) to show the probable circuitry. alv, alveus; ento, entorhinal cortex; fim, fimbria; pp, perforant path; Sch, Schaffer collateral. (Modified by J. N. P. Rawlins from Andersen et al., 1971.)

It used to be thought that CA 1 was the source of the massive efferent pathway from the hippocampal formation which travels in the fornix, But both electrophysiological and anatomical evidence now makes it clear that the CA 1 axons terminate in the subicular area. This region is transitional both in position and in morphology between the simple cortical structure of the hippocampus and the six-layered neocortical mantle encountered in the entorhinal area, on to which it abuts (Figure 33.5) and to which it projects. It is the subicular area which is also the origin of the descending fibers which travel in the fornix. These fibers go by way of the post-commissural fornix, a massive continuation of the fimbria-fornix that sweeps down in two columns, one in each hemisphere, through the posterior septal area. The targets of the descending columns of the fornix are the mammillary bodies in the hypothalamus and certain nuclei in the anterior thalamus, particularly the anteroventral nucleus. The anteroventral thalamic nucleus in turn projects to the cingulate cortex, which completes the loop by projecting again to the subicular area. The loop is also completed by a shorter route, from the anteroventral thalamus to the subicular area directly.

FIG. 33.5. Photograph of a horizontal section of the right side of the adult mouse brain as seen from above. The section passes through the posterior arch of the hippocampus and entorhinal cortex (R.E.). Caudal is up. Note the gradual transition from the six-layered entorhinal cortex through the parasubiculum (Par), presubiculum (Pres), subiculum (Sub), and prosubiculum (Pros) to the three-layered cortex of the hippocampus proper (CA 1–CA 4) and the fascia dentata (F.D.). Fi is the fimbria. (Modified by O'Keefe and Nadel, 1978, from an original by R. Lorente de No.)

The subicular cortex appears to be at a nodal point in the complex circuitry of the septo–hippocampal system (Figure 33.6). To begin with, it is in direct receipt of projections from the same regions which project also to the entorhinal area. It then also receives a projection from the entorhinal area. Next, it receives more delayed messages from the entorhinal area after these have traveled around the hippocampal circuit already described (via the dentate gyrus, CA 3 and CA 1). It then sends this information back to the entorhinal area and also out along the descending columns of the fornix, only to have the same information come back yet again after a long trip through the mammillary bodies, the anterior thalamus and the cingulate cortex. Of course, it is not "the same" information. If it were, we should need to suppose that a large part of the brain does nothing but echo back the messages that it receives. So these reverberating loops cry out for functional interpretation.

The interconnections between the hippocampal formation and septal area are simpler than this. The major septo–hippocampal projection has its

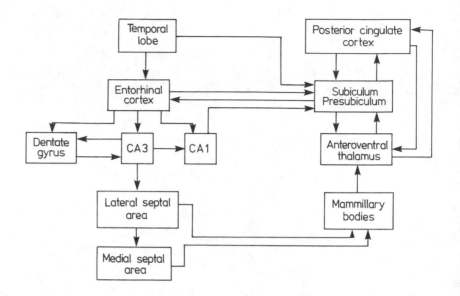

FIG. 33.6. Some efferents from, and return projections to, the septo-hippocampal system. The subicular complex also projects (not shown) to the mammillary bodies.

origin in the medial septal area (Figure 33.7), appears to be cholinergic (that is, uses acetylcholine as its neurotransmitter), and terminates throughout the hippocampal formation, as well as in the subicular area and entorhinal cortex (e.g., Lynch *et al.*, 1978). There is abundant evidence that this projection controls the hippocampal theta rhythm, that is, a quasi-sinusoidal, high-voltage, slow (4–12 Hz) EEG rhythm which can be recorded in the hippocampus and dentate gyrus under a wide variety of behavioral conditions. There is a return projection from the hippocampus to the septum, originating in CA 3 and terminating in the lateral septal area (Figure 33.7). Completion of this loop is accomplished by a projection from the medial to the lateral septal area.

Two important projections to the septo-hippocampal system originate in the brain stem and are monoaminergic, that is, they utilize as their neurotransmitter the monoamines, noradrenaline (or norepinephrine) and 5-hydroxytryptamine (also called serotonin), respectively. The noradrenergic projection originates in the locus coeruleus and travels in the dorsal noradrenergic bundle (Figure 33.8; Livett, 1973) before innervating both the septal area (medial and lateral) and the hippocampal formation (diffusely, but with a particularly high concentration of terminals within the arch formed by the dentate granule cells, an area known as the dentate hilus). The serotonergic afferents to the septo-hippocampal system originate in the dorsal and median raphe nuclei of the brainstem, mainly the median raphe. Like the noradrenergic fibers, they terminate throughout the

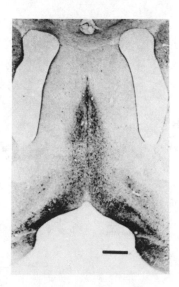

FIG. 33.7. Low-power photomicrograph of the rat's septum stained by acetylcholinesterase (AChE) histochemistry. In this frontal section the septal area is bounded by the corpus callosum at the top and the ventricles laterally. The densely stained cells in the midline constitute the medial septal area descending ventrolaterally into the nucleus of the diagonal band. Cells in the lateral septal area do not stain for AChE. Bar = 0.5 mm. (From Lynch et al., 1978.)

septal area and hippocampal formation. Both the noradrenergic (Gray, 1977; Redmond, 1979) and the serotonergic (Stein, Wise, & Berger, 1973) pathways have been attributed important roles in anxiety and in the mediation of the behavioral effects of anti-anxiety drugs.

IV. FUNCTIONS OF THE SEPTO-HIPPOCAMPAL SYSTEM

The theory (see Figure 33.9) maps the functions of the behavioral inhibition system, as outlined in Figure 33.1, onto the septo-hippocampal system and connected structures, but at the same time it proposes specific ways in which these brain regions act so as to respond appropriately to the adequate stimuli for anxiety (i.e., the inputs listed to the left of Figure 33.1). As well as the septal area and hippocampus themselves, the theory allots functions to the ascending noradrenergic and serotonergic afferents to these regions; to the Papez loop, i.e., the outflow from the subicular area (itself, as we have seen, in receipt of afferents from area CA 1 of the hippocampus) which circles, via the mammillary bodies, the anterior thalamus and the cingulate cortex, back to the subicular area; to the

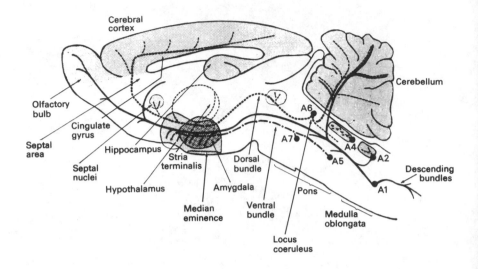

FIG. 33.8. Sagittal representation of the rat brain, showing the principal ascending and descending noradrenergic pathways. Shaded areas indicate regions of noradrenergic terminals. The dorsal ascending bundle originates in the locus coeruleus, A6. (From Livett, 1973.)

neocortical input to the hippocampus from the entorhinal area; to the prefrontal cortex, which projects to the entorhinal area and cingulate cortex; and to the particular pathways by which these structures are interrelated. The central task of this overall system is to compare, quite generally, actual with expected stimuli. The system functions in two modes. If actual stimuli are successfully matched with expected stimuli, it functions in "checking" mode, and behavioral control remains with other (unspecified) brain systems. If there is discordance between actual and expected stimuli or if the predicted stimulus is aversive (conditions jointly termed "mismatch"), it takes direct control over behavior, now functioning in "control" mode.

In control mode, the outputs of the behavioral inhibition system (Figure 33.1) are operated. First, there is an immediate inhibition of any motor program which is in the course of execution. This inhibition is not envisaged as taking place at the level of individual motor commands, but rather as an interruption in the function of higher-level systems concerned with the planning and overall execution of motor programs.

Secondly, the motor program which was in the course of execution at the time that "mismatch" was detected is tagged with an indication which, in English, might read "faulty, needs checking." This has two further consequences. (i) On future occasions the relevant program is executed with greater restraint (more slowly, more easily interrupted by hesitations

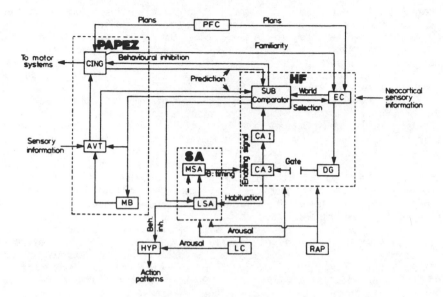

FIG. 33.9. A summary of Gray's (1982a, b) theory. The three major building blocks are shown in heavy print: HF, the hippocampal formation, made up of the entorhinal cortex, EC, the dentate gyrus, DG, CA 3, CA 1, and subicular area, SUB; SA, the septal area, containing the medial and lateral septal areas, MSA and LSA; and the Papez circuit, which receives projections from and returns them to the subicular area via the mammillary bodies, MB, anteroventral thalamus, AVT, and cingulate cortex, CING. Other structures shown are the hypothalamus, HYP, the locus coeruleus, LC, the raphe nuclei, RAP, and the prefrontal cortex. PFC. Arrows show direction of projection; the projection from SUB to MSA lacks anatomical confirmation. Words in lower case show postulated functions; beh. inh., behavioral inhibition. (For further explanation, see text.)

for exploratory behavior, more readily abandoned in favor of other programs, and so on). (ii) The tagged motor program is given especially careful attention the next time it occurs, that is to say, the system exercises particularly carefully its basic function of checking predicted against actual events. It is supposed that in performing this function the system is able to subject each stimulus in the environment to an analysis that ranges over many dimensions (e.g., brightness, hue, position, size, relation to other stimuli). The dependencies between the subject's responses and behavioral outcomes can similarly be subjected to this type of multidimensional analysis (e.g., a turn in a rat's T-maze can be classified as left-going, white-approaching, alternating, moving along the longitudinal axis of the room that surrounds the maze, and so on) (cf. Kimble, 1975).

Thirdly, the system initiates specific exploratory and investigative behavior designed to answer particular questions that arise from the operation of the comparator (e.g., is it left-going or white-approaching which gives rise to mismatch?). Like the inhibition of on-going motor programs, it is not supposed that this function is exercised directly at the point of motor output, but rather that the system modulates the control of exploratory behavior carried out by other brain systems.   Among the environmental stimuli for which the system commands a special search in this way, an important class is constituted by those which themselves predict disruption of the subject's plans, i.e., stimuli which are associated with punishment, nonreward or failure.

These, then, are the major consequences of mismatch; and, given that the behavioral inhibition system is most active under these conditions, this is the time at which anxiety will be greatest.   But the system is also continuously active at times when there is agreement between predicted and actual events; if it were not, mismatch would go undetected. Furthermore, it is supposed that the system is able to identify certain stimuli as being particularly important and requiring especially careful checking.   In this connection the role played by the ascending monoaminergic inputs to the hippocampus is of particular importance. Physiological experiments have shown that impulses coming from the entorhinal cortex to the hippocampus and thence to the subicular area must pass through a "gate" in the dentate gyrus and area CA 3 of the hippocampus proper.   With stimulus repetition such impulses either habituate quickly or become consolidated and enhanced (e.g., Vinogradova, 1975; Segal, 1977a, b; Andersen, 1978). It seems, furthermore, that impulses are more likely to pass the dentate–CA3 gate if they are associated with events of biological importance to the animal, such as those associated with the delivery of food or foot–shock (Segal, 1977b). Segal (1977a) has in addition shown that the passage of impulses in the hippocampal formation is facilitated by the simultaneous stimulation of either noradrenergic or serotonergic inputs to the hippocampus.   The action of noradrenergic afferents in particular can apparently be characterized as increasing the signal–to–noise ratio of hippocampal neurons, allowing them to respond more effectively to sensory information (of neocortical origin) entering the system from the entorhinal cortex (McNaughton & Mason, 1980).   In accordance wih these data, my theory of anxiety attributes to the noradrenergic and serotonergic afferents to the septo–hippocampal system the role of labelling such sensory information as "important, needs checking," as outlined above.

V.   AN INTERFACE BETWEEN EMOTION AND COGNITION

I hope this is sufficient to give the reader the flavor of the theory; although it is no doubt too complex to be easily summarized in this way. In the space that remains I shall try to justify the assertion that, within this theory, the septo–hippocampal system and the associated Papez circuit constitute an interface between emotional and cognitive processes.   In fact, as we shall see, there are several senses in which this assertion can properly be made.

The septo–hippocampal system and the Papez circuit are charged by the theory with the basic function of comparing actual with predicted stimuli.   As outlined above, this function is central to my analysis of anxiety; but it is, *par excellence*, a cognitive function. It is instructive to consider what kinds of information processing are necessary for such a comparator to work.

Clearly the comparator must have access to information both about current sensory events ("the world") and about expected ("predicted") events. These two classes of information must, however, themselves be closely interrelated. Predictions can only be generated (a task for which a "generator of predictions" is needed) in the light of information about the world. Thus current sensory events must be transmitted to the generator of predictions. In addition, the latter must have access to information about past environmental regularities which, in conjunction with the current state of the world, determine the content of the next prediction. But this will be sufficient only if the environment is one over which the subject has no control. If, as in most cases, the subject's behavior affects the world, the generator of predictions must also have access to information about the next intended set of movements, i.e., motor programs or plans. Prediction will now depend on the conjunction of the present state of the world, stored past environmental regularities, the next intended step in the motor program, and stored past relationships between such steps and changes in the environment.

Once made, the prediction must be tested against the world. For this to be possible, the correct sensory inputs must be chosen. It follows that the input of sensory information must be selected in the light of what is predicted. Such selection can in principle be accomplished in one of two ways. The subject's motor behavior can be left under the control of other systems, and selection accomplished by chosing among the sensory events which occur anyway as a result of that motor behavior; or the selection mechanism can take active control over motor behavior and command appropriate exploratory action. This might range from simple adjustments of sensory organs (e.g., dilating the pupils) to complex patterns of locomotion.

It follows from these arguments that the major information transactions for which we must seek neural correlates are those set out in Figure 33.10. This figure contains a simplifying assumption: that the same neural message that identifies the current state of the world, once it has been successfully matched with the relevant prediction, also serves as the information which initiates generation of the next prediction. In this way, the occurrence of a mismatch has the reasonable consequence that it automatically brings to a halt the process of generating predictions.

In attempting to match these psychological functions to the anatomy and physiology of the septo-hippocampal system and its associated structures, we first propose that sensory information enters the system via the entorhinal cortex. Figure 33.10 prompts one to seek for the comparator itself in a place where an incoming message which started out in the entorhinal cortex can both: (1) terminate a loop by identifying the current state of the world which has to be matched against a prediction that has circled round the loop; and (2) initiate a loop (to commence generation of the next prediction). The subicular area (Figure 33.6) is ideally suited for this purpose. Information reaches it from CA 1 after passage round the basic hippocampal circuit. It is then recirculated through the fornix to the mammillary bodies, the anteroventral thalamus, the

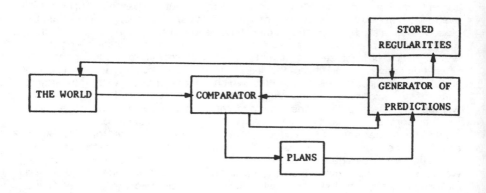

FIG. 33.10. The kinds of information processing required for the successful functioning of the hypothetical comparator. (See text for further explanation.)

cingulate cortex and back to the subicular area. Thus a plausible hypothesis accords the status of comparator to the subicular area: the input from CA 1 identifies the current state of the world; its recirculation round the Papez loop forms part of the process of prediction generation; and arrival in the subicular area of the input from the cingulate cortex (and/or the direct input from the anteroventral thalamus) constitutes the prediction.

If the subicular loop subserves in this way the generation of predictions, it must have access to stored regularities concerning the environment and concerning the relations between the organism's responses and the environment. It is plausible to see the anteroventral thalamus as a point of access for response-independent environmental regularities. Where response-dependent environmental irregularities are concerned, it is possible that the cingulate cortex plays this role. A generator of predictions must have information, not only about past dependencies between responses and outcomes, but also about current and intended responses. This information might also be available in the cingulate cortex, where it could function as part of the Papez loop. Alternatively, this region or the prefrontal cortex might send relevant motor information to the entorhinal area, to which both project. In that case informaiton about current and intended movements would form part of the description of the world. Probably both routes are used, current movements entering into the entorhinal description of the world, intended movements taking part in the generation of predictions as the Papez loop sweeps past the cingulate cortex.

In these ways, then, the septo-hippocampal system and the Papez circuit might discharge the essentially cognitive functions displayed in Figure 33.10. The fact that these functions are seen as central to an emotional state (e.g., anxiety) is therefore one reason to regard these structures as constituting an interface between cognition and emotion. But there is another, more interesting, reason so to regard them. For the present theory supposes that the functions discharged by these structures may be activated in two distinct ways. The first has already been analyzed above: it arises when the ascending monoaminergic inputs to the septo-hippocampal system facilitate passage through the dentate-CA 3 gate of sensory inputs, associated with biological reinforcement, from the entorhinal cortex. This is the process which constitutes anxiety in, say, a rat threatened by a stimulus previously associated with footshock. But the septo-hippocampal system can also be activated in a manner which is relatively free of emotional content. Thus in this sense too the system is an interface between cognition and emotion: the functions it discharges can participate in both emotional and non-emotional behavior.

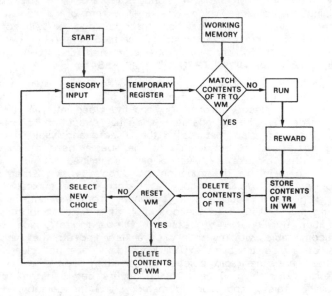

FIG. 33.11. Olton's (1978) model of the memory processes underlying performance in the radial-arm maze. (For explanation, see text.) TR, temporary register; WM, Working memory.

This (relatively) non-emotional mode of activation of the septo-hippocampal comparator can be most conveniently illustrated with reference to Olton's (1978) analysis of working memory. Olton has proposed the model shown in Figure 33.11 to account for the performance of a rat which has to find food at the end of each arm of a radial-arm maze, and the neural substrate of the model is thought to lie in the hippocampal formation (Olton, Becker, & Handelmann, 1979). There are considerable similarities between Olton's model of working memory and the functions attributed to the hippocampus in my own theory of anxiety (Gray, 1982a, p. 284) .

Olton's system

... .begins with sensory input that enters into a temporary register. A comparison process attempts to match the contents of the temporary register with each of the items in working memory. A match indicates that the choice in question (of one of the arms in the maze) has already been made and should not be repeated; information in the temporary register is deleted, a decision made as to whether to reset working memory, a search is initiated for a new choice, and new sensory input is obtained. A failure to match the contents of the temporary register and some item in working memory indicates the choice in question has not been made previously and ought to be made now. Running down the arm produces reward. The information defining the choice which was in the temporary register is stored in working memory so that the choice will not be repeated, the temporary register is cleared, the reset decision made, a search for another choice is initiated, and new sensory input is obtained (Olton, 1978, pp. 363-364).

The similarities between this system and my own analysis of hippocampal function include the central role played by the comparison of actual with stored sensory information, and the close interplay between the decisions made by the comparator and the execution of exploratory behavior. (These are also features of the O'Keefe and Nadel, 1978, model of hippocampal function, but limited to the special case of spatial tasks.) Accordingly, the circuits I have briefly described above need little modification to perform the function of Olton's working memory. The comparison process would proceed as before in the subicular area; the temporary register is equivalent to the input from the entorhinal cortex, and working memory itself to the operation of the Papez loop. But there are in addition two special processes in Olton's model which, while they are not incompatible with my theory of anxiety (indeed, they help account for some otherwise troubling experimental data), are not clearly required by it. These are: the rapid deletion of an item from the list of items awaiting further matching when a match occurs between the contents of the temporary register and working memory; and the deletion of all items of working memory at the start of a new trial in the maze (Figure 33.11), allowing that trial to proceed without interference from previous trials. The former process can be regarded as a kind of boosted habituation (a particular item no longer needs checking out); the latter, as a kind of boosted dishabituation (all items again need checking out).

Since (as briefly outlined above; but see also Gray, 1982a, Chapter 10) regulation of the habituation of stimuli passing round the hippocampal circuit lies at the heart of my theory of anxiety, it is not difficult to incorporate Olton's two special processes into the theory. Following Vinogradova (1975; and see Gray, Feldon, Rawlins, Owen & McNaughton, 1978), habituation is seen as depending on an interaction between inputs arriving at area CA 3 of the hippocampal formation from the entorhinal cortex and the septal area, respectively; and monoaminergic regulation of habituation occurs at the entorhinal–dentate–CA 3 step (see above), determining the level of anxiety. Olton's processes of boosted habituation and dishabituation may be incorporated into the same general framework, by supposing them to depend on fibers linking the subicular region to the septal area (for details, see Gray, 1982a, p. 284). In this way, habituation and dishabituation can be regulated (as Olton's model requires) by the outputs of the comparator itself, and in a flexible manner which does not depend on monoaminergic influences form the brain–stem.

It is possible, then, to give along these lines a relatively precise analysis of the way in which the septo–hippocampal system and its allied Papez circuit could function as an interface between cognition and emotion. Regulation of the number and nature of the items to be checked in the predictive and comparator circuits described above can proceed in two distinct ways. The first is relatively "emotion–rich": monoaminergic regulation at the dentate–CA 3 gate determines the level of anxiety and is susceptible to the action of the anti–anxiety drugs. The second is relatively "emotion–free": regulation by the outputs of the comparator itself (via subiculo–septal projections) allows the system to be used for such activities as solving the day's trial in familiar radial–arm maze (or equivalent list–dependent activities in our own species). In accordance with this view we have found that the anti–anxiety drug, chlordiazepoxide (which otherwise is remarkably similar in its behavioral effects to septo–hippocampal damage: Gray, 1982a), impairs performance in the radial–arm maze only during initial learning (Rawlins, Preston, & Gray, unpublished observations). Hippocampal damage, in contrast, gives rise to a substantial and enduring deficit (Olton, Becker, & Handelmann, 1979); in the terms developed here, this is because such damage eliminates both emotion–rich and emotion–free use of the comparator.

The analysis given above sees the hippocampus as an interface between cognition and emotion in the sense that its comparator function can be put either to emotional use (anxiety) or to non–emotional use (e.g., in the radial–arm maze). But, especially in man, there is a further sense in which (according to the theory) the hippocampus may act as an interface between cognition and emotion: the outputs of neocortical (and especially linguistic) cognitive processes may directly control the activities of the hippocampus. This part of the theory derives principally from a consideration of the role played by two regions whose destruction has been used successfully in the treatment of human anxiety: the prefrontal and cingulate cortices (Powell, 1979).

Within the present theory the role of these regions is twofold. First, they supply to the septo–hippocampal system information about the subject's own ongoing motor programs; this information is essential if the septo–hippocampal system (together with the Papez circuit) is to generate adequate predictions of the next expected event to match against actual events. Secondly, they afford a route by which language systems in the neocortex can control the activities of the limbic structures which are the chief neural substrate of anxiety. In turn, these structures, via subicular and hippocampal projections to the entorhinal cortex, are able to scan verbally coded stores of information when performing the functions alloted to them in the theory outlined above. In this way, it is possible for human anxiety to be triggered by largely verbal stimuli (relatively independently of ascending monoaminergic influences) and to utilize in the main verbally coded search strategies to cope with perceived threats. It is for this reason, if the theory is correct, that lesions to the prefrontal and cingulate cortices are effective in cases of anxiety that are resistant to drug therapy (Powell, 1979).

Earlier theories of anxiety based on animal experiments have usually found their most natural application to human phobias. But the elusive manifestations of the obsessive-compulsive neurosis (Rachman & Hodgson, 1979) have proved more intractable. In contrast, the present theory is most naturally applied to the symptoms of this condition, especially those that involve excessive checking of potential environmental hazards, real or imagined (Gray, 1982a, Chapter 14). The obsessions are the most cognitive of the symptoms of anxiety. It is encouraging, therefore, to see them yield some of their mystery to an analysis based on the results of behavioral and physiological experiments with animals. This progress has been possible, I believe, because the present theory of anxiety blurs two distinctions which are better blurred.

The first distinction is that between behaviorist and. cognitive psychology. The model illustrated in Figure 33.1 (the behavioral inhibition system) is essentially behaviorist; that is, it describes a pattern of input-output relationships. The more developed theory of anxiety outlined above, in contrast, is essentially cognitive; that is, it attempts to describe the internal information processing from which those relationships derive. But the later theory has not superseded the earlier one, it incorporates it: the BIS model specifies what an animal does under certain conditions, while the later theory speculates about the processes which enable the animal to do it.

The second distinction which has been blurred (but not lost, or the notion of interface would not be appropriate) is that between thought and emotion. This is surely as it should be. We do not stop thinking when we are emotionally aroused, nor use different machinery with which to think. Nor do we *only* think at such times: we also act (or interrupt action) and feel. A successful theory will need, therefore, in this as in any other branch of psychology, to bind thought, action and emotion into a single whole.

REFERENCES

Andersen, P. Long-lasting facilitation of synaptic transmission. In K. Elliott & J. Whelan (Eds.), *Functions of the septo-hippocampal system.* Amsterdam: Elsevier, 1978.

Andersen, P., Bliss, T. V. P., & Skrede, K. Lamellar organization of hippocampal excitatory pathways. *Experimental Brain Research*, 1971, *13*, 222–238.

Gray, J. A. *Elements of a two-process theory of learning*. London: Academic Press, 1975.

Gray, J. A. Drug effects on fear and frustration: Possible limbic site of action of minor tranquilizers. In L. L. Iversen, S. D. Iversen, & S. H. Snyder (Eds.), *Handbook of psychopharmacology*. (Vol. 8.) New York: Plenum, 1977.

Gray, J. A. *The neuropsychology of anxiety: An enquiry into the function of the septo-hippocampal system*. Oxford: Oxford University Press, 1982. (a)

Gray, J. A. Precis of Gray's 'The neuropsychology of anxiety: An enquiry into the functions of the septo–hippocampal system'. *The Behavioral and Brain Sciences*, 1982, *5*, 469–534. (b)

Gray, J. A., Feldon, J., Rawlins, J. N. P., Owen, S., & McNaughton, N. The role of the septo–hippocampal system and its noradrenergic afferents in behavioural responses to nonreward. In K. Elliott & J. Whelan (Eds.), *Functions of the septo-hippocampal system*. Amsterdam: Elsevier, 1978.

Hirsh, R. The hippocampus and contextual retrieval of information from memory: A theory. *Behavioral Biology*, 1974, *12*, 421–444.

Kimble, D. P. Choice behavior in rats with hippocampal lesions. In R. L. Isaacson & K. H. Pribram (Eds.), *The hippocampus, Vol. 2. Neurophysiology and behavior*. New York: Plenum, 1975.

Livett, B. G. Histochemical visualization of peripheral and central adrenergic neurones. *British Medical Bulletin*, 1973, *29*, 93–99. In *Catecholamines*, L. L. Iversen (Ed.).

Lynch, G., Rose, G., & Gall, C. Anatomical and functional aspects of the septo–hippocampal projections. In K. Elliott & J. Whelan (Eds.), *Functions of the septo-hippocampal system*. Amsterdam: Elsevier, 1978.

McNaughton, N., & Mason, S. T. The neuropsychology and neuropharmacology of the dorsal ascending noradrenergic bundle – a review. *Progress in Neurobiology*, 1980, *14*, 157–219.

O'Keefe, J., & Nadel, L. *The hippocampus as a cognitive map*. Oxford: Clarendon Press, 1978.

Olton, D. S. Characteristics of spatial memory. In S. H. Hulse, H. Fowler, & W. K. Honig (Eds.), *Cognitive processes in animal behavior*. Hillsdale, N.J.: Erlbaum, 1978.

Olton, D. S., Becker, J. T., & Handelmann, G. E. Hippocampus, space, and memory. *The Behavioral and Brain Sciences*, 1979, *2*, 313–322.

Papez, J. W. A proposed mechanism of emotion. *Archives of Neurological Psychiatry*, 1937, *38*, 725–743.

Powell, G. E. *Brain and personality*. London: Saxon House, 1979.

Rachman, S., & Hodgson, R. *Obsessions and compulsions*. Englewood Cliffs, N.J.: Prentice-Hall, 1979.

Redmond, D. E. New and old evidence for the involvement of a brain norepinephrine system in anxiety. In W. Fann *et al.* (Eds.), *Phenomenology and treatment of anxiety*. New York: Spectrum, 1979.

Segal, M. The effects of brainstem priming stimulation on interhemispheric hippocampal responses in the awake rat. *Experimental Brain Research*, 1977, *28*, 529–541. (a)

Segal, M. Changes of interhemispheric hippocampal responses during conditioning in the awake rat. *Experimental Brain Research*, 1977, *29*, 553–565. (b)

Stein, L., Wise, C. D., & Berger, B. D. Anti–anxiety action of benzodiazepines: Decrease in activity of serotonin neurons in the punishment system. In S. Garattini, E. Mussini, & L. O. Randall (Eds.), *The benzodiazepines*. New York: Raven Press, 1973.

Vinogradova, O. S. Functional organization of the limbic system in the process of registration of information: Facts and hypotheses. In R. L. Isaacson & K. H. Pribram (Eds.), *The hippocampus*, *Vol. 2. Neurophysiology and behavior*. New York: Plenum, 1975.

Winocur, G. The amnesic syndrome: A deficit in cue utilization. In L. S. Cermak (Ed.), *Human memory and amnesia*. Hillsdale, N.J.: Erlbaum, 1982.

34 BRAIN SYSTEMS AND COGNITIVE LEARNING PROCESSES

Karl H. Pribram
Stanford University

I. BEYOND ASSOCIATION

Experimental analyses of the learning process have developed into two rather different approaches. One, carried out by biologically oriented scientists, seeks to establish the locus of plasticity, and the nature of the more or less permanent changes which allow an accumulation of experience to alter behavior. The other engages computer-oriented scientists and experimental psychologists using human subjects. It focuses on the nature of processes of retrieval: Questions are asked regarding the span which can handle a store, the types of accessing which make available that which is stored, and the structures of the accessing processes. Both of these approaches are based on a model in which the memory store associates spatio-temporally contiguous experiences and accumulates the residues of such associations which then become accessible when some similar experience addresses the storage locus.

This associative model, while valuable in the analyses of simpler forms of learning such as classical and instrumental (operant) conditioning, may not encompass problems encountered when the learning of cognitive processes is involved. Nor does the associative model, as it is currently conceived, allow for the likely possibility that the memory mechanisms of the brain are content rather than location addressable.

A content addressable cognitive learning process involves coding the residues of experience in such a way that subsequent experience automatically addresses the residue on the basis of similarity rather than on the basis of location. Whereas a location addressable process traverses the same paths during acquisition and retrieval, a content addressable process operates somewhat more independently of specific pathways. Mailing a letter uses a location-addressable mechanism; broadcasting a television program utilizes a degenerate form of content addressability (the content is encoded on a carrier frequency).

The work reported here suggests that the cognitive operations of the primate brain are essentially coding operations which "label" the residues of experience so as to make them readily retrievable. In such a scheme, classification of learning processes ought, at some level, to mirror the classification of retrieval mechanisms. Thus, the evidence from experiments involving the primate forebrain, which makes up the bulk of this chapter, should overlap and be congruent with that obtained from the approaches used in memory research involving humans. But at the same time a more comprehensive understanding of the role of the primate forebrain in learning should be achieved.

The primate brain is a complex organ composed of many systems and subsystems. Damage to one system influences some learning but not all; damage to another system will affect learning processes considerably different from those influenced by injury to the first. Even when consideration is restricted to work in my laboratory, a variety of types of learning can be distinguished: the learning of perceptual or motor skills, reference learning, learning based on interest, and learning to transfer experience gained in one situational context to another. This leaves out higher-order forms of learning such as linguistic learning which in its fully-developed form is uniquely human.

In this chapter, I shall develop evidence which suggests that these varieties of cognitive learning processes can be arranged hierarchically according to the brain systems which have been identified to be involved. The specifics of the hierarchy proposed will most likely be subject to change as new evidence accrues. However, I will maintain that the complexities of cognitive learning processes will not be understood until the relationships among them and to brain systems becomes clarified. The current tendency to do no more than to serve up series of dichotomies results only in a monumental tower of Babel.

At least seven different learning processes can be identified. At the base of the hierarchy, shown in Figure 34.0, there are four processes that form two basic branches. In the first are the perceptual and motor skills, and in the second, the processing of interesting or novel (registration) and familiar (extinction) episodes. Skills are elaborated by search and sampling procedures to form the next higher level: the *referential* learning processes. Episodic learning, dependent on registration and extinction, becomes elaborated by spatio-temporal probability structures which at a higher level, frames the *context* within which the episodes occur, and allows transfer of the training obtained in one context to another. In turn, referential and contextual learning processes interact to produce declarative linguistic learning.

Each of the nodes of the hierarchy has a forebrain system identified with it. Thus perceptual learning involves the primary sensory systems; motor learning, the primary motor systems. Processing novelty involves systems converging on the amygdala; processing the familiar involves those converging on the hippocampus. (See also Chapter 33, by Gray, this volume.) Search and sampling are disturbed by resection of the posterior intrinsic, probabilistic programming by resections of the far frontal cortex. The methods and data from which these conclusions stem are described below.

II. THE MULTIPLE DISSOCIATION TECHNIQUE

The experimental analysis of subhuman primate model systems has uncovered a host of learning disturbances. The initial method by which these brain-behavior relationships were established is called the method of multiple dissociation based on an "intercept of sums" technique (Pribram,

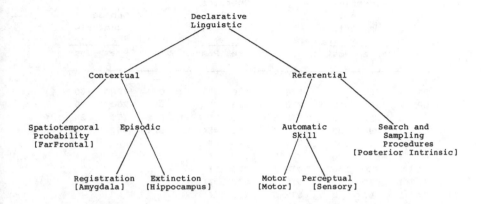

FIG. 34.0.   Hierarchical scheme of relationships between types of cognitive learning based on their involvement with particular cerebral systems.

1954) akin to what Teuber named the method of double dissociation of signs of brain trauma in humans.   The multiple dissociation technique depends on classifying the behavioral deficit produced by cortical ablations into yes and no instances on the basis of some arbitrarily chosen criterion; then plotting on a brain map the total extent of tissue associated with each of the categories *ablated: deficit*; *not ablated: no deficit*; and finally finding the intercept of those two areas (essentially subtracting the *noes* from the *yesses-plus-noes.*) This procedure is repeated for each type of behavior.   The resulting map of localization of disturbances is then validated by making lesions restricted to the site determined by the intercept method and showing that the maximal behavioral deficit is obtained by the restricted lesion. (See Table 34.0 and Figure 34.2)

Once the neurobehavioral correlation has been established by the multiple dissociation technique, two additional experimental steps are undertaken.   First, holding the lesion constant, a series of variations is made of the task on which performance was found defective.   These experimental manipulations determine the limits over which the brain-behavior disturbance correlations hold and thus allow reasonable constructions of models of the learning and retrieval processes impaired by the various surgical procedures.

Second, neuroanatomical and electrophysiological techniques are engaged to work out the relationships between the brain areas under examination and the rest of the nervous system.   These experimental

TABLE 34.1
Simultaneous Visual Choice Reaction

		Operates without deficit			Operates with deficit			Nonoperate controls	
		Pre	Post			Pre	Post		Pre	Post
OP	1	200	0	PTO	1	120	272	C 1	790	80
OP	2	220	0	PTO	2	325	F	C 2	230	20
OP	3	380	0	PTO	3	180	F	C 3	750	20
LT	1	390	190	PTO	4	120	450	C 4	440	0
LT	2	300	350	T	1	940	F			
H	1	210	220	T	2	330	F			
HA		350	240	VTH	1	320	F			
FT	1	580	50	VTH	2	370	F			
FT	3	50	0	VTH	3	280	F			
FT	4	205	0	VTH	4	440	F			
FT	5	300	200	VT	1	240	F			
FT	6	250	100	VT	2	200	F			
DL	1	160	140	VT	3	200	890			
DL	2	540	150	VT	4	410	F			
DL	3	300	240	VT	5	210	F			
DL	4	120	100							
MV	1	110	0							
MV	2	150	10							
MV	3	290	130							
MV	4	230	10							
MV	5	280	120							
CIN	1	120	80							
CIN	2	400	60							
CIN	3	115	74							
CIN	4	240	140							

Note-- Pre- and post-operative scores on a simultaneous visual choice reaction of the animals whose brains are diagrammed in Figure 34.2, indicating the number of trials taken to reach a criterion of 90% correct on 100 consecutive trials. Deficit is defined as a larger number of trials taken in the "retention" test than in original learning. (The misplacement of the score H 1 does not change the overall results as given in the text.)

procedures allow the construction of reasonable models of the functions of the areas and of the mechanisms of impairment.

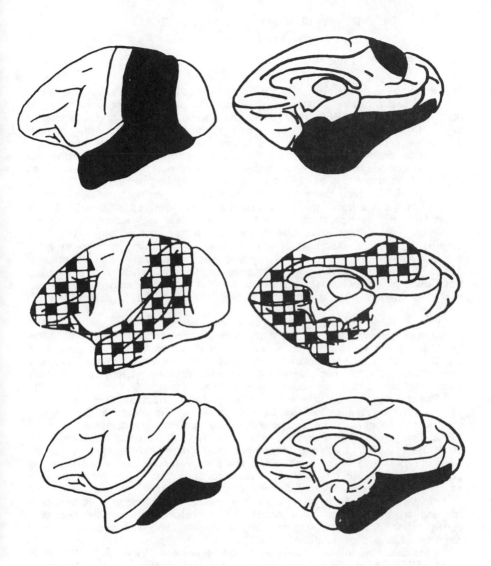

FIG. 34.2. The upper diagram represents the sum of the areas of resection of all of the animals grouped as showing deficit. The middle diagram represents the sum of the areas of resection of all to the animals grouped as showing no deficit. The lower diagram represents the intercept of the area shown in the black in the upper diagram and that **not** checkerboarded in the middle diagram. This intercept represents the area invariably implicated in visual choice behavior in these experiments.

III. PERCEPTUAL AND MOTOR SKILLS

A. PERCEPTUAL LEARNING

The impetus for work in our laboratory to study perceptual learning came from studies by Patrick Bateson, an ethologist at Cambridge University. In his thesis, Bateson (1964, see 1972) had shown that imprinting is a special case of perceptual learning. He raised newborn chicks in an environment of either horizontal or vertical stripes, and showed that this early experience dramatically influenced subsequent imprinting. Thus, it appeared that the development of an expectancy or neuronal model was as important to imprinting as to later perceptual performances (as shown by Sokolov, 1960). Once the model was established, learning took place within one or at most a very few trials. Bateson then showed, at Stanford, that a similar type of nonproblem-oriented (latent) learning occurred in young monkeys. A pattern was placed in the animals' home cage for three months. Then a discrimination task was given using this pattern in connection with a novel one and learning was compared to that obtained in a task where the novel one was matched to one which had previously been used in a problem solving situation. The "latently"-learned cue proved easily as influential in determining behavior as did the "problem" learned one.

Direct evidence from brain recordings also confirms the fact that perceptual learning can proceed without help from problem-guided learning. Records of the electrical activity evoked in the occipital (striate) cortex of monkeys shows a differentiation of wave forms even when the animal is simply exposed to two different patterns (Spinelli, 1967) and before discrimination learning has taken place (Pribram, Spinelli, & Kamback, 1967).

Sharpening of the difference in wave forms occurs over the course of several repeated exposures to the patterns. Furthermore, the cortical electrical responses either increment or decrement (Grandstaff & Pribram, 1972; Bridgeman, 1982) and the sites for these differential responses are distributed. After the initial incrementing or decrementing, which occurs over the first five or so trials, each specific electrode placement provides consistent and reliable recordings which continue unchanged from day-to-day and week-to-week. Adjacent placements show markedly different electrical response patterns, that is, the spatial arrangement of these cells appears to be random. We concluded, therefore, that at the cortex a configuration develops during perceptual learning and that perception is a function of this configuration.

Considerations which I have reviewed elsewhere (Pribram, 1966; 1969; 1974; 1982), have led me to propose that this configuration resembles a quantally organized multiplexed (strip or patch) holographic pattern. The critical evidence is the fact that extensive destructions of primary projection cortex do not interfere with pattern recognition except for the production of scotomata (i.e., holes in the sensory field). The mechanism upon which recognition is based must therefore be distributed over the primary cortex and perceptual recognition must therefore result from an operation which constructs or composes it by means of the distributed mechanism. Direct neuroelectric evidence for such distribution comes from the experiments just cited. The concept of a neural hologram, that is, of sets of interfering wave forms (or of a matrix of hyper- and depolarizations) constituted of postsynaptic and dendritic potentials, provides a reasonable model that handles many hitherto unexplainable neurobehavioral data and provides a solid base for the associative properties of recognition. These data include the lack of effect of epileptogenic lesions and cortical cross hatchings on perceptual performances (Kraft, Obrist, & Pribram, 1960; Stamm & Pribram, 1961; Stamm & Knight, 1963; Stamm, Pribram, & Obrist, 1958; Stamm & Warren, 1961; Pribram, Blehert, & Spinelli, 1966; Sperry, Miner, & Meyers, 1955; for review, see Pribram, 1982, Chapter 6). The reconstructive process is, however, complicated and will be reviewed in the section on Reference Learning.

B. MOTOR LEARNING

Our experiments on the nature of motor learning were motivated by the question posed in the neurological literature as to whether muscles or movements are represented in the motor cortex. In an experiment which was designed to replicate a study of Lashley's (1929), we found (Pribram, Kruger, Robinson, & Berman, 1955) that resection of large extents of motor cortex did not produce weakness of any muscle group, nor did the resection interfere with any specific movement (defined as a sequence of muscle contractions and studied by examining progressive frames of cinematographic records obtained in different behavioral situations). What we did find was a marked delay in acquisition, and a change in the fluency of performance in opening a latch box and retrieving a peanut (reaction times doubled or tripled). Since no such change in reaction times was seen in other practiced situations, I interpreted the change in performance to indicate that a specific problem-solving act had been impaired. Thus, it appeared that the issue of representation was even more complex than had been stated. Not only muscles but movements and actions (defined as the consequences of movements) had to be considered. The resolution of the problem came when it was realized that the representation of muscles was anatomically determined, the representation of movements resulted from physiologically-oriented studies, while the concept of a representation of actions came from neurobehavioral experimentation. All three types of representation were, in fact, tenable: the issue is not which, but how the representations interact. Other ways of stating the problem are: How (by what physiological process) are the anatomical organizations which characterize the motor system mapped into the behaviors of the organism? How do movements relate muscles to the environment upon which they operate?

A clue to how such mappings might occur came from the work of Bernstein (1965) in which he made cinematographic analyses of the actions of humans performing tasks such as hammering nails, jogging on a spring-supported platform, or writing on a blackboard. His subjects were dressed in black leotards but had white spots marking their joints. The photographic film therefore recorded the movements of the joints as the actions were carried out. The record consisted of a pattern of continuous waveforms, one for each joint. By performing a frequency analysis on the waveforms, Bernstein was able to correctly predict the amplitudes and locations of the next movements in the sequence.

It seemed plausible to me that the analyses which served Bernstein so well might similarly serve the motor systems of the brain, especially as there is considerable evidence (noted above) that the sensory systems operate by way of such waveform analytic processes. We therefore undertook some experiments to determine whether single neurons in the basal ganglia and cerebral motor cortex were frequency selective. The results of the experiment showed that a 20% portion of a total of 308 cells sampled resonate (i.e., increase or decrease their activity at least 25% over baseline spontaneous activity) to a narrow (1/2 octave) band of the range of cycle frequencies.

Tuning could be due to a spurious convergence of factors relating to the basic properties of muscle as discussed in the introduction: metric displacement and tonicity or tension. An examination was therefore undertaken of variables related to these basic properties, variables such as velocity, change in velocity (acceleration), as well as tension, and change in tension. These factors in isolation were found not to account for the frequency selective effects. This does not mean that other cells in the motor system are not selectively sensitive to velocity and tension. But it does mean that the frequency selectivity of the cells described is dependent on some higher order computation of the metric and tonic resultants imposed by the foreleg musculature and by the external load.

The other variable investigated was position in the cycle of movement. Position *is* encoded by cortical cells (and not by caudate nucleus cells) but only at the site of phase shift and only for a particular frequency. The result thus supports the hypothesis that the cortical cells are in fact frequency selective, in that any sensitivity to phase shift presupposes an encoding of phase and therefore frequency. Furthermore, the fact that the cortical cells respond to position suggests that they are directly involved in the computation of the vector space coordinates within which actions are achieved.

There is thus no question but that an approach to analysis of the functions of the motor system in frequency terms is useful not only in studying the overall behavior of the organism, but in studying the *neural* motor mechanisms involved in the acquisition of motor skills. Motor learning, just like perceptual learning, appears to depend on computations involving the networks of the primary sensory and motor cortexes, computations which are readily carried out in the frequency domain but which can be specified as well in terms of quantal matrix characteristics.

C. AUTOMATIC (SKILLED) VERSUS CONTROLLED (PROCEDURAL) PROCESSING

There is additional evidence that, for some tasks at least, learning needs only the primary projection, input–output systems of the brain. Shiffrin and Schneider (1977; Schneider and Shiffrin, 1977) and Treisman (1977) have developed tasks which differentiate between automatic and controlled processing. They differ in that tasks which can be automatically processed involve over-learned skills in which a choice can proceed without serial search. Thus, the number of alternatives from which a cue is chosen has no effect on reaction time since all are processed in parallel. Controlled processing involves an earlier stage of skill and requires a serial search with reaction time dependent on the number of alternatives.

To determine what brain systems were involved in these two types of reference learning tasks, we used a modification of Treisman's displays and measured the event-related electrical activity recorded from the striate and peristriate cortex, the inferior temporal lobe, far frontal and precentral cortex of monkeys. The subject had to select a green square from a set of a colored squares and diamonds, each of equal contour and luminance when compared to the rewarded cue.

The following display combinations were used in the experiment described here: a) a simple disjunctive display in which the green square had to be identified in a background of eight red diamonds; b) a more complicated disjunctive display in which the green square had to be identified in a background of red diamonds, white circles, and blue triangles not held identical); c) the conjunctive display in which the green square had to be identified in a background of green diamonds, red diamonds and red squares. The results showed that differences in the electrical recordings made from the primary sensory areas reflected differences in distinct features of the displays. Conversely, changes in potentials recorded from the posterior intrinsic association cortex reflected the difficulty of the task as determined by the number of alternatives and the conjunctive/disjunctive dimension. When the task was novel, the far frontal intrinsic cortex was shown also to be involved.

Other experiments have allowed us to make a dissociation between the brain electrical activity evoked in the primary sensory projection cortex and the posterior intrinsic association cortex of the temporal lobe (Rothblat & Pribram, 1972; Nuwer & Pribram, 1979). These earlier studies, as well as the current ones, showed that the brain electrical activity evoked in the primary sensory receiving areas was largely determined by the features in the stimulus display, irrespective of whether they were being reinforced, whereas the electrical potential changes evoked in the temporal cortex were primarily related to the cognitive operations, i.e., the choices involving categorizing or pigeon holing (Broadbent, 1974). Clear and consistent involvement of the frontal cortex was found only on occasions when the task was novel or the reinforcing contingencies were shifted between runs. These relationships to categorizing and novelty are consonant with the results described below.

IV.   REFERENCE LEARNING AND THE POSTERIOR CORTICAL CONVEXITY

A. SENSORY SPECIFICITY

Between the sensory projection areas of the primate cerebral mantle lies a vast expanse of parieto–temporo–preoccipital cortex. Clinical observation has assigned disturbance of many cognitive and language functions to lesions of this expanse. Experimental psychosurgical analysis in subhuman primates of course, is limited to nonverbal behavior; within this limitation, however, a set of sensory–specific agnosias (losses in the capacity to categorize cues) have been produced. Distinct regions of primate cortex have been shown to be involved in each of the modality–specific cognitive functions: anterior temporal in gustation (Bagshaw & Pribram, 1953), inferior temporal in vision (Mishkin & Pribram, 1954) midtemporal in audition (Weiskrantz & Mishkin, 1958; Dewson, Pribram, & Lynch, 1969) and occipitoparietal in somesthesis (Pribram & Barry, 1956; Wilson, 1955). In each instance, categories learned prior to surgical interference are lost to the subject postoperatively and great difficulty (using a "savings" criterion) in reacquisition is experienced, if task solution is possible at all.

The behavioral analysis of these sensory–specific agnosias has shown that they involve a restriction in sampling of alternatives, a true information processing deficit, a deficit in reference learning. Perhaps the easiest way to communicate this is to review the observations, thinking, and experiments that led to the present view of the function of the inferior temporal cortex in vision.

B. SEARCH AND SAMPLING PROCEDURES

All sorts of differences in the physical dimensions of the stimulus, for example, size, are processed less well after inferior–temporal lesions (Mishkin & Pribram, 1954) but the disability is more complex than it at first appears – as illustrated in the following story:

One day when testing my lesioned monkeys at the Yerkes Laboratories at Orange Park, Florida, I sat down to rest from the chore of carrying a monkey a considerable distance between home–cage and laboratory. The monkeys, including this one, were failing miserably at visual tasks such as choosing a square rather than a circle. It was a hot, muggy, typical Florida summer afternoon and the air was swarming with gnats. My monkey reached out and caught a gnat. Without thinking I also reached for a gnat – and missed. The monkey reached out again, caught a gnat, and put it in his mouth. I reached out – missed! Finally the paradox of the situation forced itself on me. I took the beast back to the testing room: He was still deficient in making visual choices, but when no choice was involved, his visually–guided behavior appeared to be intact. On the basis of this observation the hypothesis was developed that *choice* was the crucial variable responsible for the deficient discrimination following infero–temporal lesions. As long as a monkey does not have to make a choice, his visual performance should remain intact.

To test this hypothesis, monkeys were trained in a Ganzfeld made of a translucent light fixture large enough so the animal could be physically inserted into it (Ettlinger, 1957). The animal could press a lever throughout the procedure but was rewarded only during the period when illumination was markedly increased for several seconds at a time. Soon response frequency became maximal during this "bright" period. Under such conditions no differences in performance were obtained between infero-temporally lesioned and control animals. The result tended to support the view that if an infero-temporally lesioned monkey did not have to make a choice he would show no deficit in behavior, since in another experiment (Mishkin & Hall, 1955) the monkeys failed to choose between differences in brightness.

In another instance (Pribram & Mishkin, 1955), we trained monkeys on a task in which they had to choose between easily discriminable objects: an ashtray and a tobacco tin. These animals had been trained for two or three years prior to surgery and were sophisticated problem-solvers. This, plus ease of task, produced only a minimal deficit in the simultaneous choice task. When given the same cues successively, the monkeys showed a deficit when compared with their controls, despite their ability to differentiate the cues in the simultaneous situation.

This result gave further support to the idea that the problem for the operated monkeys was not so much in "seeing" but in being able to *refer* in a useful or meaningful way to what had been reinforced previously. Not only the stimulus conditions but an entire range of response determinants appeared to be involved in specifying the deficit. To test this more quantitatively, I next asked whether the deficit would vary as a function of the *number of alternatives* in the situation (Pribram, 1959). It was expected that an informational measure of the deficit could be obtained, but something very different appeared when I plotted the number of errors against the number of alternatives (see Figure 34.3).

If one plots repetitive errors made before the subject finds a peanut — that is, the number of times a monkey searches the same cue — *vs.* the number of alternatives in the situation, one finds there is a hump in the curve, a stage where control subjects make many repetitive errors. The monkeys do learn the appropriate strategy, however, and go on to complete the task with facility. What intrigued me was that during this stage the monkeys with infero-temporal lesions were doing better than the controls! This seemed a paradox. However, as the test continued, the controls no longer made so many errors, whereas the lesioned subjects began to accumulate errors at a greater rate than shown earlier by the controls.

When a stimulus sampling model was applied to the analysis of the data, a difference in sampling was found (Figure 34.4). The monkeys with infero-temporal lesions showed a lowered sampling ratio; they sampled fewer cues during the first half of the experiment. Their defect can be characterized as a restriction on the number of alternatives searched and sampled. Their sampling competence, that is, their competence to process information, had become impaired. The limited sampling restricted the ability to construct an extensive memory store and to reference that memory during retrieval.

Search during novel cue presentation

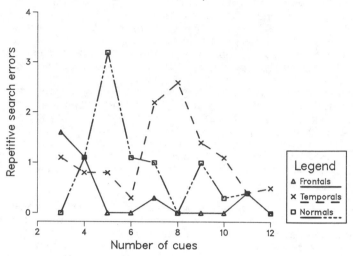

FIG. 34.3. The average number of repetitive errors made in the multiple object experiment during those search trials in each situation when the additional, that is, the novel, cue is first added.

C. ELEMENT LEARNING

The multiple object task had been administered in a Yerkes testing apparatus operated manually. Because administration was tedious and time consuming and because inadvertent cueing was difficult to control, an automated testing device was developed (Pribram, Gardner, Pressman, & Bagshaw, 1962; Pribram, 1969b). The resulting computer controlled Discrimination Apparatus for Discrete Trial Analysis (DADTA) proved useful in a large number of studies, ranging from testing one-element models of learning (Blehert, 1966) to plotting Response Operator Characteristic (ROC) curves to determine whether bias was influenced toward risk or toward caution by selected brain resections (Spevack & Pribram, 1973; Pribram, Spevack, Blower, & McGuinness, 1980).

To investigate whether learning proceeds by sampling one element at a time, eight monkeys were trained on a two choice and a five choice sample displayed on the screen of the DADTA panels of which only one was rewarded when pressed. The choices of individual monkeys were plotted for each of the cues sampled by panel pressing. As can be seen from the accompanying figure (Figure 34.5), sampling of cues is initially random, producing prolonged periods of stationarity. Behavior then becomes concentrated on the rewarded cue in steps, each of which is preceded by another period of stationarity and the elimination (i.e., choice

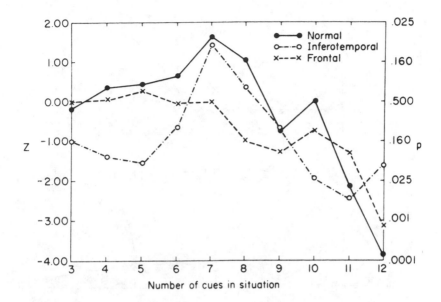

FIG. 34.4. The average proportion of objects (cues) that are sampled (except novel cue) by each of the groups in each of the situations. To sample, a monkey had to move an object until the content or lack of content of the food well was clearly visible to the experimenter. As was predicted, during the first half of the experiment the curve representing the sampling ratio of the posteriorly lesioned group differs significantly from the others.

drops to zero) of one of the unrewarded cues.

The study was undertaken in order to determine whether cross-hatching (with a cataract knife) of the inferior temporal cortex would produce subtle effects which would otherwise be missed. No such effects were observed. By contrast, restricted under-cutting of the inferior temporal region, which severed its major input and output connections, produced the same severe effects as extensive subpial resection of the cortex *per se*. Sampling was severely restricted as in the multiple object experiment (Pribram, Blehert, & Spinelli, 1966).

Subtle effects are obtained, however, when abnormal electrical foci are induced by implanting epileptogenic chemicals in the cortex. In such preparations, the period of stationarity in a two-choice task is increased five-fold. Despite this, the slope of acquisition, once it begins, remains unaffected. Obviously during the period of stationarity something is going on in the nervous system – something which becomes disrupted by the process which produces the electrical abnormality. Perhaps that something devolves on distributing the effects of trial and error over a sufficient reach of the neural net until an adequate associative structure is attained.

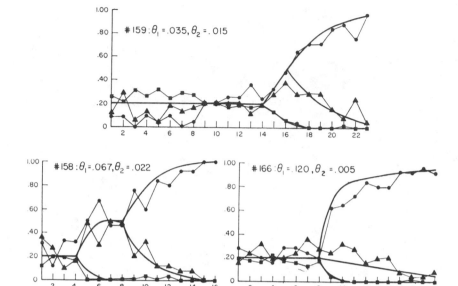

FIG. 34.5. Observed and predicted proportion responses to each stimulus for individual subjects on multiple discrimination in 25-trial blocks. Each curve represents a different stimulus. Solid lines represent predictions based on restricted sampling. The ordinate shows the proportion of responses, the abscissa shows 25-trial blocks.

D. REFERENCE LEARNING

How do the search and sampling systems interact with the perceptual and motor systems to produce skilled performance? We have shown that recovery functions in the primary visual and auditory systems have been influenced by electrical stimulations of the sensory specific intrinsic association areas and the frontolimbic systems (Spinelli & Pribram, 1966). This influence is a function of the attentive state of the monkey (Gerbrandt, Spinelli, & Pribram, 1970). Visual receptive fields have also been shown to become altered by such stimulation (Spinelli & Pribram, 1967). Finally, the pathways from the sensory specific intrinsic association and frontolimbic formations to the primary input systems have been to some extent delineated (Reitz & Pribram, 1969). Perhaps the most surprising findings of these studies is that input control is to a large measure effected through structures which had hitherto been thought of as regulating motor function.

This brings me to a consideration of the brain as the instrument with which we develop learning skill. The brain as we know it now is considerably different from the one that early learning theorists thought they were working with. Most formulations of learning depended heavily

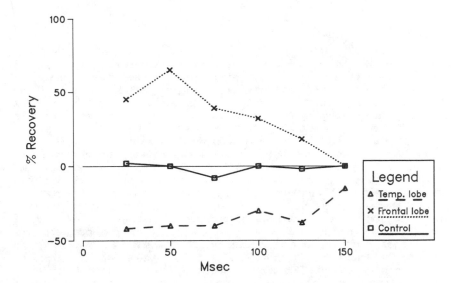

FIG. 34.6. The percentage recovery during stimulation compared to prestimulation response. Control stimulations were performed on the parietal cortex. Records were made immediately after the onset of stimulation and weekly for several months. Stimulations one month later showed no significant change.

on the concept of associative strength based on contiguity and number. Configural variables were relegated to perception and the existance of perceptual learning was, until the past two decades, denied or ignored. Further, the configural and sampling aspects of perceptual learning had not been teased apart.

An even more pervasive difficulty with classical learning theory is its dependence on the reflex-arc, stimulus -> organism -> response model of brain function. We now know that the brain is organized along servo-mechanism principles. The discovery of the function of the gamma efferent fibers of motor nerves made it necessary to modify our conceptions of the organization of the reflex and therefore of behavior. Thus, sensory functions are controlled by output systems; behavior is regulated not by a piano keyboard control over muscle contraction but by servo-control of the setting of muscle receptors (see Pribram, Sharafat, & Beekman, 1983). In such a brain, learning is hierarchic and constructional: the brain must build up programs to organize perceptions and to compose a behavioral repertoire. Instead of simple "association" by contiguity, learning proceeds by matching configurations; and the accretion of skils through practice (the develoment of subroutines) occurs by dropping out unnecessary actions and movements, not by forming new associative connections.

V.   EPISODIC LEARNING AND THE LIMBIC FOREBRAIN

A. CONTEXTUAL LEARNING

The second major division of the cerebral mantle to which learning functions have been assigned by clinical observation lies on the medial and basal surface of the brain and extends forward to include the poles of the frontal and temporal lobes. This frontolimbic portion of the hemisphere is cytoarchitecturally diverse. The expectation that different parts might be shown to subserve radically different functions was therefore even greater than that entertained for the more uniform posterior cortex. To some extent this expectation was not fulfilled: Lesions of the frontolimbic region, irrespective of location (dorsolateral frontal, caudate, cingulate–medial frontal, orbitofrontal, temporal polar–amygdala, and hippocampal) disrupted "delayed alternation" behavior. The alternation task demands that the subject alternate his responses between two cues (for example, between two places or between two objects) on successive trials. On any trial the correct response is dependent on the outcome of the previous response. This suggests that the critical variable which characterizes the task is its temporal organization. In turn, this leads to the supposition that the disruption of alternation behavior produced by frontolimbic lesions results from an impairment of the process by which the brain achieves its temporal organization. This supposition is only in part confirmed by further analysis: It has been necessary to impose severe restrictions on what is meant by "temporal organization" and important aspects of spatial organization are also severely impaired.

For instance, *skills* are not affected by frontolimbic lesions, nor are discriminations of melodies. Retrieval of long–held memories also is little affected. Rather, a large range of short–term memory processes are involved. These clearly include tasks which demand matching from memory the spatial location of cues (as in the delayed response problem) (Anderson, Hunt, Vander Stoep, & Pribram, 1976) as well as their temporal order of appearance (as in the alternation task) (Pribram, Plotkin, Anderson, & Leong, 1977). A similar deficit is produced when, in choice tasks, shifts in which cue is rewarded are made over successive trials (Mishkin & Delacour, 1975). The deficit appears whenever the organism must fit the present event into a "context" of prior occurrences, and there are no cues which address this context in the situation at hand at the moment of response.

B. THE REGISTRATION OF EVENTS AS EPISODES

As noted, different parts of the frontolimbic complex would, on the basis of their anatomical structure, be expected to function somewhat differently within the category of contextual memory processes. Indeed, different forms of contextual amnesia are produced by different lesions. In order to be experienced as memorable, events must be fitted to context. A series of experiments on the orienting reaction to novelty and its registration have pointed to the amygdala as an important locus in the "context-fitting" mechanism. The experiments were inspired by results obtained by Sokolov (Sokolov, 1960).

Sokolov presented human subjects with a tone beep of a certain intensity and frequency, repeated at irregular intervals. Galvanic skin response (GSR), heart rate, finger and forehead plethysmograms, and electro-encephalograms were recorded. Initially, these records showed the perturbations that were classified as the orienting response. After several repetitions of the tone, these perturbations diminish and finally vanish. They habituate. Originally it had been thought that habituation reflected a lowered sensitivity of the central nervous system to inputs. But when Sokolov decreased the intensity of the tone beep, leaving the other parameters unchanged, a full-blown orienting response was reestablished. Sokolov reasoned that the central nervous system could not be fatigued in general but that it was less responsive to sameness: when any difference occurred in the stimulus the central nervous system became more sensitive. He tested this idea by rehabituating his subjects and then occasionally omitting the tone beep, or reducing its duration without changing any other parameter. As predicted, his subjects now oriented to the unexpected silence.

The orienting reaction and habituation are thus sensitive measure of the process by which context is organized. We therefore initiated a series of experiments to analyze in detail the neural mechanisms involved in orienting and its habituation. This proved more difficult than we imagined. The dependent variables – behavior, GSR, plethysmogram, and electro-encephalogram – are prone to dissociate (Koepke & Pribram, 1971).

Forehead plethysmography turned out to be especially tricky, and we eventually settled on behavior, the skin conductance (GSR), heart and respiratory responses, and the electrical brain manifestations as most reliable.

The first of these experiments (Schwartzbaum, Wilson, & Morrissette, 1961) indicated that, under certain conditions, removal of the amygdaloid complex can enhance the persistence of locomotor activity in monkeys who would normally decrement their responses. The lesion thus produces a disturbance in the habituation of motor activity (Figure 34.7).

The results of the experiments on the habituation of the GSR component of the orienting reaction (Bagshaw, Kimble, & Pribram, 1965) also indicate clearly that amygdalectomy has an effect (Figure 34.8). The lesion profoundly reduces GSR amplitude in situations where the GSR is a robust indicator of the orienting reaction. Concomitantly, deceleration of heartbeat, change in respiratory rhythm, and some aspects of the EEG indices of orienting also are found to be absent (Bagshaw & Benzies, 1968). As habituation of motor activity (Pribram, 1960a, b) and also habituation of earflicks (Bateson, 1972) had been severely altered by these same lesions, we concluded that the autonomic indicators of orienting are in some way crucial to subsequent behavioral habituation. We identified the process indicated by the autonomic components of the orienting reaction as "registering" the novel event.

However, the registration mechanism is not limited to novelty. Extending the analysis to a classical conditioning situation (Bagshaw & Coppock, 1968; Pribram, Reitz, McNeil, & Spevack, 1979) using the GSR as a measure of conditioning, we found that normal monkeys not only

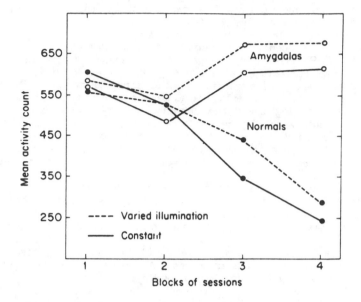

FIG. 34.7. Postoperative activity scores of normal and amygdalectomized monkeys for successive blocks of three sessions under conditions of constant illumination and more intense, varied illumination.

condition well but produce earlier and more frequent anticipatory GSR's as time goes by. Amygdalectomized subjects fail to make such anticipatory responses. As classical conditioning of a striped muscle proceeded normally, it is not the conditioning *per se* which is impaired. Rather, it appears that registration entails some active process akin to rehearsal – some central mechanism aided by viscero-autonomic processes that maintains and distributes excitation over time.

Behavioral experiments support this suggestion. Amygdalectomized monkeys placed in the 2-cue task described above fail to take proper account of reinforced events. This deficiency is dramatically displayed whenever punishment, that is, negative reinforcement, is used. For instance, an early observation showed that baboons with such lesions will repeatedly (day after day and week after week) put lighted matches in their mouths despite showing obvious signs of being burned (Fulton, Pribram, Stevenson, & Wall, 1949). These observations were further quantified in tasks measuring avoidance of shock (Pribram & Weiskrantz, 1957). The results of these two experiments have been confirmed in other laboratories and with other species so often that the hypothesis needed to be tested that amygdalectomy produces an altered sensitivity to pain. Bagshaw and Pribram (1968) put this hypothesis to test and showed that the threshold of GSR to shock is *not* elevated as it would be if there were an elevation of the pain threshold. Rather the threshold is, if anything, reduced by the ablation. This experimental result suggests that

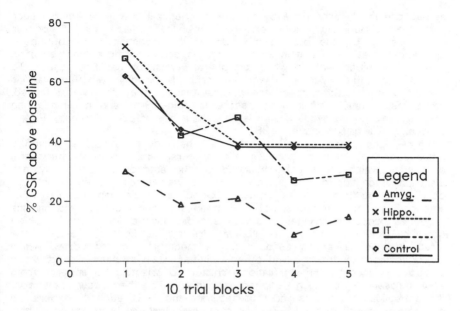

FIG. 34.8. Curves of percent GSR response to the first 50 presentations of the original stimulus for the control and three experimental groups (Hippo., IT, Amyg.), i.e., hippocampal, inferior temporal, and amygdala resected monkeys.

amygdalectomy produces its effects by way of a "loss of fear" defined as a disturbance in "registering" the noxious event by placing it in context. In other words, the animal does not *remember* the noxious event; its recurrence is experienced as novel and not fear–producing.

Another set of data are relevant to this issue of registration. These data were gathered within an entirely different frame of investigation. Around the turn of the century, the observation was made that after severe head injuries, patients could not remember what had happened to them for a period of time prior to the injury. The duration of such retrograde amnesia varied as a function of the severity of the injury. This suggested that the process of registering an experience in memory took some time and that the injured brain could not carry out this process. The process was labeled "consolidation."

During the 1960s and 1970s, James McGaugh and his collaborators (McGaugh & Hertz, 1972) carried out a series of experiments during which they interfered with, or enhanced, consolidation by injecting rats with different chemical substances immediately after they had experienced shock. The times of injection were varied in order to chart the course of the consolidation process. Once McGaugh had accomplished this he set out to locate the brain systems involved in the process. The amygdala seemed a good choice as a starting point in the search. Consolidation was now successfully manipulated by electrical and chemical stimulations much as had previously been done by peripheral chemical injections. In any such series of experiments, however, the possibility remains that all one is accomplishing by the brain stimulation is the boosting of a peripheral chemical secretion so that in essence one is doing no more than repeating the original experiments in which peripheral stimulation had been used. To control for this, Martinez, working with McGaugh, removed various peripheral structures such as the adrenal gland. They found that indeed, when the adrenal medulla which secretes epinephrine and norepinephrine was absent, the amygdala stimulations had no effect (Martinez, Rigter, Jensen, Messing, & Vasquez, 1981).

McGaugh's experiments indicate, as had ours, that the amygdala influences the learning process via visceral and glandular peripheral processes which are largely regulated by the autonomic nervous system. Electrical excitation of the amygdala -- as well as of the entire anterior portion of the limbic cortex: anterior cingulate, medial and orbital frontal, anterior insula, and temporal pole -- in anesthetized monkeys and humans produces profound changes in such visceroautonomic processes as blood pressure and respiratory rate (Kaada, Pribram, & Epstein, 1949). The amygdala thus serves as a focus for a mediobasal motor cortex which regulates visceroautonomic and other activities (such as head-turning which is also produced by the stimulations) related to orienting. It appears from all this research that such peripheral activities when they occur, can boost the consolidation process and thus facilitate the registration of experience in memory. Vinogradova (1975) has suggested that the boost given by this visceroautonomic system stands in lieu of repetition of the experience. As noted above, the experiments on conditioning suggest that visceroautonomic arousal acts somewhat like internal rehearsal. One can take visceroautonomic arousal as an indication that interest and emotions have been engaged: thus the mechanism has been tapped which accounts for the well-known fact that emotional involvement can dramatically influence learning.

C. PROCESSING THE FAMILIAR

Context is not composed solely of the registration of reinforcing and reinforced events. As important are the errors, the non-reinforced aspects of a situation, especially if on previous occasions they had been reinforced. It is resection of the primate hippocampal formation (Douglas & Pribram, 1966) which produces relative insensitivity to errors, frustrative non-reward (Gray, 1975; and see Chapter 33, this volume, by Gray) and more generally to the familiar, non-reinforced aspects of the environment (the SΔ of operant conditioning; the negative instances of mathematical psychology). In their first experience with a discrimination learning situation subjects with hippocampal resections show a peculiar retardation provided there are many nonrewarded alternatives in that situation: For example, in an experiment using the computer-controlled automated testing apparatus (DADTA), the subject faced 16 panels; discriminable cues are displayed on only two of these panels and only one cue is rewarded. The cues are displayed in various locations in a random fashion from trial to trial. Hippocampectomized monkeys were found to press the unlit and unrewarded panels for thousands of trials, long after their unoperated controls ceased responding to these "irrelevant" items. It is as if in the normal subject, a "ground" is established by enhancing "inattention" to all the negative instances of those patterns that do not provide a relevant "figure." This "inattention" is an active, evaluating process as indicated by the behavior shown during shaping in a discrimination reversal task, when the demand is to respond to the previously nonreinforced cue: Unsophisticated subjects often begin by pressing on various parts of their cage and the testing apparatus before they hit upon a chance response to the now-rewarded cue.

These and many similar results indicate that the hippocampal formation is part of an evaluative mechanism that helps to establish the "ground," the familiar aspects of context.

D. THE SPATIOTEMPORAL STRUCTURE OF CONTEXT

In some respects the far frontal resection produces memory disturbances characteristic of both hippocampectomy and amygdalectomy, though not so severe. Whereas medial temporal lobe ablations impair context formation by way of habituation of novel and familiar events, far frontal lesions wreak havoc on yet another contextual dimension, that of organizing the spatial and temporal structure of the context (Pribram, 1961; Anderson, Hunt, Vander Stoep, & Pribram, 1976; Pribram, Plotkin, Anderson, & Leong, 1977). This effect is best demonstrated by an experiment in which the normal scallop produced by a fixed interval schedule of reinforcement fails to develop and another in which the parameters of the classical alternation were altered. Instead of interposing equal intervals between trials (go right, go left every 5 seconds) in the usual way, couplets of R/L were formed by extending the intertrial interval to 15" before each R trial (R 5" L 15" R 5" L 15" R 5" L 15". . .). When this was done, the performance of the far frontally lesioned monkeys improved immediately and was indistinguishable from that of the controls (Pribram & Tubbs, 1967; Pribram, Plotkin, Anderson, & Leong, 1977). This result suggests that for the subject with a bilateral far frontal ablation, the alternation task is experienced similarly to reading this page without any spaces between the words. The spaces, like the holes in doughnuts, provide the contextual structure, the parcellation or parsing of events by which the outside world can be coded and deciphered.

E. CONTEXT AS A FUNCTION OF REINFORCING CONTINGENCIES

Classically, disturbance of "working" short-term memory has been ascribed to lesions of the frontal pole. Anterior and medial resections of the far frontal cortex were the first to be shown to produce impairment on delayed response and delayed alternation problems. In other tests of context-formation and fitting, frontal lesions also take their toll. Here also impairment of conditioned avoidance behavior and of classical conditioning and of the orienting GSR is found. Furthermore, error sensitivity is reduced in an operant conditioning situation. After several years of training on mixed and multiple schedules, the monkeys were extinguished over 4 hours. The frontally lesioned monkeys failed to extinguish in the 4-hour period, whereas the control monkeys did (Pribram, 1961).

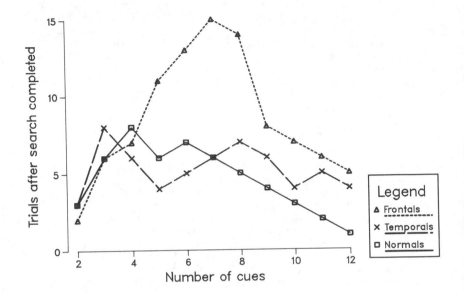

FIG. 34.9.   The average number of trials to criterion taken in the multiple object experiment by each group in each of the situations after search was completed, that is, after the first correct response. Note the difference between the curves for the controls and for the frontally operated group, a difference that is significant.

This failure in extinction accounts in part for poor performance in the alternation already described: the frontally-lesioned animals again make many more repetitive errors. Even though they do not find a peanut, they go right back and keep looking (Pribram, 1959).

This result was confirmed and amplified in studies by Wilson (1962) and by Pribram, Plotkin, Anderson, & Leong (1977) in which we asked whether errors followed alternation or non-reinforcement. We devised a situation in which both lids over two foods wells opened simultaneously, but the monkey could obtain the peanut only if he had opened the baited well. Thus, the monkey was given "complete" information on every trial and the usual correction technique could be circumvented. There were four procedural variations: correction-contingent, correction-noncontingent, noncorrection-contingent, and noncorrection-noncontingent. The contingency referred to whether the position of the peanut was altered on the basis of the monkey's responses (correct or incorrect) or whether its position was changed independently of the monkey's behavior. We then analyzed the relationship between each error and the trial that preceded that error. Table 34.1 shows that for the normal monkey, the condition of reinforcement and non-reinforcement of the previous trial makes a difference. For the frontally-lesioned monkey this is not the case. Change in location, however, affects both normal and frontal subjects about equally. In this situation, as well as in an automated computer-controlled version of the alternatives problem, frontal subjects are simply uninfluenced by rewarding or nonrewarding consequences of their behavior.

TABLE 34.2
Percentage of Alternation as a Function of Response and Outcome of Preceding Trial

S	A-R	A-NR	NA-R	NA-NR
		Preceding Trial		
Normal				
394	53	56	40	45
396	54	53	36	49
398	49	69	27	48
384	61	83	33	72
Total	55	68	34	52
Frontal				
381	49	51	41	43
437	42	46	27	26
361	49	48	38	35
433	43	39	31	32
Total	46	46	33	33

Note-- S = Subject; A-R = Alternation and Reinforcement; A-NR = Alternation and No Reinforcement; NA-R = No Alternation and Reinforcement; NA-NR = No Alternation and No Reinforcement.

In the original multiple choice task (Pribram, 1959) (see Figure 34.9) the procedure called for a strategy of returning to the same object for five consecutive times, that is, to criterion, and then a shift to a novel item. The frontally-lesioned animals are markedly deficient in doing this. Again, the conditions of reinforcement are relatively ineffective in shaping behavior in animals with frontal lesions and the monkeys' behavior becomes nearly random when compared to that of normal subjects (Pribram, Ahumada, Hartog, & Roos, 1964). Behavior of the frontally-lesioned monkeys thus appears to be minimally controlled by the expected outcome.

F. TRANSFER LEARNING

When we take a monkey who has learned to choose between circles of different sizes and ask him to transfer his experience to a situation in which he must choose among ellipses of different sizes (Bagshaw & Pribram, 1965) he will quickly master the new task unless he has a lesion of the limbic forebrain. This is not due to faulty generalization (Hearst & Pribram 1964a, b) -- generalization is impaired by lesions of the posterior cortical convexity. Rather, the difficulty stems from an inability to transfer what has been learned in one situation to another which is more or less similar. If his hippocampus has been resected bilaterally, the familiar cue will be normally effective only if it had previously been the rewarded one. The previously unrewarded cue will be reacted to as if it also were novel -- as if it had been completely ignored in the original discrimination problem. Just the opposite occurs when a monkey has been amygdalectomized. Now effective familiarity relates to non-reward (SΔ; negative instances); the previously rewarded cue is treated as novel in the transfer situation (Douglas & Pribram, 1966).

A variety of other problem situations have demonstrated this relationship between the hippocampus and the previously non-reinforced (non-salient) aspects of a situation and between the amygdala and prior reinforcement. Multiple choice (Douglas, Barrett, Pribram, & Cerny, 1969) and distraction (Douglas & Pribram, 1966) experiments have been especially illuminating. In all instances, as in the reversal situations, whenever the probability structure of reinforcement becomes insufficiently distinct, or the distractions sufficiently powerful, limbic-lesioned subjects fail to persist in a strategy that had proved useful in prior situations. Attention and search are no longer directed (programmed) by previous experience; hypotheses are no longer pursued (Pribram, Douglas, & Pribram, 1969). The monkeys no longer expend the effort to maintain useful strategies and relapse to position habits which assure them a constant, if not a maximum number of reinforcements. In short, the monkeys become biased to caution. By contrast, resections of the inferotemporal cortex bias monkeys to risk (Pribram, Spevack, Blower, & McGuinness, 1980).

VI.   CONCLUSION: LINGUISTIC LEARNING

The evidence presented here makes it not unlikely that one function of the posterior intrinsic and frontolimbic formations of the forebrain is to code events occurring within the input systems. As noted, the distribution of information (dis-membering) implies an encoding process that can reduplicate events. Regrouping the distributed events (re-membering) also implies some sort of coding operation - one similar to that used in decoding binary switch settings into an octal format and that into assembly and still higher-order programming language. An impaired coding process would be expected to produce grave memory disturbances. Lesion-produced amnesias, reference and contextual, therefore reflect primary malfunctions of coding mechanism and not the destruction of localized engrams.

Concretely, the intrinsic cortex is thus conceived to program, or to structure, an input channel. This is tantamount to saying that the input in the projection systems is coded by the operation of the intrinsic cortex. In its fundamental aspects, computer programming is a coding operation: The change from direct machine operation through assembler to one of the more manipulable computer languages involves a progression from the setting of binary switches to conceptualizing combinations of such switch settings in "octal" code and then assembling the numerical octals into alphabetized words and phrases and, finally, parcelling and parsing of phrases into sentences, routines, and subroutines. In essence, these progressive coding operations minimize interference among the configurations of occurrence and recurrence of the events.

This, then, is a sketch of the model derived from analyzing the effects on cognitive learning processes which resections and stimulations of the non-human primate brain have produced. What then distinguishes man's brain, identifies him as human? The psychopathology of human learning processes has almost universally been interpreted in terms of transcortical connections. All we have learned from experiments on non-human primate brains (e.g., the data described above) is evidence against the importance of such connections. Either the interpretation of the basis for the learning deficiencies in man is in error or else we have, through our efforts, stumbled on the difference between man's brain and that of his primate relatives. Thus it becomes paramount to review and test out once again, from this new vantage, the clinical evidence.

The converse of this approach has also proved fruitful. Experiments have tested the linguistic abilities, one-by-one, of non-human primates. The results have shown marked differences in syntactic competence which depends for its development on procedures which determine perceptual, motor and referential skills. In turn, such development depends on the construction of contexts from episodes and flexibly shifting these contexts in accordance with the spatiotemporal probabilities of reinforcement. These results indicate that the difference between non-human and human primates encompass a great deal of their forebrain and that these differences may well be due to an increase in transcortical connectivity.

ACKNOWLEDGMENTS

The author is the recipient of Research Career Award M.H. 15214-22 from the National Institutes of Health.

REFERENCES

Anderson, R. M., Hunt, S. C., Vander Stoep, A., & Pribram, K. H. Object permanancy and delayed response as spatial context in monkeys with frontal lesions. *Neuropsychologia*, 1976, *14*, 481–490.

Bagshaw, M. H., & Benzies, S. Multiple measures of the orienting reaction to a simple non-reinforced stimulus after amygdalectomy. *Experimental Neurology*, 1968, *20*, 175–187.

Bagshaw, M. H., & Coppock, H. W. GSR conditioning deficit in amygdalectomized monkeys. *Experimental Neurology*. 1968, *20*, 188–196.

Bagshaw, M. H., & Pribram, K. H. Cortical organization in gustation (*macaca mulatta*). *Journal of Neurophysiology*, 1953, *16*, 499–508.

Bagshaw, M. H., & Pribram, K. H. Effect of amygdalectomy on transfer of training in monkeys. *Journal of Comparative and Physiological Psychology*, 1965, *59*, 118–121.

Bagshaw, M. H., & Pribram, J. D. The effect of amygdalectomy on stimulus threshold of the monkey. *Experimental Neurology*, 1968, *20*, 197–202.

Bagshaw, M. H., Kimble, D. P., & Pribram, K. H. The GSR of monkeys during orienting and habituation and after ablation of the amygdala, hippocampus, and inferotemporal cortex. *Neuropsychologia*, 1965, *3*, 111–119.

Bateson, P. P. G. An effect of imprinting on the perceptual development of domestic chicks. *Nature*, 1964, *202*, 421–422.

Bateson, P. P. G. Ear movements of normal and amygdalectomized monkeys. *Nature*, 1972, *237*, 173–174.

Bernstein, N. *The co-ordination and regulation of movement.* New York: McGraw-Hill, 1965.

Blehert, S. R. Pattern discrimination learning with rhesus monkeys. *Psychological Reports*, 1966, *19*, 311–324.

Bridgeman, B. Multiplexing in single cells of the alert monkey's visual cortex during brightness discrimination. *Neuropsychologia*, 1982, *20*, 33–42.

Broadbent, D. E. Division of function and integration. In *Neurosciences study program, III.* Cambridge, Mass.: MIT Press, 1974.

Dewson, J. H., III., Pribram, K. H., & Lynch, J. Ablations of temporal cortex in the monkey and their effects upon speech sound discriminations. *Experimental Neurology*, 1969, *24*, 579–591.

Douglas, R. J., & Pribram, K. H. Learning and limbic lesions. *Neuropsychologia*, 1966, *4*, 197–220.

Douglas, R. J., Barrett, T. W., Pribram, K. H., & Cerny, M. C. Limbic lesions and error reduction. *Journal of Comparative and Physiological Psychology*, 1969, *68*, 437–441.

Ettlinger, G. Visual discrimination following successive unilateral temporal excisions in monkeys. *Journal of Physiology: London*, 1957, *140*, 38–39.

Fulton, J. F., Pribram, K. H., Stevenson, J. A. F., & Wall, P. D. Interrelations between orbital gyrus, insula, temporal tip, and anterior cingulate. *Transactions of the American Neurologists Association*, 1949, *74*, 175.

Gerbrandt, L. K., Spinelli, D. N., & Pribram, K. H. The interaction of visual attention and temporal cortex stimulation on electrical recording in the striate cortex. *Electroencephalography and Clinical Neurophysiology*, 1970, *29*, 146–155.

Grandstaff, N., & Pribram, K. H. Habituation: Electrical changes in visual system. *Neuropsychologia*, 1972, *10*, 125–132.

Gray, J. A. *Elements of a two-process theory of learning*. London: Academic Press, 1975.

Hearst, E., & Pribram, K. H. Facilitation of avoidance behavior in unavoidable shocks in normal and amygdalectomized monkeys. *Psychological Reports*, 1964, *14*, 39–42. (a)

Hearst, E., & Pribram, K. H. Appetitive and aversive generalization gradients in normal and amygdalectomized monkeys. *Journal of Comparative and Physiological Psychology*, 1964, *58*, 296–298. (b)

Kaada, B. R., Pribram, K. H., & Epstein, J. A. Respiratory and vascular responses in monkeys from temporal pole, insula, orbital surface and cingulate gyrus. *Journal of Neurophysiology*, 1949, *12*, 347–356.

Koepke, J. E., & Pribram, K. H. Effect of milk on the maintenance of sucking in kittens from birth to six months. *Journal of Comparative and Physiological Psychology*, 1971, *75*, 363–377.

Kraft, M., Obrist, W. D., & Pribram, K. H. The effect of irritative lesions of the striate cortex on learning of visual discrimination in monkeys. *Journal of Comparative and Physiological Psychology*, 1960, *53*, 17–22.

Lashley, K. S. *Brain mechanisms and intelligence*. Chicago: University of Chicago Press, 1929.

Martinez, J. L., Jr., Rigter, H., Jensen, R. A., Messing, R. B., & Vasquez, B. J. Endorphin and enkephalin effects on avoidance conditioning: The other side of the pituitary–adrenal axis. In J. L. Martinez, Jr. (Ed.), *Endogenous peptides in learning and memory processes*. New York: Academic Press, 1981.

McGaugh, J. L., & Hertz, M. J. *Memory consolidation*. San Francisco: Albion, 1972.

Mishkin, M., & Delacour, J. An analysis of short–term visual memory in the monkey. *Journal of Experimental Psychology: Animal Behavior Processes*, 1975, *1*, 326–334.

Mishkin, M., & Hall, M. Discrimination along a size continuum following ablation of the inferior temporal convexity in monkeys. *Journal of Comparative and Physiological Psychology*, 1955, *48*, 97–101.

Mishkin, M., & Pribram, K. H. Visual discrimination performance following partial ablations of the temporal lobe: I. Ventral vs. lateral. *Journal of Comparative and Physiological Psychology*, 1954, *47*, 14–20.

Nuwer, M. R., & Pribram, K. H. Role of the inferotemporal cortex in visual selective attention. *Electroencephalography & Neurophysiology*, 1979, *46*, 389–400.

Pribram, K. H. Toward a science of neuropsychology: Method and data. In R. A. Patton (Ed.), *Current trends in psychology and the behavioral sciences*. Pittsburgh: University of Pittsburgh Press, 1954.

Pribram, K. H. On the neurology of thinking. *Behavioral Science*, 1959, *4*, 265–287.

Pribram, K. H. The intrinsic systems of the forebrain. In J. Field & H. W. Magoun (Eds.), *Handbook of physiology: Neurophysiology.* (Vol. 2.) Washington, D.C.: American Physiological Society, 1960. (a)

Pribram, K. H. A review of theory in physiological psychology. *Annual Review of Psychology,* 1960, *11,* 1–40. (b)

Pribram, K. H. A further experimental analysis of the behavioral deficit that follows injury to the primate frontal cortex. *Experimental Neurology,* 1961, *3,* 432–466.

Pribram, K. H. Some dimensions of remembering: Steps toward a neuropsychological model of memory. In J. Gaito (Ed.), *Macromolecules and behavior.* New York: Academic Press, 1966.

Pribram, K. H. Four R's of remembering. In K. H. Pribram (Ed.), *The biology of learning.* New York: Harcourt, Brace, & World, 1969. (a)

Pribram, K. H. DADTA III: An on-line computerized system for the experimental analysis of behavior. *Perceptual and Motor Skills,* 1969, *29,* 599–608. (b)

Pribram, K. H. The isocortex. In D. A. Hamburg & H. K. H. Brodie (Eds.), *American handbook of psychiatry.* (Vol. 6.) New York: Basic Books, 1974.

Pribram, K. H. *Languages of the brain: Experimental paradoxes and principles in neuropsychology.* New York: Brandon House, 1982.

Pribram, K. H., & Barry, J. Further behavioral analysis of the parietotemporo-preoccipital cortex. *Journal of Neurophysiology,* 1956, *19,* 99–106.

Pribram, K. H., & Mishkin, M. Simultaneous and successive visual discrimination by monkeys with inferotemporal lesions. *Journal of Comparative and Physiological Psychology,* 1955, *48,* 198–202.

Pribram, K. H., & Tubbs, W. E. Short-term memory, parsing and the primate frontal cortex. *Science,* 1967, *156,* 1765.

Pribram, K. H., & Weiskrantz, L. A comparison of the effects of medial and lateral cerebral resections on conditioned avoidance behavior in monkeys. *Journal of Comparative and Physiological Psychology,* 1957, *50,* 74–80.

Pribram, K. H., Ahumada, A., Hartog, J., & Roos, L. A progress report on the neurological processes distributed by frontal lesions in primates. In J. M. Warren & K. Akert (Eds.), *The frontal granular cortex and behavior.* New York: McGraw-Hill, 1964.

Pribram, K. H., Blehert, S., & Spinelli, D. N. The effects on visual discrimination of crosshatching and undercutting the inferotemporal cortex of monkeys. *Journal of Comparative and Physiological Psychology,* 1966, *62,* 358–364.

Pribram, K. H., Douglas, R., & Pribram, B. J. The nature of non-limbic learning. *Journal of Comparative and Physiological Psychology,* 1969, *69,* 765–772.

Pribram, K. H., Gardner, K. W., Pressman, G. L., & Bagshaw, M. H. An automated discrimination apparatus for discrete trial analysis (DADTA). *Psychological Reports,* 1962, *11,* 247–250.

Pribram, K. H., Kruger, L., Robinson, F., & Berman, A. J. The effects of precentral lesions on the behavior of monkeys. *Yale Journal of Biology & Medicine,* 1955, *28,* 428–443.

Pribram, K. H., Plotkin, H. C., Anderson, R. M., & Leong, D. Information sources in the delayed alternation task for normal and "frontal" monkeys. *Neuropsychologia*, 1977, *15*, 329–340.

Pribram, K. H., Reitz, S., McNeil, M., & Spevack, A. A. The effect of amygdalectomy on orienting and classical conditioning in monkeys. *Pavlovian Journal*, 1979, *14*, 203–217.

Pribram, K. H., Sharafat, A., & Beekman, G. J. Frequency encoding in motor systems. In H. T. A. Whiting (Ed.), *Human motor actions*. North–Holland Publishing, 1983. In press.

Pribram, K. H., Spevack, A. A., Blower, D., & McGuinness, D. A decisional analysis of the effects of inferotemporal lesions in the rhesus monkey. *Journal of Comparative and Physiological Psychology*, 1980, *94*, 675–690.

Pribram, K. H., Spinelli, D. N., & Kamback, M. C. Electrocortical correlates of stimulus response and reinforcement. *Science*, 1967, *156*, 94–96.

Reitz, S. L., & Pribram, K. H. Some subcortical connections of the inferotemporal gyrus of monkeys. *Experimental Neurology*, 1969, *25*, 632–645.

Rothblat, L., & Pribram, K. H. Selective attention: input filter or response selection? *Brain Research*, 1972, *39*, 427–436.

Schneider, W., & Shiffrin, R. M. Controlled and automatic human information processing: I. Detection, search and attention. *Psychological Review*, 1977, *84*, 1–66.

Schwartzbaum, J. S., Wilson, W. A., Jr., & Morrissette, J. R. The effects of amygdalectomy on locomotor activity in monkeys. *Journal of Comparative and Physiological Psychology*, 1961, *54*, 334–336.

Shiffrin, R. M., & Schneider, W. Controlled and automatic human information processing: II. Perceptual learning, automatic attending, and a general theory. *Psychological Review*, 1977, *84*, 129–190.

Sokolov, E. N. Neuronal models and the orienting reflex. In M. A. Brazier (Ed.), *The central nervous system and behavior*. New York: Josiah Macy, Jr. Foundation, 1960.

Sperry, R. W., Miner, N., & Meyers, R. E. Visual pattern perception following subpial slicing and tantalum wire implantations in the visual cortex. *Journal of Comparative and Physiological Psychology*, 1955, *48*, 50–58.

Spevack, A. A., & Pribram, K. H. A decisional analysis of the effects of limbic lesions on learning in monkeys. *Journal of Comparative and Physiological Psychology*, 1973, *82*, 211–226.

Spinelli, D. N. Evoked responses to visual patterns in area 17 on the rhesus monkey. *Brain Research*, 1967, *5*, 511–514.

Spinelli, D. N., & Pribram, K. H. Changes in visual recovery functions produced by temporal lobe stimulation in monkeys. *Electroencephalography and Clinical Neurophysiology*, 1966, *20*, 44–49.

Spinelli, D. N., & Pribram, K. H. Changes in visual recovery function and unit activity produced by frontal cortex stimulation. *Electroencephalography and Clinical Neurophysiology*, 1967, *22*, 143–149.

Stamm, J. S., & Knight, M. Learning of visual tasks by monkeys with epileptogenic implants in temporal cortex. *Journal of Comparative and Physiological Psychology*, 1963, *56*, 254–260.
Stamm, J. S., & Pribram, K. H. Effects of epileptogenic lesions in inferotemporal cortex on learning and retention in monkeys. *Journal of Comparative and Physiological Psychology*, 1961, *54*, 614–618.
Stamm, J. S., & Warren, A. Learning and retention by monkeys with epileptogenic implants in posterior parietal cortex. *Epilepsia*, 1961, *2*, 220–242.
Stamm, J. S., Pribram, K. H., & Obrist, W. The effect of cortical implants of aluminum hydroxide on remembering and on learning. *Electroencephalography and Clinical Neurophysiology*, 1958, *10*, 766.
Treisman, A. Focused attention in perception and retrieval of multidimensional stimuli. *Perception & Psychophysics*, 1977, *22*, 1–11.
Vinogradova, O. S. Functional organization of the limbic system in the process of registration of information: Facts and hypotheses. In R. L. Isaacson & K. H. Pribram (Eds.), *The hippocampus, Vol. 2. Neurophysiology and behavior*. New York: Plenum, 1975.
Weiskrantz, L., & Mishkin, M. Effects of temporal and frontal cortical lesions on auditory discrimination in monkeys. *Brain*, 1958, *81*, 406–414.
Wilson, M. Effects of circumscribed cortical lesions upon somesthetic and visual discrimination in the monkey. *Journal of Comparative and Physiological Psychology*, 1955, *50*, 630–635.
Wilson, M. Alternation in normal and frontal monkeys as a function of responses and outcome of the previous trial. *Journal of Comparative and Physiological Psychology*, 1962, *55*, 701–704.